Pags 75 ¦ 76

Anatomy and Physiology
for Physiotherapists

Anatomy and Physiology for Physiotherapists

D. B. MOFFAT
VRD, MD, FRCS
Professor of Anatomy
University College, Cardiff

R. F. MOTTRAM
BSc, PhD, MB, BS, LMSSA
Senior Lecturer in Physiology
University College, Cardiff

SECOND EDITION

OXFORD
Blackwell Scientific Publications
LONDON   EDINBURGH   BOSTON
MELBOURNE   PARIS   BERLIN   VIENNA

© 1979, 1987 by
Blackwell Scientific Publications
Editorial Offices:
Osney Mead, Oxford OX2 0EL
25 John Street, London WC1N 2BL
23 Ainslie Place, Edinburgh EH3 6AJ
3 Cambridge Center, Cambridge,
   MA 02142, USA
54 University Street, Carlton
   Victoria 3053, Australia

First published 1979
Reprinted 1982, 1983
Second edition 1987
Reprinted 1990

Printed in Great Britain
at the University Press, Cambridge

DISTRIBUTORS

Marston Book Services Ltd
PO Box 87
Oxford OX2 0DT
(*Orders*: Tel: 0865 791155
          Fax: 0865 791927
          Telex: 837515)

USA
   Mosby-Year Book, Inc.
   200 North LaSalle Street
   Chicago, Illinois 60601
   (*Orders*: Tel: 312 726-9733)

Canada
   Mosby-Year Book, Inc.
   5240 Finch Avenue East
   Scarborough, Ontario
   (*Orders*: Tel: 416 298-1588)

Australia
   Blackwell Scientific Publications
   (Australia) Pty Ltd
   54 University Street
   Carlton, Victoria 3053
   (*Orders*: Tel: 03 347-0300)

British Library
Cataloguing in Publication Data
Moffat, D.B.
   Anatomy and physiology for
   physiotherapists ——2nd ed.
   1. Human physiology 2. Physical
   therapy
   I. Title   II. Mottram, R.F.
   612'.0024616   QB34.5

   ISBN 0-632-01464-4

Contents

Preface to Second Edition vii

Preface to First Edition ix

1 Introduction 1

2 The Structure and Function of Living Cells 10

3 Basic Histology 33

4 The Nervous System, Introduction 44

5 General Anatomy and Physiology of the Nervous System 75

6 The Autonomic Nervous System 125

7 The Microscopic Structure and the Function of Muscle 134

8 Effects of Peripheral Nerve Damage 152

9 Bone and Bones 162

10 Joints 218

11 The Muscular System 270

12 The Peripheral Nerves 368

13 Special Senses 399

14 The Structure of the Cardiovascular System 411

15 The Cardiac Cycle and Regulation of the Heart's Activity 438

16 The Circulation of the Blood and its Regulation 455

17 Blood, Lymph and Other Body Fluids 473

Contents

18	Gas Transport in the Blood	491
19	The Respiratory System	502
20	The Lymphatic System	536
21	The Urinary System	546
22	The Skin and Body Temperature Control	556
23	Effects of Exercise	567
24	The Digestive System	586
25	Regional Anatomy	608
26	The Endocrine Glands	621
27	Reproduction, Growth and Development	636
	Index	652

Preface to the Second Edition

For this new edition the whole text has been reviewed with the intention of improving the presentation of the original material and bringing it up to date where necessary. In particular, we have revised the sections on the nervous system, on posture, on the roles of calcium in the body, and on the effects of exercise. Chapter 27, on Growth, Reproduction and Development has been amplified and largely rewritten.

Some of the original diagrams have been revised and a number of new ones added. These are the work of Catherine Hemington and it is a pleasure to acknowledge her skill in preparing them from our own rough sketches. We should also like to thank our colleagues and friends in various physiotherapy departments who have made many useful suggestions and we very much hope that they will give us the benefit of further constructive criticism of the new edition so that this book may continue to be helpful to physiotherapy students.

D. B. MOFFAT
R. F. MOTTRAM

Preface to First Edition

Physiotherapists need to know a great deal about bones, joints and muscles; enough, indeed, to require them to read one of the more ponderous tomes on anatomy. They also require a detailed knowledge of the physiology of parts of the nervous system, of the muscular system, and of the effects of exercise on the cardiovascular and respiratory systems, so that one of the rather advanced textbooks of physiology is needed. Their knowledge of other systems of the body can easily be met by a study of one of the smaller textbooks. This book has been written in the hope that it will replace, in one volume, the small library of suitable textbooks which the physiotherapist needs in the study of these basic subjects and, indeed, may be an improvement on them since it deals with the subjects purely from a physiotherapist's point of view. It is essentially a student textbook and there is thus a certain amount of repetition where it is thought that this might be helpful in driving home a point and there are also a number of mnemonics and *aides memoires* which, childish though some of them may be, have been found useful, often by the authors themselves, in remembering details. It is perhaps superfluous to add that the diagrams are diagrammatic—they, again, are meant purely as an aid to learning and do not necessarily present an accurate representation of the human body, which may better be studied in an anatomical atlas or in a living subject. We hope that this book will also be helpful after student days are over and that the practising physiotherapist will continue to find it useful as a reference book in her everyday work and in understanding some of the disease processes that she will encounter.

Contrary to the belief of many students, anatomy and physiology are one and the same subject, even though they employ different methods of research. Structure and function are mutually interdependent—our knowledge of the mechanism of muscle contraction is based on structural studies with the electron microscope while a description of the attachments of latissimus dorsi makes sterile reading without a knowledge of the use that can be made of this muscle by paraplegics. For this reason, rather than dividing the book into two sections— anatomy and physiology—the subjects are integrated as far as possible so that, for example, the chapters on the nervous system deal both with the structure and function of this rather difficult subject.

We have tried, as far as possible, to keep the terminology up to date, using S.I. units and the Paris anatomical nomenclature, but in places we have also used other terminology since this is more likely to be met in a clinical context. Thus, blood pressure is recorded in mm Hg rather than kilopascals, while we have preferred to use the term 'posterior root ganglion' rather than the modern (and less sensible) 'spinal ganglion'.

We would like to express our sincere thanks to all our colleagues and co-examiners in the Chartered Society of Physiotherapy who have, perhaps unconsciously, taught us so much about the applications of anatomy and physiology to physiotherapy. In particular we would like to thank Miss Dilys Gronow, who, in addition to giving us much invaluable advice, has read parts of the manuscript and suggested numerous improvements. We would also like to express our gratitude to Miss Mary Jo Drew who prepared all the illustrations from our own crudely drawn efforts and whose patience and tolerance must have been stretched to the utmost as drawing after drawing was returned to her many times for minor alterations. We also owe a debt of gratitude to Miss Pamela Lee, Miss Janice Gunn and Mrs. Joan Moffat, who between them have typed the manuscript at least twice over. Finally, we have the greatest pleasure in thanking the publishers, and in particular Mr. Per Saugman, for their help, advice and encouragement during the preparation of this book.

D. B. MOFFAT
R. F. MOTTRAM

1 · Introduction

The human body, in elementary textbooks, is often compared to an engine and analogies are drawn between food and fuel, energy output and horsepower and expired air and exhaust gases. The physiotherapist is at some disadvantage compared to the engineer, however, in that the subject of his or her attentions is almost completely covered with skin and other tissues through which all manipulations and examinations have to be carried out. Furthermore, each patient is an individual with his own likes and dislikes and he may often be reluctant to carry out instructions, either through anxiety or pain. This book will attempt to give you a theoretical knowledge of the structure and function of the body, but much of the real learning will have to be done with the hands and eyes from the living body. Only by careful and accurate palpation and manipulation can the normal structure of the tissues and their normal functioning be understood and the limitations imposed by disease be appreciated.

Although anatomy and physiology are often regarded as separate subjects, they are much better thought of as different aspects of the same study. A knowledge of pure anatomy is meaningless without understanding how the structures involved contribute to the functioning of the body and, similarly, function cannot be fully understood without a knowledge of structure, either as seen with the naked eye or on a microscopic level. We have not, therefore, divided this book up into 'anatomical' and 'physiological' sections but have dealt rather with the different systems of the body from a combined viewpoint. Naturally, some chapters will have a preponderance of description of structure while others will be concerned mainly with physiology but in the majority of chapters a combined approach has been used.

Much of your future work will be concerned with bones, muscles and joints, and you will therefore need a very good knowledge of their structure and function and be able to identify individual muscles and also to put joints through their full range of movement. You will need to have a good idea of the mechanisms by which the nervous system controls muscular activity and the effects of such activity on the body as a whole. The methods by which the body produces the necessary energy for muscular activity are also important. On the other hand, many aspects of anatomy and physiology need only be understood in outline and these will be dealt with quite briefly. The organs and tissues of the body can only function efficiently when the environment in which they work is satisfactory, and it is essential that the internal environment of the body is kept very constant with regard to temperature, chemical composition, acidity or alkalinity and oxygenation. You will therefore find that you will need to study at some length the means whereby the internal environment of the tissues can compensate for changes in the external environment and for the changes produced by body activity.

CELLS AND SYSTEMS

The basic unit of living matter (not only animals but also plants) is the cell, and the whole body is composed of cells, extracellular material and water, both intra- and extracellular. The idea of the cell was first introduced in 1839, by Schleiden and Schwann (one a botanist and the other a zoologist), and the difficulty of understanding the body mechanisms before this time can be well understood by reading an account of the process of conception which was written before it was realised that the essential step here was the fusion of two cells—a spermatozoon and an ovum. In a book published in 1821 a Dr. T.Bell M.D., wrote: 'Now the animal mucilage of semen is nearly pure albumen . . . it will powerfully attract the carbon which the blood rejects . . . Now albumen united with carbon forms fibrin . . . hence as the blood enters the albuminous drop, the union of its carbon with the surrounding albumen will form a muscular ring; successive rings will of course be formed as the carbonated blood passes onward . . . Thus muscular canals will be formed and vessels will shoot through the mass.'

It is difficult for us, nowadays, to imagine trying to work out the process of fertilisation without a knowledge of cells and their genetic structure, but Dr. Bell's account must have sounded quite convincing in his time. Once the idea of the cell was appreciated, however, it became possible to understand how cells could become modified to carry out particular functions, how they could be put together to form tissues and how damaged tissues can repair themselves.

This book will therefore start with a description of the basic components of a typical cell and the structure of cells which have become specialised to carry out particular functions. Most cells do not exist as isolated units (except for blood cells and other cells which have a nomadic type of existence) but are grouped together with other cells of similar or differing types together with varying amounts of extracellular material, to form *tissues and organs*. These will be described in the next chapter.

The body as a whole consists of a number of different tissues and organs which, for the sake of convenience, can be grouped into various *systems*. Thus the muscular system consists of a number of separate muscles, together with their tendons and other accessories, whose main functions are to maintain the posture of the body and to carry out movements, including locomotion. The nervous system consists of the brain, spinal cord and peripheral nerves which receive information from the tissues and send nerve impulses out to the tissues to stimulate their activity. A number of other systems are concerned with other aspects of bodily function and the greater part of this book will deal with each system in turn. In general, it will be found most suitable to describe the anatomy and histology of each system and then to go on to show how the structures which have been described are able to carry out their functions. In this way the two subjects of anatomy and physiology will be integrated.

INTRODUCTION TO ANATOMY

The anatomy of certain systems is of special concern to physiotherapists and will have to be learnt in considerable detail. Thus descriptions of the structure and functions of the bones, joints and muscles will occupy a great deal of the

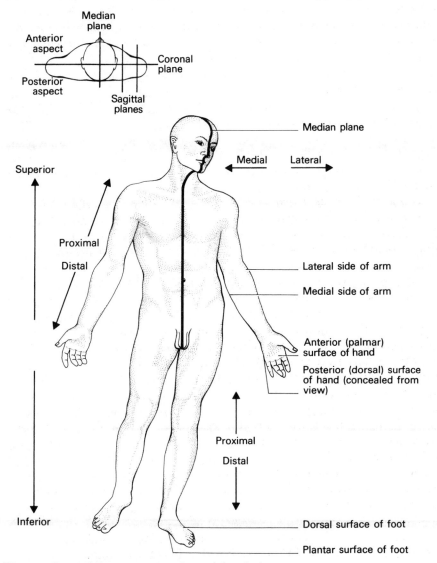

Fig. 1.1. Some of the most commonly used descriptive terms.

book, but since anatomy is basically a descriptive study you will find it helpful to read these chapters in conjunction with a practical study of the living body and of the dried bones. In this way you should try to obtain a mental picture of the structures which lie under the skin so that you will be able easily to understand what you can feel in patients in whom the structures may be abnormal. In order to describe these structures, however, it will be necessary to use a standard terminology and to refer to their position relative to each other and to the body as a whole. In order to do this, the body is imagined to be standing upright with the feet pointing towards the front and the palms of the hand facing forwards—this standard position is known as the *anatomical position* and descriptive terms are always used as though the body were in such a position. Thus the wrist is always said to be 'below' or 'inferior to' the elbow, even though when the arm is held above the head it is above it.

The terms used to describe the positions of structures are shown in Figure 1.1. The anterior surface of the body is that which faces forwards in the anatomical position and is alternatively known as the *ventral* surface. The corresponding surface of the hand is usually called the *palmar* surface. The posterior surface of the body is also known as the *dorsal* surface or *dorsum*. A plane parallel to this is called the *coronal* plane and a plane at right angles to the coronal plane is the *sagittal* plane. A structure which is nearer to the midline of the body than another structure is said to be *medial* to it, while if it is farther away it is *lateral*. Thus, in the anatomical position the thumb is on the lateral side of the hand and the big toe is on the medial side of the foot. Distance from the root or origin of a structure is indicated by the terms *proximal* and *distal*, proximal meaning nearer to and distal meaning farther away from. Hence the wrist is distal to the elbow and a hair has its root at the proximal end. The terms *superficial to* and *deep to* are often used in referring to the relative position of structures and these are particularly useful in describing regions such as the sole (or plantar surface) of the foot which is normally examined upside down. Thus, it is tempting to say that, in this region, the muscles lie beneath the skin whereas in the anatomical position they are actually above it—confusion may thus be avoided by saying that they are *deep to* the skin.

Terms which are used to indicate movements are shown in Figure 1.2. Bringing together two ventral surfaces is known as *flexion* (as in bending the elbow) and movement in the opposite direction is *extension*. Bringing a structure nearer to the midline is *adduction* (as in lowering the arm to the side) and the opposite movement is *abduction*. In *rotation*, a limb or the trunk is rotated about its long axis—in the case of the limbs, rotation may be defined as *medial* or *lateral*. A special form of rotation takes place in the forearm, where the radius, carrying the hand with it, rotates around the ulna. This type of rotation in which the palm of the hand is turned to face backwards is called *pronation*, while the opposite movement is called *supination*—thus the forearm is supinated in the anatomical position. A special type of movement also occurs in the foot. Turning the foot so that the sole is directed medially is known as *inversion*, the opposite movement being *eversion*. In the case of the hip and shoulder and at certain other joints, a compound movement occurs which is known as *circumduction*. In this movement the limb describes a cone whose apex is at the joint; circumduction at the shoulder joint may be demonstrated by drawing a circle on the blackboard while the elbow is kept extended.

A comparison of the upper and lower limbs may, at first, be puzzling when these terms are used. The big toe and the thumb are obviously comparable and yet the big toe is medial and the thumb lateral. Similarly, one would expect the back of the knee region to correspond to the front of the elbow. These apparent contradictions may be understood by reference to the intra-uterine development of the limbs. In the fetus, at an early stage of development, the palms of the hands and the soles of the feet both face medially so that the thumb and the big toe occupy a corresponding position; they are said to lie at the *pre-axial* border of the limb—i.e. the border nearest to the head end of the fetus. Similarly the dorsal surfaces of the hand and of the foot face in the same

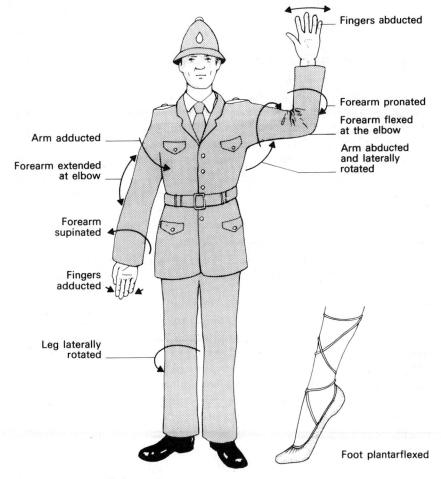

Fingers abducted

Forearm pronated

Forearm flexed at the elbow

Arm abducted and laterally rotated

Arm adducted

Forearm extended at elbow

Forearm supinated

Fingers adducted

Leg laterally rotated

Foot plantarflexed

Fig. 1.2. Terms applied to movements.

direction, as do the knee and the elbow. Bringing the ventral (palmar) surface of the hand nearer to the ventral surface of the forearm is flexion and bringing the sole of the foot nearer to the calf of the leg is also flexion. During development the upper and lower limbs rotate in opposite directions so that the thumb becomes lateral and the big toe medial, but flexion at the ankle joint, by definition, is still a movement of straightening the foot on the leg (Fig. 1.2). Flexion does not therefore necessarily mean 'bending'. Confusion can be avoided by using the terms *plantarflexion* (or true flexion) and *dorsiflexion* (extension). Once this explanation is understood you will see why the muscles on the calf of the leg and on the front of the forearm are flexors; you will also find it easier to understand the distribution of dermatomes (p. 87).

If you have studied zoology you may find these terms confusing, after being accustomed to describing quadrupeds, in which the head is at the anterior end of the body rather than at the upper end. Another feature of the biped which you will need to think about is the position of the centre of gravity. This is not important in quadrupeds since the centre of gravity normally lies well within the quadrangle formed by the four feet; but in man, a perpendicular dropped from the centre of gravity (the *line of gravity*) must fall between the two feet for a position of stable equilibrium to be maintained. If the weight is mainly taken on one foot, as usually occurs in a relaxed standing position, the line of gravity must pass through the weight-bearing foot and this calls for adjustment of the position of the rest of the body, particularly the pelvis, spine and shoulders. This principle is well understood by artists, and a study of Michelangelo's David or the Venus de Milo will demonstrate it to perfection. You may observe it yourself whenever you stand up from a sitting position—the first movement to take place is not extension of knee and hip joints but flexion at the hip joint so that the centre of gravity comes to lie over the feet; only then can the weight be transferred to the feet so that the body can stand upright.

The line of gravity passes slightly behind the hip joint, towards the front of the knee and just in front of the ankle. In this way the upright posture can be maintained with the minimum of muscular effort, although if the muscular system is observed closely, small adjustments can be seen from time to time, particularly in the lower limb. You can see these even more clearly if you watch the leg muscles of someone standing on one leg. One of the functions of the lower limb and trunk muscles, therefore, is to balance the body on the lower limb and this function will be referred to from time to time in Chapters 5 and 11. The head too, is balanced on top of the spine so that only a very slight muscular contraction is needed to hold it steady—the effect of relaxation of these muscles can be seen during a boring lecture, when heads are seen to drop forward suddenly before being wearily hauled back to the upright position.

Finally, you must realise that the anatomy and physiology which will be described in the book refers only to the most frequently encountered individuals. It is well known that no two persons have identical fingerprints and a brief examination will reveal that the pattern of veins on the back of the hand varies from individual to individual. Similar variations occur in the more important structures in the body so that it is not uncommon, for example, to find an extra rib attached to the last cervical vertebra, a sciatic nerve which divides in the pelvis rather than behind the knee and a biceps muscle with three heads instead of two. Similarly, the 'normal' blood pressure is very difficult to define since it varies within such wide limits, and the state of muscle tone (p. 106) may vary enormously so that one person may be 'droopy' with poor posture and another alert and upright. The more important and common of these variations will be described in this book but you must be prepared to meet many other variations, and at times you will find it difficult to decide whether a particular feature is normal or diseased. Similar variations will be met in psychological as well as in physical make-up; and whereas one patient will refer to a pain as 'excruciating' another patient with a similar condition may merely admit to it 'hurting a bit'.

INTRODUCTION TO PHYSIOLOGY

In the preceding section, the 'normal' blood pressure was used as an example to discuss the importance of normal biological variation, an extremely important concept which you must understand thoroughly. We can also discuss the subject of blood pressure in this section in order to show some of the ways in which physiologists think about the functions of the body. The pressure which the blood exerts on the walls of its containing vessels is important in any consideration of the circulation of the blood and of its adequate distribution to all parts of the body. The physiology of blood and of the circulatory system is concerned with a study of these phenomena and the ways in which they are adjusted to fulfil precisely the requirements of the body in all circumstances. Such a study will involve understanding the composition of the blood, the origin of its various components and how their formation is regulated to maintain optimal amounts of each of them within the circulation. We shall need to study the functions of the components of the blood. Then we must study the nature of the heart and its beating which maintain the circulation of the blood and how the beat is regulated so as to ensure that output from the heart at all times balances the return of blood to it from the veins. The 'blood pressure', by which we mean the pressure in the distributing arteries, must be studied in its various aspects. What factors affect blood pressure? These include output from the heart, elasticity of arteries, viscosity of blood, the body's position and the rate at which blood can 'run off' from the arteries into the small vessels in the tissues. These studies must be clearly distinguished from those concerning the regulation or control of arterial blood pressure.

Virtually all physiological functions, as well as growth and all structural development, are regulated so that they perform optimally for the body's needs. It is usually possible to look at a regulatory process from the point of view of the purpose to be served. Since the brain controls so many of the body's activities and since it is the first part of the body to be affected by a fall in blood pressure, we can say: 'The blood pressure *is regulated so as to ensure* an adequate supply of blood to the brain.' We can take this a stage further. The principal regulatory mechanism for arterial blood pressure is the alteration of the resistance to flow of blood from arteries to the tissues. This is achieved by altering the calibre of the muscular branches of the smallest arteries (*arterioles*) in response to nerve impulses reaching them from the brain. Our purposive statement can now read, 'The blood pressure is regulated, so as to ensure an adequate supply of blood to the brain, *by nerve impulses which cause alterations in arteriolar resistance elsewhere in the body*. But this statement is still incomplete. It fails to tell us the final vital link in the whole picture. How does the nervous system know the blood supply to the brain or the blood pressure are not correctly adjusted? Signals must reach it to inform it that the pressure is incorrect. In this instance it is the level of blood pressure itself that is the source of the signals. In the walls of some arteries (aorta and carotid arteries) are pressure detectors (*baroreceptors*) that signal the pressure level. If this rises the receptors produce more impulses. If it falls they produce fewer. It is these impulses, varying with changes in pressure, that then change the output of nerve impulses to the arterioles. This change will always be such that the alteration in blood pressure is reversed. Thus a fall in blood pressure (such as that produced by a blood dona-tion, about half a litre) causes reduced signalling from the baroreceptors. This *reduction* in number of incoming impulses to the brain results in an *increase* in the number of impulses going out from the brain to the arterioles. *This causes an increase in resistance to flow of blood from the arteries and thus a rise again in the arterial blood pressure towards normal so that the perfusion of the brain with blood is adequately maintained.*

This whole description is one example of what physiologists call *homeostatic control* or *homeostasis,* which means 'maintaining a constant state'. Here a change in blood pressure is detected by the baroreceptors and these bring about changes in the nervous system and its control of blood flow to tissues that reverse the changes in blood pressure. Engineers refer to such regulatory processes as *negative feedback loop controls*. This expression reminds us of two related phenomena that form part of all control systems. The first is that alteration from the 'normal' or 'set' level in the aspect of the system being studied itself brings about the response that affects the system—hence the term, 'feedback loop'. Thus changes in blood pressure activate the blood pressure regulating system. The second is that the loop always operates *negatively*, a fall in blood pressure setting off a response that causes a rise in blood pressure.

Homeostatic control systems are purely automatic, in that they are 'built in' to the body, just as thermostats (consisting of a temperature detector and a switch) are built into domestic electric water heaters. The functions of detecting a change and of producing the response that reverses the change are constant features of such control systems. It is always possible, and much to be preferred, then, to replace a purely purposive description of a change in function by a cause–effect one. This describes a change as a response (the effect) caused by some sort of receptor being stimulated and this stimulus (the cause), directly or indirectly, produces the change.

This cause–effect description of events does not only cover homeostatic mechanisms which maintain the constancy of state of a living body, but it must also be used when change of state leads to changes in some physiological function. Let us take another example. When muscles contract, they transform chemical energy into kinetic energy and a limb, it may be, moves. The ultimate source of the chemical energy comes from the oxidation of materials obtained from the blood, the oxygen itself also coming from the blood. One might use a purposive form of explanation. 'Blood flow in exercising muscle increases *so that the muscle may receive* the increased oxygen, etc., that it needs.' But such a statement tells us nothing about the mechanism involved. The correct cause–effect description would be: 'The exercising muscle *liberates materials which act on arterioles* and cause them to relax. *More blood can then flow into the muscle capillaries* and increase the oxygen, etc. supplied to the muscle.' Although in this example the identity of the materials released by exercising muscle which act on the arterioles is not known for certain, the statement of cause and effect is far more precise than the purposive statement. It tells us what mechanism causes the blood flow increase and leads us to the point of further research and perhaps to a rational understanding of disease processes and their treatment.

Such cause–effect processes, whether concerned with homeostasis or not, are found throughout the body whenever any function is being started, maintained at a constant level, or stopped. Attention has recently been directed to intracellular control processes, both of the manufacture of proteins (p. 13) and of many of a cells's chemical activities. In the study of the function of organs and systems of the whole body, physiologists continually study the cause–effect links between events. Where these links use the nervous system as the mediating pathway we call them *reflexes*. The control of growth and development of the body's structures must also be largely dependent upon similar cause–effect processes for its full description. Finally, all disease states, by disturbing normal function, will set off trains of cause–effect changes which must be described and understood before the disease process itself can be accurately defined and treatment logically planned.

2 · The Structure and Function of Living Cells

INTRODUCTION

Since the human body is entirely composed of cells and of extracellular material which is produced by cells, it is logical to begin a study of the body by describing the structure of an individual cell. This can be done by using as an example one of the unicellular plants or animals which are capable of maintaining an independent existence, such as the Amoeba or the yeast cell and, indeed, such cells are often used for basic research. It would be more realistic, however, in this account, to describe a typical mammalian cell, even though many of the components are very similar to those found in plants and lower animals. It is rather difficult to select a 'typical cell' since all the cells of the body are special-ised in structure in order to carry out their particular function: this specialisation occurs during development of the body and the process is known as *differentiation*. Thus, the fertilised ovum, which is less than 150 micrometres (μm) in diameter, must be capable of giving rise to cells as different in appearance as nerve cells, muscle cells and connective tissue cells. (A micrometre is one millionth of a metre and is the unit of length most often used by workers with microscopes. A human hair is about 100 μm across.) This it does by repeated divisions and as division occurs the daughter cells become more and more differentiated and less and less able to give rise to cells of other types. Development of the human body therefore involves two processes: *differentiation* (mostly occurring in the first two months of gestation) and *growth*, which involves the division of cells to give rise to more cells of the same type. In general it may be said that the more specialised a cell becomes, the less able it is to change into other forms of cell, or in some cases, even to reproduce itself. Thus, as will be described in Chapter 3, some cells of connective tissue remain undifferentiated and can therefore, when the occasion demands, divide to give rise to specialised cells of various types. To go to the other extreme, nerve cells are so highly specialised that they cannot divide to replace other nerve cells that have been affected by disease or injury so that destruction of a part of the brain or spinal cord leaves permanent damage. (It is very important to under-stand, however, that nerve cell *processes* outside the brain and spinal cord can regenerate as long as the cell body is not affected—see Chapter 8.)

THE STRUCTURE OF CELLS

LIGHT MICROSCOPY

This began in the 17th century and was the only method of examination avail-able until relatively recently. The most important advances were made in the

19th century when the technique of cutting thin sections was developed, and this, combined with various staining techniques, led to the discovery of the basic structure of the various tissues (the science of *histology* or microscopic anatomy) and also enabled something of the internal structure of the cells themselves to be observed. Originally, two principal types of stain were used (and still are, for routine histological purposes). These are the *acidic stains* (often pink in colour, such as eosin) and the *basic* stains (often blue in colour, such as haematoxylin). Portions of the cells which are essentially acidic in nature take up basic stains and are said to be *basophilic* while other parts take up acidic stains and are said to be *acidophilic* or *eosinophilic*. Thus the nucleus, with its high content of acid (DNA), stains blue with haematoxylin while the cytoplasm stains pink with eosin, unless a large quantity of the acid RNA is present, when it will stain a bluish colour (DNA and RNA are discussed later in this chapter).

These simple stains really represent a chemical reaction (acid-base interaction) on a microscopic scale, and other more complex reactions can similarly be carried out on tissue sections which, if the final reaction product is insoluble, remains *in situ*, and is preferably coloured, can be used to investigate the chemical composition of the tissues. This method of study is known as *histochemistry*.

ELECTRON MICROSCOPY

In the last twenty five years our knowledge of cell structure has been greatly extended by the use of the electron microscope (e.m.) which can produce an *effective* magnification of several hundred thousand, compared to the 1,500–2,000 which is the best that can be attained with the light microscope. Most of the description of the cell which appears below is therefore based on e.m. and also on histochemistry, which can be applied to e.m. if the final reaction product is electron-opaque. Electron microscope workers use nanometres (nm), each one of which is one thousandth of a micrometre, or one thousand millionth of a metre. Their instruments give about one thousand times more useful magnification than do the light microscopes.

THE CELL

As has been said, cells differ in structure according to the particular function which they have to carry out but these differences are quantitative rather than qualitative so that most of the intracellular structures or *organelles* described here can be found in any cell (Fig. 2.1). (The chemical make-up of different types of cell may, of course, be very different.) The most obvious component of the cell is the nucleus which can be easily seen by light microscopy when it is found to be basophilic. It is surrounded by a double-layered *nuclear membrane* which separates it from the remaining material in the cell which is the *cytoplasm*. The nucleus contains, among

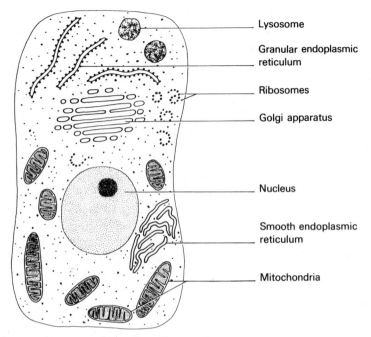

Lysosome

Granular endoplasmic reticulum

Ribosomes

Golgi apparatus

Nucleus

Smooth endoplasmic reticulum

Mitochondria

Fig. 2.1. Diagram of a typical cell and its most important components.

other things, *deoxyribonucleic acid* (DNA) which, because of its staining properties, is commonly called chromatin. This complex acid contains a large number of components whose arrangement along the molecule, as will be seen later, constitutes the *genetic code* which is able to pass information, from nucleus to cytoplasm, about the types of protein which the latter has to manufacture in order to carry out the particular function of the cell. The DNA can also reproduce itself during cell division so that the two daughter cells are exactly similar to the original cells. It would obviously be disastrous if a liver cell, for example, divided to form two nerve cells. The nucleus is thus essential to the life of the cells and if it is lost or removed experimentally, the cell will eventually die. Red blood cells (Chapter 17) obviously fall into this category and therefore have only a limited life-span.

The *cell membrane*, sometimes called the *plasma membrane*, covers the cytoplasm and separates it from the extracellular tissue. It is composed essentially of a bimolecular layer of lipid in which proteins (principally enzymes) are embedded and appears in electron micrographs as two very narrow black lines with a white line between, i.e. it is trilaminar (Fig. 2.10). It must not be thought of as a simple plastic bag holding the cytoplasm since it is an active living membrane with very important functions that will be described later.

The cytoplasm of the cell contains a number of organelles, each of which has specific functions. *Mitochondria* are relatively large structures which may be spherical or rod-shaped (Fig. 2.2). They are surrounded by a double membrane, like the nucleus, and the inner layer of this is projected into the inside of the

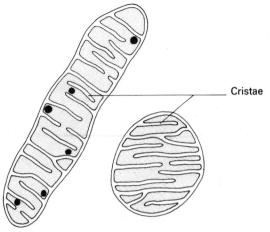

Cristae

Fig. 2.2. Mitochondria.

mitochondrion to form a series of shelves or *cristae*. This increases the internal surface area of the membrane, which is important because it is upon this inner membrane that a large number of enzymes are carried (p. 25).

Ribosomes are very small particles (15–25 nm in diameter) which are found scattered throughout the cytoplasm, either individually in small groups or attached to rough endoplasmic reticulum (see below). As will be seen later, they are responsible for stringing together the amino acids which form the proteins of the cell. The *endoplasmic reticulum* (Fig. 2.1) consists of a system of small tubules and flattened sacs or *cisternae* which are very prominent in some cells but hardly represented in others. There are two types of endoplasmic reticulum. In the *smooth* variety the tubules and vesicles consist of membrane only so that it is usually relatively inconspicuous. The *rough* or *granular* variety is similarly composed of membrane but its outer surface is studded with large numbers of attached ribosomes. The two types of endoplasmic reticulum have different functions. The smooth variety is involved in carbohydrate and fat metabolism and is also present in cells which are associated with the production of certain hormones. The precise function of the smooth variety has not yet been worked out however. The rough endoplasmic reticulum is involved in the manufacture of 'protein for export' which is carried out by the attached ribosomes. Thus, while the function of ribosomes in general may be described as protein manufacture, the free ribosome (i.e. those unattached to rough endoplasmic reticulum) produce protein for use by the cell itself, while the attached ribosomes produce protein for use outside the cell, the manufactured protein being passed, in the first instance, into the cavity of the cisternae. The subsequent fate of this protein will be outlined in the next paragraph which describes the Golgi apparatus. As an example of cells which contain numerous free ribosomes one may cite actively growing cells such as those of the embryo or those forming tumours, since a large-scale production of protein is obviously necessary for growth. Cells of the exocrine glands, such as the salivary glands, contain much

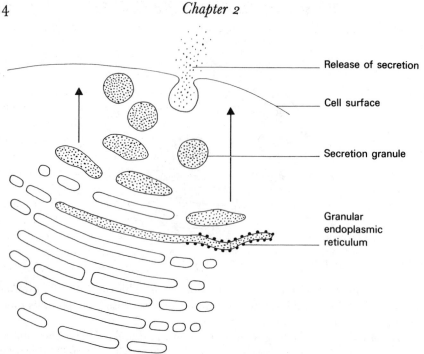

Release of secretion

Cell surface

Secretion granule

Granular
endoplasmic
reticulum

Fig. 2.3. Mode of secretion of 'protein for export'. The Golgi apparatus is at the bottom of the diagram (see text for description).

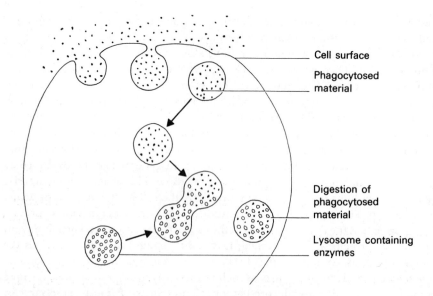

Cell surface

Phagocytosed
material

Digestion of
phagocytosed
material

Lysosome containing
enzymes

Fig. 2.4. Lysosomes and their digestion of phagocytosed material.

granular endoplasmic reticulum since they are producing large quantities of enzymes.

The Golgi apparatus was discovered by Camillo Golgi in 1898 but he was widely disbelieved until the advent of electron microscopy and histochemistry, when the Golgi apparatus was revealed as consisting of a stack of flattened saccules, together with a number of roughly spherical membrane-bound vesicles (Fig. 2.3). It is particularly prominent in cells which are secreting enzymes or other forms of protein and in these it normally lies between the nucleus and the secreting surface. The first function of the Golgi apparatus to be discovered was that of packaging 'proteins for export'. The granular endoplasmic reticulum, having manufactured and accumulated protein, passes it on from its cisternae into the saccules of the Golgi apparatus. Here it accumulates in large quantities until, from the face of the Golgi apparatus nearest to the secreting surface of the cell, large vacuoles are formed containing dense granules of secretory product (*zymogen granules*). These are passed to the surface of the cell and, when they reach it, their contents are discharged by a form of reversed phagocytosis. The Golgi apparatus is also involved in the production of some carbohydrate and of lysosomal enzymes (see below).

Lysosomes are roughly spherical membrane-bound bodies which are found in the cytoplasm. They contain a very large number of destructive enzymes, capable of breaking down many components of the cell such as proteins, fats, carbohydrates, DNA and RNA and many others. The membrane which surrounds the lysosome and prevents its contents from mingling with the cytoplasm is thus extremely important. After phagocytosis has taken place at the cell surface, the engulfed material lies in the cytoplasm enclosed by a detached portion of cell membrane (Fig. 2.4). Lysosomes can fuse with these structures, the two membranes becoming continuous, so that the lysosomal enzymes can break down the phagocytosed material without their coming into contact with the cytoplasm. Lysosomal enzymes can also be used to break down old and worn-out components of the cell itself.

THE CELL NUCLEUS AND CELL DIVISION

MITOSIS

The nucleus of the dividing cell, ever since it was first described almost 100 years ago, has aroused great and continuing interest among biologists. The dark staining material of the nucleus (chromatin) breaks up into a small number of short rods. These are called *chromosomes* and their number, size and shape are always the same for each single species of animal or plant but they vary from species to species. For example, human cells contain 46 chromosomes, those of a species of fruit-fly only 8 (this species has thus been extensively used for research purposes). The nuclear membrane has in the meantime disappeared and the chromosomes line up in the equatorial plane of the cell along a series

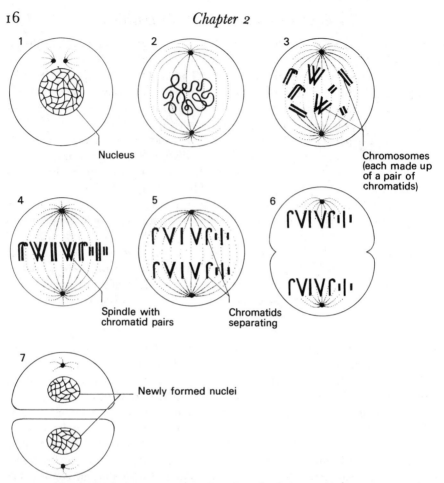

Fig. 2.5. Mitosis. Each chromosome consists of a pair of chromatids and the chromosomes themselves are also paired. When the chromatids separate, each becomes a new chromosome so that each of the daughter cells has the same number of chromosomes (8) as the original cell.

of threads which form a spindle shape (Fig. 2.5). Each chromosome is now seen to be divided lengthwise into two identical *chromatids*. The members of each pair of chromatids migrate to opposite poles of the cell. Two new nuclei are then reconstituted from the exactly shared chromosomal material. The cytoplasm constricts between the two nuclei and then the whole cell divides and the two new cells commence growing. At some time during cellular growth there is a sudden spurt of activity within the nucleus so that the chromosomal material becomes doubled in amount. In some cells chromosome doubling can actually be seen to occur at this time. Thus the appearance of the paired chromatids during nuclear division is merely a delayed revealing of a process that has already taken place during the growth of the original cells. It should be finally pointed out that when the chromosomes appear in cell division, they are always recog-

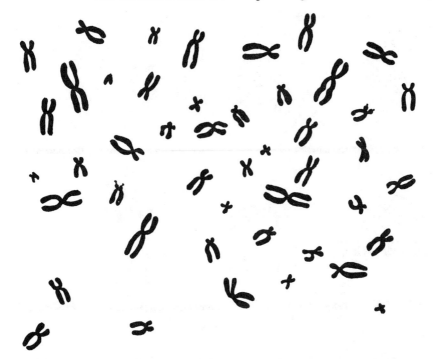

Fig. 2.6. The normal human karyotype. (From Beck, Moffat and Lloyd: *Human Embryology and Genetics*, Blackwell, Oxford.)

nisable as matched pairs. Thus the 46 that appear in human cells can be sorted into 23 pairs and the 8 in the fruit-fly into 4 pairs. One pair, however, known as the *sex chromosomes*, differ from the others in that they show a sex difference. In the female the two sex chromosomes are represented by a pair of fairly large chromosomes which are known as X chromosomes. In the male, however, one of the pair is very much smaller than the other and is known as the Y chromosome. These differences can be seen in Fig. 2.6 which shows the array of chromosomes from a human male cell—such a display of chromosomes is known as a *karyotype*.

MEIOSIS (REDUCTION DIVISION)

A modified form of cell division occurs in the cells which form the gametes, that is, the ova and the spermatozoa. In this, instead of the chromosomes, each divided into its chromatids, lining up at the cell's equator, the two members of each pair of chromosomes come to lie together and then each chromosome of the pair travels to opposite ends of the cell (Fig. 2.7). The altered process results in each gamete, when its nucleus is reconstituted, having 23 (in the human) or 4 (in the fruit-fly) chromosomes only. This is, of course, only half the number of chromosomes in normal cells. Fertilisation of an ovum by a

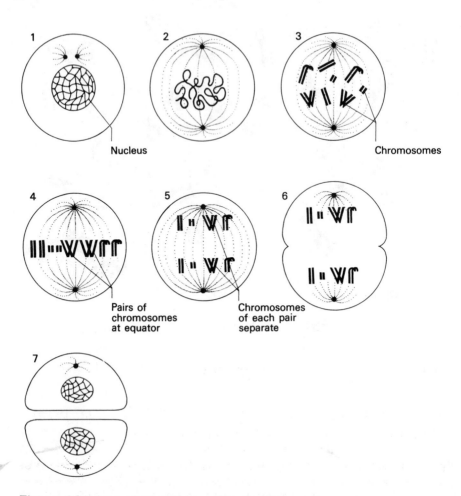

Fig. 2.7. Meiosis or reduction division. The chromosomes of each pair migrate to opposite poles of the cell so that each daughter cell has only half the number of chromosomes as the original cell. This process is normally rapidly followed by a mitosis, so that the end result is 4 daughter cells.

spermatozoon, which is completed by a fusion of their nuclei, results again in a cell containing a full complement of chromosomes, 46 in man, 8 in the fruit-fly.

While, as a result of the chromosomal reduction (Meiosis) type of cell division, all ova are identical in chromosome make up (each containing an X chromosome) the spermatozoa will be of two kinds, half of them containing an X chromosome, the others containing a Y chromosome (Fig. 2.8). At fertilisation when the ovum and spermatozoon fuse, there is thus a 50 % chance of producing cells containing two X chromosomes and a 50 % chance of cells containing an X and a Y. The former will be female and the latter male. In subsequent development of the fertilised ovum it is the presence of the Y chromosome rather

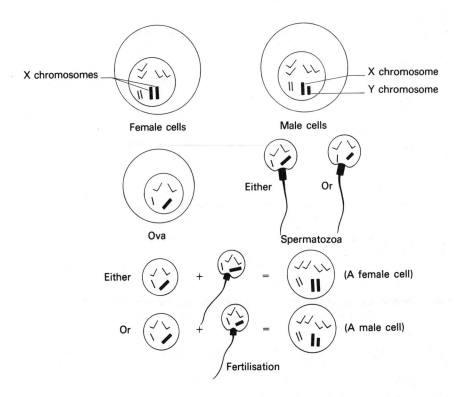

Fig. 2.8. After meiosis, each ovum possesses one X chromosome while each spermatozoon may possess one X or one Y chromosome. After fusion, the daughter cells (so called) will have two X chromosomes or one X and one Y.

than the absence of two X chromosomes that determines the development of the male gonads and thus the secondary sexual characteristics of the male in most animal species. This is shown by one of the conditions caused by chromosome abnormalities in which the patient appears to be a male although his cells contain two X chromosomes and one Y.

CELL PHYSIOLOGY

Shortening is one special function of muscle cells. Conveying impulses is the principal function of the nerve cells. All cells, however, share many common activities which are essential to them if they are to remain alive. To begin with, their chemical composition frequently differs from that of their surroundings. Maintaining this difference requires continual activity on the part of the cell. For example, sodium ions leak into many cells and have to be actively extruded. The term 'ion' indicates that the metallic sodium atom (a fiercely reactive substance) has lost its outermost electron, thereby gaining a unit of positive electrical charge and becoming chemically quite inactive. All cells need, at all times or at least some of the time, to manufacture certain chemicals, either for their own use or for that of the whole body. Nerve cells conduct impulses and muscle cells contract. All of these processes need energy, just as a car or any other machine needs energy. One of the most important functions of cells is making energy available for these processes.

ENERGY PRODUCTION

The source of the cell's energy is found in the energy contained in many chemical compounds. Chief among these are the sugar and fat molecules. By a series of oxidation reactions (combining these substances with oxygen from the air) this energy can be liberated, just as the energy contained in petroleum molecules is set free when these are burnt with oxygen in the internal combustion engine of a car. In fact, the amount of energy liberated when sugar or fat is oxidised in the body is always exactly the same as the amount liberated when the same substance is burnt in chemical apparatus, so immutable are the laws of all matter, whether living or dead. In the car engine, however, the energy is liberated in an explosion that forces the piston down, turns the crankshaft and so drives the car forwards. About four-fifths of the explosively liberated energy is wasted and emerges from the engine as heat. In the living body a glucose molecule is gently and gradually split up, each part being combined with oxygen under controlled conditions; about 50 % of the energy set free is trapped in chemical links known as *high energy phosphate bonds* in a special group of compounds which will be described later.

No cell could live for a moment in the explosive release of energy and resultant high temperature in a car engine, but all cells, by virtue of their elaborate chemical machinery, can achieve the same end result in an entirely controlled manner. How is this done? The sugar or fat molecule is fairly inert. It can be left exposed to oxygen at the temperature of the body and nothing will happen to it. In a cell, though, the glucose molecule will combine with an *enzyme*. (All enzymes are protein molecules, sometimes combined with smaller, more reactive materials or co-enyzmes. Each enzyme is specific for one chemical reaction in which it acts as a catalyst.) This combination then subtly distorts the glucose molecule so that a phosphate molecule can be combined with it. This phosphate comes from one of the high energy phosphate bonds, and the energy in the bond helps to distort the glucose molecule. The glucose molecule is now 'activated' and this activation is the essential first step for a whole series of 40 or 50 reactions during which the glucose molecule is gradually broken down into its carbon and hydrogen atoms and these are then combined with oxygen and the energy released is trapped in further high energy phosphate bonds. At each stage the sugar fragment has first to be combined with a particular enzyme which alters it in its own way and so enables the next reaction of the series to occur. Further enzymes are concerned with the trapping of energy when hydrogen combines with oxygen to form water. Some enzymes are found floating freely in the cytoplasmic water, but most of the energy-producing enzymes are found in an orderly array within the mitochondria. These organelles have truly been called the 'powerhouse of the cell'. They are found in virtually all living creatures and in fact in almost all—from amoeba and yeast to man— energy production in the cell proceeds by exactly the same series of chemical reactions. Furthermore, the enzymes associated with these reactions are similar if not identical in composition in all living tissues.

HIGH ENERGY PHOSPHATE MOLECULES

This group of compounds are also found in almost all living beings and in all are the means whereby energy is collected from oxidation reactions, stored, and used in energy-requiring reactions. The essential chemical link in all is the bond formed between two phosphate (or phosphoric acid) units (Fig. 2.9). When this bond is split by water, large amounts of energy are set free as heat. It is thought that when this link is formed in living tissues energy is stored in it and when it is destroyed, as in the addition of phosphate to glucose, the energy is absorbed in making this reaction proceed. But the phosphates do not exist free in the cytoplasm. They are far too reactive in that form. Instead they form part of larger compounds, of which the first studied, and perhaps the most important, is *adenosine triphosphate* (ATP). In this, there is an organic alkali or base, adenine, joined to a sugar molecule, ribose. On to the sugar are attached a chain of three phosphate groups (Fig. 2.9). The undulating lines joining these indicate the putative sites wherein the energy is stored. Normally it is the ter-minal phosphate group only that is involved in energy reactions, but under

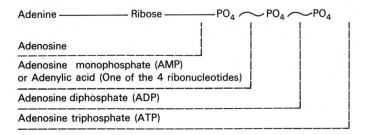

Fig. 2.9. Structures and names of adenosine triphosphate and some of its derivatives.

certain circumstances the second bond may also be split. The new compound formed after the loss of the terminal phosphate and transfer of energy is called *adenosine diphosphate*. This is reconverted to ATP in the mitochondria as oxidation liberates further energy.

Adenine may be replaced by related organic bases and phosphate may combine by a similar high-energy bond with creatine. Creatine phosphate (CrP) is plentiful in muscle tissue and is used there as an emergency energy store which is drawn upon at the onset of vigorous contraction. Other types of bonds are also described as high energy bonds. These are found in several of the intermediate compounds formed as sugars or fats are broken down and oxidised. Their presence indicates an intramolecular rearrangement and they are formed and destroyed as part of the process whereby the stored energy of the foodstuff molecule is transferred to its final place, the rebuilt ATP. For the purpose of this book these will be ignored, but ATP and CrP will be repeatedly mentioned, the latter in muscular contraction and the former in any energy-needing cellular reaction.

PROTEINS

Enzymes, which have already been mentioned, are one form of protein. They are of the class of materials known to chemists as *catalysts*. These enable chemical reactions to occur that would otherwise not happen at all, or would happen at a very much slower speed, but they are not themselves consumed. Enzymic catalysts usually act by combining with the reacting substance(s), altering their shape so that new compounds can be formed, and then allowing these reaction products to escape from the enzyme. In the living body, proteins are also found as part of the structure of cells, for example in their cell membranes (Fig. 2.10) and intracellular organelles. Hair, nails and the dead scales shed from the skin surface are all protein. The thick and the thin filaments of muscle are each composed of two or more different proteins, and these are, of course, the chief components of the protein that we eat in meat. All proteins are complex chemical compounds with large molecules, being built up as chains of smaller compounds called *amino acids*. There are about 20 of these and the precise order in which each is incorporated into a protein molecule containing from 50 to 200 or more of them determines the eventual shape and chemical activity of the protein. Altering one of the amino acids in a compound containing 100 of them can completely alter both the shape and therefore the chemical activity of the protein. This may in turn completely alter the function of the cell in which this protein is an essential enzyme. How the proteins are made and how their

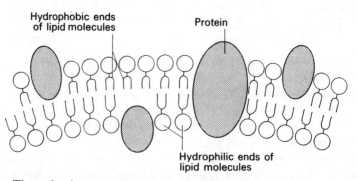

Fig. 2.10. The molecular composition of a cell membrane.

production is regulated will be described later. We must now look at some of the functions of proteins.

The production of energy in a usable form is one of the most important of these. Manufacture of adenine from simpler substances can be another (we normally get enough of this from the food). Cells have to make many new compounds as they grow and protein enzymes are involved in all of these activities, sometimes many of them being needed to make a single compound. In their complex activities cells inevitably produce poisonous waste products. Some of these are immediately converted by one or more enzymes into innocuous substances which can later be safely discarded from the body. Apart from making new materials during growth, many cells have to make other materials continually if they are to perform their specific functions. Again, enzymes are required in making these materials, often several working one after another to convert the raw materials into the finished product. In many of these reactions of protein enzymes, just as in the chemical reactions that occur as the filaments of contracting muscle slide past each other, energy is required. This energy is often needed to activate a somewhat inert compound so that it will take part in some reaction that would otherwise not proceed at all. The source of the energy is almost invariably ATP, for which the appropriate enzyme must have a special binding site.

Proteins are also very much concerned with defence of the body against various forms of damage. The antitoxins, such as are produced when one is immunised against diphtheria or tetanus, are proteins that neutralise the toxins of these bacteria. So also are the 11 compounds, collectively known as Complement, the most important of which attack a foreign cell and, literally, punch holes in its surface membrane, thus effectively destroying it. There are also protein enzymes that attack other proteins and break up their molecules. The blood clotting process is activated by at least one such process, though in the final stage enzyme thrombin helps the soluble fibrinogen molecules to clump together into the insoluble, sticky fibrin threads (p. 482). When the clot has served its purpose, another protein, fibrinolysin, appears which can break up the fibrin again.

Finally, proteins play various roles in the structure of the cell membrane and in its function. This is such an important topic that it is considered in the next section.

THE CELL MEMBRANE

The cell membrane could not be seen before the advent of the electron microscope, though its presence was never doubted. It was realised that the cell must have some sort of limit to prevent its contents mixing with the environment. It was also well known that the interior of the cell differed in many ways from its surroundings, that these differences were essential for normal function of the cell and that cells and their limiting membranes actively ensured that these differences were maintained. The e.m. shows a triple-layered membrane (see above) and it is now known that it consists of fat and protein.

Fat molecules are, at one end, miscible with water (*hydrophilic*), but the other end will not mix with water (*hydrophobic*). The triple-layered membrane of cells is thought to consist of two layers of these fat molecules with the hydrophobic ends in the middle of the membrane and two surfaces of hydrophilic ends, one facing the interior of the cell and one facing its external surroundings. Protein molecules are also present in the membrane, bound to either surface of the fat molecules as a complete or a discontinuous layer, or stretching right through its full thickness. They may help to bind the fat molecules together, but in many cases can be shown to act as membrane carriers or enzymes, enabling substances to enter or leave the cell. Figure 2.10 portrays this concept of the cell membrane but this is only one of a number of different ideas about the structure of the membrane and it is probable that different cells have quite different types of membrane.

The chemical composition of the cell membrane has important results. Hydrophilic molecules and any but the smallest of hydrophobic molecules have difficulty in passing through it, though to some small molecules like water it is freely permeable. Proteins folded so that hydrophobic side chains on the amino acids form a fat-soluble surface, shuttle to and fro through the hydrophobic interior of the membrane and thus transport substances through it that the cell requires or that it needs to extrude from its interior. One such substance is the sodium ion, since more than a trace of sodium is harmful to the proper functioning of the cell. In the tissue fluid surrounding cells however, sodium is present in about 3 parts per 1,000 parts of water. Sodium ions continually leak passively through cell membranes. So we find in cell membranes a sodium-transporting system of proteins which carry the sodium ions back outside again. These proteins need ATP for their normal working, for to extrude sodium 'uphill' from where there is only a low concentration to a region of high concentration requires energy. This energy comes immediately from ATP and ultimately from the oxidation of foodstuffs. It has been estimated that one-third of the total energy used by the living body is involved in keeping sodium out of the cells! A further result of this active extrusion of sodium from cells is the appearance of positive electrical charges on the cells' exterior and a corresponding negative charge in their cytoplasm. We will have much more to say about cell membranes, sodium transport and the electrical charges when nerve cell function is described.

REGULATION OF CELLULAR ACTIVITY

The foregoing will have given some idea of what cells look like and of some of the things that they do. In all cells, though, these must be considered along with the ways in which their activities are regulated. In fact, much of both physiology and developmental anatomy (how growth, for instance, is controlled) are concerned with regulatory mechanisms.

The fields of regulation that should be mentioned or discussed here, since they apply generally to all cells of a multicellular organism, are:

1   Intrinsic metabolic regulation
2   Regulation by the nervous system
3   Regulation by 'chemical messengers' or hormones
4   The genetic mechanism
5   Regulation of cell development

INTRINSIC METABOLIC REGULATION

Many examples of this process are found operating in such systems as the entry into and breakdown within cells of substances like glucose or fatty acids that are the source materials of the energy required for so many of the functions of cells. Two only will be considered as examples of the general process.

The mitochondria contain the enzymes that perform the major oxidative process of all living tissues, the combination of oxygen with hydrogen to form water. Along with these are the further processes that trap the energy released and retain it in the terminal inter-phosphate bond of ATP. In simple terms the following two reactions are catalysed in the mitochondria:

$$2H_2 + O_2 \longrightarrow 2H_2O + \text{Energy}$$
$$\text{ADP} + \text{Inorganic phosphate} + \text{Energy} \longrightarrow \text{ATP}$$

The first of these *can* proceed independently of the second (as can be seen in cells poisoned by the weed-killer di-nitro-ortho-cresol, or exposed to too much of the thyroid gland's secretion). In normal life, however, the oxygen uptake by mitochondria is controlled by ADP. They start using oxygen only when ADP is present and stop, even though all the other materials for their many reactions are still available, as soon as the ADP is used up. Thus ADP acts as the regulator of mitochondria, switching oxidation on when it is formed, and off again as it is all converted to ATP. Many more of the metabolic processes of the cell are switched on and off by this sort of process. It is fashionable nowadays to employ terms from the branch of engineering known as *cybernetics* or control theory. This ADP–oxidation control is known as a *servo mechanism*, in which output (the energy available to form the terminal interphosphate bond of ATP) follows signal (the quantity of ADP present) and the result is always to maintain a very low concentration of ADP and a maximum concentration of ATP.

Equilibrium conditions, in which a specific optimal concentration of a material is to be maintained, require a different regulator system that is known to engineers as *negative feedback control*. An example is that of glycogen storage in muscles. Glucose can be stored in an inactive form as glycogen, for use as an energy source in an emergency (for more about this, see Chapter 7). A glucose transfer system exists in the muscle cell membrane to enable the hydrophilic glucose molecules to cross the hydrophobic membrane. Evidence suggests that the amount of stored glycogen affects the transfer system, so that the more glycogen there is, the slower the rate of transfer of glucose into the cell. Ultimately the glycogen concentration will reach the level at which no glucose enters the cell at all—the equilibrium state. In other words, the increasing glycogen content of the cell has an inhibiting effect on glucose transport. Both servo mechanisms and negative feedback control mechanisms can be seen operating in many different physiological systems. These paragraphs merely serve as an introduction, at the cellular level, to the ways in which they operate. Servo mechanisms, by *stimulating a response*, return the concentration of a substance (input signal) to zero. Negative feedback, in which a product of a response *inhibits that response*, maintains the optimal concentration of a substance.

REGULATION BY THE NERVOUS SYSTEM

This does not apply to all cells. Its most important aspect, regulation of muscle contraction, will be discussed at length in Chapter 4. Many glandular cells discharge their secretions when stimulated by the nerves supplying them. Examples of this are the production of saliva and other digestive tract secretions. The nerve supply to the adrenal gland medulla regulates its activity, impulses causing adrenaline to be released into the blood. The adrenaline produces reactions that raise the blood pressure. This happens when a fall in blood pressure causes an increase in the number of impulses in the nerves to the adrenal gland. Again we have an example of both servo mechanisms and of feedback control. An increase in nerve impulses causes an increase in adrenaline secretion (servo). An increase in blood adrenaline causes an increase in heart artery and vein muscle activity (servo). The resultant rise in blood pressure reduces the stream of impulses to the adrenal gland, this reduces the output of adrenaline and hence the blood pressure (negative feedback).

REGULATION BY 'CHEMICAL MESSENGERS' OR HORMONES

Many of a cell's metabolic functions are controlled from a distance, by the liberation into the blood stream, by endocrine glands of chemical substances known as *hormones* (Chapter 26), which then travel in the circulation to reach their *target cells*, i.e. the cells which are affected by the hormone.

Insulin is one of the best known of these hormones. It is formed in certain cells of the pancreas and is liberated from these cells when the blood concentraion of glucose rises after a meal. When carried by the blood to muscle tissue, insulin increases the rate at which this tissue extracts glucose from the blood, presumably by raising the activity of the membrane transfer system. It has a similar action on fat cells, where it further depresses the rate of fat liberation into the blood.

Another hormone, *adrenaline*, can cause the muscle present in the walls of some blood vessels to relax but that of most to contract. It causes the heart muscle to contract with more force. In many tissues it increases the rate of glyco-gen breakdown and oxidation to provide more energy.

Less widespread in action is a group of hormones that help to regulate the formation of digestive juices. *Gastrin* appears in the lining of the distal part of of the stomach when food is present. It enters the blood and thus reaches and stimulates the acid-producing cells of the stomach. In a similar way other hormones are formed in the wall of the small intestine by the partially digested food when it leaves the stomach for the duodenum. These hormones pass in the blood to the pancreas and liver where they stimulate the production of the pancreatic digestive secretion and expulsion of bile from the gall bladder.

THE GENETIC MECHANISM

The long history of the study of genetics is a fascinating one, but beyond the scope of this book. It begins with the work of Darwin and Wallace in the 1840s and 1850s on the existence of variation within an animal species and with Mendel's discovery in the 1860s of how the variations breed true. At the beginning of the 20th century the curious similarity between Mendel's 'characters' or *genes* and the cell nucleus in division was recognised. The genes, or units of inherited information which govern a cell's activity, became generally acknowledged as being carried on the chromosomes of the dividing cell. In the last twenty-five years the chemical nature of the gene has been determined, as have the ways in which the genes affect cellular function. One can say that each gene provides the blueprint for the formation of one of the enzyme proteins that the cell needs. We also know how the blueprint is copied and, in some cases, how its presentation to the protein-building machinery in the cell is itself regulated. Genes can be switched on and off or kept in a latent state, according to the stimuli that reach them within a cell's nucleus. For example, the cells in the root of a hair on the chin are stimulated to produce keratin, the hair protein and thus to build a hair, by the hormone from the testis of the adolescent and adult human male. They are inhibited by the hormones from the female ovary. The genes responsible for producing the enzymes that make keratin are thus inactive in the presence of female sex hormone, becoming active if this is absent and male hormone present. In all other cells of the body these same genes will be present, but they are held inactive by some means at present unknown, since hair formation would be inappropriate in either sex at any site other than skin!

Another example of gene regulation, presumably, is in the series of changes that occur in repeatedly exercised muscle. Increases in bulk, in the amount of the oxygen-storing compound myoglobin, and in the amount of glucose-splitting and -oxidising enzymes have all been described. In all, it must be assumed that the genes have been given some signal to produce more blueprint copies which are then transferred to the protein-forming machinery so that more of the requisite proteins will be made.

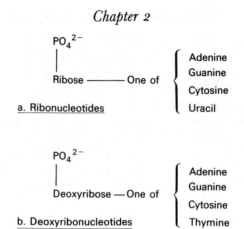

a. Ribonucleotides

b. Deoxyribonucleotides

Fig. 2.11. Structures of ribonucleotides and deoxyribonucleotides.

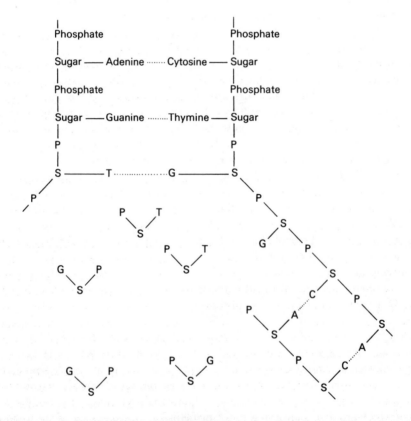

Fig. 2.12. A deoxyribonucleic acid double-chain (top) splits into two and one half (bottom right) is rebuilding its complementary chain from free nucleotides.

In order to explain how these processes work, one has to consider their chemical aspects. Crude chemistry by the 1940s had shown the chromosomes to be made of large chain-like molecules called *nucleic acids*. These could be broken down and were found to be made up of large numbers of smaller molecules of four types, called *nucleotides*, each being repeated many times over in a single nucleic acid molecule. The nucleotides themselves were divisible further into three still smaller molecules. One of these is *phosphoric acid*, another a sugar, *deoxyribose* (hence the name 'deoxyribonucleic acid', or DNA for short). The third is any one of the four organic base molecules, *adenine, guanine, thymine* and *cytosine*. Other nucleic acids were found in the cell cytoplasm. These had ribose for their sugar and three out of the four bases, but had *uracil* instead of thymine. They are known as *ribonucleic acids* (RNA). Figure 2.11 shows the general formula for a nucleotide. It will at once be seen that ATP (Fig. 2.9) is simply the adenine nucleotide with two extra phosphate groups added to the one of the nucleotide.

It was in 1952, after patient years of work by a number of people in building models of chemicals and in X-ray analysis of their structures, that the structure of the nucleic acid of chromosomes was finally worked out by Sir Francis Crick and Dr. James Watson. Two coils of alternating sugar and phosphate molecules were found, linked to each other by pairs of the bases. Thus the base adenine always linked with thymine and guanine always with cytosine. Furthermore, it was found that the double coil or helix could readily be pulled apart into its two halves, giving single chains of nucleotides. If this happened when free nucleotides were also present then each single chain rapidly built up its 'other half' and two identical double coils were made, replacing the original single molecule (Fig. 2.12). Thus we find that a longitudinal splitting of nucleic acid molecules can occur in a manner very like that of the chromosomes at cell division, and that each new nucleic acid molecule is as exact a replica of the original as is the chromosome.

It is now possible to identify the chromosomal nucleic acid molecule, deoxyribonucleic acid (or DNA) as the gene. A typical DNA molecule may contain 400–900 nucleotide units. A nucleotide containing any one of the four bases will occur in each place along the chain, so there can be many millions of nucleic acids; in a nucleic acid containing only 5 nucleotides, for instance, there are $4 \times 4 \times 4 \times 4 \times 4 =$ about 1,000 possible permutations. The complete sum gives an answer greater than 1×10^{99}. It is not surprising that there are in fact enough genes, and then some to spare, to account for all the myriads of ways cells can function and can differ from one another.

Work since the 1950s has helped to confirm the assumption that each gene controls the production of a single enzyme protein. It is known that the amino acid sequence along the protein chain is absolutely vital to the function of the protein. A single one out of place in a molecule containing 100 or more may completely wreck that protein's function. If a gene were to provide the information for protein building, then there must be some correspondence between nucleotide sequence along the DNA chain and amino acid sequence along the protein molecule. But while there are 20 amino acids, there are only

4 nucleotides, so a one-to-one correspondence is not possible. What if the nucleotides were combined in pairs and each pair corresponded to one amino acid? To permutate 4 nucleotides by pairs allows only of 16 possible arrangements. It is only when we take nucleotides in groups of three at a time, with a total of 64 possible nucleotide triplets, that we have sufficient. nucleic acid material to devise a 'code' for the 20 amino acids (most amino acids have more than one code triplet). Such a code exists and has by now been largely broken down. For each of the 20 amino acids the bases of the corresponding nucleotide triplet are known (the sugar and phosphate are the same for all nucleotides). However, the proteins are not themselves put together in the cell nucleus. On the DNA of the gene molecule a ribonucleic acid molecule is built, with uracil (instead of thymine), pairing with adenine, guanine pairing with cytosine and with ribose in the sugar-phosphate backbone. This is a *messenger RNA* molecule, for it leaves the nucleus and travels through the cell's cytoplasm to meet there one or more ribosomes. These may be free in the cytoplasm or attached to rough endoplasmic reticulum. The messenger RNA (mRNA) slots into one end of the ribosome, on the surface of which are two further binding sites for yet another type of RNA molecule. These are small soluble molecules. Because they pick up amino acids and transfer them to a growing protein chain, they are called *transfer- or tRNA*. All these processes take place in the cytoplasm. The tRNA with the activated amino acid on one end now approaches the ribosome mRNA complex. At the end opposite to the amino acid on the tRNA molecule there is a group of three nucleotides that can 'fit' one group of three nucleotides on the mRNA molecule (adenine to uracil, guanine to cytosine). Only when the correct three molecules are present in the adjacent bit of mRNA will a particular tRNA attach itself to the first binding site on the ribosome. The mRNA then moves through its slot on the ribosome, carrying with it the tRNA and amino acid to the second binding site. The two binding sites for tRNA are so positioned that the next three nucleotides of the mRNA molecule are now in position at the first tRNA binding site. The appropriate tRNA, carrying its own amino acid, can now attach itself to the ribosome. The two amino acids are placed so that they will combine using the energy originally supplied by the ATP molecules. The first amino acid is freed from its tRNA which leaves the ribosome. This leaves the binding site free for the second tRNA and the mRNA moves another three nucleotides through its slot, carrying with it the second tRNA and amino acid. A third tRNA now moves into the first binding site, fitting correct bases on to the mRNA, and carrying its specific amino acid, which is now positioned to join the amino acid chain in its own appropriate place. In this way the sequence of base triplets of the mRNA nucleotides becomes translated into the amino acid sequence of the corresponding protein molecule. One mRNA molecule can become slotted into several ribosomes at the same time, on each of which the protein molecules will be built up, always strictly in accordance with the base triplet sequence of the mRNA molecule. Figure 2.13 illustrates this process. Special base sequences on the mRNA molecule signal the start and stop points for building the protein chain. The mRNA molecules are only

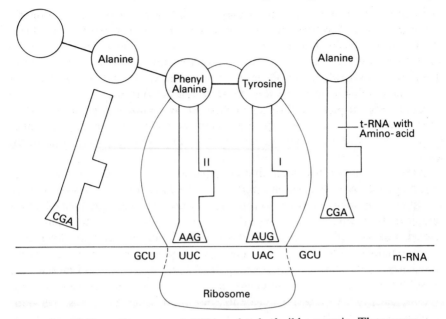

Fig. 2.13. mRNA, a ribosome, and tRNA molecules build a protein. The process moves from right to left.

formed by the genes when appropriate signals penetrate the cell's nucleus. Other signals can suppress the gene – mRNA transcription. Some of the stimulating and inhibiting signals are known (as in the example of hair growth on the chin) but many are not.

REGULATION OF CELL DEVELOPMENT AND OF GROWTH

All cells of a living body, originating by repeated cell division from a fertilised ovum, ought, if nuclear division has proceeded correctly on every occasion, to possess all the genetic information contained in that original cell. In each and every cell all the chromosomes have reduplicated themselves exactly and at each division the daughter cells have each received their full complement of chromosomes and therefore of genes. Each cell ought, therefore, to be capable of performing all the functions that all other cells perform. It contains the DNA molecules in its nucleus and the protein-forming equipment in its cytoplasm. We must look at some of the aspects that affect the differing development and behaviour of the cells in different organs of the body.

It is necessary to say at first that very little is known about organ differentiation in the developing embryo organism. Why one cell group should form a liver and another a gastrocnemius muscle seems in part to be due to their respective positions in the embryo. The 'messenger' which instructs some cells to become a liver, and others to form a muscle, is quite unknown. There are similar mysteries surrounding cell growth. We know that if half the liver is cut out then the remaining half grows until the total liver mass is restored. Growth then ceases. If muscles are exercised vigorously in certain ways their bulk similarly increases. In this case the opposite is also true—paralysed muscles or muscles rendered inactive through splinting will actually waste away. If widespread paralysis occurs early in life a limb may never reach its full size. We know very little about how these processes of growth and its cessation are themselves regulated and we have to take it on trust that the pattern, being laid down in the beginning, will be followed without fail throughout normal life. This concept is obviously of the greatest importance to physiotherapists.

When we turn to the functional activities of cells, there is now just a little information about how these are regulated. We have already stressed that the activity of a cell depends upon its protein enzymes, that these are formed on the template of the mRNA molecules by the ribosomes, this mRNA template being copied from the DNA gene molecules in a cell's nucleus. But the liver cell's genes actually provide templates for one set of enzymes and other proteins, and a muscle cell's genes for a different group of proteins. Both liver and muscle cells make the materials needed for the energy-producing machinery already described. Neither cell produces digestive enzymes, insulin or other hormones. Yet both contain an identical share of all the genes that were originally present in the chromosomes of the fertilised ovum. The wonder is not how genes work but how they are kept from working!

The concept of 'repression' has been developed. Some factors normally prevent mRNA from being formed on the DNA molecules and it is only when a gene has been de-repressed that it can then produce its mRNA which then enters the cell's cytoplasm, finds ribosomes and tRNA with amino acids and can start producing the relevant protein (enzyme) molecule. In most cases the nature of repressors or de-repressors is not known, though in some they have been identified. One example is seen in the hormones that determine secondary sexual characters and also the menstrual cycle. These, called *oestradiol* and *progesterone*, belong to the chemical class of compounds known as steroids. It is thought that all the steroid hormones act upon genes in cell nuclei. Thus oestradiol represses hair formation on the female human face, but de-represses growth in the lining of the uterus, while progesterone initiates its secretory activity. In an analogous way, one must think of nerve stimulation and exercise as acting as de-repressor agents on muscle cells, allowing then the formation of extra contractile proteins and metabolic enzymes of these cells, production of all these having been long known to depend upon muscular activity. In this case no specific chemical substance has been shown to have de-repressor activity.

3 · Basic Histology

As was mentioned in Chapter 1, although the cell is the basic unit of living matter, cells only exist as isolated independent elements in relatively few situations such as in the blood and, as will be seen in this chapter, as wandering cells in connective tissue. More often they are grouped together in an orderly fashion to fulfil a particular function and they are usually combined with extracellular material such as collagen or elastic fibres and diffuse ground substance in which the cells are embedded. Such combinations of cells and extracellular material form tissues of various types such as muscular tissue, nervous tissue and so on. Tissues do not usually exist in a pure unmixed form but are mostly combined with other forms of tissue to form a whole organ. Thus a muscle such as the biceps consists not only of muscular tissue (which itself is formed of muscle cells and extracellular material) but also of blood vessels and nerve fibres. It is important to remember that examination of a dissected body or of a microscope slide cannot give a true impression of the appearance of tissues during life since in both cases they have been treated with chemicals to preserve them ('fixation') and, in the case of histological specimens, they have been stained. The process of fixation hardens the tissues and precipitates many components such as proteins. Most tissues (except for those of the skeletal system) are soft and pliable and have a high water content. The latter, as well as the pH, the osmotic pressure, the temperature, the oxygen pressure (pO_2) and numerous other factors have to be kept very constant to form an internal environment (*milieu interne*—p. 549) which is essential for the proper functioning of the tissues.

THE BASIC TISSUES

The body is composed of four basic tissues which are (a) *Epithelia,* (b) *Connective tissues,* (c) *Muscular tissue* and (d) *Nervous tissue.* The latter two tissues are highly complex and of very great importance to physiotherapists, so they will be dealt with in later chapters concerned with the muscular and nervous systems.

EPITHELIA

Epithelia (Sing. *epithelium*) consist mainly of cells, with relatively little extracellular material, and they form a covering or lining layer for the body surfaces, both internal and external. In a few cases cells may become detached from the surface to form clumps of cells within a tissue, but this is unusual and for the most part the above definition is true. Epithelia are thus very often protective in funtion, serving

33

to prevent damage from external sources, to prevent bacterial invasion, to keep internal surfaces moist and slippery and to prevent moisture loss from external surfaces. In addition, many epithelia have very specialised functions and their structure is modified accordingly. Perhaps the best example of this is the skin, whose functions are so varied and important that this tissue is dealt with in a separate chapter. In many epithelia, glands are present. These are formed by an invagination of cells from the surface and the glands themselves and their ducts are thus lined by epithelium (Fig. 3.3).

Simple epithelia (Fig. 3.1)

These consist of a single layer of cells arranged on a very thin flat membrane of extracellular material known as the *basement membrane*. *Squamous epithelium* consists of a single layer of extremely thin cells, so thin that the nucleus forms a bulge on the surface. They are thus found in situations where it is desirable to allow diffusion of gases and the passage of salts and water across the layers of cells, such as occurs in the lining of the alveoli of the lungs. A special type of squamous epithelium known as *endothelium* forms the thin lining of the vascular system and is particularly important in the capillaries where such exchanges of gases and solutions take place. *Cubical epithelium* again consists of a single layer of cells arranged on a basement membrane, but here the cells are cubical in shape with the nucleus usually placed in the centre. *Columnar epithelium* is similar except that the cells are relatively taller. These two types of epithelium are really two varieties of the same type and sometimes it is impossible to say whether a particular epithelium is columnar or cubical. The surface of the cell may be specialised in some way—for example, it may discharge a secretion on to the surface or it may carry very fine motile hairs known as *cilia*. These have quite a complex internal structure when examined with the electron microscope

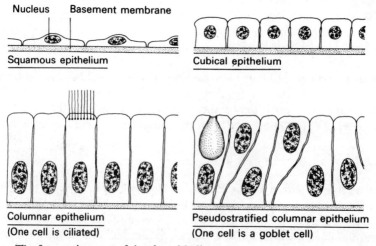

Fig. 3.1. The four main types of simple epithelium.

and they move to and fro in such a way that particles on the surface of the cell are moved in a particular direction. In several regions, columnar epithelium appears to be made up of more than one layer of cells (see 'stratified epithelia' below) but on closer examination it will be found that all cells maintain contact with the basement membrane even though the nuclei are all at different levels. This type of epithelium is therefore called *pseudostratified*. As has been mentioned already, the surface of both cubical and columnar epithelia may secrete material on to the surface and such epithelium is therefore often found lining exocrine glands (see below). In these the secretion granules may be visible with the light microscope after suitable staining and, in the case of glands which secrete mucus, the secretion may form a large mass within the cell—such cells are known as *goblet cells* (Fig. 3.1). From this account of cubical and columnar epithelia it is not difficult to guess their distribution. Secreting epithelium is found, for example, in the cells of the digestive tract and its associated glands such as the pancreas, in glands of the skin, including the sweat glands and, of course, the mammary gland, and in many other situations. Goblet cells are found in regions where the surface is covered by a protective layer of mucus, such as in the digestive tract and in parts of the respiratory tract. In the latter, the mucus which is secreted on the surface is wafted up to the top of the trachea by ciliated epithelium, which is also found in various other similar situations.

Stratified epithelia

These are epithelia in which the cells are arranged in two or more layers. In *stratified squamous epithelium* most of the cells are flattened, although the cells nearest the basement membrane may be cubical or columnar (Fig. 3.2). It is in these basal cells that active cell division takes place and as the cells move nearer and nearer to the surface they become more and more flattened until eventually they are shed, to be replaced by more cells produced by the basal layer. Such an epithelium is therefore found in regions which are subject to much wear and tear and where continuous replacement is needed. It is thus found lining the mouth, pharynx and oesophagus, and, particularly, in the skin. The stratified squamous epithelium which forms the superficial layer of the skin is, in fact, rather specialised in that it is a *keratinised* epithelium. That is, the cells, as they

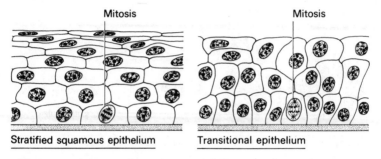

Fig. 3.2. The two main types of stratified epithelium.

pass towards the surface, become loaded with a special protein known as *keratin*. These cells ultimately die and form the insensitive horny superficial layer of the skin which is shed as minute flakes as replacement proceeds. This layer forms the very tough and waterproof surface of the body and is particularly thick in regions where friction occurs such as on the palms of the hands and the soles of the feet.

A special type of stratified epithelium is found lining the urinary tract. This is *transitional epithelium* and it is formed of a basal layer in which cell division occurs, one or more intermediate layers consisting of oval or polyhedral cells and a superficial layer of oblong cells. This description is necessarily vague since the appearance of the epithelium changes with the state of distension of the organ, particularly in the bladder. This epithelium is relatively waterproof and allows for extreme distension.

GLANDS

Glands are organs of secretion and are of two types. The first type do not have ducts, the active cells of the gland passing their secretions directly into the blood stream which transports them to other parts of the body. Such glands are known as *endocrine (or ductless) glands* and they are so important to the general functioning of the body that they will be dealt with separately in Chapter 26. The other type are known as *exocrine* glands and their secretions pass through ducts on to an epithelial-covered surface, internal or external. It is a general rule (although there are a few exceptions) that such glands develop as invaginations of the surface epithelium. The invagination may or may not branch and the epithelial cells which have been invaginated form the secreting cells of the gland and also the duct of the gland, as shown in Fig. 3.3. In the branching type of glands, the secreting cells usually become arranged round a central lumen to form *alveoli* or *acini*. The glandular cells become surrounded by connective tissue

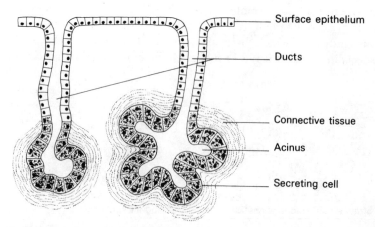

Fig. 3.3. Simple and compound glands.

to form the complete gland, which does, of course, receive a blood supply and a nerve supply. Examples of such exocrine glands are the sweat glands and the mammary glands which open on to the skin, and the salivary glands which open into the mouth. The pancreas is peculiar in that it is a mixed endocrine and exocrine gland, the former component secreting into the blood stream (p. 195) and the latter secreting via the pancreatic duct into the intestine.

CONNECTIVE TISSUES

The connective tissues comprise a very wide variety of tissues in which the cellular components are embedded in a relatively large quantity of extracellular material which may itself take various forms. The connective tissues thus comprise connective tissue itself (see below), cartilage, bone (and, perhaps surprisingly to the student, blood, in which the extracellular material is fluid.) Bone and blood are highly specialised tissues and are of particular importance to the physiotherapist so they will be described in later chapters. This chapter will therefore concern itself only with *connective tissue* and cartilage. The terminology here is highly misleading since connective tissue, as described below, is one of the connective tissues, as defined above, but this is standard terminology and must, unfortunately, be used.

Connective tissue

This takes a number of different forms. The basic form is known as *areolar connective tissue* (Fig. 3.4) which is a soft, loose, moist tissue, found as a 'packing

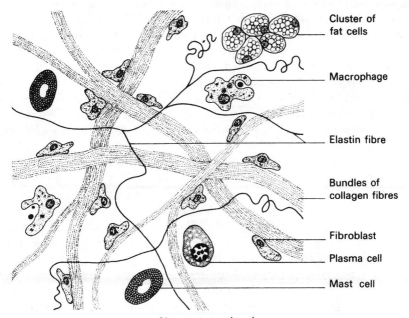

Cluster of fat cells

Macrophage

Elastin fibre

Bundles of collagen fibres

Fibroblast

Plasma cell

Mast cell

Fig. 3.4. Some of the components of loose connective tissue.

material' around viscera, muscles and other structures, particularly when mobility is needed so that one structure is enabled to slide over another. It is found in rather large quantities, for example, between the scapula and the chest wall so that the scapula and its attached muscles can move freely over the ribs. In the dissecting room, after fixation, it does not greatly resemble living connective tissue and a better idea of its form during life can be obtained by inspecting a joint of veal, which contains a large quantity of areolar connective tissue. It consists of cells of various types, extracellular substance (or matrix) and collagen and elastic fibres. The other types of connective tissue are variants of areolar tissue since the basic components are similar for the most part but one or more of them predominate, as will be seen later. The cells of areolar connective tissue include: (a) *Fibroblasts*—which are cells capable of laying down collagen fibres (see below) and, if necessary, of proliferating if a great deal of collagen is needed as occurs, for example, in scar formation. (b) *Macrophages*—these are rounded or stellate cells which are actively phagocytic so that they contain many lysosomes which may be loaded with phagocytosed material. When necessary, for example in infections, they migrate to the site where they are required. (c) *Plasma cells*—these cells are characterised by a very large quantity of granular endoplasmic reticulum which, you will remember, indicates the manufacture of 'protein for export'. The protein in this case consists of antibodies which combat bacterial toxins and other foreign proteins. (d) *Mast cells*—granular cells which produce both heparin and histamine as well as certain other substances. (e) *Undifferentiated cells*—which are capable, when necessary, of giving rise to other types of connective tissue cells so that the tissue is readily capable of reacting promptly and effectively to any abnormal conditions. (f) *Fat cells*—these are cells which become so loaded with fat that the nucleus becomes displaced to one side and the whole cell is little more than a fat-filled envelope. These may be scattered throughout the tissues but may also be accumulated together to form large masses of fatty or *adipose tissue*. Such tissue is found, for example, in the superficial fascia (see below), particularly in females. In histological sections, adipose tissue looks like wire netting, since the reagents used in preparing sections remove fat unless special precautions are taken. The cells therefore appear empty except for the nucleus which is compressed against one wall of the cell.

These miscellaneous cells, which vary in number in different types of connective tissue, are embedded in an extracellular matrix composed of protein-carbohydrate compounds which have a complex chemical structure. Also running through this matrix are a variable number of fibres. *Collagen fibres* (also known as white fibres) are made of protein, are produced by fibroblasts and form interlacing bundles. As will be seen later, when they are present in large quantities they make an extremely strong tissue, e.g. in ligaments and tendons. One particular type of collagen fibres are known as *reticular fibres* which form a supporting network in some tissues. *Elastic fibres* are found singly rather than in bundles and they may branch and reunite. They are elastic in the sense that they are highly extensible.

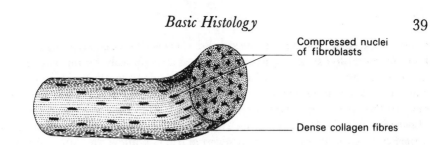

Compressed nuclei
of fibroblasts

Dense collagen fibres

Fig. 3.5. Tendons and ligaments are composed of tightly packed collagen fibres with scattered fibroblasts.

Areolar connective tissue (sometimes called 'loose' connective tissue) thus has a number of specialised components. The other forms of connective tissue, in which one or more components predominate are as follows. *Dense connective tissue*, or *white fibrous tissue*, consists almost entirely of dense masses of collagen fibres with scattered fibroblasts. The cells are compressed by the bundles of fibres so that they are stellate in cross sections (Fig. 3.5). Such tissue is immensely strong and forms ligaments, tendons and deep fascia. Ligaments and tendons thus have precisely the same histological structure and are only different in that tendons join muscles to bone while ligaments unite two bones. *Elastic tissue* contains principally elastic fibres and often has a yellowish colour. It is strong but stretchable, and is found, for example, in the ligamenta flava (p. 227) and in the lung (p. 519). *Reticular tissue* contains mainly reticular fibres and various types of cells. It will be described later in connection with the lymphatic system, with which it is particularly involved.

Superficial and deep fascia

Since the former consists mostly of adipose tissue and the latter of dense connective tissue, they will be described here, although they are more closely related to the skin and the muscles respectively. Superficial fascia is the layer of tissue which is found directly under the skin. It consists of areolar tissue and adipose tissue, principally the latter. It varies in thickness in different parts of the body, being thick, for example, over the abdomen and buttocks and more or less absent in the eyelids and the scrotum. It is very much thicker in the female than the male and is principally responsible for the characteristic shape of the former, although the shape of the skeleton also contributes to some extent. The breasts consist of a localised accumulation of fat in the superficial fascia, in which the components of the mammary gland are embedded. In places it contains a great deal of white fibrous tissue which binds down the skin to the underlying tissues. This occurs particularly in areas which are exposed to pressure, for example in the buttocks, the palm of the hand and the soles of the feet. The reason for this is that both fibrous tissue and fat are relatively inactive and do not need a rich blood supply—they are thus able to withstand long periods of compression, unlike very vascular tissue such as muscle.

Superficial fascia is an important tissue with a number of functions. Firstly fat is an excellent thermal insulator and superficial fascia helps to maintain body temperature. Channel swimmers are thus invariably plump and supplement their own fatty layer with 5-6 lb of lanolin (wool-fat). It serves as a convenient layer in which nerves and blood vessels can branch and anastomose before travelling to their destination in the skin. It forms a mobile layer which allows the skin to move freely on the underlying tissues—destruction of the superficial fascia, for example by a severe burn, can allow the skin to fuse with the underlying muscle and deep fascia and cause severely restricted movement. In later life, and sometimes in youth, the amount of fat tends to increase, particularly around the thorax, abdomen and proximal parts of the limbs, and may be an annoying barrier between the physiotherapist and the underlying structures which she wishes to feel.

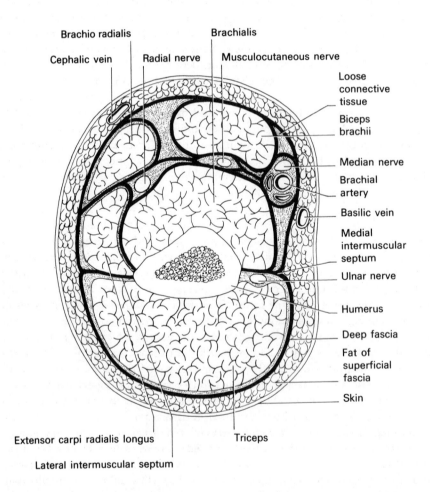

Fig. 3.6. Diagrammatic cross-section of the upper limb above the elbow to show how the deep fascia divides the limb up into compartments.

The *deep fascia* is primarily a layer of dense connective tissue lying deep to the superficial fascia, and thus investing the limbs, and parts of the trunk and the head and neck. It is best marked in the limbs, where it increases in thickness distally. It is thicker in the lower limb than in the upper limb and in the thigh it is given the special name of the *fascia lata*. Over the chest it is very thin and it is absent over the ventral abdominal wall. The reason for this is that it is a relatively inelastic layer so that if it were as thick in the chest and abdomen as in the thigh, a large meal would cause discomfort and breathing would be very difficult. As well as forming an investing layer, it extends between groups of muscles, to form intermuscular septa, and it also ensheaths individual muscles (Fig. 3.6). In places it may be especially thick; for example in the wrist and ankle regions such thickenings form strap-like bands called *retinacula* which hold the tendons in place and prevent them from 'bowstringing'. Other thickenings are found around individual tendons where they change direction, so that they act as slings or pulleys. Deep fascia also forms sheaths around bundles of nerves and blood vessels (*neurovascular bundles*) such as the axillary sheath and the femoral sheath.

Deep fascia thus has a wide distribution and has important functions to perform. In the limbs it not only helps to keep muscles, tendons and other structures in place but also forms a relatively inelastic sheath around the whole limb, so that although muscle contraction causes a bulge in the deep fascia which is visible on the surface, the muscles are nevertheless sufficiently contained to cause a rise in pressure within the fascial compartment. This compresses the deep veins and plays an extremely important part in the return of venous blood to the heart (p. 458). The muscle sheaths and intermuscular septa form slippery planes so that muscle groups and individual muscles can contract independently and their relations to each other can change in different positions of the limb. A practical example of the importance of this mobility: when a surgeon is exploring the track of a bullet or a knife wound, the limb is placed in a position similar to that in which the injury occurred so that the perforations in successive layers of muscle and fascia lie in a straight line rather than being displaced. The formation of retinacula, slings and pulleys which help to prevent displacement of tendons has already been mentioned. Finally, the deep fascia in many situation serves as an attachment of muscles—the gluteus maximus (the largest muscle in the body) is partly attached to the femur but about 75% of its fibres are attached distally to part of the deep fascia of the thigh. Attachment of muscles to deep fascia occurs to a greater extent in the lower than in the upper limb. This is important from a functional point of view since in the upper limb precise and often delicate movements are required, so that muscles and tendons tend to have precise and localised attachments. In the lower limb, movements are much more ponderous and, in particular, many of the muscles are as much concerned with balance and the maintenance of posture as with movement. They thus have rather diffuse attachments in general, and are often attached to wide bands of fascia rather than to precise bony points.

CARTILAGE

Cartilage is another form of connective tissue in which the cells are situated in a relatively large amount of extracellular material which may itself contain collagen fibres or elastic tissue. There are three types of cartilage: *hyaline cartilage, fibrocartilage* and *elastic cartilage* which, as will be seen, are named according to the composition of the extracellular material or matrix.

The cells of cartilage are called *chondrocytes* and they lie in cavities or lacunae in the matrix, often forming small groups or columns. In histological preparations they are often distorted and shrunken and they may even fall out, leaving empty lacunae. Near the surface of the cartilage the cells become more and more flattened until they blend with the fibrous layer which covers the surface of cartilage and is called the *perichondrium* (Fig. 3.7). The cells of the deepest layer of the perichondrium can form new chondrocytes and so are important for growth. No perichondrium is found at the surface of articular cartilage, which covers the ends of the bones in synovial joints. Cartilage contains no blood vessels or nerves and must receive all its nourishment by diffusion from the surrounding tissues or, in the case of articular cartilage (p. 223), from synovial fluid.

In all types of cartilage the intercellular matrix is composed of protein-carbohydrate compounds which are related to, but not identical with, the ground substance of connective tissue. In hyaline cartilage this matrix contains very fine collagen fibres but these cannot be seen in ordinary preparations. This type of cartilage is semitransparent in appearance, is tough and springy and is found where some rigidity but also some resilience and flexibility are required. It is formed, for example in the trachea, which it holds open, preventing collapse, and also in the costal cartilages at the ends of the ribs, so that the ribs are better able to move during respiration. With increasing age, hyaline cartilage tends to become calcified (i.e. calcium salts are deposited in it) so that it becomes rigid and, unlike normal cartilage, is visible in X-rays. Fibro-cartilage is similar

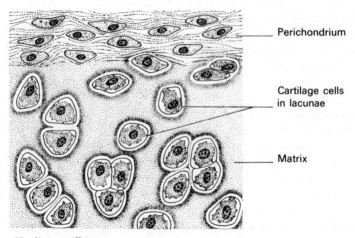

Perichondrium

Cartilage cells in lacunae

Matrix

Fig. 3.7. Hyaline cartilage.

to hyaline cartilage but the matrix contains large bundles of densely packed collagen fibres, which make it even more resistant to stresses than hyaline cartilage. It is therefore found in places where cartilage is subject to pressure and wear, such as in the intervertebral discs.) COPY

In *elastic cartilage* the matrix contains branching elastic fibres so that it is very resilient and after deformation it rapidly springs back to its original shape.) You can get an idea of the consistency of this type of cartilage very easily since it forms the 'skeleton' of the auricle of the ear.)) — STOP

In the embryo, the skeletal system is composed entirely of hyaline cartilage. The process by which this becomes transformed into bone is complicated but important, so it will be described in some detail in Chapter 9.

4 · The Nervous System—Introduction

The nervous system is the main signalling and regulating system of the living body. It also stores memories and, particularly in Man, enables communication between individual members of the species. It receives stimuli at its many and varied receptor endings. These are each adapted to respond to a particular physical or chemical stimulus, be this light rays entering the eye, vibrations in the ear, stretch in the muscles and blood vessels, touch, pressure, heat and cold in the skin or the many different blood-borne chemicals. Each receptor converts its own specific stimulus into nerve impulses which then pass to the central nervous system along *sensory* or *afferent* fibres. The nervous system also controls effector organs or tissues via *motor* or *efferent* fibres. By means of these it makes muscle contract and glands secrete. It speeds up or slows down the heart beat. It does all these things because muscles, glands, the heart and many other organs are sensitive to the nerve impulses reaching them in their nerve supply. Without the arrival of nerve impulses at their effector cells, muscles will not contract and many glands will not secrete. The heart, however, will beat, but at one fixed rate only. Nerve impulses may speed it up and make it beat with greater force, or they may do the opposite, but the heart, unlike the muscle of limbs, will continue to beat since it possesses its own inherent rhythmicity.

Before we study the more general structural and functional aspects of the nervous system, we must briefly consider the basic units of which it is made up.

The basic *structural* unit is the nerve cell which, together with all its processes, is known as a *neurone* (Fig. 4.1). In each neurone one of the processes (*the axon*) may be extremely long and bundles of these, together with various associated structures, form the nerves of the body. They are usually loosely called 'nerve fibres' and may end in a specialised effector ending or in a receptor ending. They may also end by coming into very close relationship with another nerve cell, the junctional region being called a *synapse*.

The basic units of *function* are the nerve impulse itself and the reflex response. The nerve impulse is a wave of chemical and electrical activity that travels along a nerve fibre. It is not comparable with an electrical impulse that travels along a copper wire at 300,000 kilometres per second, for nerve impulses travel at 1–100 metres per second only. Their electrical size is measured in millivolts (up to 120 mV), and some nerve fibres can conduct up to 1,000 impulses per second. In any one nerve fibre all impulses are of the same size and they always travel at the same speed. Variations in strength of stimulus at a receptor produce variations in the number of impulses it discharges per unit of time. Similarly, variations in response strength are produced by variations in the number of impulses per unit time that reach an effector organ through its nerve fibres. Also, more nerve fibres can be brought into action when an increased response is

44

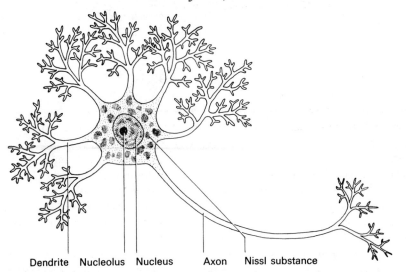

Dendrite Nucleolus Nucleus   Axon   Nissl substance

Fig. 4.1. A nerve cell. This cell has a relatively short axon.

required. As a general rule, then, we can say that in the nervous system alterations in signal strength are produced by altering impulse frequency (number per second) or altering the number of units (receptors, nerve cells or effectors) involved, and *never* by altering the size of the individual nerve impulse.

The reflex has already been briefly mentioned in Chapter 1. It is best described as an unconscious activity involving:

1 The initiation of nerve impulses at a sensory receptor organ.
2 Their transmission along sensory (afferent) fibres, through connector neurones and ultimately down motor fibres.
3 Their modulation at the synapses.
4 The production of a response by an effector organ, gland or muscle.

Some reflexes are involved in the maintenance of the constancy of the internal environment or of body posture. Others are involved in producing and controlling changes in the body in response to change in its external environment. But in all reflexes the nerve impulses start at a receptor, pass through one or more synapses and end by producing a response in gland or muscle tissue.

Two other terms are often linked with the word, reflex. The first is *reflex arc*. This is the anatomical pathway composed of nerve fibres, cells and synapses along which the impulses pass in a reflex. The second is *reflex response*, sometimes known as *reflex action*. This is the alteration in the effector organ or in the activity of a tissue that occurs consequent upon the receptor stimulation. The term reflex response is to be preferred, for it implies that the effector's change of activity is an effect caused by some prior change affecting a receptor. All reflexes are responses to stimulation of receptors. Further details about their nature are given on page 68.

After these brief introductory statements we can now look at the structure and behaviour of a 'typical' neurone in greater detail.

THE STRUCTURE OF THE NEURONE

A diagram of a typical neurone is shown in Fig. 4.1. All neurones have a similar basic structure but they differ in detail such as in the number of processes they possess. Each neurone has a central cell body containing a nucleus with a prominent nucleolus, mitochondria and other organelles common to all types of cells. In particular, however, the cell body contains a great deal of granular endoplasmic reticulum as well as free ribosomes in its cytoplasm, indicating the manufacture of a great deal of protein. These granules can be seen very easily by light microscopy when they are stained by basic dyes such as thionin. The blue granules which can be seen after such a procedure are known as *Nissl substance* (or Nissl granules) and they give some idea of the state of the cell since under certain conditions such as fatigue, or after section of the axon, the Nissl substance breaks up into fine fragments, a process known as *chromatolysis*. The cell body also contains numerous bundles of filaments which also extend into the processes of the cell and are called *neurofilaments*.

From the cell body extend one or more processes. These may be branched or run direct from cell body to termination without branches. In general, processes

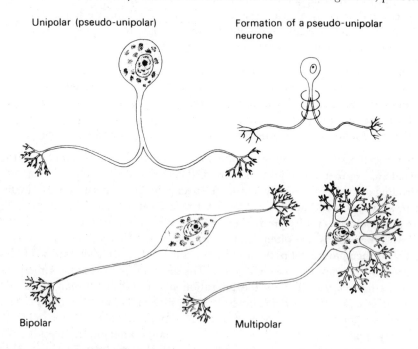

Unipolar (pseudo-unipolar)

Formation of a pseudo-unipolar neurone

Bipolar

Multipolar

Fig. 4.2. Three types of nerve cell.

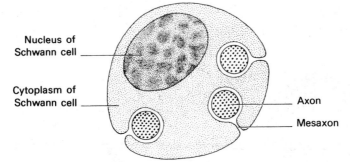

Nucleus of
Schwann cell

Cytoplasm of
Schwann cell

Axon

Mesaxon

Fig. 4.3. A Schwann cell surrounding three axons in a non-myelinated nerve.

conveying impulses towards the cell body are called *dendrites*. They are usually relatively short except for those running from sensory receptors. Processes that carry impulses away from the nerve cell body are called *axons*. They, again, may be quite short or very long (up to a metre in some cases) and there is usually only one axon to each cell body. The cytoplasm within the axon is known as *axoplasm* and the cell membrane which surrounds it is called the *axolemma*. Dendrites and axons may vary in diameter from less than 1 μm to about 10 μm (1 μm is 1/1,000,000 of a metre). Nerve cells are classified, according to the number of processes they have, as unipolar, bipolar or multipolar (Fig. 4.2). In the important cells of the posterior root ganglia of the spinal nerves (see Chapter 5) and in the ganglia of the cranial nerves, the axon and dendrite fuse with each other for a short distance and these cells are known as *pseudo-unipolar cells*.

THE MYELIN SHEATH

Most, but not all, nerve fibres are covered by a *myelin sheath*, a layer of lipid material which has long been recognised by light microscopists in sections stained by special methods for fats. Electron microscopy has shown that it is more complicated than was previously thought. It is best studied in the peripheral nerves. In those peripheral nerves which do not have a myelin sheath, the axon is embedded in a special type of supporting cell called a *Schwann cell*. More than one axon may be embedded in the same Schwann cell and as may be seen from Fig. 4.3 the axons invaginate the cell membrane of the Schwann cell to produce a *mesaxon* which resembles the mesentery of the digestive tract. Such nerve fibres are said to be *non-myelinated* and most nerves end in this way even though they may have myelin sheaths over the greater part of their course.

In myelinated nerves the axon is similarly embedded in a Schwann cell, but cytoplasm of the latter apparently wraps itself around the axon a number of times to form a sheath which, on cross section is seen to be composed of a spiral consisting mostly of fused cell membranes with very little enclosed cytoplasm (Fig. 4.4). A mesaxon can be recognised and so can the nucleus and the outer layer of cytoplasm of the Schwann cell. This outer layer, together with the

Myelin sheath Axon Nucleus of Schwann cell Node of Ranvier

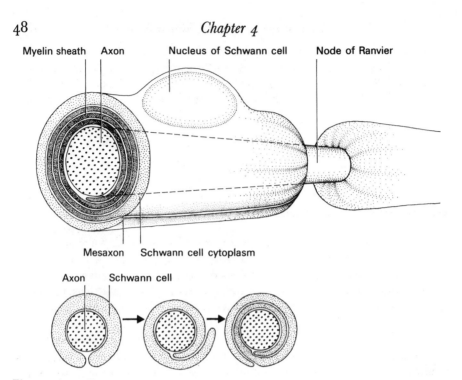

Mesaxon Schwann cell cytoplasm

Axon Schwann cell

Fig. 4.4 A myelinated nerve. The lower figure shows how the myelin sheath is formed by the wrapping of a Schwann cell around an axon.

surrounding tissues, used to be called the *neurilemma sheath* before electron microscopy showed the true nature of the sheath which surrounds nerve fibres. Along the course of a long axon there may be a very large number of Schwann cells, each cell wrapping itself around the axon to form the myelin sheath of one segment. Between each of the Schwann cells there is a very short length of the axon which is devoid of a myelin sheath. This constriction in the sheath, which can be very easily seen by light microscopy (Fig. 4.4) is known as a *node of Ranvier;* these are extremely important in speeding up the conduction of a nerve by a process known as *saltatory conduction* which will be described later in this chapter. It should be remembered that in sections prepared by routine histological methods, fatty substances are dissolved out by the reagents used in processing the tissue so that, as is the case with fat cells, only an empty space is visible where the myelin should be. Special techniques are used to show the myelin sheath and also to show degenerating myelin after damage to nerves (Chapter 8).

Within the central nervous system (p. 75), myelin sheaths are again found around the majority of fibres but here Schwann cells are absent and the sheath is produced by a quite different type of supporting cell oligodendrocytes. This is important because the presence of Schwann cells is essential for the regeneration of nerves after they have been divided or damaged. Thus, while a lesion of a peripheral nerve may be followed by a process of regeneration with complete or almost complete regain of function, regeneration of nerve fibres within the central nervous system does not occur. The paralysis which follows section of the spinal cord in spinal injuries is therefore permanent and the physiotherapist's job is to teach the patient to compensate for his disability as far as is possible by the development of remaining functions.

PERIPHERAL NERVES

Peripheral nerves, such as the sciatic or radial nerves, are composed mainly of bundles of axons whose cell bodies are in the spinal cord, the posterior root ganglia or in sympathetic ganglia (see Chapters 5 and 6). The majority of these axons are myelinated but the smaller ones may be non-myelinated and surrounded only by Schwann cell cytoplasm. Around each nerve fibre is a delicate sheath of connective tissue known as the *endoneurium* (Fig. 4.5). The individual nerve fibres are gathered together in bundles or *fasciculi* and each fasciculus is surrounded by a rather thicker sheath of connective tissue, the *perineurium*. Finally, the whole nerve is surrounded by a fairly thick connective tissue sheath called the *epineurium*. Nerve tissue, in general, needs a good blood supply, and small arteries enter nerves at intervals. These are usually too small to be noticeable, but in the case of a large nerve like the sciatic, the supplying artery is large enough to be dignified with its own name. Slight pressure on a nerve for several minutes may deprive it of its blood supply, interfere with its powers of conduction and give rise to the condition commonly known as 'pins and needles' but known clinically as *paraesthesia.*

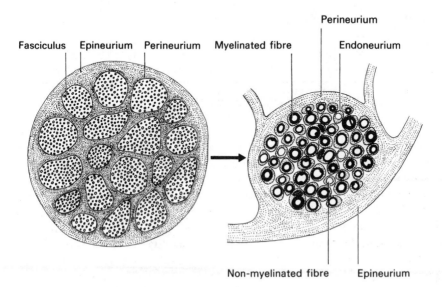

Fig. 4.5. Cross-section through a myelinated nerve. A single fasciculus, highly magnified, is seen on the right.

THE SIZE OF NERVE FIBRES

Nerve fibres are classified by size and by the presence or absence of a myelin sheath. Unfortunately two size classifications have been developed by those working separately with sensory and motor fibres. As is shown in Table 4.1 these overlap and many physiologists now refer to fibre diameter only.

Table 4.1 A classification of peripheral nerve fibres

Motor nerves		Diameter	Conduction velocity	Sensory nerves	
Function	Type			Type	Function
		μ	m/sec		
Innervation of ordinary muscle fibres (extrafusal fibres)	⎧ Aa	12–20	70–120	Ia	Annulo-spiral spindle endings
				Ib	Golgi tendon organs
	⎩ Aβ ⎫				
Innervation of muscle spindle fibres	Aγ ⎬	5–12	30–70	II	Flowerspray spindle endings, skin touch & pressure
	Aδ	2–5	5–15	III	*Pain and temperature from skin, chemoreceptors
Autonomic preganglionic	B	1–3	3–15		
Autonomic postganglionic	C†	0.5–1	0.5–2	IV†	*Pain & chemoreceptors

*Commonly now referred to as δ and C fibres respectively.
†C or Type IV fibres are unmyelinated fibres.

B and C fibres are typically found in the autonomic portion of the nervous system (see Chapter 6). C fibres are the only non-myelinated ones, and they are also present in sensory nerves, coming from pain and perhaps some chemo-receptors.

This classification was developed for the peripheral nerves outside the brain and spinal cord but could also be applied to nerve fibres within these structures. It is unfortunate that the two classifications are often used together, but as this convention has become generally adopted we shall use it in this book.

RECEPTORS

All activity in the nervous system must begin at a receptor and each receptor

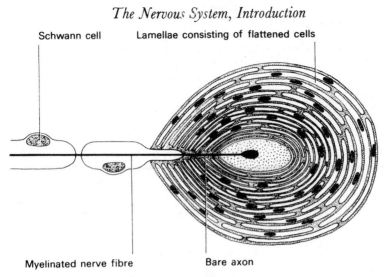

Schwann cell   Lamellae consisting of flattened cells

Myelinated nerve fibre   Bare axon

Fig. 4.6. A Pacinian corpuscle—a sensory receptor.

is designed to respond to one specific sort of physical or chemical stimulus. There are therefore many different kinds of receptor. In many of these, non-myelinated terminal nerve fibres or a mass of branching nerve fibres can be seen: for example, in the taste bud, in the Pacinian corpuscle (Fig. 4.6) and in the ear. The function of the specialised cells in receptors is to convert the chemical or physical stimulus into an electrical potential that then initiates the nerve impulse. This process will be described more fully later. It is sufficient to point out here that specialised receptor structures exist to perform this function, each 'transducing' the appropriate stimulus into nerve impulses. (A transducer is a structure that changes energy from one form to another.) The different form of different receptors is linked to their specific transducing function. In a few cases we know something of how this is done and can then relate the appearance of the receptor to its function. In many cases we do not understand the links between the specific structures and the functions of receptor organs. A typical receptor nerve ending is shown in Fig. 4.6. This is a Pacinian corpuscle such as may be found in the deeper layers of the skin. It is sensitive to deep pressure. (Note that the expression 'nerve ending' is used for both motor and sensory specialised endings of nerve fibres although in the latter case, strictly speaking, it is a nerve beginning.)

SYNAPSES

A synapse is the name given to the junction between two neurones. This usually takes the form of a junction between an axon termination of one cell and the cell body or a dendrite of another cell. In no case in the mammalian nervous system is there actual cytoplasmic continuity between one cell and another—the axoplasm is always separated from the cytoplasm of the next neurone by a *synaptic gap* or *cleft* which electron microscopy has shown to be about 20 nm in

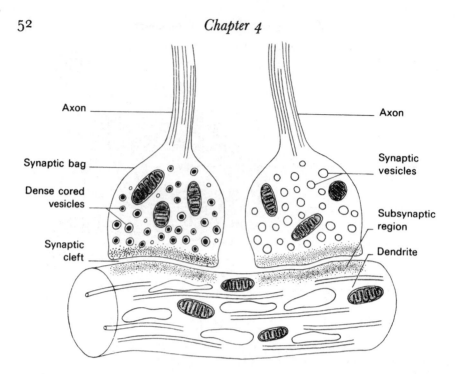

Fig. 4.7. Two axons ending by means of synapses with a dendrite. The axon on the left contains dense cored vesicles in its synaptic bag.

width (1 nm — one-thousandth of a μm). An axon usually breaks up into as many as a hundred branches before forming synapses with an adjacent neurone. Each branch loses its myelin sheath and finally terminates in a swelling called the *synaptic bag*. These were originally recognised by light microscopy and were (and still often are) known as *boutons terminaux*. The synaptic bag is, of course, bounded by the cell membrane (*axolemma*) and the cytoplasm within it contains mitochondria, occasional lysosomes and, in particular, a large number of *synaptic vesicles*. The vesicles may be spherical or flattened and, in the case of synapses containing adrenaline or noradrenaline, contain densely staining bodies; the vesicles themselves, in this case, are called *dense-cored* vesicles. The synaptic vesicles contain the appropriate transmitter substance—acetylcholine, noradrenaline or any one of a number of other substances (see p. 65). The surface of the synaptic bag may be flattened against the side of a dendrite or a cell body or it may be moulded around irregularities in the wall of the adjacent cell. In either case there is a synaptic cleft between the synaptic bag and the cell membrane of the next cell (Fig. 4.7). On both sides of the cleft there are one or more zones of particularly dense cytoplasm. The portion of the cell lying immediately deep to the synaptic cleft is called the *subsynaptic region*.

Synaptic bags are applied all over the surface of the cell body and dendrites of the receiving cell. A large neurone may have as many as 30,000 synapses on its surface, these coming from many hundreds or thousands of different pre-

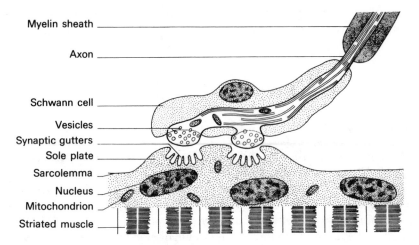

Myelin sheath

Axon

Schwann cell

Vesicles
Synaptic gutters
Sole plate
Sarcolemma

Nucleus
Mitochondrion
Striated muscle

Fig. 4.8. An axon ending in a motor end plate on a skeletal muscle fibre.

synaptic axons. In this way, one neurone may be affected by the activity of a very large number of adjacent neurones, some of which may have cell bodies which are a very long distance away in other regions of the nervous system.

NERVE-EFFECTOR ORGAN JUNCTIONS

These share many features in common with synapses (Fig. 4.8), there being no cytoplasmic continuity between the cytoplasm of the two cells involved. Although there are minor differences between effector nerve endings in various situations, a description of a typical motor nerve ending at a skeletal muscle fibre will suffice to give a general idea. Such endings are called *motor end plates;* each muscle fibre has but one such junction but a single axon may supply, by means of its terminal branches, between 5 and 150 muscle fibres. When an axon of a motor nerve is followed towards its ending it first splits up into a number of branches and each branch approaches a muscle fibre and loses its myelin sheath. The axon is not bare, however, since it is still enclosed within a Schwann cell. The axon then ends by forming an expanded branched termination which is applied to the surface of the muscle fibre. The expansion contains numerous mitochondria and spherical vesicles similar to those of a synaptic bag. The terminal expansion is separated from the surface of the muscle fibre by a gap of 30–40 nm similar to a synaptic cleft and beyond this the surface layers of the muscle fibre are modified to form a *sole plate*. In this region the sarcolemma is deeply folded to form a series of gutters and the sarcoplasm is thickened and contains numerous mitochondria.

The terminations of autonomic axons are similar but the gap between the axon termination and the smooth muscle is wider and there is no sole plate—the expression motor end plate is not, therefore, applied to such nerve endings. In the case of the smooth muscle of the blood vessel walls, there may be a double nerve supply. One pathway causes the muscle to contract and the other makes it relax, there being two sets of neuromuscular junctions. A similar situation is found in the heart.

NERVE CELL FUNCTION

IMPULSE CONDUCTION

In Chapter 2 we saw that many living cells have a mechanism that actively extrudes sodium ions as these slowly diffuse through the slightly permeable cell membrane into their interior. Within the cells' cytoplasm is found the rather similar element, potassium. The positively charged sodium ions in the tissue fluid accompany an almost equal number of negatively charged chloride ions, the two together in this fluid forming a solution of sodium chloride in the water. Inside the cell there is a small amount of chloride, some phosphate and some sodium but most of the potassium is present as a potassium-protein salt. As a result of the different amounts of sodium, potassium, chloride and protein on the two surfaces of the membrane, and of their different abilities to pass through it, we find a permanent electrical charge on the cell membrane. In human nerve cells this is around 70 or 80 millivolts (1 mV = 1/1,000 volt) with the positive charges on the outside and the negative charges on the inside of the membrane (Fig. 4.9). This electrical charge persists as long as the sodium 'pump' extrudes most of the sodium ions from the cell as fast as they leak in, and the pump will do this work throughout life so long as it is provided with energy in the form of ATP. The charged state of the membrane is often referred to as a polarised condition. Loss of charge is called *depolarisation*.

Both the relatively high concentration of sodium ions on the outside of the membrane compared with the inside, together with the positive charges outside the membrane compared with the negative charges in the cytoplasm, result in forces which tend to drive sodium ions inwards through the membrane. It is

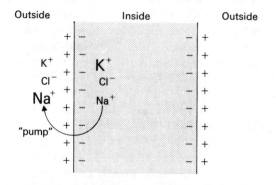

Fig. 4.9. The distribution of ions and charges on the cylindrical membrane of a nerve fibre. The size of the symbols indicate their concentrations.

only the resistive power of the membrane and perhaps the shell of water molecules that gather around the sodium ions that keeps them out. In the case of potassium ions, the high concentration inside compared with that outside tend to drive them out, but the negative charge on the inside holds them back. They have only a small inclination to move out from the cell. Similarly the balance of concentration difference and electrical force give the chloride ions a slight push in the inward direction. For each of the three substances a theoretical electrical potential can be determined which would just balance the concentration gradient and result, if the membrane were freely permeable to the substance, in there being no net movement of the substance. For sodium ions this *equilibrium potential* is 60 mV positive inside, very different from the potential actually present on the resting membrane. For potassium it is 85 mV and for chloride also about 85 mV, with the negative charge on the inside. For these two ions, then, the equilibrium potential is close to the resting membrane potential. The values of the equilibrium potentials, of course, are putting in a numerical way what we have already said about the strong tendency of sodium to move into the cell, as well as the relative lack of any tendency to move on the part of potassium and chloride.

The nerve cell membrane has another peculiar property with respect to sodium. If for any reason its electrical charge is reduced beyond a certain value,

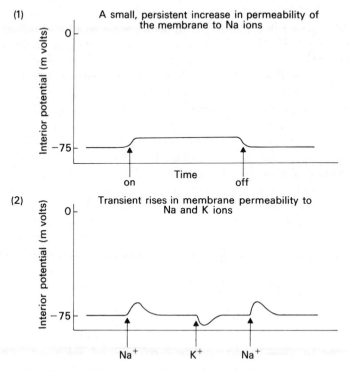

Fig. 4.10. The effect on the transmembrane potential of small changes in permeability of the membrane to sodium and potassium ions.

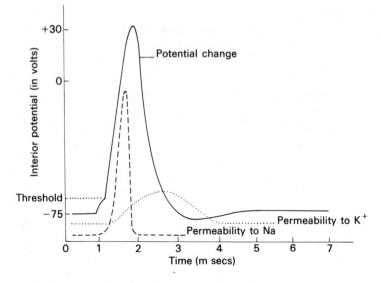

Fig. 4.11. The changes in electrical potential, sodium and potassium permeability and flow that occur in a nerve impulse.

known as the *threshold*, then it becomes more permeable to sodium ions. These can then move in more freely, carrying with them further positive electrical charges and thus further reducing the membrane potential. If nothing else were to happen this would obviously develop into a runaway state so that more and more sodium ions would pile in as the membrane's resistance to their passage falls (the fall being due to the positive charges carried in by the sodium ions themselves). Eventually the concentration gradient and electrical force would balance each other, at the sodium equilibrium potential of 60 mV, positive inside. Two different things can happen to prevent this state from ever being reached. When the membrane potential falls, potassium ions move out from the cell both because the membrane becomes more permeable to potassium ions and since the new actual potential is further from the potassium equilibrium potential. If the change in sodium permeability and fall in potential is small but persistent this movement of potassium will offset it and the membrane will now be held at a new, lower level for as long as the cause of the potential fall persists and the membrane is said to be slightly *depolarised* (Fig. 4.10.1). If, on the other hand, the change in sodium permeability is small and brief, then the outward movement of potassium ions restores the membrane potential to its original resting level in 3–10 milliseconds (ms) after the fall occurs (Fig. 4.10.2).

However, there is a threshold change of membrane potential, about 8 mV below the resting potential, beyond which the restorative movement of potassium is ineffective. Now we do have the runaway state and many sodium ions flood into the interior of the nerve cell. Even as this happens, however, potassium ions are beginning to flow out since there is no negative charge inside to hold

(1) Non myelinated axon

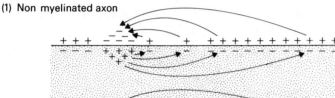

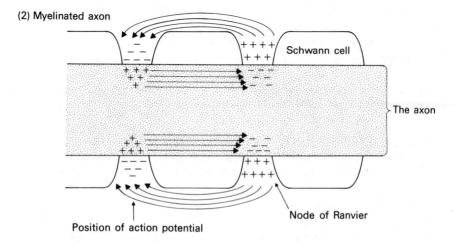

Position of action potential

(2) Myelinated axon

Position of action potential

Node of Ranvier

Fig. 4.12. Electrical current flow that precedes a nerve impulse in (1) a non-myelinated and (2) a myelinated fibre. The impulse travels from left to right.

them back. At the same time the membrane reacquires its resistance to movement of sodium ions, though before this happens the cell's interior may have become 30 mV positive on the inside. Continuing movement of potassium ions now restores the electrical charge to its original state of 80 mV positive outside. This whole process takes about 3 ms in human nerve cells (Fig. 4.11). It results in a minute increase in Na^+ and fall in K^+ within the fibre's axoplasm. The original state is restored by the Na–K ATP-splitting pump system. This takes several ms to be completed and in no way does it interfere with the passage of further impulses.

The next point to note is that when this process occurs at one site on the cell, the altered electrical state at this site will cause electrical currents to flow over the cell surface towards the site (Fig. 4.12). These currents will so reduce the positive charge at adjacent regions of the cell as to initiate the increase in sodium permeability, sodium influx, potential fall, etc., that we have already described at the original site. Thus the process will spread successively over the whole surface of the nerve cell including its processes. This is what we call a nerve impulse or a *propagated action potential*. The action potential is thus best described as a spreading wave of temporary reversal of electrical charge over the surface of the nerve cell. It can travel at speeds of between 1 and 100 metres per second, depending on the size and type of the nerve cell. The presence of the myelin sheath affects the rate of conduction. It is only at the exposed regions of the nerve fibres (i.e. the nodes of Ranvier between adjacent Schwann cells) that the phenomena just described can occur. Electrical currents move between nodes of Ranvier at the speed of electrical currents in any conductor (300 million metres per second). The other changes, especially sodium entry, take much longer. In myelinated nerve fibres, then, the impulse leaps from node to node down the fibre, and the farther apart are the nodes, the longer are the leaps and the faster the impulse travels down the fibre. This mode of impulse conduction is called *saltatory conduction* (from the Latin 'saltare', to leap). Cells can convey between 10 and 1,000 separate impulses in a second. The limit to this number is set by the time required after an impulse has passed before the recovery of the fibre allows a second impulse to pass. This time interval is known as the *refractory period*, since the fibre is inert and cannot be made to respond to artificial stimulation or to conduct another impulse. The maximum rate and frequency of impulse conduction along a nerve fibre are diminutive when compared with those of an electrical current passing along a copper wire, or with those exhibited by radio waves. However, nerve cells gain as conductors by their small size and the signal does not become attenuated as it passes along the fibre but is reamplified to full strength at each node of Ranvier.

RECEPTOR FUNCTION; THE INITIATION OF NERVE IMPULSES

Now that the nature of the nerve impulse has been described, we must see how they are produced and how they are handed on from cell to cell. We should begin this study at the sensory receptor for it is almost always possible to trace any nerve impulse back to a receptor organ, except when we are dealing with the highest levels of neurophysiological activity and their mental correlates. In technical terms, receptors are 'transducers'. They transform different sorts of physical or chemical energy into the energy of nerve impulses, in a way analogous to the transformation of a radio wave into the electrical voltage in a radio receiving set's tuning circuit. Each modality of physical or chemical stimulus must have a receptor specially designed to transform it into nerve impulses. For a few stimuli the chain of events linking stimulus to nerve impulses has been identified, but for many it is only the final event that is known. The appropriate stimulus causes a reduction in the voltage charge of the nerve terminal contained in the receptor. This electrical change is proportional in size to the size of the stimulus, and persists as long as the stimulus persists (Fig. 4.13). If the voltage reduction is great enough (about 8 mV) then an action potential is set up and this passes up the nerve fibre towards the CNS. If the stimulus still continues, the voltage charge (which probably returned to the resting level immediately after the nerve impulse left the receptor) again falls and a second impulse follows the first. This process goes on, for a while, so long as the stimulus is constantly applied to the receptor (Fig. 4.13). If the strength of the stimulus alters, then the speed at which the resting electrical charge falls becomes altered also. The stronger the stimulus, the more rapidly the charge drops, and the sooner is the threshold reached for producing another impulse (Fig. 4.13). Thus, once a stimulus is strong enough to produce impulses from a receptor, the latter signals variation in strength of stimulus by producing a corresponding variation in the frequency of nerve impulse production. The size of the individual impulses remains unchanged. If you are used to the computer analogy, it may be clear to you that the receptors in the nervous system are signalling in a digital and not in an analogue manner, although the first stage, the speed of reduction of the resting membrane potential or voltage charge, is related in an analogue manner to the stimulus strength. The rate of impulse production is not simply related to the strength of stimulus (stretch in the case of Fig. 4.13) but to a mathematical function of stimulus strength, as is shown in Fig. 4.14. This allows nerve fibres and perceptual mechanisms to respond meaningfully to very wide ranges of stimulus strength.

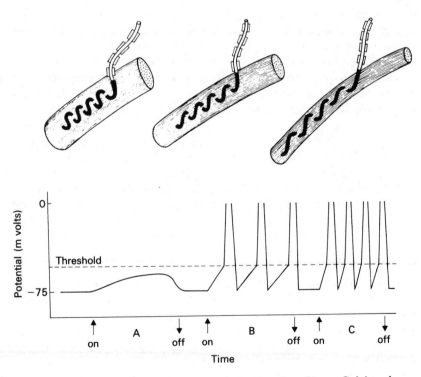

Fig. 4.13. The effect of increasing the stretch applied (A to B to C) to a Golgi tendon organ on the membrane potential of its nerve fibre.

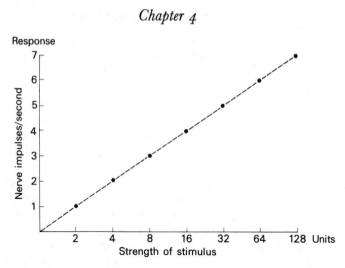

Fig. 4.14. The relationship between the strength of a stimulus at a receptor and the size of the receptor's response.

If a stimulus is applied at constant strength for some time, the response is seen to become steadily weaker, the number of impulses decreasing with time. This process of attenuation occurs rapidly in some receptors, more slowly in others. In some an initial rapid fall off in response is followed by a constant stream of impulses that does not vary, no matter how long the stimulus is applied. This alteration in receptor function is called *adaptation*, and receptors are classed as fast-, slow- or non-adapting. Adaptation is not the same as fatigue, for an adapted touch receptor will respond fully if the stimulus is removed and then immediately reapplied to the receptor. Touch and temperature receptors in the skin, the eyes and the ears are all fast adapting, so they respond best to a change in the stimulus. Joint receptors and Pacinian corpuscles are similar. The length and tension receptors of muscles and ligaments are, however, among the most slowly adapting of receptors. If a man is to remain upright his muscle and ligament stretch receptors must be capable of signalling continually the state of these structures.

SYNAPTIC TRANSMISSION

The nerve fibres bearing impulses from receptors come to an end by making synaptic connections with one or more of the vast number (perhaps 10 million million or thereabouts) of nerve cells in the brain or spinal cord. As the nerve fibre approaches its destination, it breaks up into as many as 100 fine branches, each ending in a synaptic bag, the vesicles of which contain one or another of a group of specific chemicals called *neurotransmitters*. While the identity of these has been proven in a few sites, in most cases we do not know which substance might be present in any particular synaptic bag. We do know, though, fairly precisely, what happens when a nerve impulse reaches, for example, an excitatory synaptic bag. A small amount of transmitter substance is liberated into the gap between the two nerve cells. This diffuses across the gap and becomes attached to the surface (subsynaptic) membrane of the second (post-synaptic) nerve cell immediately adjacent to the synaptic bag. For a brief period, probably less than 1 msec, the subsynaptic membrane becomes permeable to sodium ions. The transmitter is rapidly destroyed and sodium permeability falls back to normal. While it was raised, however, enough sodium entered to cause the membrane potential of the post-synaptic neurone to fall by 2 mV. This is not enough to cause the development of an action potential in the post-synaptic neurone (Fig. 4.15A). Movement of potassium and perhaps of chloride ions restore the *status quo* in a further few ms. This small and brief fall in electrical charge is known as the *excitatory post-synaptic potential* (ESP). Here, once again, we have an analogue phenomenon. The total

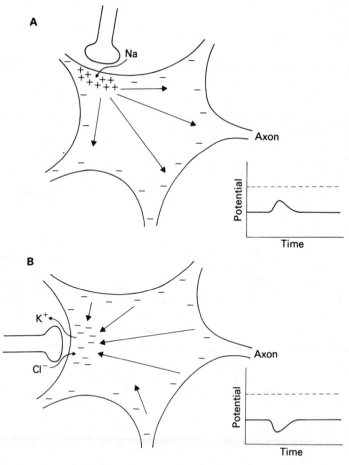

Fig. 4.15. The presumed effects of the arrival of an impulse at an excitatory (A) and an inhibitory (B) synapse.

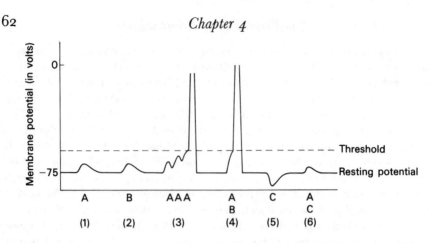

Fig. 4.16. Summation of post-synaptic potentials. A and B are formed by excitatory synapses and C by an inhibitory synapse. Synapse A, stimulated repetitively (3) or A and B stimulated simultaneously, (4) cause a nerve impulse. Stimulation of synapse C, simultaneously with A (6) reduces the potential change caused by A alone, and so reduces the ability of A to cause impulse production.

EPSP of a cell is related in size to the number of synapses liberating transmitter at one and the same time i.e. it is the sum of the separate synapses' EPSPs. This is known as *spatial summation* (i.e. summation over an area). Since EPSPs last a few msecs, rapid repetitive release of transmitter can also cause EPSP to build up, i.e. the resting nerve cell potential will fall progressively. This is known as *temporal summation* (i.e. summation in time). Eventually, either form of summation of EPSP (the simultaneous stimulation of many synapses—spatial summation, or impulses arriving in rapid succession at one or more synapses—temporal summation) will cause the membrane potential to fall so far that the threshold for action potential is reached. A nerve impulse is then generated in the post-synaptic neurone and passes down its axon. As impulses continue to arrive at the presynaptic terminal and liberate transmitter, EPSP is repeatedly developed in the post-synaptic membrane, and the post-synaptic neurone produces further impulses (Fig. 4.16). The rate of production of these impulses in the post-synaptic cell will also depend, through the rate of build-up of EPSP, on the frequency of impulses arriving at any synapse and on the number of synapses involved. Once again, as in the receptor, the strength of the effective stimulus (here the number of incoming impulses) determines the frequency of the impulses generated.

A further point about synaptic excitation is that not all of the post-synaptic cell's membrane has the same threshold for impulse generation. The place of origin of the axon, known as the *axon hillock*, is more sensitive to the fall in potential than is any other part of the cell's membrane. It is therefore in this region that the impulses are first produced. From this region they not only pass along the axon, but they also pass over the membrane of the cell body and its dendrites. Immediately after the passage of the impulses, the membrane potential returns to the original resting value before fresh EPSP is developed.

SYNAPTIC INHIBITION

Nerve cells can not only be excited, they can also be switched off or inhibited. While the transmitter released at some synapses can cause a fall in membrane potential or epsp, other transmitters, released at other synapses when impulses arrive at them, increase the permeability of the subsynaptic membrane to potassium ions, and perhaps to chloride ions. Movement of either of these down their concentration gradients will cause the interior of the cell to become more negative, or will raise the membrane potential (Fig. 4.15B). Again, the transmitter is rapidly destroyed and after a few msecs the normal potential is restored. Since this potential change is in the opposite direction to the EPSP, it will subtract from any EPSP produced at the same time. It is therefore called the *inhibitory post-synaptic potential*, or IPSP, for short. The final output from any nerve cell is the resultant of the opposing actions of excitatory and inhibitory synapses, producing respectively EPSP and IPSP (Fig. 4.16).

As well as this form of inhibition, another form exists. This is known as presynaptic inhibition. Many sensory neurones' axons, shortly before their terminal bags, have applied to them secondary synapses. These, effectively, are inhibitory for, by partial depolarisation of the primary synapse terminal, they reduce the amount of excitatory transmitter it can release, and thus the ability of the terminal to excite the post-synaptic neurone, while this neurone remains readily excitable to all other inputs since presynaptic inhibition has raised no IPSP in the cell.

Each nerve cell may have impinging on it the axon terminals of as many as 1000 other cells (although each axon, by its division, may contribute many separate synapses so that the total number of synaptic terminals may be several thousand). Some convey excitatory and some inhibitory impulses. It is not then possible to liken a nerve cell to a single transistor or valve in electronic equipment. The cell is the point where up to 1000 separate input signal trains meet and are integrated into a single output. Each input signal may similarly be delivered directly, or through a chain of intermediate nerve cells, to as many as 100 output cells. It is small wonder, then, that neurophysiologists are only at the beginning of establishing some sort of 'wiring diagram' of only the simplest forms of nerve networks. Excitatory and inhibitory synapses can sometimes be distinguished structurally. The former are typically between axon terminals and dendritic 'spines', short outgrowths from a dendrite which give it a mossy appearance. The latter (inhibitory synapses) are found between the spines, near to the bases of dendrites or on the cell body of the neurone. To confirm functionally whether any particular synapse is excitatory or inhibitory one needs to know what transmitter is liberated and what its action is on that particular post-synaptic cell. As many as eight different substances may be released at different synapses and while two of these are probably always inhibitory, the function and identity of the others at specific sites is not known. The only exception, on the 'looking at it' basis, is the case of presynaptic inhibition. Here it is not the post-synaptic cell that is inhibited, but only one of its input channels, and this is achieved, as already described, without any interference with the cell's power of responding to other inputs.

CHEMICAL TRANSMITTERS

HISTORICAL SIGNIFICANCE

In 1905 it was first suggested that *adrenaline*, first discovered ten years earlier in the adrenal glands, might also be liberated from sympathetic effector nerve fibres when impulses reached their terminals. That they released, into the surrounding fluid, an adrenaline-like substance was amply demonstrated in the 1920s and 1930s. Finally, in 1946 the closely related compound *noradrenaline* was detected in the nerve terminals of many sympathetic effector nerve fibres. In most sites the effect of noradrenaline is to stimulate the structure innervated, but in some, noradrenaline decreases their activity. The latter are examples of peripheral inhibition, which is not widely employed in control mechanisms in the human body. Adrenaline, coming from the adrenal medulla, has actions which are largely similar to sympathetic nerve stimulation. The exceptions to this remained puzzling until recently, and were the cause of much speculation. The solution of the problem will be described later, as will the precise role of adrenaline.

Dale, in 1910, began to study the active compounds present in the fungus of rye and wheat commonly known as ergot and which had borne, from the middle ages, an evil reputation. Among other materials he identified the chemical compound *acetylcholine*. This had various actions. Like small doses of nicotine, it stimulated all autonomic nervous system (ANS) ganglia (p. 75). It also mimicked the action of another fungal poison, Muscarine, which was extracted from the mushroom, *Amanita muscaria*. This substance causes the heart beat to become very weak and slow among its effects. So also did acetylcholine, and also electrical stimulation of the vagus nerve fibres supplying the heart. In 1921 it was found that, like the sympathetic cardiac nerve fibres, the vagal nerve fibres also liberated a chemical that mimicked their action. This substance was soon shown to be acetylcholine. It is now known that all parasympathetic effector nerve terminals release acetylcholine. So also do some sympathetic effector nerve terminals, all autonomic ganglion synapses or connector nerve terminals and all the neuromuscular junctions of skeletal muscle. Evidence is now being accumulated to show that acetylcholine is also found at some sites within the CNS acting as a transmitter.

Thus the concept of chemical transmission of nerve impulses between nerve cells was first proposed for adrenaline and the sympathetic nerves in 1905, first proved for acetylcholine and parasympathetic nerves in 1921 in one site, and subsequently elsewhere including neuromuscular junctions in the middle 1930s. The place of noradrenaline in the ANS was discovered in the late 1940s and the search for chemical transmitters in the CNS is still being actively pursued at the present time. Both noradrenaline and acetylcholine are implicated in these studies, as are numerous other chemical substances (see Chapter 4).

RELEASE, DISPOSAL AND SYNTHESIS OF TRANSMITTERS

Both acetylcholine and noradrenaline are present in nerve terminals in micro-

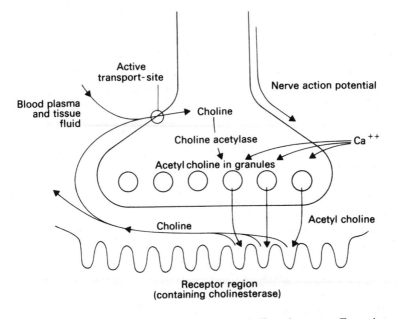

Fig. 4.17. The release and fate of acetylcholine at a cholinergic neuro-effector junction.

scopic vesicles (see Chapter 4). Also present in the terminals are the enzymes required to make fresh transmitter or to reabsorb it from the surrounding tissue fluid. The arrival of a nerve impulse allows calcium ions to enter the nerve terminal, a fixed amount entering for each impulse. This calcium then promotes the release of transmitter substance from the vesicles into the gap between the nerve terminal and the ganglion cell or innervated effector organ. Since the amount of transmitter released is proportional to the amount of calcium that enters the terminal, it is also proportional to the number of nerve impulses reaching the terminal. The synaptic gap or cleft is 15–50 nm across. The transmitter molecules diffuse across this gap and combine with specialised receptor sites on the far side, just as in the synapses in the CNS which have been already described. The disposal of the two transmitters, acetylcholine and noradrenaline, follows different routes. Most of the acetylcholine is hydrolysed by an enzyme, *acetyl cholinesterase*, forming free choline and acetic acid, both of which are inactive so far as the post-synaptic cell membrane is concerned. The choline is actively reabsorbed by the presynaptic terminal and then reacetylated to form fresh acetylcholine. Both these processes require energy. Most of the noradrenaline is reabsorbed by the presynaptic terminal unaltered and is returned forthwith to the storage vesicles. Noradrenaline reabsorption also requires energy. The processes of synthesis, release and disposal of the transmitters are summarised in Figs. 4.17 and 4.18. Similar figures can be devised for the other transmitters.

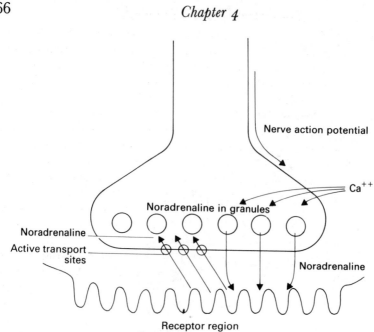

Nerve action potential

Ca^{++}

Noradrenaline in granules

Noradrenaline

Active transport
sites

Noradrenaline

Receptor region

Fig. 4.18. The release and fate of noradrenaline at a noradrenergic neuro-effector junction.

INTERFERENCE WITH TRANSMISSION

Transmission can be artificially influenced both at autonomic ganglia and at effector nerve fibre terminals. Since all ganglion cells are stimulated by the same process, viz. release of acetylcholine from the connector neurone terminals, it is not surprising to find that all can be affected by the same materials. The best known of these, and the most widely used, or abused and therefore dangerous, is *nicotine*. In small amounts this stimulates all ANS ganglion cells, but in larger amounts it paralyses them. Most of the physical and the potentially harmful effects of nicotine result from this ganglion cell stimulation, though the pleasure derived by some from it is presumably due to some action within the CNS. *Atropine*, which is found in plants of the Nightshade family, blocks the action of acetylcholine released from autonomic effector nerve terminals, whether these are in the parasympathetic or sympathetic nerves. It was formerly used as an aid to beauty, when the pupillary dilation that it causes was thought to look attractive (hence the name belladonna for this substance).

In the case of the noradrenaline-liberating sympathetic effector nerve terminals (usually shortened to *noradrenergic terminals*), the position is more complex. On the post-synaptic side two types of receptor exist. One group of drugs blocks one type of noradrenaline receptor and another group blocks a second type. These are known as α and β *receptors*. The former are mainly excitatory in action, while β type receptors may be excitatory (on the heart) or inhibitory (muscle of the intestine). It is therefore possible by choosing the correct drug, to affect selectively some noradrenergic terminals, while leaving others unaffected. A further point has emerged with the discovery of the two sorts of noradrenergic receptors. Noradrenaline reacts more powerfully with the α receptors, while adrenaline reacts equally with both types of receptor. The synthetic material, *Isopropyl noradrenaline* (Isoprenaline), acts almost exclusively on β receptors. Finally, as well as receptor blockers of these two types, a third group of compounds exists which prevents the release of noradrenaline from the nerve terminals when impulses reach them.

The action of acetylcholine can be potentiated by compounds which prevent its destruction by acetyl cholinesterase. *Physostigmine* and *Neostigmine* have been used for many years now in treating *myasthenia gravis*. Similarly, amongst its other actions, *cocaine* can potentiate the action of noradrenaline by blocking its re-uptake into nerve terminals and thus allowing its effect to be prolonged.

Within the central nervous system, which is much less accessible than autonomic ganglia, effector ANS fibre terminals or neuromuscular junctions, it has only been since the 1960s that essentially similar processes of chemical synaptic transmission have been analysed. Acetylcholine and noradrenaline have both been identified as CNS transmitters, but so also have several different other molecules. These include dopamine (chemically like noradrenaline), 5-hydroxytryptamine (made in the body from the amino acid tryptophan and frequently known as 5-HT), gamma-amino butyric acid (made from glutamine and called GABA), the amino acids glutamic acid and glycine (both of which are inhibitory transmitters) and numerous peptides. Of these, Substance P and the Enkephalins may be true transmitters affecting the sensation of pain (see below, p. 107), while others, often being released from the same terminal as a transmitter, may act on special post-synaptic receptors which then alter some aspect of the post-synaptic cell's metabolism on a long-term basis and so its responsiveness to excitatory or inhibitory inputs. Such peptides are called 'co-transmitters' or 'neuromodulators'. Their action may be the basis of learning and memory formation. Compounds that potentiate, mimic, or block the action of transmitters in the CNS have been found in the same way as for the ANS ganglia and effector terminals. These are developed by the pharmaceutical industry for the relief of many neurological and psychiatric conditions and for the relief of pain. Unfortunately, since a transmitter may be active at many different sites, unwanted side-effects of these compounds may outweigh the advantages sought by their use. Furthermore, the CNS has a truly remarkable ability to develop resistance to any extraneous substance, which then gradually loses its effectiveness.

TRANSMISSION AT NEUROMUSCULAR JUNCTIONS

Neuromuscular junctions are very similar to synapses. As has been seen, in skeletal muscle each muscle fibre receives an axon branch terminal from only one nerve cell but through its terminal branches a single nerve cell can stimulate as many as 150 muscle fibres. All these junctions are excitatory in function, there being no inhibitory nerve fibres to human skeletal muscle (though the heart and the muscle of some blood vessels receive both excitatory and inhibitory nerve supplies).

The transmitter at skeletal neuromuscular junctions is *acetylcholine* (A.Ch.). When an impulse arrives at the motor end plate, A.Ch. is liberated into and diffuses across the gap between nerve and muscle membranes. It then causes sufficient depolarisation of the muscle cell's membrane to set off a propagated action potential in the muscle fibre. Unlike synaptic excitation in nerve cells, excitation of the muscle fibre requires only one impulse in its controlling nerve fibre. There is a 1:1 relationship between impulses in the nerve fibre and impulses in the muscle fibre membrane. How these impulses are converted into contraction within the muscle will be left to a later chapter. Here we must note that as soon as the A.Ch. has done its work it is destroyed by an enzyme known as *acetylcholinesterase* that is found in the membranes of the muscle cell. Occasionally the effect of the A.Ch. becomes defective due to a reduction in the number of acetylcholine receptors in the subsynaptic region of the muscle fibre and the patient then suffers from an inordinate degree of fatigue, muscle weakness and paralysis. This condition is called *myasthenia gravis* (—severe weakness of muscles). It can be substantially relieved by giving a drug physostigmine that blocks acetylcholinesterase (which normally destroys the A.Ch.). The persistence of acetylcholine at the small number of receptors then has a more nearly normal action in generating propagated action potentials in the muscle cell's membrane.

REFLEX RESPONSES

The term reflex has already been defined. What follows here is a detailed description of the nature and properties of reflex activity. It will become apparent how closely reflex activity depends upon the behaviour of synaptic transmission.

Reflexes are automatic, occurring without one 'taking thought' about them, frequently without conscious awareness that they have even happened. For the most part they are unlearned, being determined by anatomical pathways laid down, in many instances, before birth. They are always purposive and frequently found as the means of maintaining the normal physiological state of the living body, or causing it to respond adequately to altered environmental states, both internal and external. Thus the withdrawal of the bare foot from a tack left on the floor, or the hand from a hot stove are both protective reflexes. The 'knee jerk' (p. 116) is a response to a brief stretch of the muscles on the front of the thigh produced by tapping their common tendon at the knee. It is a momentary excitation of a stretch reflex—one of the many that maintain us in our normal upright position and help to keep balance as we move about. The heart rate and the arterial blood pressure are reflexly controlled in such a way as to ensure an adequate blood supply to the organs of the body. Reflexes partly control the secretion of the digestive juices that are liberated when food is eaten but not at other times. In most of these cases, while the primary reflex arc mediates the response that is appropriate to the particular stimulus, at various points along its path nerve fibres branch and synaptically connect with many other parts of the nervous system so that the original reflex is carefully blended in with all the other bodily activities and any alterations in these are all automatically detected and the required adjustments made. The pathway through the various nerve fibres and synapses along which nerve impulses pass to produce a reflex response is called a *reflex arc*. The simplest reflex arc, producing the most constant response contains only one synapse; it is called a *monosynaptic arc*. Such arcs are found only in stretch reflexes (pp. 115–118). All other reflexes are mediated through *polysynaptic arcs*, the impulses passing through two or more synapses (Figs. 4.16 and 4.17). Detailed description of many regulatory reflexes will be given at their appropriate places in this book, but we would like here to follow through some of the consequences of one of the simplest of reflexes, that of withdrawal of the foot when a sharp object perforates the skin of the sole.

At first, impulses travel from the pain receptors along sensory nerve fibres to the spinal cord. After passing through at least one intermediate neurone, the impulses reach excitatory synapses on the motor nerve cells whose axons go to muscles in the same leg. But not to all muscles—only those motor neurones are excited that supply the muscles which, when they contract, will withdraw the leg from the drawing pin. These are the flexor muscles of the hip, knee and ankle joints. The antagonistic extensor muscles are not excited by the impulses from the pain receptors, but their motor neurones are actually inhibited from activity. In the case of the extensor motor neurones, the impulses from the pain receptors, after passing through interneurones, terminate at inhibitory synapses, so that the extensor muscles become relaxed at the same time as the flexor muscles are contracting in response to the painful stimulus. The extent to which the flexor muscles are excited and the extensor muscles inhibited depends upon the strength of the stimulus, i.e. the number of pain receptors stimulated and the intensity

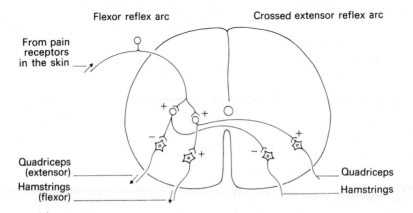

Fig. 4.19. The simplest arrangement of interneurones that could form the arcs for flexor and cross-extensor reflex responses to cutaneous receptor stimulation.

with which each is stimulated. Both of these factors will determine the number of nerve impulses reaching the motor neurones in a given time and will then determine the rate at which epsp builds up and causes impulses to leave the motor neurones. This response to painful stimulation is called a *flexor reflex* (Fig. 4.19).

Branches of the incoming axons of the sensory neurones or of axons of intermediate neurones, cross the midline of the spinal cord and stimulate the motor neurones controlling the extensor muscles of the opposite leg and, at the same time, inhibit those controlling flexor muscles. This limb, which is now

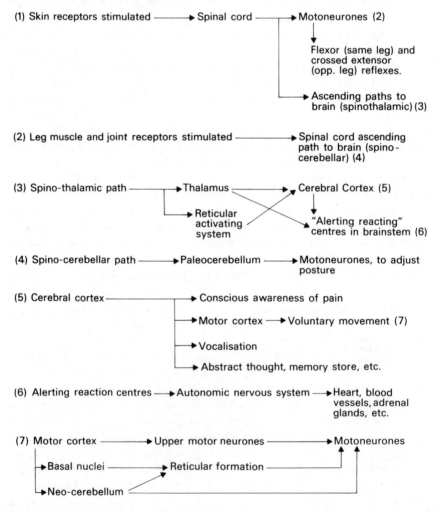

When a foot treads on a drawing-pin:

(1) Skin receptors stimulated ⟶ Spinal cord ⟶ Motoneurones (2)
⟶ Flexor (same leg) and crossed extensor (opp. leg) reflexes.
⟶ Ascending paths to brain (spinothalamic) (3)

(2) Leg muscle and joint receptors stimulated ⟶ Spinal cord ascending path to brain (spino-cerebellar) (4)

(3) Spino-thalamic path ⟶ Thalamus ⟶ Cerebral Cortex (5)
⟶ Reticular activating system
⟶ "Alerting reacting" centres in brainstem (6)

(4) Spino-cerebellar path ⟶ Paleocerebellum ⟶ Motoneurones, to adjust posture

(5) Cerebral cortex ⟶ Conscious awareness of pain
⟶ Motor cortex ⟶ Voluntary movement (7)
⟶ Vocalisation
⟶ Abstract thought, memory store, etc.

(6) Alerting reaction centres ⟶ Autonomic nervous system ⟶ Heart, blood vessels, adrenal glands, etc.

(7) Motor cortex ⟶ Upper motor neurones ⟶ Motoneurones
⟶ Basal nuclei ⟶ Reticular formation
⟶ Neo-cerebellum

Fig. 4.20. How the whole central nervous system becomes involved in a flexor reflex response.

responsible for support since the other is flexed, becomes stiff and straight, and better able to bear the body's weight (the *crossed extensor reflex*) (Fig. 4.19.) Further impulses go to the appropriate muscles of the trunk, which adjust its position so as to bring the centre of gravity over the supporting limb. At the same time, impulses are leaving the receptors in the muscles and tendons (length and tension receptors respectively) and the joints of the affected limb. These go up the spinal cord to the cerebellum, which controls and co-ordinates the movements so that they are smoothly performed and stop when their purpose is achieved. The cerebellum also controls, following the information from appropriate receptors, the contraction of the trunk and other muscles needed to maintain the normal body posture. All these controls are essentially reflex in nature, being automatically initiated by receptor stimulation and fulfilling all the other criteria already described for reflex activity. Following upon this muscular activity and following the eventual arrival at the 'alerting reaction centres' in the brain of the impulses from the pain receptors, a further set of reflexes arises which adjusts the activity of the heart, blood vessels and respiration in accordance with the actual or expected muscular activity. Thus it can be seen that there is no such thing as a 'simple' reflex response. At all stages each response triggers off further receptor stimulation and so further reflex responses occur, all being perfectly co-ordinated and arranged so that the adjustments required by the altered state of the body are continually and automatically performed (Fig. 4.20).

PROPERTIES OF REFLEX ACTIVITY

Some of these have already been described, such as the way in which inhibition is just as much part of reflex responses as is excitation. Also we have seen how the effects of the initial receptor stimulation and its primary response set off other reflexes which spread, like ripples over the surface of a pond, until it seems that the whole nervous system is involved in the effects of the initial stimulus. We should now, though, look in greater detail at many of the phenomena of reflex activity and see how these depend upon the properties of synaptic transmission and the possible anatomical arrangements of synapses upon neurones.

The response in a muscle to painful stimulation varies with the strength of the stimulus. This is true whether the stimulus is an episode such as a jab with a needle or tack, or a continuous painful stimulus arising, perhaps, from a torn ligament. In the former, the more powerful the stimulus the larger the number of afferent nerve fibres stimulated. More synapses liberate transmitter on the motor neurones so that the epsp reaches threshold in more of them. This causes a larger number of muscles fibres to contract. Thus spatial summation of synaptic activity (described above) causes summation of the reflex response, which follows, in a graded manner, the strength of the initial stimulus to the receptors. Temporal summation of epsp (see above) also increases the size of the reflex response. The more powerfully any receptor is stimulated, the more frequently it produces nerve impulses. The more frequently these reach motor neurones,

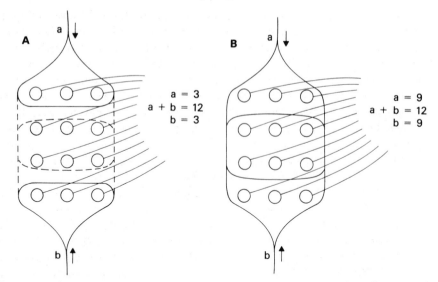

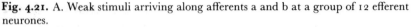

Fig. 4.21. A. Weak stimuli arriving along afferents a and b at a group of 12 efferent neurones.
B. Powerful stimuli reaching the same 12 neurones along the same two afferents.

the more rapidly does the epsp build up to threshold value in them and so the more frequently do they produce impulses and the greater the consequent contraction of the muscle they supply.

Spatial summation can also be seen to operate when two sensory nerves are stimulated simultaneously. If both stimuli, when applied separately, produce a response, two possible results may be observed when both are applied together. The response may be either larger or smaller than expected (Fig. 4.21). Both **of these results** can be explained by supposing that each afferent nerve supplies part of the whole pool of motor neurones supplying a muscle, and that there is some overlap between them. If the stimuli are weak (Fig. 4.21A), then each alone only produces an impulse in a few neurones, but some EPSP (insufficient to produce an impulse) in several more, which are said to be in the *sub-threshold* or *subliminal fringe* of the afferent fibre. When both are stimulated to-gether, those neurones that are in the subliminal fringes of both afferent fibres now have sufficient epsp to cause action potentials to be set off in them. The resultant contraction is greater than the sum of the two separate ones. If the stimuli were powerful (Fig. 4.21B) then all neurones, including those in the fringes, would respond to each separate stimulus. But those in the overlap area cannot respond more to the two stimuli than to one alone. Therefore the resultant contraction is less than the sum of those produced by the two stimuli applied separately. These phenomena then, known to physiologists respectively as *subliminal fringe activity* and *occlusion*, are dependent upon the anatomical arrangement of the afferent fibres' synaptic terminals, upon the summatory property of epsp and upon the all-or-none nature of the evoked nerve impulse.

Facilitation and *reinforcement* are also due to subthreshold levels of epsp in motor neurones. The former term is used when a stimulus, previously too weak to evoke a reflex response, can do so when a quite different stimulus has, by producing a continuous but subthreshold level of activity, rendered a group of motor neurones more readily excited. Facilitation is thus a special sort of subliminal fringe activity. In reinforcement, a given reflex is rendered more powerful by muscular activity in another part of the body. For instance, the knee jerk response (Chapter 5) may only be produced (facilitation) or may be greatly increased in strength (reinforcement) by voluntary effort involving the arms. This has the effect of raising the level of excitation of the quadriceps motor neurones.

Some reflexes, like the crossed extensor response, show a gradual build-up in strength during a period of stimulation, and an equally gradual decline in strength when the stimulus ceases. This means that, while the sensory stimulus is constant, more and more neurones are gradually being stimulated sufficiently for their level of EPSP to reach the threshold for impulse production. Impulses must be arriving at these neurones through several delaying pathways, these impulses having farther to travel and passing through additional synapses (Fig. 4.22). Similarly, after the sensory stimulation has ceased, impulses continue to arrive for some time at these motor neurones. These two linked phenomena, dependent upon a network of delaying and *reverberatory circuits* (Fig. 4.22) in which nerve impulses can travel repeatedly through a circuit of neurones, are known respectively as *recruitment* and *after discharge*.

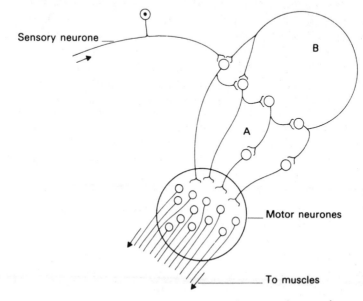

Fig. 4.22. Postulated neurone networks that cause both progressive recruitment and after discharge from a group of motor neurones.

Very similar to recruitment is *irradiation*. A stimulus to the sole of the foot, sufficient initially to cause ankle dorsiflexion only, will, if maintained at a constant strength, gradually come to affect first the knee then the hip flexor muscles also. Again, delay and reverberatory circuits are gradually delivering more impulses to motor neurones until eventually their excitatory state (EPSP) is sufficient to cause them to generate nerve impulses and so cause the muscles they supply to contract.

We have already described how inhibitory synapses work, producing inhibitory post-synaptic potentials (IPSP) which are the arithmetic opposite of epsp. Whereas epsp is a reduction of cell membrane potential (caused by the entry of sodium ions), ipsp is an elevation of the potential and it is caused by either the entry of negatively charged chloride ions, or the loss from the cell of positive potassium ions. Presynaptic inhibition also occurs, blocking the access of impulses in particular sensory nerve impulses to their synaptic terminals. In reflexes, inhibition is at work as much as is excitation. In the flexor reflex, the extensor muscles become relaxed (inhibited). In the crossed extensor reflex, it is the flexor muscles of the extended limb that are inhibited. In a stretch reflex, which occurs in any muscle when its length detectors (the spiral endings of the muscle spindle—see Chapter 5) are stimulated, the antagonist muscle is always inhibited. If the tension receptors (Golgi organs) of a muscle tendon are powerfully stimulated, then that same muscle is inhibited by the impulses travelling in sensory nerve fibres from these receptors. As a general rule one can say that in a reflex response, whenever a muscle's motor neurones are excited and cause that muscle to contract, the motor neurones of the muscle's antagonist are inhibited. This phenomenon is known as *reciprocal innervation* or *reciprocal inhibition*. Inhibition will also be seen to play a vital role in regulation of the activity of the heart, the control system of the blood vessels and in many other regulatory functions of the nervous system. The simple models that have been here described can be directly applied to all forms of reflex activity. These will be described later in the various chapters dealing with different systems but particularly in Chapter 7, concerned with movement and muscle tone.

5 · General Anatomy and Physiology of the Nervous System

Chapter 4 described, among other things, the extraordinary shape of the nerve cells, particularly those with long axons. This may be best appreciated by considering one of the neurones which supplies the muscles of the sole of the foot. The cell body lies at approximately the level of the first lumbar vertebra (see below), so that the axon extends from here to the foot, a distance of perhaps 1 metre. If the cell body were to be drawn to occupy one page of this book, say 10 cm across, and was actually 100 μm in diameter, the magnification would be × 1,000 and the axon would have to be drawn 1 km long—from Trafalgar Square to Buckingham Palace or half the distance from Queen Street to Bloor Street in Toronto. The importance of the axon in transmitting nerve impulses from one part of the body to another can thus be easily appreciated and, as will be seen later in this chapter, relatively few neurones may be needed to convey impulses all the way from the brain to the foot or vice versa.

The nervous system may be subdivided into two basic parts: the *cerebro-spinal* and the *autonomic* nervous systems. The latter is that part which is autonomous, or self-governing, and it controls unconscious or involuntary activities which are essential to the proper functioning of the body—such activities as the secretion of digestive juices, the movements of the viscera and the regulation of blood pressure. It will be dealt with in Chapter 6. The nervous system may also be subdivided in another way, that is, into the *central nervous system* (CNS), consisting of the brain and spinal cord, and the *peripheral nervous system* which consists of the *cranial* and *spinal nerves* and the peripheral autonomic nerves and ganglia (see Chapter 6). The twelve pairs of cranial nerves arise from the brain itself and include the nerves of the special senses such as the optic nerve. The 31 pairs of spinal nerves arise from the spinal cord and are distributed to all parts of the body.

The brain and spinal cord consist of nerve cells and their processes, together with certain supporting cells which are not concerned with the transmission of nerve impulses and which are known collectively as *neuroglia*. They fulfil the function of connective tissue in other parts of the body, since true connective tissue is not found in the nervous system. Collections of nerve cells in the central nervous system form the *grey matter* since they are visible to the naked eye as a greyish-pink substance. Nerve fibres form the *white matter*. In the spinal cord there is a central column of grey matter which is completely surrounded by white matter, but in most of the brain there is a thin covering of grey matter (the cortex) and a central mass of white matter. In the latter, however, are many isolated collections of grey matter which in this situation are known as *nuclei* and are mostly named—see, for instance, the gracile and cuneate nuclei below. Outside the central nervous system, collections of nerve cells are called *ganglia*

75

and they are found associated with the spinal and some of the cranial nerves and also with the autonomic nervous system.

The nerve fibres which form the white matter may follow various courses. They may run from one side of the brain to the other (*commissural fibres*) for the purpose of co-ordination; they may run from one part of the brain to another on the same side (*association fibres*) or from one part of the brain to a distant collection of neurones in the brain or spinal cord (*projection fibres*). The latter may be extremely long since they may have to travel from the cerebral hemisphere to the spinal cord or vice versa. Collections of axons which are conveying impulses over a long distance are usually grouped together in bundles to form *tracts* which follow well-defined pathways through the brainstem and spinal cord. There are various tracts associated with the motor system, such as the *pyramidal tract*, and others with the sensory system, such as the *spinothalamic tracts*. Interference with these anywhere along their path will cause disturbances of motor or sensory activity respectively so that it is most important that you should know the whereabouts of the principal tracts in the brain and spinal cord. They will be described in detail later in this chapter.

The brain and spinal cord are covered by three membranes known collectively as the *meninges*. They are called, from within outwards, the *pia mater*, *arachnoid mater*, and *dura mater*. The narrow space between the pia and arachnoid (the *subarachnoid space*) is occupied by a clear fluid called *cerebrospinal fluid* (CSF) and this fluid is also contained within a series of cavities or ventricles inside the brain and in a tiny *central canal* of the spinal cord. The whole central nervous system is well protected by bone (namely, by the skull and vertebrae) although the spinal cord and its surrounding meninges can be reached by a long needle passed between the vertebrae.

THE BRAIN

The general plan of the adult brain may best be understood by surveying very briefly its development in the embryo. The very early nervous system consists only of a long tube (the *neural tube*) which runs the whole length of the embryo and round which the skull and the vertebrae develop. At the front end of the tube three slight dilatations develop which are known as the *forebrain*, the *midbrain* and the *hindbrain* (Fig. 5.1). The forebrain rapidly develops a pair of outgrowths, one on each side, which form the *cerebral hemispheres*. The cavity of the tube persists, so that each cerebral hemisphere contains a cavity which will later be called the *lateral ventricle* of the brain, while the central part of the forebrain contains the *third ventricle*. The cerebral hemispheres will later grow enormously until they completely overshadow the rest of the brain. The midbrain undergoes very little change except for a marked thickening in its walls. Its cavity remains relatively small and forms the adult *aqueduct*. The hindbrain forms the *pons* and the *medulla* (oblongata), and, at a later stage of development, the *cerebellum*. The cerebellum grows quickly backwards until it overhangs and

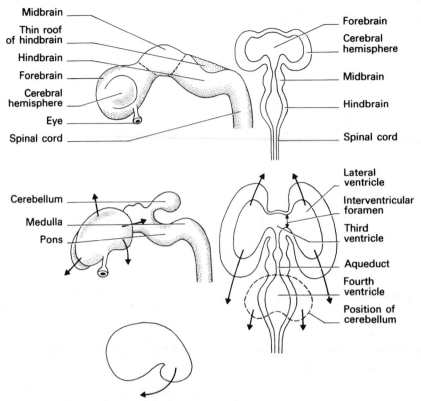

Midbrain

Thin roof of hindbrain

Hindbrain

Forebrain

Cerebral hemisphere

Eye

Spinal cord

Forebrain

Cerebral hemisphere

Midbrain

Hindbrain

Spinal cord

Cerebellum

Medulla

Pons

Lateral ventricle

Interventricular foramen

Third ventricle

Aqueduct

Fourth ventricle

Position of cerebellum

Fig. 5.1. The development of the brain. The arrows indicate the direction of growth of the cerebral hemispheres and the cerebellum.

conceals the dorsal part of the hindbrain but is itself very largely concealed by the backward growth of the cerebrum. The original cavity of the hindbrain widens out and becomes the *fourth ventricle* of the brain.

The adult brain may therefore be described as consisting essentially of a central axis or *brain-stem*, formed by the midbrain, pons and medulla, which is continuous anteriorly with the forebrain, whose enormous cerebral hemispheres (called collectively the *cerebrum*) overshadow the rest of the brain (Fig. 5.2). Under cover of the cerebral hemispheres, the much smaller cerebellum may be seen, and this is connected to the brain-stem by three pairs of *cerebellar peduncles*. These consist of masses of nerve fibres (i.e. white matter); the *superior peduncle* connects the cerebellum to the midbrain, the *middle peduncle* connects it to the pons and the *inferior peduncle* to the medulla.

The cerebral hemispheres are not particularly large in most animals but in the human brain this region is concerned with peculiarly human characteristics such as intelligence, memory, speech and the carrying out of delicate manipulations with the hands. It is therefore enormously developed in man, and in order to increase its surface area it is thrown into folds and wrinkles and is

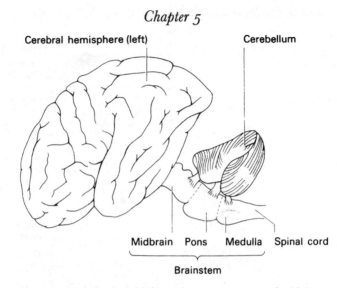

Cerebral hemisphere (left)   Cerebellum

Midbrain Pons Medulla Spinal cord

Brainstem

Fig. 5.2. A side view of the brain with its various parts separated widely (and artificially) to show its components.

much the largest part of the brain (Fig. 5.4). The brain-stem is relatively inconspicuous but this is not to say that it is unimportant since it is in the pons and medulla that the so-called *vital centres* are found. These are rather diffuse collections of nerve cells which control many unconscious activities such as respiration and the beating of the heart. Furthermore, you will appreciate that any bundles of nerve fibres (tracts) which are passing from the cerebrum to the spinal cord or vice versa, must pass through the brain-stem, while tracts to or from the cerebellum must pass through one of the peduncles and through part of the brain-stem. Lesions of this region, for example by a haemorrhage from one of the brain arteries, can therefore cause extensive disturbances of sensory or motor activity in various parts of the body. The structure of the brain will now be discussed in a little more detail.

THE CEREBRAL HEMISPHERES

As has been mentioned, these form by far the largest part of the brain (Fig. 5.2). Each consists essentially of a central core of white matter, within which are a number of nuclei consisting of grey matter, and outside the whole is a thin layer of grey matter called the *cerebral cortex*. It is in the latter that most of the particularly human functions are represented, so its surface area is increased by its being thrown into folds or *gyri* (sing. gyrus) which are separated by quite deep clefts called *sulci*. Laterally, the surface is so folded that one portion of the cortex called the *insula* is completely buried below the general surface of the brain (Fig. 5.3). The gyri and sulci form useful landmarks and the most important of them are shown in Figure 5.4. The cerebral hemispheres are subdivided also into four lobes, whose names correspond to the names of the skull bones which overlie them and they are therefore called the *frontal, parietal, occipital* and

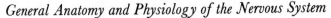

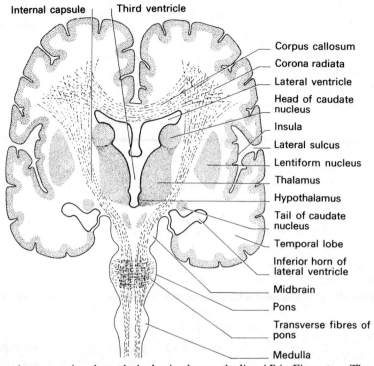

Fig. 5.3. A cross-section through the brain close to the line AB in Figure 5.4. The dotted lines indicate the pyramidal (corticospinal) tract running longitudinally and the corpus callosum and pontine fibres running transversely.

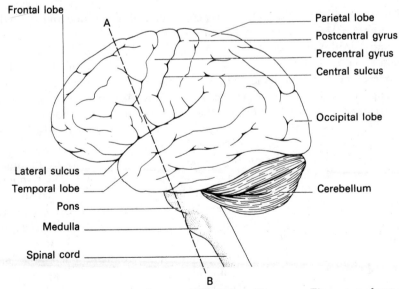

Fig. 5.4. A side view of the brain, more lifelike than Figure 5.2. Figure 5.3. shows a section through the brain close to the line AB.

temporal lobes. As will be seen later, for many functions one cerebral hemisphere controls the opposite side of the body but some co-ordination between the two sides is necessary so the two hemispheres are connected by a number of fibres which cross the midline. These are known as commissures and the largest is the *corpus callosum* which can be seen cut in cross section in Figure 5.5.

While it is difficult to ascribe a definite function to all parts of the cerebral cortex, some parts are definitely associated with particular functions (Fig. 5.6). Thus the gyrus in front of the central sulcus (the *precentral gyrus*) is the main motor area which apparently initiates purposeful movements of the body; in this area the body is represented upside-down so that the motor functions of the face and upper part of the body are represented at the lower end of the precentral gyrus. It is not surprising, therefore, to find the motor speech centre situated close to this (Fig. 5.6). This speech centre, however, is usually one-sided, being found only in the dominant hemisphere — the left hemisphere in right-handed persons. Damage (due to defective blood supply, for example) to this region on the dominant side causes a difficulty in articulation. The patient knows what he wants to say, but cannot produce the right movements to get it out; speech becomes hesitant, staccato and telegraphic in nature (aphasia). The facial and other muscles involved are, however, perfectly able to take part in movements of expression, mastication and swallowing. It cannot be stressed too strongly that particular cells of the cerebral cortex do not control particular individual muscles—it is whole movements which are represented, movements which may involve the simultaneous contraction of a number of different muscles and the relaxation of others. In fact the representation of the body on the cortex is quite complicated but it is of such importance that further details will be given when motor activity is considered (see below).

The post-central gyrus is particularly concerned with sensation and again the body is represented upside-down so that the sensory impulses from the face are appreciated in the lower part of the gyrus. The occipital lobe of the brain is associated with vision, particularly on the medial aspect of the cerebral hemispheres, i.e. the areas which face each other. In the temporal lobe an area in the uppermost of its gyri is associated with hearing. As will be seen from Figure 5.6, there are many areas unlabelled. Many of these areas have, however, been allocated functions but they are not very well understood and are poorly defined. For example, a large area of the frontal lobe is concerned with behaviour, and patients who have had part of the frontal lobe affected by disease or injury show characteristic changes in behaviour, usually for the worse. The temporal lobe is associated with memory. Some of these areas will be referred to later when motor and sensory functions are discussed, but much of the cerebral cortex consists of the so-called 'association areas', to which the incoming sensory information is relayed and further processed, being matched with stored memories, or incorporated into the memory store as a new item. These areas also initiate appropriate motor responses and are perhaps collectively also the sites of consciousness, thinking and emotional feeling.

This account of the cerebral cortex is grossly oversimplified as its functions

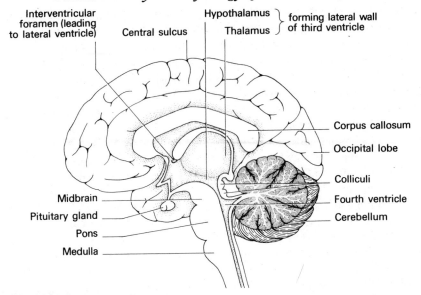

Fig. 5.5. A midline sagittal section through the brain.

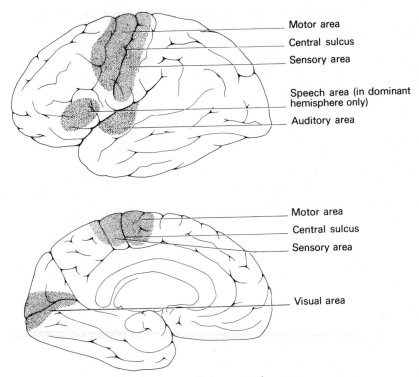

Fig. 5.6. Some of the functional areas of the cerebral cortex.

are so complex that we are not yet in a position to elucidate them with what are, at present, relatively clumsy and unrefined experimental methods. Furthermore, the areas are not static but can change from time to time as a result of the acquiring of new skills and experiences so that electrical stimulation of a particular area may give a response on one occasion which cannot be obtained again on subsequent stimulation.

From the cells of the cerebral cortex, axons pass down to the brainstem, some of them being destined for nuclei in the brainstem, some for the cerebellum and some for the spinal cord. Since the uppermost part of the midbrain is small compared to the cerebrum these fibres must converge towards a 'bottleneck' and the very prominent fan-shaped mass of fibres which results is called the *corona radiata* (Fig. 5.3). The fibres are finally collected together to form a compact mass of white matter called the *internal capsule* which has important relationships to some masses of grey matter (nuclei) in the cerebrum which will be described below. From the internal capsule the fibres pass into the ventral part of the midbrain. The internal capsule and corona radiata contain fibres other than those passing down from the cerebral cortex. There is here a two-way traffic and fibres which have come up the brain-stem (i.e. sensory and other fibres) also pass into the internal capsule after forming synapses in the thalamus (see below) and then fan out to form part of the corona radiata before reaching various parts of the cerebral cortex. The internal capsule is supplied with blood by a number of small arteries (the *lenticulostriate arteries*) from the base of the brain. These vessels are rather vulnerable and in patients with cardiovascular disease they are liable to bleed or to become thrombosed. A relatively small lesion in this region can cause a very widespread effect, particularly on motor function, and such a cerebral vascular accident or 'stroke' produces a condition in which physiotherapy is of great value.

In order to examine the relations of the internal capsule, it will now be necessary to study the central part of the forebrain which, in the midline, contains the third ventricle. As can be seen in a vertical section through the brain (Fig. 5.3), this is a very narrow cavity because the grey matter in its lateral walls bulges inwards and, in fact, may meet in the midline to form a bar of grey matter across the cavity. The largest mass of grey matter in this region is the *thalamus* (Fig. 5.5), which consists of cells that form an important relay station for sensory fibres ascending from the spinal cord. These form synapses with the cells of the thalamus and from here axons ascend, via the internal capsule and the corona radiata, to the sensory areas of the cortex. Various other ascending fibres also synapse in the thalamus and some of these will be referred to later.

Below the thalamus, in the lateral wall and floor of the third ventricle, is the *hypothalamus* (Fig. 5.5). This contains a number of nuclei. These are concerned with the autonomic control of many functions including regulation of body temperature, food intake and water intake and loss from the body (see Chapter 21). The hypothalamus also controls the 'flight, fear and fight' reactions, through the autonomic nervous system (see Chapter 6) and many of

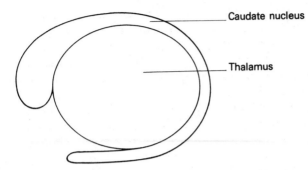

Fig. 5.7. The caudate nucleus and the thalamus.

the endocrine glands of the body. The latter function is carried out in conjunction with the pituitary gland, with which the hypothalamus has important connections.

Wrapped around the thalamus is the *caudate nucleus* (caudate: 'having a tail'). This is a mass of grey matter shaped rather like a tadpole with the head overhanging the anterior end of the thalamus, and a long thin tail which is wrapped around the posterior part (Figs. 5.3 and 5.7). The internal capsule is directly lateral to the mass of grey matter formed by the thalamus and the caudate nucleus. Finally, lateral to the internal capsule, is another group of nuclei which, since they are approximately the shape of a biconvex lens, are known collectively as the *lentiform nucleus*. The lentiform nucleus lies immediately deep to the insula, the buried part of the cerebral cortex (Fig. 5.3). The lentiform and caudate nuclei (often called collectively the *corpus striatum* or, along with neighbouring nuclei, the *basal nuclei*) are part of the *extrapyramidal system;* this is the complex of nuclei and tracts which are concerned with certain aspects of motor activity but lie outside the main motor pathway in the brain. The latter is known as the pyramidal or *corticospinal tract*.

THE MIDBRAIN

The appearance of cross-sections through the midbrain are shown in Figure 5.8. The original cavity of the neural tube remains very small and there are no complicated outgrowths such as the cerebrum or cerebellum. The midbrain is therefore relatively simple. The tiny cavity, or canal, in the dorsal part of the midbrain is called the *aqueduct*. Ventral to it lies the cerebral peduncle and dorsal to it are four little elevations called the *corpora quadrigemina* or *colliculi*. The superior colliculi are seen in the section of the upper midbrain in Figure 5.8 and are concerned with visual reflexes such as dilatation and constriction of the pupil of the eye, and the inferior colliculi are in the lower part of the midbrain and are concerned with reflexes associated with the auditory system.

The cerebral peduncle is divided into two main parts by a peculiar pigmented zone called the *substantia nigra*. Dorsal to this, the midbrain contains a

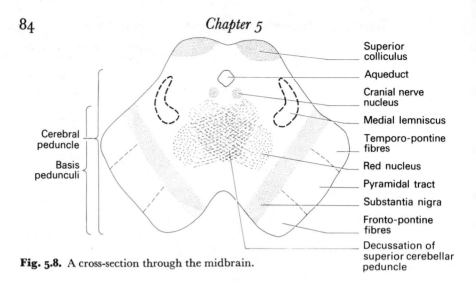

Superior colliculus

Aqueduct

Cranial nerve nucleus

Medial lemniscus

Temporo-pontine fibres

Red nucleus

Pyramidal tract

Substantia nigra

Fronto-pontine fibres

Decussation of superior cerebellar peduncle

Cerebral peduncle

Basis pedunculi

Fig. 5.8. A cross-section through the midbrain.

number of nuclei, including the large *red nucleus* and also the main ascending (sensory) tracts. These include a prominent mass of fibres called the *medial lemniscus*, which is the main pathway for sensory nerve impulses, as well as a number of other ascending tracts. Ventral to the substantia nigra on each side is the *basis pedunculi*. This contains the main descending tracts. The central part (three-fifths approximately) of the basis pedunculi contains the main motor pathway (the *pyramidal (corticospinal) tract*) which is descending from the motor area of the cerebral cortex to the spinal cord. The medial and lateral one-fifth of the basis pedunculi consist of fibres which are travelling from other parts of the cerebral cortex to some nuclei in the pons—the *pontine nuclei*. The medial one-fifth consists of *frontopontine* fibres and the lateral one-fifth are *temporopontine* fibres coming from the corresponding lobes of the cerebrum. The further course of these tracts will be described later. The cerebral peduncle, like the cerebellar peduncles, therefore consists mainly of descending and ascending tracts passing to and from the brain-stem—it is in fact the 'stalk' of the cerebrum.

THE PONS AND MEDULLA

The pons and medulla form the hindbrain component of the brain-stem. The original cavity of the neural tube widens out here; the ventral part of its wall becomes considerably thicker but its roof remains extremely thin and, in the adult brain, is concealed by the overhanging cerebellum (Fig. 5.5). The cavity here forms the *fourth ventricle* which is connected to the third ventricle by the aqueduct and which is continuous behind with the tiny central canal of the spinal cord. Thus the dorsal surface of the pons and of the medulla form the floor of the fourth ventricle and the structures in the pons and the medulla which lie in this region are exposed to the cerebrospinal fluid which fills the ventricle. This relationship is important—see, for example, the effect of carbon dioxide on the respiratory centre, p. 532. In the lateral part of the floor of the fourth

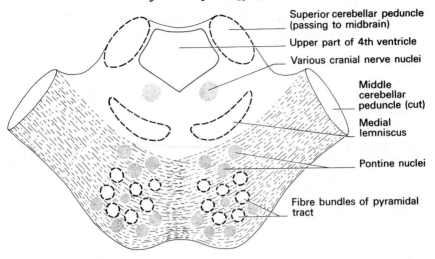

Fig. 5.9. A cross-section through the pons.

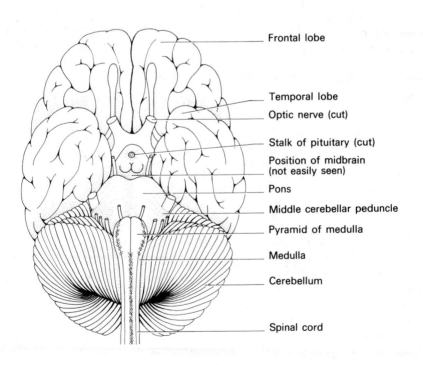

Fig. 5.10. The ventral surface of the brain.

ventricle (i.e. in the dorsal part of the pons and medulla), are a number of nuclei known collectively as the *vestibular nuclei*. These receive afferent fibres from the *vestibulocochlear* (eighth) cranial nerve and communicate with the cerebellum, the eye muscles and the anterior horn cells in the spinal cord. The vestibular impulses from the inner ear can therefore affect movements of the eyes, head, limbs and trunk which they do, under cerebellar control, to maintain balance (see Chapter 13).

The most prominent feature of the pons, especially noticeable if you have previously studied animal brains, is the large and prominent ventral region (Fig. 5.9). This again is a peculiarly human characteristic and it is associated with the enormous cerebral hemispheres and the large neocerebellum (see below). In order to effect the intricate and co-ordinated movements which the human body can carry out, it is necessary that there shall be close liason between the cerebrum and the cerebellum, the latter being particularly concerned with the co-ordination of muscle activity. A number of nerve fibres from the cerebrum (*corticopontine fibres*), including the frontopontine and temporopontine tracts (see above) therefore terminate in the ventral part of the pons by synapsing with the cells of the extensive grey matter in this region which forms the *pontine nuclei*. (Fig. 5.9 and 5.10). The axons of the cells of these nuclei cross the midline and travel to the neocerebellum (see below) via the middle cerebellar peduncle. Passing longitudinally between the numerous pontine nuclei are the pyramidal (corticospinal) tracts which are therefore rather dispersed in this region. The dorsal part of the pons contains a number of cranial nerve nuclei and various longitudinal tracts; in particular it contains a large bundle of fibres which ascend to the thalamus. This bundle is the *medial lemniscus* and it

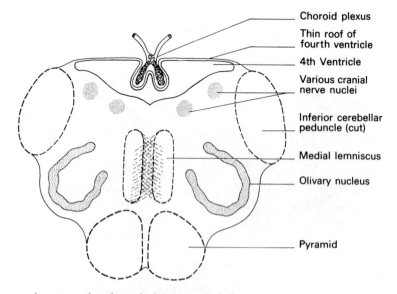

Choroid plexus

Thin roof of fourth ventricle

4th Ventricle

Various cranial nerve nuclei

Inferior cerebellar peduncle (cut)

Medial lemniscus

Olivary nucleus

Pyramid

Fig. 5.11. A cross-section through the upper medulla.

contains all the sensory fibres from the whole body, which form synapses in the thalamus with neurones whose axons ascend in the corona radiata to reach the cerebral cortex.

The medulla is much smaller in cross-sectional area than the pons, since the large bulge caused by the pontine nuclei is absent. The fibres of the corticospinal tract therefore converge to form a discrete bundle of fibres on each side of the midline. This produces a bulge on the surface known as the *pyramid* (Figs. 5.10 and 5.11). A little dorsal to this, another surface bulge is produced by an underlying nucleus called the *olivary nucleus*. On either side of the midline is the large bundle of sensory fibres which forms the medial lemniscus. Other longitudinal fibre tracts also traverse the medulla, and there are also various cranial nerve nuclei. The dorsal surface of the medulla, as has been mentioned already, forms the floor of the fourth ventricle. A number of fibre tracts pass from the medulla to the cerebellum and vice versa, via the inferior cerebellar peduncle.

In the lowermost part of the medulla, which projects through the foramen magnum to become continuous with the spinal cord, the shape becomes almost circular in cross-section (Fig. 5.12). Both the corticospinal tracts and the medial lemniscus cross the midline so that there is a considerable change in the internal structure at this level. Furthermore, the fourth ventricle narrows right down to become continuous with the minute central canal of the spinal cord. The corticospinal tracts decussate to form a prominent crossover of white matter in the lower medulla known as the *decussation of the pyramids*, although some fibres remain uncrossed (p. 104). The fibres which ascend from the spinal cord in the posterior white columns synapse in two prominent nuclei in the dorsal region of the lower medulla known as the *nucleus gracilis* (nearer the midline) and *nucleus cuneatus*. The fibres from these nuclei then cross the midline in the *decussation of the lemnisci* or *internal arcuate fibres* to form the medial lemnisci.

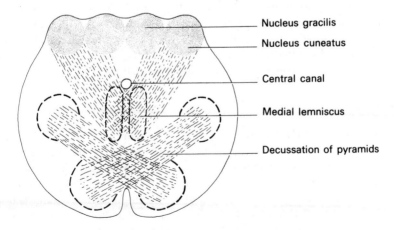

Nucleus gracilis

Nucleus cuneatus

Central canal

Medial lemniscus

Decussation of pyramids

Fig. 5.12. A cross-section through the lower medulla.

Throughout the whole length of the brain-stem various cranial nerves are given off, the last (twelfth) nerve being just above the first spinal nerve. There is also, in the brain-stem, an extremely complex network of white matter and scattered islands of grey matter which is known as the *reticular formation*. It is connected to many parts of the nervous system including the cerebellum, the spinal cord (reticulo-spinal tracts), the cerebrum and the cranial nerve nuclei. Its functions are obscure but there is no doubt that it fulfils an important role. In man, it is important in control of the lower motor neurones *via* the spinal cord, and has other functions concerned with the autonomic nervous system and with the rhythms of sleep and waking. The reticular formation is also concerned with incoming pathways to the CNS from sensory receptors, controlling the access of their impulses to higher centres. Some of its neurones serving this function contain the opioid peptide transmitters, the enkephalins. This may partly explain the effectiveness of the opiate alkaloids in pain relief.

The cerebellum is much smaller than the cerebrum (Fig. 5.5) but the structure of the two parts of the brain shows some similarities. Thus the cerebellum consists of a right and a left lobe, connected across the midline by a central portion. It has a thin cortex of grey matter and it contains a number of nuclei of grey matter within its central white matter. Anatomically, it consists of a number of named parts but from a functional point of view it may best be described as consisting of primitive parts (the archicerebellum and the palaeocerebellum) which are concerned with basic functions in relation to the motor system such as the muscle tone and the maintenance of posture, and a more recently developed part (the *neocerebellum*) which, in man, is extremely well developed and is concerned with the co-ordination of voluntary muscle action. Some of the incoming tracts to the last-named part of the cerebellum have already been described and the functions of this and the other parts of the cerebellum will be discussed later. Most of the outgoing tracts from the cerebellum travel in the superior cerebellar peduncle, decussate in the midbrain (Fig. 5.8) and then reach the thalamus (and thence the cortex) and various nuclei in the brain-stem such as the red and recticular nuclei (and thence the spinal cord). The fibres of the superior peduncle decussate in the midbrain and the fibres which then descend to the spinal cord also decussate so that, unlike the cerebrum, one side of the cerebellum controls the *same* side of the body.

THE SPINAL CORD

Once again it is helpful to refer to prenatal development. In the early fetus the spinal cord is the same length as the vertebral column so that it extends from the foramen magnum to the end of the sacrum. The vertebral column, however, grows rather faster than the spinal cord so that, as development proceeds, the lower end of the spinal cord retreats gradually up the vertebral canal, its upper end being fixed in position by the brain within the skull. At birth, the spinal cord ends at about the level of the third lumbar vertebra while in the adult it ends at the lower border of the first lumbar vertebra. This process also affects

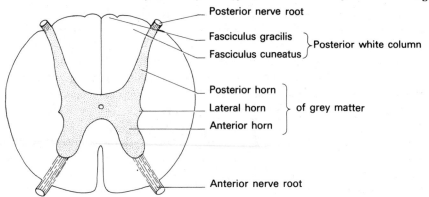

Fig. 5.13. A cross-section through the spinal cord.

the direction of the roots of the spinal nerves so that in the cervical region they leave the spinal cord approximately at right angles, but their direction becomes more and more oblique until they are vertical. That is to say, the roots of the lumbar nerves below the first, and those of all the sacral nerves come off close together near the level of the first lumbar vertebra and descend in a bundle to the levels at which they respectively leave the vertebral canal. Since this bundle of nerve roots bears a remote resemblance to a horse's tail it is called the *cauda equina*. Thus a needle can be introduced into the vertebral canal between the spines of the lower lumbar vertebrae without the danger of its impaling the spinal cord. This minor operation is called *lumbar puncture* and is used, for example, to obtain a sample of cerebrospinal fluid (see below).

The spinal cord is circular in cross-section near its upper end and in the thoracic region (look at the shapes of the vertebral canals in the vertebrae from these regions), but in the lower cervical region it expands and becomes more oval, while a similar enlargement occurs in the lower thoracic and upper lumbar regions. These *cervical* and *lumbar enlargements* indicate the sites of attachment of those nerves which form the limb plexues. The spinal cord needs a good blood supply and so various longitudinally running vessels may be seen along its length.

In a cross-section of the spinal cord (Fig. 5.13) it can be seen that the grey matter is centrally placed around the central canal. It takes the form of a long fluted column so that in cross-section it is somewhat H-shaped, with *anterior* and *posterior horns* and, in some regions, a small *lateral horn*. In general, it may be said that the anterior horn cells are concerned with motor activity and the posterior horn with sensory. The lateral horn contains cell bodies belonging to the autonomic nervous system.

THE SPINAL NERVES

A series of spinal nerves arise from the spinal cord and pass out between the vertebrae through the intervertebral foramina (see Chapter 9).

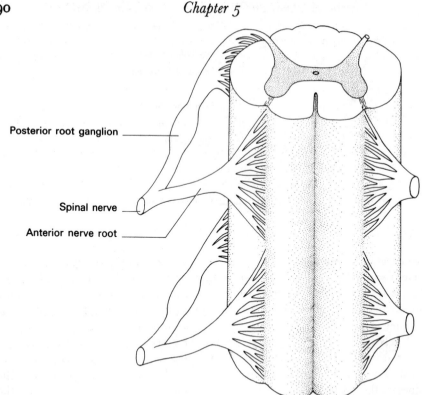

Fig. 5.14. Part of the spinal cord viewed from in front to show the origins of the spinal nerves.

Posterior root ganglion

Spinal nerve

Anterior nerve root

The first cervical nerve (C1) passes between the first cervical vertebra and the base of the skull, the second nerve between the first and second vertebrae, the third between the second and third, and so on. The last cervical nerve, however, passes between the seventh cervical vertebra and the first thoracic vertebra so that there are eight cervical nerves but only seven cervical vertebrae. This peculiar situation is caused by changes which occur during development. The first thoracic nerve (T1) lies between the first and second thoracic vertebrae, and from here down, each spinal nerve passes between its own vertebra and the one below. There are thus the same number of nerves as vertebrae, i.e. twelve thoracic (T1–12), five lumbar (L1–5) and five sacral (S1–5). There is also a very small coccygeal nerve (Co 1).

The spinal nerves arise by *anterior* and *posterior* (ventral and dorsal) *nerve roots* which unite in the intervertebral foramen to form a single nerve (Fig. 5.14). Just before the point of junction, the posterior nerve root is enlarged to form the *posterior root ganglion* (spinal ganglion); a ganglion, you will remember, consists of a collection of nerve cells. The posterior nerve root is associated with sensation and therefore contains ingoing or *afferent* fibres while the anterior root is associated with motor activity and contains outgoing or *efferent* fibres.

The nerve fibres which make up the anterior and posterior nerve roots will have to be studied in more detail later in this chapter but for the moment it will suffice to say that the posterior root ganglion contains the cell bodies of pseudo-unipolar cells of which one process (the axon) passes into the spinal cord and either proceeds straight up towards the medulla or forms synapses in the posterior horn of grey matter and then passes upwards (or occasionally downwards). The other process (the dendrite) passes outwards along the spinal nerve and ends as a specialised sensory nerve ending (see below). The nerve cell bodies whose axons form the anterior nerve roots lie in the anterior horn of grey matter and are known as *anterior horn cells*. Since these axons supply all the muscles of the neck, trunk and limbs they are of the greatest importance to physiotherapists and their connections will be described in detail later. The anterior nerve roots also contain fibres of the autonomic nervous system and these will be discussed in Chapter 6.

It may be helpful to make a small digression here in order to point out the similarity between many of the cranial nerves and the spinal nerves. This does not apply to the nerves of the special senses (smell, hearing, vision) since these are, in every way, special. The other cranial nerves, however, may be mainly motor, mainly sensory, or mixed. In any nerves which have a motor component, the nerve cell bodies of the axons in the nerve will lie in a collection of cells in the brain-stem called the *motor nucleus* of that particular nerve, so that this nucleus corresponds to a group of anterior horn cells. Similarly, cranial nerves with a sensory component will have cell bodies of the pseudo-unipolar type lying in a ganglion along the course of the nerve which thus corresponds exactly to a posterior root ganglion.

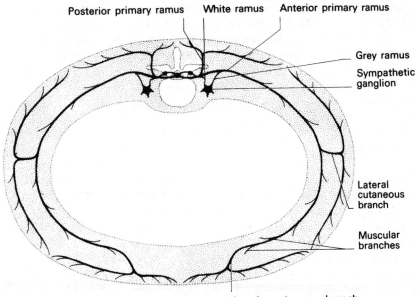

Fig. 5.15. Diagram of the course of a typical spinal nerve.

Spinal nerves are all mixed nerves, containing both afferent (sensory), efferent (motor) and connector and effector (see below) autonomic nerve fibres, the latter travelling with the spinal nerves as 'passengers', later leaving them to run to the smooth muscle of blood vessels and of other viscera and also to glands. The ultimate distribution of spinal nerves depends on the region to which they belong. Thus the upper cervical, the thoracic nerves (except for T_1) and the first lumbar nerve have a fairly typical course. This may best be described by considering a nerve from the mid-thoracic region (Fig. 5.15). Shortly after its formation by the junction of the two roots, the nerve gives a *white ramus communicans* to the corresponding ganglion on the sympathetic trunk. This contains connector fibres whose cell bodies are situated in the lateral horn of grey matter (see Chapter 6). Effector fibres leave the ganglion and pass back to the nerve in a *grey ramus communicans* and these will be distributed to structures in the skin, to blood vessels, and to other structures which lie along the course of the spinal nerve. The spinal nerve divides into two branches: the primary rami. The *posterior primary ramus* passes backwards, giving off branches to the muscles of the back and it ends by reaching the superficial fascia where it divides into medial and lateral branches which are sensory to the skin of the back. The *anterior primary ramus* passes around the chest wall between the ribs, where it is known as an intercostal nerve, and gives branches to supply the intercostal muscles. As it passes round the side of the chest, the nerve gives off a *lateral cutaneous branch* which divides up to supply the skin in this region. At the front of the chest, close to the side of the sternum, the nerve terminates by passing forwards to form an *anterior cutaneous branch* which supplies the skin of the front of the chest. The nerve thus supplies a band of muscle and skin which forms a narrow girdle running around the chest or, in the case of the lower thoracic nerves, around the abdomen. The area of skin supplied by a single spinal nerve is known as a *dermatome* (see below). In the thoracic and abdominal region, therefore, the dermatomes are regularly arranged, but it must be remembered that there is a great deal of overlap between one dermatome and the next so that damage to a single thoracic nerve may cause surprisingly little sensory impairment.

THE LIMB PLEXUSES

The nerves which travel to the limbs are arranged in a much more complicated fashion than are those mentioned in the previous section. This is because the individual spinal nerves do not innervate the limb in simple segmental fashion but join with each other in a complicated way to form a *limb plexus* from which the nerves of the limb are derived. Thus the spinal nerves C5 to T1 form the *brachial plexus* from which the named nerves of the upper limb arise. The lumbar plexus supplies nerves mainly to the front and sides of the thigh and is formed by L1 to L4, while the sacral plexus (L4–S3) supplies nerves mainly to the back of the leg and the sole of the foot. You will notice that L4 is common to both plexuses— this nerve bifurcates, one division going to the lumbar and one to the sacral plexus.

Because of the limb plexuses, individual named nerves of the limbs may contain fibres derived from a number of different spinal nerves. For example, the femoral nerve in the thigh is formed by contributions from L2, 3 and 4 while the radial nerve in the upper limb is derived from C5, 6, 7, 8 and T1. The numbers of the spinal nerves which contribute to a particular named nerve are referred to as its *root value*. The cervical nerves also form a plexus of sorts but this is a comparatively simple affair.

DERMATOMES

As has been mentioned already, a dermatome is the area of skin supplied by a particular spinal nerve, and in the trunk they form a series of regular segments

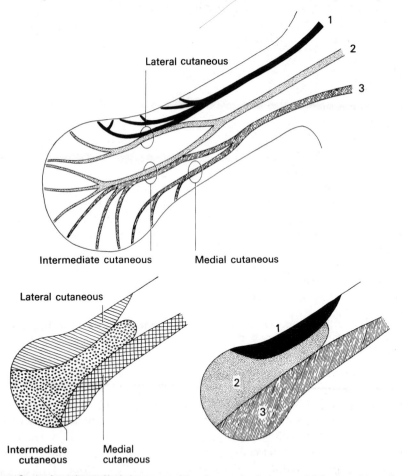

Fig. 5.16. An imaginary limb innervated by three spinal nerves which give rise to three cutaneous nerves, two of which contain components of 2 spinal nerves. Note that the 'map' of the area supplied by each cutaneous nerve differs from the 'map' of the dermatomes.

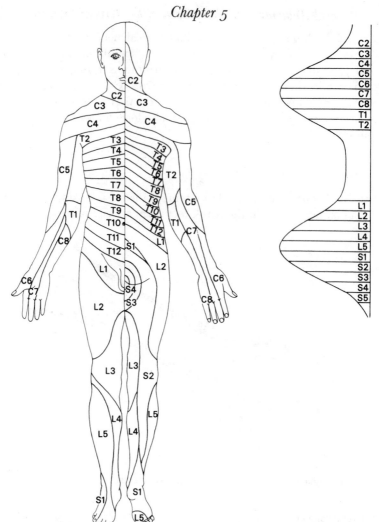

Fig. 5.17. The dermatomes of the front and back of the body. The small diagram shows the regular arrangement of dermatomes in the upper and lower limbs of the embryo.

running around the body (Fig. 5.17). In the limbs, however, it is difficult, though not impossible, to see the segmental arrangement of the nerves since a single cutaneous nerve may carry components of several spinal nerves. It is very important that you should understand clearly the difference between the distribution of a named cutaneous nerve (such as the posterior cutaneous nerve of the thigh) and a dermatome. This is explained in Figure 5.16. This illustrates a purely hypothetical limb which is innervated by 3 spinal nerves, 1, 2 and 3. Suppose that the cutaneous nerves of the limb are called lateral, intermediate and medial cutaneous nerves with root values 1 and 2, 2 and 3, and 3 respectively.

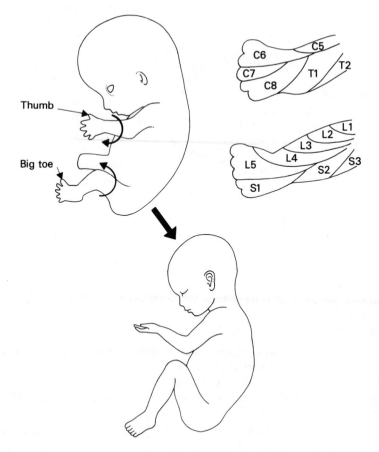

Fig. 5.18. Diagram to show rotation of the limbs during development.

The diagram shows that a map of the areas innervated by the three nerves will be very different from a map of the dermatomes.

Figure 5.18 shows how the arrangement of the limb dermatomes comes about. If it is remembered that the embryonic limbs are simple in shape and face in the same direction (see also p. 5), it is clear that the adult arrangement of dermatomes is the result of the elongation of the limb with the carrying of some of the dermatome areas to the ends of the limbs. Again it must be stressed that there is a wide area of overlap between adjacent dermatomes and also between areas supplied by individual cutaneous nerves.

THE VENTRICLES OF THE BRAIN

The four brain ventricles have already been briefly mentioned. Each of the two *lateral ventricles* lie within a cerebral hemisphere. The bodies of the ventricles lie near the medial sides of the hemispheres but an *inferior horn* of each diverges

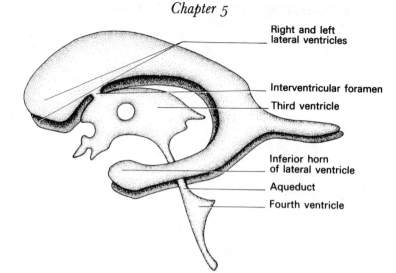

Right and left
lateral ventricles

Interventricular foramen

Third ventricle

Inferior horn
of lateral ventricle

Aqueduct

Fourth ventricle

Fig. 5.19. A cast of the four ventricles of the brain. Leonardo da Vinci was the first person to make such a cast.

laterally and downwards to lie within the temporal lobe (Figs 5.3 and 5.19). Each lateral ventricle communicates via an *interventricular foramen* with the anterior part of the *third ventricle*. This is a very narrow cavity lying in the midline, with its lateral walls formed by the thalami and the hypothalamus. It leads backwards through the *aqueduct* of the midbrain to open into the *fourth ventricle* which lies in the dorsal part of the pons and medulla.

Each of the four ventricles has one part of its wall which is extremely thin. This thinned-out region is invaginated into the cavity of the ventricle and the invagination contains a dense plexus of blood vessels known as the *choroid plexus* (Fig. 5.20). The choroid plexus secretes cerebrospinal fluid into the cavity of the ventricles so that the whole ventricular system is filled with fluid,

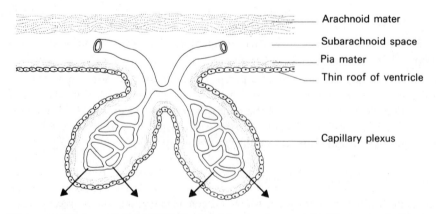

Arachnoid mater

Subarachnoid space

Pia mater

Thin roof of ventricle

Capillary plexus

Fig. 5.20. Diagram of the choroid plexus of the fourth ventricle. The arrows indicate the secretion of the cerebrospinal fluid into the cavity of the ventricle.

which is in continuous production. The thin part of the fourth ventricle is its roof and in the region which is not invaginated to form the choroid plexus are three holes or *foramina*—one in the midline and one at each side. These allow the exit of CSF from the ventricular system so that it passes directly into the subarachnoid space. There is thus a continuous circulation of CSF from the choroid plexuses into the subarachnoid space. From here it passes around the brain and spinal cord and finally enters the venous blood through the *arachnoid villi*. These are small projections of the arachnoid mater into the wall of the large venous channel which passes over the top of the skull in the midline (the *superior sagittal sinus*). The physiology of the CSF will be described in Chapter 17.

THE PHYSIOLOGY OF THE NERVOUS SYSTEM

As has already been described in Chapter 4, the reflex is the basic unit of activity in the nervous system. In reflexes, nerve impulses pass from sensory receptors along nerve fibres to the CNS, where the 'information' is transferred, perhaps to several different regions, before being fed to motor nerve cells so that motor responses to sensory stimuli may be performed in a purposive and controlled manner and achieve exactly the requirement signalled by the receptors' stimulation. The need has also to be met without excessive disturbance to the body's position or, for instance, to the blood supply of its vital organs.

Thus a study of the functions of the CNS must begin with a study of receptors and the incoming sensory fibres' pathways whereby information is brought to the brain. (In the healthy body, it must be remembered, reflex activity proceeding without the intervention of the brain is of minimal importance.) We can then continue with what we know about how the brain initiates and controls muscular movements in response to sensory stimulation. Finally, in this chapter we will continue description of the way the position of the resting body, its posture, is maintained by a continual state of partial muscular contraction, or *muscle tone*.

SENSORY MODALITIES

The special senses of vision, hearing, balance, taste and smell will be considered in Chapter 13. In this chapter we will be concerned with sensation from skin, muscles, tendons, joints and the visceral organs of the thorax and abdomen. The sensations considered are touch, pressure, pain and temperature from the skin; position and movement sense (*proprioception*), and pain from muscles and joints and from viscera. Although not specifically covered in this chapter, pressure and stretch are also detected in blood vessels and in the lungs, and many different chemical components of blood and other body fluids are detected by specific *chemoreceptors*.

SKIN RECEPTORS AND SENSATIONS

At least six different kinds of ending occur in the skin as shown in Fig. 5.21. These include relatively specialised nerve endings which are encapsulated in some sort of a sheath (including Meissner's corpuscles and Pacinian corpuscles) and others in which no sheath is present. Some axons end by dividing into a number of branches and ending in the dermis or in the deeper layers of the epidermis. The superficial part of the epidermis, however, is devoid of nerve fibres and therefore can be pricked or cut with no feeling of pain. The roots of the hairs are surrounded by a network of nerves, many of which end in the vicinity of the root and since the hair can act as a lever, hairy skin is, in general, very sensitive, so that you can feel the lightest touch on your hair, at least near the scalp. Hairs themselves are, of course (fortunately) devoid of nerve fibres. Four basic types of sensation are detected by sensory nerve endings: touch-pressure, pain, cold and warmth. In addition to simple touch, we have the power to discriminate between neighbouring points on the skin when they are touched simultaneously. This *two-point discrimination*, forms the basis of our ability to recognise and handle objects by touch alone. It is readily diminished by disease affecting the density of sensory nerve endings and fibres, being dependent on the number of discrete endings and fibres present. It has not generally been possible to identify any one sensation with any one receptor-type. However, whatever the histological appearance of a nerve ending, each ending of whatever type is specific for only one of the four modalities of sensation. An exception to the lack of relationship between appearance and sensation mediated is seen

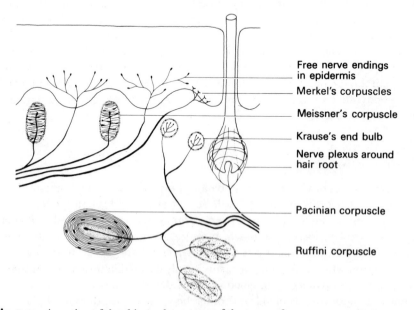

Free nerve endings in epidermis
Merkel's corpuscles
Meissner's corpuscle
Krause's end bulb
Nerve plexus around hair root
Pacinian corpuscle
Ruffini corpuscle

Fig. 5.21. A section of the skin to show some of the types of sensory nerve endings.

in the Pacinian corpuscles which detect changes and fluctuations of pressure applied to them. These have been used to study the way in which the stimulus, i.e. pressure sufficient to deform the receptor, is transduced into nerve impulses, as has been described for the Golgi tendon organ on pp. 101–102.

ADAPTATION

Most receptors become *adapted* to a continuing stimulus and soon cease to respond if the stimulus is maintained unchanged in strength (Fig. 5.22). The receptor is not fatigued, only 'bored', for at the slightest change in stimulus strength, the receptor responds again as vigorously as before, only to lapse again into inactivity once it has become adapted to the new level of stimulus. Many of the cutaneous touch receptors readily adapt, so we soon become unaware of the clothes we are wearing. Others, particularly the length receptors in muscle spindles are for ever vigilant and continually produce impulses even though the length of the muscle is unchanged (Fig. 5.22). We will have more to say about these receptors in the next section of this chapter, in addition to what has already been said in Chapter 4.

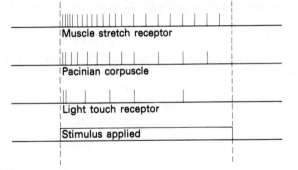

Muscle stretch receptor

Pacinian corpuscle

Light touch receptor

Stimulus applied

Fig. 5.22. The decay of the response of different receptor types when a constant stimulus is applied to the receptors.

MUSCLE, TENDON AND JOINT RECEPTORS

NEUROMUSCULAR SPINDLES

These structures, sometimes known simply as muscle spindles, are the receptor organs of the muscles themselves, being situated in amongst the muscle fibres. You should remember that when talking of the 'nerve supply of a muscle' you are not only referring to motor or efferent fibres but also to sensory or afferent fibres which originate in these receptor organs.

The spindles are, as their name implies, fusiform in shape (i.e. like a straight banana!) and they are enclosed by a sheath of connective tissue (Fig. 5.23). Within the sheath there are 5–15 modified skeletal muscle fibres which are known as *intrafusal fibres* and are of two types. As will be described later, skeletal muscle fibres are cells, each of which contain a very large number of nuclei and this is also true of the intrafusal fibres. In one type of intrafusal fibre the nuclei are almost all found clustered in a slight dilatation of the fibre and these are known as *nuclear bag fibres*. In the other type—*the nuclear chain fibres*—the nuclei form a longitudinal chain along the centre of the fibre. The nuclear chain fibres are entirely contained within the connective tissue spindle sheath but the nuclear bag fibres are longer and project at either end of the spindle to be attached to the connective tissue sheath (*endomysium*) of the ordinary muscle fibres which form the main mass of the muscle. The ordinary muscle fibres are sometimes called *extrafusal fibres* to distinguish them from the fewer, smaller and specially innervated intrafusal fibres. The extrafusal fibres are not, of course, part of the muscle spindle, so this term is a little misleading.

The nerve supply of the neuromuscular spindles is both motor and sensory. The intrafusal fibres are supplied by small diameter or γ efferents whose cell bodies are in the anterior horns of grey matter of the spinal cord and whose axons end on both types of intrafusal fibre, particularly in the region away from their centres. Impulses travelling along the γ efferents can cause contraction of the intrafusal fibres. The sensory nerve supply of the spindles is of two types. Group Ib fibres enter the spindle and break up to form a number of branches which wrap themselves around the nuclear bag types of intrafusal fibres in a spiral fashion to form the *annulospiral* nerve endings. Smaller Group IIb fibres enter the spindle and branch out into a number of beaded nerve endings which are particularly related to the nuclear chain fibres. They are known as *flower-spray* endings.

The structure of the spindles gives some clues to their function. Stretching a muscle as a whole will obviously stretch the intrafusal fibres, particularly the nuclear bag fibres which are attached outside the spindle to the endomysium of the muscle fibres themselves. Thus the annulospiral nerve endings will be stimulated. They can also be stimulated by contraction of the intrafusal fibres themselves as a result of stimulation of the γ efferent fibres since the latter cause contraction particularly of the ends of the intrafusal fibres so that the central portions are stretched. In this way the sensitivity of the annulospiral nerve endings can be varied since if the central portions of the intrafusal fibres are already taut, a very small elongation of the muscle will produce an immediate response. It seems likely that the nuclear bag fibres, with their extensive annulospiral nerve endings, respond rapidly to active changes in length of the whole muscle, whereas the flower-spray endings on the nuclear chain fibres respond to slower and less dynamic changes in length which occur in the maintenance of posture. These functions will be discussed in more detail later; they play an important part in the mechanisation of stretch reflexes.

THE TENDON AND JOINT RECEPTORS

Golgi first described the net-like collection of nerve endings within tendons. They are therefore often called *Golgi tendon organs* (Fig. 5.23). Their afferent fibres are Group Ib in type. Impulses pass along these fibres as a muscle contracts and tension develops in its tendon. They are thus properly described as the muscle's tension receptors. just as the spindles are its length receptors.

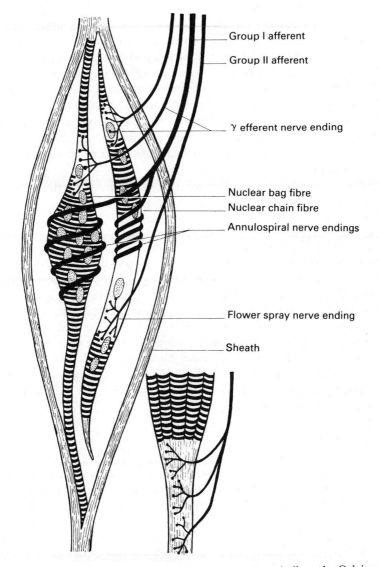

Group I afferent

Group II afferent

γ efferent nerve ending

Nuclear bag fibre
Nuclear chain fibre
Annulospiral nerve endings

Flower spray nerve ending

Sheath

Fig. 5.23. A diagrammatic representation of a neuromuscular spindle and a Golgi nerve ending in a tendon (small diagram).

Various types of receptor in joint capsules and ligaments signal movement and position of the respective joints. Little is known about them in comparison to the detailed knowledge of the muscle spindle, but they must be as intimately concerned in the subjects of control and awareness of body position and movement as are the muscle and tendon receptors.

PROPRIOCEPTION

This is the term used for our conscious awareness of the position of the parts of the body, particularly the limbs. Without looking to see where it is we know the position of our hand or foot, how fast and in what direction it is moving, so we can hit or kick an object towards a target from which we do not take our eyes. Proprioception is the result of integrating sensory information from many receptors including Pacinian corpuscles and other touch receptors, both the length and the tension receptors of muscles and tendons, and the joint receptors. All these receptors signal not just where a limb is in a space, but how fast the joints are altering in position, thus supplying information comparable to that coming from the annulospiral endings of the muscle spindles. We do not know how the combining of all the information is performed by the brain, but find life very difficult when the sensory pathways are damaged by disease.

VIBRATION SENSE

Finally, all tissues, especially bone, are sensitive to vibration. If a 50 cycles/sec tuning fork base be applied to a bony point, a 'buzzing' sensation is experienced as the fork vibrates. What purpose is served by this sensation is not known, but as the afferent nerve fibres run in the CNS with those from muscles, tendons and joints, it is sometimes of value to the physician to determine whether vibration sense is present in a limb. Vibration sense disturbance occurs at the same time as disturbance of proprioception, so serves as a useful indicator for real loss of this important sensory function. The naïve patient, feigning loss of proprioception, can thus be detected when he declares he can perceive the vibration.

PAIN FROM MUSCLES, JOINTS AND LIGAMENTS

Muscles, tendons and joints may be the sites of origin of painful sensation, brought about by many causes. The whole subject of pain, of such importance to physiotherapists, has been given a special section (p. 105), so nothing more will be said here.

SENSORY PATHWAYS

The first nerve cells along the sensory pathways are found in the posterior root ganglia or the ganglia of cranial nerves, particularly the trigeminal nerve. As has already been mentioned, these cells are pseudo-unipolar cells whose peripheral process ends as a specialised sensory nerve ending and whose central process enters the dorsal part of the spinal cord. The subsequent pathway taken by sensory nerve impulses varies according to the type of sensation involved and it will therefore be necessary to consider these pathways individually.

THE SPINOTHALAMIC PATH

Sensory nerve fibres, from pain, temperature and some touch receptors end in the posterior horn of the spinal cord grey matter as soon as they enter the cord in the posterior spinal nerve roots. The cells on which these fibres terminate include local interneurones supplying anterior horn cells of the same segment, intersegmental neurones which run from one segment of the spinal cord to another, and finally neurones whose axons pass up through the spinal cord to the brain. The axons of these neurones cross obliquely upwards through the spinal cord to form the *spinothalamic tracts* of the opposite side on the anterolateral aspect of the cord, between the anterior fissure and the anterior nerve roots (Fig. 5.24). Pain and temperature fibres travel separately from the touch fibres and lie behind them in the spinothalamic tracts. Fibres from the lower limb are placed laterally and cervical fibres placed medially. These afferent fibres end in the posterior ventral part of the thalamus, whence they are relayed to the sensory cortex.

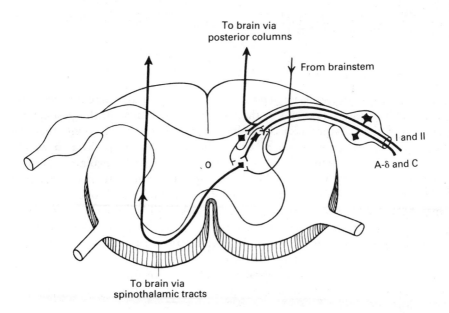

Fig. 5.24. The paths taken by sensory nerve fibres entering the spinal cord, and the pathways which can influence them.

THE POSTERIOR COLUMN PATH

Some of the fibres concerned with touch and two-point discrimination, and those concerned with position and vibration sense enter the posterior columns of the spinal cord, without making any synaptic connections in the posterior horn (Fig. 5.24). They travel on the same side as their entry through posterior nerve roots. In the medulla oblongata these fibres end on nerve cells of the *nuclei cuneatus and gracilis* (Fig. 5.12). The axons of these cells cross the midline and pass up through the brain-stem on the opposite side to reach the posterior ventral part of the thalamus. During their pathway through the brainstem these fibres form a well-defined tract on each side of the mid-line known as the *medial lemniscus*.

As was explained earlier in this chapter, the ganglia on the cranial nerves correspond to the posterior root ganglia of the spinal nerves. The sensory nerve supply to the face and much of the scalp is derived from one of the cranial nerves, the trigeminal nerve, and the axons of its ganglion cells synapse with cells in various nuclei in the brain-stem whose axons cross the midline and then proceed, like the fibres from the rest of the body, to the posterior ventral part of the thalamus.

Thus all general body sensation reaches the thalamus on the side of the central nervous system opposite to that part of the body from which it originated. Shortly before entering the thalamus all the sensory fibres, whether originally from the spinothalamic path, from the posterior column path, or from the trigeminal path, send branches into the adjacent *reticular formation*, which consists of scattered masses of grey matter in the brain-stem, the precise anatomy of which is impossible to map out.

THE SENSORY CORTEX

From the cells of the posterior ventral part of the thalamus, where the ascending sensory fibres end, axons run to the sensory cortical areas in the post-central gyrus and the adjacent lateral cerebral sulcus (Fig. 5.6). It seems that the reticular formation neurones may act on this direct sensory pathway in such a way that one 'attends to' or is 'aware of' only a portion of the incoming neuronal activity. Unless the recticular formation 'activates' the thalamocortical part of the path, the incoming impulses do not, apparently, get through. Thus one may be quite unaware of the ticking of a clock until it suddenly stops.

In the sensory cortex the body is represented upside down. The foot area is close to the mid-line, at the top of the post-central gyrus, followed by leg, hip, trunk, shoulder, arm and hand, after which come the face, lips, teeth and tongue and pharynx. The sizes of cerebral cortical areas for different bodily regions are related to the density of receptors in those regions, being small for the trunk and largest for hands and oral regions where sensory information is so important.

Adjacent and posterior to the cortical area for the reception of impulses from the body's sensory receptors is an area known as the *sensory association area*.

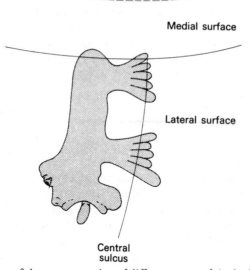

Medial surface

Lateral surface

Central
sulcus

Fig. 5.25. The size of the representation of different parts of the body on the motor cortex.

It is suggested that this association area correlates the pattern of impulses originating various receptor types in adjacent skin areas and thus identifies an object placed, for example, in the hand. Nothing more precise than that can yet be said about what goes on in this part of the brain. It is clear that impulses must travel to many other brain areas, those concerned with memory retention and emotional expression being important. If the post-central gyrus is affected by disease, touch and position sense are most affected, temperature sensation is less affected and pain only slightly disturbed.

However little we actually know for certain about the brain and sensation, it is clear that most movement originates from sensory input, either immediately, or as a result of recall from short- or long-term memory. So it is appropriate now to turn from sensory input to the brain to the motor output and see how the brain initiates and controls voluntary movement.

PAIN

RECEPTORS AND PERCEPTION

There are no identifiable specific pain receptors. All we know, so far as man is concerned, is that impulses travelling towards the CNS in fine myelinated (A-delta) or unmyelinated (C) fibres may be associated with a sensation of pain. Impulses in A-delta fibres are linked with 'fast' pain, rapid in onset and stinging or pricking in quality, whereas impulses in C fibres produce 'slow' pain, which is dull, aching or burning in quality. Impulses in these fibre types may also be associated with sensations of heat and cold, itch and touch. Some of the fibres come from recognisable receptor structures and some from branching free nerve endings in the tissues. Many, although typical sensory nerve fibres, have collateral endings on neighbouring arterioles. Impulses in these cause the vessels to dilate and produce the triple response to injury, arteriolar and capillary dilatation and wheal formation.

STIMULATION OF THE RECEPTORS

A variety of stimuli produce a sensation of pain, but the most important are those that result in actual tissue damage. This may be chemical, thermal or mechanical. In any muscle, striated, cardiac or smooth, powerful contraction, or contraction in excess of the circulatory system's power to maintain metabolism, may cause intense pain, presumably due to the build-up of damaging chemical products of the raised metabolic activity. The search for a common chemical stimulus that, when applied to the tissue damage receptors, results in a sensation of pain, has produced several different chemicals. These include acetylcholine, 5 HT, histamine, bradykinin, K^+ and H^+. Some of these may be liberated in any sort of tissue damage, for example, histamine and bradykinin in the inflammatory process, mechanical disruption or temperature extremes. Hydrogen and potassium ions may be the stimuli in ischaemic muscle exercise. Thus, experiments on healthy people can produce exactly the same sort of pain in the muscles of a limb exercised with a tourniquet on it as is produced when the blood flow is limited by arterial disease, and these experiments suggest that when metabolism is increased out of proportion to the blood supply, a metabolic product is the stimulus to the pain receptors.

Pain arising from a joint or other deep tissue may be felt to be coming from the skin—but not necessarily the skin overlying the site of the pain. It will come from skin supplied by the same spinal nerve segment (dermatome) as the affected organ, and is said to be *referred* to that region. Thus pain from the diaphragm between thorax and abdomen is referred to the shoulder and pain from the hip to the region over the knee joint. Pain from damaged or diseased joints, or from the torn periosteum over a broken bone always results in powerful and sustained contraction (spasm) of the muscles around the damaged joint, with its resultant immobilisation.

VISCERAL SENSATION

The viscera are supplied by the autonomic nervous system. The only sensations of which we are conscious as having arisen directly from the viscera are unpleasant ones—usually painful! Pain may be associated with inflammation, distension or powerful muscular contraction of the walls of hollow viscera, but not with cutting or pricking with a sharp instrument. Distension and inflammation produce a sustained pain, while contraction of smooth muscle in a viscus typically causes an intermittent pain known as a *colic*. Distension may cause excessive contraction, so colicky pains may be superimposed upon a sustained pain, thus conditions causing obstructions to the intestine may result in colic as the intestinal muscle contracts powerfully. Visceral pain is typically referred to skin areas, sometimes distant from the site of the disease. An inflamed appendix causes pain around the umbilicus (both sustained and colicky), until its inflammatory state involves the lining of the abdominal wall (the parietal peritoneum which is supplied by the spinal nerves, not the autonomic system); whereupon the pain becomes more accurately loca- lised and protective muscle contraction in the abdominal wall can also be de- tected. Pain from the heart is felt in the mid-line of the chest, whence it spreads up into the neck and to the inner side of the left arm.

SPINAL CORD CONNECTIONS

Most of the A-delta and C fibres, on entering the posterior horn of grey matter in the spinal cord, terminate there in the substantia gelatinosa, now known as lam- inae I and II. The excitatory transmitter at these fibres' terminals is now thought to be a polypeptide of 11 amino acids, appropriately called *substance P*. These synaptic terminations are not simple. The incoming sensory fibres receive inhibi- tory pre-synaptic terminals, and the cells of the posterior horn onto which they terminate may have inhibitory terminals on them, as well as the excitatory ter- minals of the sensory fibres. The origins of these pre- and post-synaptic inhibitory terminals, affecting transmission through the 'pain' pathway, are both other sen- sory fibres entering the same segment of the spinal cord (via local interneurones) and fibres coming down the spinal cord from the brainstem's reticular formation. The former may be the basis of the 'gate control' theory of pain perception. The latter are of special interest for at some synapses in this pathway the opiate pain- relieving drugs have been found to be active, and the opioid transmitters, small peptide molecules called *enkephalins* have been detected. Figure 5.24 illustrates these connections.

ASCENDING AND DESCENDING PATHWAYS

The 'second order' axons involved in the conscious perception of pain pass up the spinal cord in the anterior and lateral columns of white matter in the spinothalamic tracts. These are crossed tracts, fibres on the left side conveying information from the right side of the body. The cross-over is obliquely arranged; the axons travel- ling some segments up the spinal cord from their origin in the posterior horn on one side before taking up their final position in the spinothalamic tract on the other side. The fibres from the sacral region, entering the column first, come to lie most superficially, and as more fibres enter the tract as one passes up the spinal cord they join the deeper aspect of the tract, those entering in the cervical region lying close to the anterior horn of grey matter. Activity in these axons is the resultant, as has been described, of both stimulation of the 'pain' receptors and of the various inhibitory influences operating at their origin in the spinal cord. Brain-stem nuclei, within the midbrain periaqueductal grey matter, via synaptic connections with the raphe nuclei of the pons and medulla can inhibit the cells of origin of the spinothalamic axons, by impulses passing down the raphe-spinal path in the lateral columns of the spinal cord. It is in these brain-stem nuclei that the opioid transmitters, the enkephalins, have been detected. Two of these have been found, leucine and methionine containing peptide molecules of five amino acids only. Their physiological action is mimicked by morphine and is blocked by drugs that block the pain-relieving properties of morphine. Now that their chemistry is known, it is possible that a new range of opium-like pain killing drugs may be developed, which do not have the notorious harmful side-effects of morphine.

The spinothalamic fibres enter the thalamus, whence the 'third order' axons pass to the post-central gyrus of the cortex. They also enter the reticular formation, from where, through many neurones, they connect with the hypothalamus and the 'emotional' regions of the cerebral cortex. It should be noted that artificial electrical stimulation of the cortex in a conscious patient has not, so far, been accompanied by a sensation of pain. Some doubt, therefore, must remain about that part of the brain with which we actually experience pain.

VARIATIONS IN PAIN SENSATION

Patients will experience pain where no peripheral cause can be detected, and pain may not be experienced when tissue damage is known to be present. Both of these states may be attributed to varying activity of the various inhibitory connections in the pain pathway. Stimulating the midbrain's periaqueductal grey matter, in a region known to be linked to the alerting or alarm reaction centre, reduces the responses of an animal to mild tissue damaging stimuli. This may be the mechanism whereby soldiers in battle, for example, become unaware of the severity of the injuries they have suffered. It is also possible that the otherwise inexplicable pains following herpes zoster or Coxsackie virus infections of posterior root ganglia may be due to a loss of inhibition from local group I and II afferents, releasing the A-delta and C fibres from their inhibitory influence. Rubbing a painful area or heating it gently may reduce the sensation of pain, presumably because these alternative afferent pathways, when stimulated, partially inhibit the passage of the 'pain' impulses into the spinothalamic tract axons, through the hypothetical gate control mechanism.

HYPERALGESIA

This means enhanced pain sensation in response to relatively trivial stimulation. In the presence of inflammation, or in conditions in which other sensory modalities are depressed but the pain mechanism remains intact, pain sensation is increased in response to mild stimuli. This is most often seen in the region of the arteriolar flare of the *triple response* (p. 469) to cutaneous injury, where it may persist for many days after the initial injury. It may also be experienced in the skin in the dermatome corresponding to the site of spinal cord damage.

REFERRED PAIN

When changes occur in viscera that stimulate their pain receptors, the resultant sensation may be felt in regions other than the actual site of the affected viscus. Thus pain from the heart receiving an inadequate blood supply is felt under the sternum, in the neck on the left side and down the medial surface of the left arm to the elbow. Pain from the inflamed appendix is felt around the umbilicus, until it affects also the local parietal peritoneum (p. 107) when it is correctly localised to the right iliac fossa, while pain from the diaphragm is felt at the tip of the shoulder. The simple anatomical reason for these effects is that the innervation of these viscera is determined by their embryonic sites of origin, so that pain from them is felt in the skin supplied by the spinal nerves of these same sites.

THE MOTOR CORTEX

About 100 years ago electrical stimulation of the brain in animals showed an area where stimulation produced movements, usually of the opposite limbs. This area was found by histologists to contain large pyramid-shaped cells and was called by them area 4. A corresponding area was found in monkeys, apes and finally in man. In the last-named species it is located in the central sulcus and spreads from there on to the precentral gyrus. In this *motor cortex* the body is represented precisely as in the adjacent general sensory cortex—feet at the top or actually on the medial surface of the cerebral hemisphere, followed by ankle, leg, knee, etc., until low down, at the end of the precentral gyrus adjacent to the lateral sulcus, is the area for laryngeal and tongue movements (Fig. 5.25). Once again, most space is given to those regions of the body with the largest repertoire of movements—hand, fingers (particularly the thumb) and the face.

It should be noted once again, that *movements*, not individual muscles, are localised in the cortex. Stimulation at one point produces, say, flexion of the fingers, which will be caused by contraction of flexor digitorum profundus and superficialis, relaxation of the finger extensor muscles, synergic contraction of the wrist extensors and relaxation of the antagonistic wrist flexors. All of these muscles groups are involved in this way in the conscious act of gripping an object and the co-ordinated action of the four groups of muscles is controlled by one area of the cortex. In addition to the motor cortex, stimulation of which produces discrete movements, there are other cortical regions, stimulation of which produces movement. An accessory motor cortex lies in the medial surface of each cerebral hemisphere wherein the body is regionally represented. A roughly triangular area, known to the histologists as area 6, lying in front of area 4, when stimulated, initiates more generalised and less discrete movements than does area 4. In front again of area 6 is a region where stimulation causes *both* eyes to move so that they look towards the opposite side, and a similar area exists in the occipital lobe of the cortex.

Apart from the eye movements, it should be emphasised that each cerebral hemisphere controls movements of the *opposite side* of the body. As a result of this, and of the crossing over of the sensory fibres running towards the brain, the left hemisphere receives sensory information from the right half of the body and discharges impulses which cause movements also in the right half of the body. Some mid-line structures, like the jaws and tongue or, even more surprisingly, the anus, are still divided functionally into right and left halves when it comes to their motor nerve supply, but in the very important function of speech, one hemisphere seems to take complete control of both sides of the body. It seems here that a region in the premotor cortex or area 6 has assumed control of the appropriate portions of area 4 of both hemispheres. The speech area lies in the left hemisphere in right-handed persons and vice versa—in right-handed persons, the left hemisphere is said to be *dominant*. We are, unfortunately, unable to tell you which is the dominant hemisphere in parrots or mynah birds.

MOTOR PATHWAYS

From the brain, two sets of nerve fibres run to the motor neurones of the brain-stem and the spinal cord.

THE CORTICOSPINAL OR PYRAMIDAL PATHWAY (Fig. 5.26)

This was given the latter name for two reasons. First, it appears as a discrete bundle of about a million fibres on each side of the anterior surface of the medulla oblongata in a body known as the *pyramid*. When traced back to the cerebral cortex, these fibres are found to originate in an area of the brain containing large pyramid-shaped cells (*Betz cells*). However, there are in the motor .cortex only 35,000 Betz cells on each side. In fact, only some 3% of pyramid tract fibres come from the Betz cells and only 60% of the remainder come from areas 4 and 6 of the cortex.

Before appearing on the surface of the medulla, these corticospinal fibres have left the cortex, converged in the corona radiata, passed between the cerebral basal nuclei in the internal capsule, then through the midbrain in the right and left bases pedunculi (Fig. 5.8). In the pons the corticospinal fibres become dispersed as they are interrupted by the pontine nuclei (Fig. 5.9) but soon after they come together again to form the pyramids of the medulla. They then mostly decussate, crossing the mid-line and coming to lie in the lateral columns of white matter in the spinal cord. (Fig. 5.12). About 30% of fibres remain on the same side, but at least some of these cross eventually. The corticospinal axons finally terminate on motor neurones or interneurones in the anterior horn of the spinal cord grey matter. It will thus be realised that the pathway from the cerebral cortex to the muscles may involve only two neurones. The *upper motor neurone* has its cell body in the cortex, its axons passing down to the anterior horn cells. The *lower motor neurone* consists of the anterior horn cell and its axon which passes to the muscle.

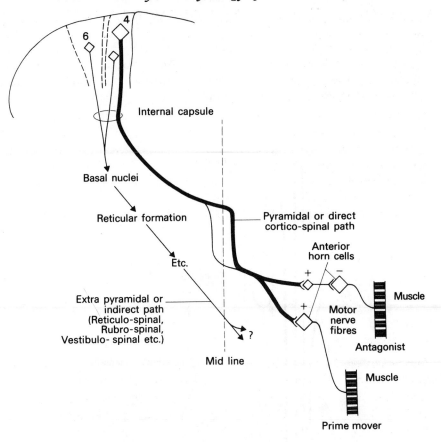

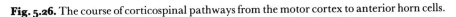

Fig. 5.26. The course of corticospinal pathways from the motor cortex to anterior horn cells.

THE EXTRAPYRAMIDAL PATHWAYS (Fig. 5.26)

Areas 4 and 6 and other parts of the cortex contribute axons which run to the basal nuclei and various other brain-stem nuclei. There are many complex interconnections between these nuclei, and fibres from this *extrapyramidal system* run to the anterior horn cells of the spinal cord grey matter. The influence of the motor cortex regions on the extrapyramidal system, whatever it may be, does not directly affect the motor neurones as it may do via the direct corticospinal path. For this reason it is also called the indirect corticospinal pathway.

It is possible, but probably not wholly accurate, to sum up the functions of these two routes as follows: the direct path from cortex to anterior horn cell initiates voluntary movement, while the indirect paths control both voluntary movements and the postural state in which movement begins and ends. This function of the latter path is described in detail in the following section.

CONTROL OF VOLUNTARY MOVEMENT

As nerve impulses pass from the cortex to the anterior horn cells, other impulses also pass to controlling systems in the brain. An important part of the brain concerned with these systems is the cerebellum, which also receives impulses from the muscles, tendon and joint receptors.

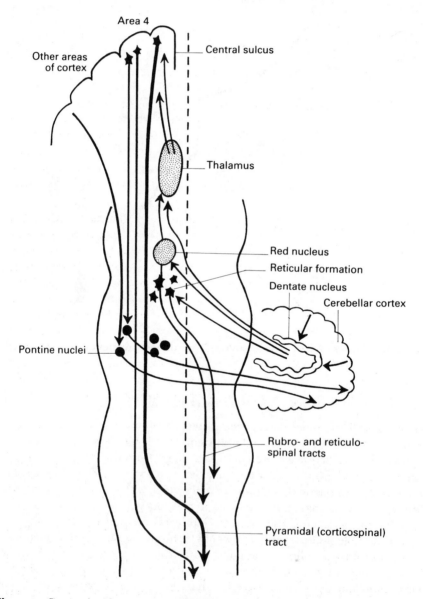

Fig. 5.27. Connections between motor cortex and the neocerebellum.

THE CEREBELLUM

The cerebellum can be divided physiologically into three parts. The first, the *archicerebellum* is concerned with resting posture. Its function will be outlined later. The second is the *palaeocerebellum* which deals with the control of reflex limb movements and the third is the *neocerebellum* which is involved in controlling the voluntary movements. The former two components are relatively primitive and form the main part of the cerebellum in lower animals. In man, the neocerebellum is the most extensive part of the cerebellum, and it is naturally linked closely with the motor cortex and with the basal part of the pons (Fig. 5.27). One important feature is that cerebellar function is uncrossed. It is connected, unlike the cerebral cortex, with its own side of the body. It receives fibres from the cortex via the pontine nuclei. These fibres enter the cerebellar cortex together with the *spinocerebellar* tract fibres which carry impulses from the muscle spindles, tendon organs and from movement receptors in joint capsules and ligaments. The output from the cerebellum is principally through the thalamus and back to the motor cortex, and also via the red nucleus and reticular nuclei to the anterior horn cells. The cerebellar cortex 'synthesises' the incoming information, perhaps comparing the command signal (from cerebral cortex via pons) with the incoming signals from muscle, tendon, and joint, that are altered consequent upon muscular contraction, and then altering the command signal from cerebral cortex to anterior horn cells in accordance with the returning information from the moving part and modifying directly the anterior horn cells. This, at any rate, is how an engineer might design a control system and it is noticable that loss of the cerebellum causes loss of control of, especially, fine movement. Damage to the cerebellum causes a loss of muscle *tone*, the continual state of partial contraction of all skeletal muscles which forms the basis of normal posture, as well as of control and co-ordination of voluntary movements. The speed, force and direction of all movements are all affected. Gait becomes 'drunken' and speech slurred. If such a patient attempts to touch an object, the finger will oscillate back and forth past the object, the state being known as *intention tremor*. This tremor is, of course, absent at rest.

THE BASAL NUCLEI OF THE CEREBRAL HEMISPHERES

These are sometimes misleadingly called the basal ganglia. Connections between cerebral cortex, the basal nuclei, substantia nigra, and thalamus are complex (see Fig. 5.28). Two important regions are the caudate nucleus and the substantia nigra. Damage to the former leads to rapid or slow purposeless and irregular movements (*chorea* or *athetosis* respectively). Damage to the nigral pathway leads to the tremor and the rigidity of *Parkinson's disease*. Tremor appears first in a resting limb and is abolished when voluntary movements take place, but the gradual development in muscles of a steady state of partial contraction leads to a progressive difficulty in movement and of facial expression. Parkinson's disease has attracted attention because the axons of the substantia nigra cells employ

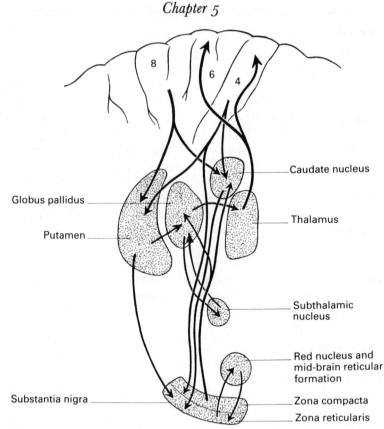

Fig. 5.28. Connections between the motor cortex and the basal nuclei of the cerebral hemisphere.

dopamine as their transmitter, and a partial loss of their activity can be opposed by oral administration of L-dopa, which is converted into dopamine in the brain, or by the giving of a substance that blocks an antagonistic pathway employing acetylcholine (perhaps the axons from the caudate nucleus to globus pallidus). Apart from the consequences of their damage, we have little knowledge of what the basal nuclear circuits actually do to control voluntary movements. As well as the L-dopa and anti-acetylcholine treatments, operative interference may return the patient's state towards normal. It seems that while loss of one part leads to trouble due to the unopposed actions of the remaining parts, loss of all improves things again! A final curiosity about the basal nuclei is their sensitivity to some metabolic poisons. Chronic mild carbon monoxide poisoning is associated with Parkinson's disease, copper and bile pigments may get deposited in the lentiform nucleus and cause chorea or athetosis. Chorea is also seen in some cases of hypersensitivity to streptococci causing throat infections and, in middle age, as the presenting symptom of an inherited presenile dementia known as Huntington's chorea.

CONTROL OF MUSCLE TONE AND POSTURE

Muscle tone and posture both depend upon the stretch reflex which is found in all muscles. While they were first studied experimentally in single muscles and isolated spinal cord segments, it must be realised that the basic stretch reflexes so revealed are extensively modified and controlled by a variety of influences from brain-stem, cerebellum, basal nuclei and cerebral cortex. Some of these higher centres exert their control in response to impulses in afferent paths coming originally from the muscles themselves. In other cases, it is impulses from joints, from the inner ears, or from the eyes that modify the stretch reflex controlling mechanism. Finally, presumably through the motor cortex, thoughts and emotions have a profound effect upon the stretch reflex and posture controlling mechanisms.

THE BASIC STRETCH REFLEX

As with all reflexes, we begin with the receptors. These are the spiral endings in muscle spindles. As has already been described, these respond to elongation. The more they are pulled out, the more impulses they produce. They are stimulated in two ways. One is by passive stretch of the muscle—the usual experimental technique for activating the receptors. The other is stimulation of the γ-efferent motor neurones. This causes contraction of the muscle fibres within the spindle structure (the intrafusal fibres) and thus extension of their non-contractile parts around which the spiral endings are wound. This mode of stimulating the spiral endings is probably the one that occurs most frequently in real life states.

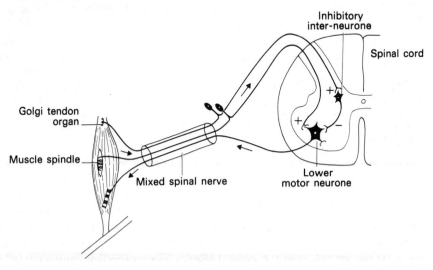

Fig. 5.29. The reflex arcs of the excitatory stretch reflex and the inhibitory reflex caused by strong stimulation of the Golgi tendon organ.

In the spinal cord the sensory fibres from the spindles end by synapsing with the motor neurones of the same muscle as their origin, thus forming a 2-neurone, monosynaptic reflex arc (Fig. 5.29). The synapses are excitatory, causing the muscle to contract when the spiral endings of its spindles have been stretched. The contraction will persist so long as the muscle or its spiral endings are stretched, these endings showing little adaptation to the continuing stimulus. The muscle relaxes, however, immediately the afferent impulses cease, as happens naturally when the spiral endings are released from the stretch, or experimentally if the sensory nerve roots are cut, thus preventing the impulses from gaining access to the motor neurones. Within a muscle the individual cells may be found in a state of alternating contraction and relaxation, and they may work in relays, so that no single cell becomes excessively fatigued.

As well as the spindle afferent fibres synaptically connecting directly with motor neurones, branches from these fibres end in synapses upon a type of interneurone, called the Golgi *bottle neurones*. Axons from these run to motor neurones of antagonist muscles and inhibit them, causing relaxation of the antagonist.

A powerful passive stretch of a muscle may cause its reflex contraction response to be suddenly overcome by a complete relaxation. This has been found to be caused by impulses coming from the Golgi tension receptors in the tendon of the muscle. Just as the muscle length receptors are stimulated by elongation, the tendon receptors are stimulated by an increase in tension. In an experimental situation, increase in length always accompanies an increase in tension, the tissue itself having elastic properties. In the body, however, an increase in tension is not always associated with a change in muscle length (see p. 146 on isometric and isotonic muscle contraction). Apart from these considerations, the tension receptor impulses can override the length receptor impulses and cause motor neurone inhibition if the tension is great enough. The inhibitor activity of the Golgi tendon organs is most dramatically seen in athletes with a muscular cramp. If its contraction is deliberately opposed by a stronger external force, then the muscle will suddenly relax. In the calf muscles, which plantarflex the ankle joint, a deliberate dorsiflexion of the foot by a team-mate or trainer will 'break' the cramp in a moment. In the intact animal or man, both length and tension signals are used in the automatic control of muscle contraction, carried out, it is believed, in the cerebellum.

TENDON JERKS

A brief elongation of a muscle, produced by tapping its tendon, may elicit a correspondingly brief stretch reflex response. This technique can easily be carried out in the conscious human being. If a person sits with his legs crossed and with one foot clear of the ground, then a tap on the ligamentum patellae (which is really the tendon of the knee extensor muscles, see Fig. 11.37) will momentarily lengthen these muscles. They briefly contract and then relax again.

The foot will kick out and return to its original position. This reflex response is known as the 'knee jerk'.

Similar responses can be obtained from the tendo calcaneus, if the calf muscles are stretched by straightening the knee and dorsi-flexing the ankle joints. One of the present authors can obtain a similar response from his tibialis anterior muscle. Responses are also readily obtained from triceps, biceps and brachioradialis muscles of the arm and from the temporalis and masseter muscles that close the jaw.

All these stretch reflexes can be detected (except the jaw jerk!) in a patient with his spinal cord isolated from the brain, but their behaviour is extensively modified when the higher levels of the CNS retain some of their connections to the spinal cord.

γ-EFFERENT CONTROL OF STRETCH REFLEXES

The contraction of the intrafusal fibres of the neuromuscular spindle is brought about by the small motor neurone axons, type A-γ (known for short as the γ-efferents) whose cell bodies, like those of any other lower motor neurone, are in the anterior horn of grey matter. Any alteration in the contractile state of these intrafusal fibres will alter the responsiveness of the spiral nerve ending and thus alter the amplitude of stretch reflexes, including tendon jerks. The γ-efferents are controlled by both excitatory and inhibitory fibres descending from the brain. Shivering results from a synchronous firing of many γ-efferents together, which thus results in intrafusal fibre contraction and the resultant reflex contraction of the rest of the muscle mass. Hooking the fingers of the two hands together and then pulling apart the arms stimulates all γ-efferents in the body. This therefore enhances the tendon jerk response (*reinforcement*). In some people emotional excitement enhances γ-efferent activity, while more stable people may have diminished activity. All these changes are reflected in corresponding changes in the tendon jerk responses (and in other stretch reflexes, were we able to observe them). It has been suggested that the altered muscle tone of Parkinson's disease, i.e. the increased rigidity of this condition, is also due to increased γ-efferent activity.

THE BRAIN AND STRETCH REFLEXES

Quite apart from the modifications produced by higher centres affecting the γ-efferents, these centres are themselves stimulated by impulses coming from the stretch receptors. The tendon jerk when the spinal cord is isolated from the brain, e.g. in spinal injuries, consists of a brief twitch-like contraction followed by an immediate and complete relaxation of the muscle. After a knee jerk has been produced, the foot will swing to and fro like a pendulum. In the healthy person, the contraction is held for a moment and then followed by a controlled relaxation of the quadriceps so that the foot gently returns to its resting position.

If the corticospinal pathway is interrupted (most commonly in the internal capsule) producing a paralysis of the opposite side of the body (*hemiplegia*), then the response is even more powerful and sustained than in the normal person and *clonus* (alternate contraction and relaxation) may be seen.

STRETCH REFLEX AND MUSCLE TONE

Even when a person's muscle is relaxed, it still feels moderately firm to the touch and is not absolutely flaccid. Similarly, if you seize a person's forearm and vigorously shake it, although the hand will 'flap' it is still under some muscular control and will not move altogether freely, however hard your subject tries to relax her muscles. In other words, even when a muscle is voluntarily fully relaxed and at rest, it still retains some degree of involuntary contraction. This is known as *muscle tone*, which is only completely abolished in deep surgical anaesthesia, death, or by cutting the sensory nerves from the muscle. Tone depends upon impulses from the neuromuscular spindles, being thus a manifestation of sustained stretch reflex activity. You might think that fatigue would develop in muscles exhibiting tone. This does not happen for three reasons. First, only a few of a muscle's cells need to contract to produce the tone; second, each cell is alternately contracted and relaxed; and third, a shift system may operate, the impulses from the spindles going first to the spinal motoneurones supplying one group of muscle cells and then to other groups. Muscle tone depends not only upon the spindle sensory fibres (the afferent limb of the stretch reflex arc, entering the spinal cord by the posterior nerve roots) but also by the axons from the brain-stem and the cerebellum that control the γ-efferents. Naturally, tone is also affected by damage to the motoneurones themselves.

POSTURE

This is the position your body may assume when you are not carrying any specific movement. Some may think the natural posture for man is lying down! But in fact, we automatically assume postures, which are maintained at any time that we are not actually moving under conscious control. The stretch reflex is the foundation of posture. Remove the spindle afferents in experiment or by disease and the maintenance of posture is lost. The oldest part of the cerebellum, the archicerebellum, is largely concerned with maintaining normal posture in man. Not only does it receive afferent information from muscle spindles, tendon organs and joint receptors, but it also is informed of bodily position and movement in space by the labyrinth portion of the inner ear and by the eyes. A person who has lost his tendon and joint afferents to the brain, due to spinal cord disease affecting the posterior columns of white matter, falls over when he washes his face, for he is now deprived of his last remaining source of information. Movements, which are, after all, a change from one postural state to another, can, in extreme, be thought of as γ-efferent induced changes in spindle afferent activity causing a subsequent change in α-efferent activity and thus in the contractile state of the whole muscle. Like all extreme views, this one is not wholly true. In fact, movements are caused directly by α-efferent activity, together with such changes in γ-efferent drive as to maintain a constant level of spindle afferent stimulation.

Posture can be altered by many diseased states. If a muscle or group of muscles become paralysed, the pull of the antagonist muscles is unopposed and they can draw limbs or the trunk into abnormal postures. If a joint is damaged, then it will be held in the posture that puts the minimal amount of stress upon the damaged region. Damage to the internal capsule (p. 122 of this chapter) in man results in a very characteristic alteration in posture of the limbs of the affected side. The shoulder joint is adducted and medially rotated, elbow, wrist and fingers flexed and the forearm is pronated. The hip is adducted and extended. The knee is extended and the ankle plantar-flexed. Furthermore the tone of the affected muscles in the limbs is increased, so they are powerfully and continually held in the abnormal posture.

PROPRIOCEPTIVE NEUROMUSCULAR FACILITATION (PNF)

The reflex activity of the neuromuscular spindles and other proprioceptive receptors is used in the technique of *proprioceptive neuromuscular facilitation*. In this technique the muscle, or more commonly muscles, to be exercised are first put into the position of greatest length or full stretch so that the spindles are stimulated. From this position the patient is told to contract the muscles towards the position of greatest shortening. For maximal effect, this is done by initiating a chain of movements in which a whole series of muscles contract in turn, each being fully stretched at the beginning of its contraction (remember that it is *movements* rather than individual muscles that are represented in the cerebral cortex). At the same time, of course, the antagonists will pass from their position of greatest shortening to full stretch. Certain patterns of movement, usually spiral and diagonal in nature, are most effective in facilitation of this type and the muscles concerned are grouped together on one aspect of a limb or the trunk. For example, to put the hip joint successively through the movements of flexion, adduction and lateral rotation, starting from a position of extension, abduction and medial rotation will affect the anterior and medial musculature of the thigh, viz. iliopsoas, adductors, sartorius, pectineus, etc. In practice, the whole limb may be used for manoeuvres in which hip, knee, ankle and foot joints are all successively moved, the movements proceeding from distal to proximal.

SOME EFFECTS OF THE DISEASES OF THE CENTRAL NERVOUS SYSTEM

In this section we do not wish to describe particular disease states but to outline the more general types of functional loss that may follow any structural damage to the brain and spinal cord. Whether this damage causes death of cell bodies, death or disruption of conducting nerve fibres, or loss of their surrounding myelin sheaths, the result is always loss of nerve impulse conduction from or through the diseased area. In a later chapter we will take up in more detail the structural and physiological changes that follow damage in peripheral nerves because these are, to some extent, different from those seen in disease of the central nervous system. They are more accessible to study, and the presence of the myelin sheath and the endoneurial sheath surrounding each fibre is associated with considerable powers of regrowth of nerve fibres. This regeneration is not seen within the CNS. Here, once a fibre has been cut or damaged, while the portion between the site of damage and its cell body will survive, there is no regrowth into the part severed by the damage and therefore no restoration of the route taken originally by the nerve impulses.

LESIONS OF THE SENSORY SYSTEM

Damage at any point along the path between receptor and cerebral cortex results in loss of normal sensation when the normal stimulus is applied to the receptor. Often, though, the person affected will describe sensations apparently derived from the isolated region. This may be appreciated, whenever you have a local anaesthetic for minor dental conservative or restorative work, by those of you who suffer from attacks of migraine causing temporary and partial loss of vision, and more dramatically in the person who has lost a limb or suffers from 'post-herpetic neuralgia' following shingles—but more about this later. Loss of normal sensation together with a lesser or greater degree of abnormal sensation always occur whenever sensory pathways from receptors to brain are damaged.

Damage to the spinal cord may disrupt the fibres connecting the sensory neurones in posterior root ganglia (posterior columns) and posterior horns (spinothalamic fibres) to the brain. All sensation is then lost below the site of the damage, with, perhaps a zone of increased or abnormal sensation at the level of the damaged part of the cord.

It is not uncommon for one side only of the spinal cord to be damaged. If this happens, position sense, fine touch and 2-point discrimination are disturbed on the same side and up to the same level as the lesion, while pain and temperature sense are disturbed on the opposite side since fibres carrying the latter sensations cross over in the spinal cord. Owing to the oblique nature of this cross-over the loss may begin 2 or 3 segments *below* the site of the lesion. These findings were first described by the nineteenth century neurologist Brown-Séquard.

SENSORY NEURONE DISEASE

Apart from being involved in other diseases or injury, there are some diseases affecting sensory neurones specifically. These may be within the spinal cord itself, in the posterior root ganglia or more peripherally in sensory nerve fibres (*peripheral neuropathy*). Both peripheral arterial disease and diabetes mellitus (or arterial disease caused by diabetes) can damage sensory nerves or receptors. The lack of pain sensation leads to neglect of trivial injuries which readily become infected, producing *trophic ulcers* which are slow to heal. Joints may be damaged due to lack of pain sensation and loss of muscle tone which protects the joints from abnormal movements. The extreme form of this is known as a *Charcot's joint*, after the first description by the French physician Charcot about 100 years ago. The patient with peripheral neuropathy may also experience abnormal sensations of heat, cold or pain rising from the affected area.

The varicella (chicken pox) virus may affect the posterior root ganglion of usually only one segmental nerve perhaps many years after an attack of the general disease, or perhaps as a new infection with the virus, the condition being known as *herpes zoster* or *shingles*. Not only does the virus travel down the axons to the skin, producing the characteristic rash of chicken pox, confined to the affected segmental nerve's distribution but it also causes stimulation of the nerve fibres passing through the ganglion, with a resultant sensation of pain, which may be intense, usually pricking or stabbing in quality. For the academically minded, the rash in a case of herpes zoster provides an excellent demonstration of the extent of a dermatome. Occasionally after the inflammatory effects of the illness have completely regressed, attacks of pain may persist in the affected region. This is the condition called *post-herpetic neuralgia* and it may be trivial or in some cases incapacitating. A similar disease, though more generalised and with no rash but with persistence of abnormal sensation, may be caused by Coxsackie B virus infections.

LESIONS OF THE MOTOR SYSTEM

Before examining actual conditions affecting motor neurones, it is best to review generally the types of lesion that are seen in the nervous system, and their characteristic features.

The upper motor neurone

Lesions of the direct corticospinal path lead to loss of voluntary movement, but persistence and exaggeration of muscle tone and of reflexly produced movement. No wasting of the muscles occurs, however, in spite of the paralysis. The condition is therefore said to be a *spastic paralysis*. One consequence of such lesions is the appearance of the 'positive' Babinski response, of which a fuller description is given later in this chapter.

Extrapyramidal pathways

Lesions of these pathways are characterised by increases of muscle tone, so that limbs appear stiff, and by the emergence of abnormal movements. These may be tremors (Parkinsonism) or purposeless jerky or writhing movements (chorea and athetosis respectively). Voluntary movements can still occur, but may be interfered with by the disturbances of tone and abnormal movements.

The lower motor neurone

When these are damaged, then complete *flaccid* paralysis of the muscles they innervate ensues. The reflexes are diminished or lost, muscle tone is diminished and the muscles feel flabby—very different from the 'spastic' condition in upper motor lesions. Deprived of nerve impulses in the motor axons, the muscle fibres eventually degenerate and become wasted, as will be described in Chapter 8. Another feature of a flaccid paralysis is *fasciculation* or periodic contractions of small groups of muscle fibres which can be seen on the surface of superficial muscles.

Now that these phenomena have been described, we can look briefly at some of the disturbances of physiological function that are commonly seen in the motor system.

Hemiplegia

This most commonly results from thrombosis in or bleeding from, arteries in the internal capsule. Both the direct corticospinal path and fibres running from the cerebral cortex to the extra-pyramidal system are damaged. The resultant state is thus one of loss of voluntary movement (upper motor neurone damage) and the emergence of abnormal tone (extrapyramidal damage) on the opposite side of the body. This latter has already been described in the section on posture. *Quadriplegia* and *paraplegia* result when the motor pathways in the spinal cord are divided above or below the cervical enlargement respectively. In the former, all four limbs are isolated from the brain, and in the latter the lower limbs only are isolated. The results will be described in detail in the later section on spinal shock. If only one side of the cord is damaged, some voluntary control of the isolated region may persist as 70% only of the pyramidal tract fibres cross over in the medulla and some, at least, of the remainder finally supply lower motor neurones of the opposite side of the spinal cord.

DISEASES OF THE LOWER MOTOR NEURONE

Poliomyelitis is the best known of these. This is a virus disease, primarily of the gut and it occurs in areas of poor standards of cleanliness in food preparation where it is extremely common as a gastro-intestinal infection. A small minority of those infected suffer a further invasion of the CNS. One period of G.I. tract infection confers a lengthy (perhaps life-long) immunity. It is small wonder, then, that where health standards are low, it is called 'infantile paralysis', for the first attack by the disease will occur early in life. The virus has a predilection for the anterior horn cells (i.e. the lower motor neurones), particularly those that are active at the time of its CNS invasion. It destroys permanently the neurones attacked and paralysis inevitably results from this destruction. Not only is voluntary movement lost but also all possibility of the muscles being reflexly stimulated. Deprived of any nerve impulses the muscles atrophy (see Chapter 8). Although the widespread adoption of oral vaccination in the 1950s has led to a near disappearance of the disease in Britain and North America, there are many people still alive who were attacked before that time, and the disease is still common in many countries of the world and may easily infect the un-vaccinated person, whether native or visitor. Such reappearance of the disease is reported from time to time in different parts of Great Britain.

Apart from this infective disease, lower motor neurones may be the main site of certain ill-understood degenerative diseases. The consequence of all is a progressive weakening of voluntary movements, loss of tone and finally atrophy of skeletal muscles as their controlling motor neurones cease to transmit impulses. The causes of these conditions are unknown and little can be done for those suffering from them, for dead motor neurones cannot be replaced. However, a sex-linked inherited variety, Duchenne's muscular dystrophy, has been extensively studied, and its presence can now be both predicted and detected early in pregnancy in male fetuses.

SPINAL CORD SECTION

Originally, many of the properties of reflexes were studied in experimental animal preparations. In particular, the reflex activity of the spinal cord was studied after its severance from the brain. As soon as this had been done, a quiescent state was observed very similar to that seen in human patients in whom an accident has severed the spinal cord. All regions below the damage cease activity for a varying length of time and all reflexes are lost. This is a few minutes in a cat or a dog, some days in a monkey and several weeks in a human being. The state is called *spinal shock*.

After 2–3 weeks, some degree of sustained or tonic contractile activity begins to appear, at first and more markedly in the flexor limb muscles. This tone is reflexly generated, being dependent upon impulses coming from the muscle spindles. It is, in fact, a continuously maintained low-level stretch reflex. Subsequent to this, flexor reflex activity appears. When the skin of a limb is pricked or scratched, all the physiological flexor muscles contract and the ex-tensors relax. Even an extension (*crossed extensor reflex*) of the other (contralateral) limb, and contraction of the anterior abdominal muscles may be seen. One such reflex response has been particularly studied in patients with an injury of the spinal cord. It is called after its original describer, Babinski. The Babinski responses are obtained when the skin of the sole of the foot is firmly stroked with a blunt instrument. In a normal person the toes adduct and curl towards the sole (plantar flexion). Other muscles that have been seen (sometimes) to contract include the tensor fasciae latae, on the lateral side of the thigh. When nerve pathways from the brain to spinal cord are severed or are not conducting impulses, then the same stimulus produces fanning (abduction) and extension of the toes (particularly the big toe), dorsiflexion at the ankle joint and sometimes contraction of the muscles at the back of the thigh (the hamstring muscles) which cause flexion of the knee (*withdrawal* responses). The abnormal Babinski response is usually termed a positive response or 'up-going toe' by clinicians.

The knee and ankle tendon jerks reappear 3–6 weeks after the spinal cord injury. Although they are more powerful than the general level of tone (itself a form of stretch reflex), it may be many months before the tendon jerks regain their normal strength. They may then become exaggerated and a reflex contraction elicited in an extensor muscle group, e.g. by sudden flexion of the knee joint, might irradiate to the neighbouring flexor muscles also. This can result in a rigid but straight limb, even capable of supporting the body's weight. Differences in the quality of the tendon jerks are usually seen. The contraction may be as brisk as in a normal person, but relaxation is rapid and complete. In the normal person, relaxation of the thigh muscles (quadriceps) after the knee jerk proceeds smoothly and the limb returns to its resting position. In the isolated spinal cord the immediate and complete relaxation leaves the distal part of the limb free to swing to and fro like a pendulum.

Occasionally, either flexor or extensor (withdrawal or stretch) reflexes may become grossly exaggerated, even when evoked by the trivial stimuli. Mass flexor responses may be seen after minor skin stimulation. Widespread extensor spasm also occasionally occurs.

Thus, while considerable independent reflex activity may develop in the spinal cord severed from the brain, this lacks the degree of co-ordination and control usually seen in the intact nervous system. It is also true that the many elaborate experimental studies made on the isolated cord's reflex activity all demonstrate phenomena that may be considerably modified in the intact organism.

6 · The Autonomic Nervous System

The word 'autonomic' means that this part of the nervous system deals with functions of which we may have no conscious awareness and over which we can exert no conscious control. Like all definitions, this one has many exceptions and it might be better to consider the anatomical definition as being preferable. This rests on the fact that, in its efferent or motor aspect, the autonomic nervous system (ANS) is clearly distinguishable from the remainder of the nervous system which supplies the skeletal or voluntary muscle. The distinctive feature of the ANS is that there are two neurones on the pathway from the central nervous system to the organs or tissues innervated. Along the path of autonomic nerves are found collections of nerve cells (*ganglia*). The axons leaving the CNS terminate in the ganglia, forming synaptic connections with the ganglionic neurones. It is the axons of these ganglionic neurones that then innervate the structures controlled by the ANS.

Traditionally, the axons leaving the CNS and passing to the ANS ganglia were called 'preganglionic' axons. They are type B, small myelinated fibres, the myelin giving them a whitish appearance. The fibres leaving the ganglion cells were called 'post-ganglionic' axons. They are type C, small and non-myelinated fibres. The lack of myelin gives them a greyish colour. The terms pre- and post-ganglionic are now often replaced by *connector* and *excitor* respectively. We can see nothing wrong with the former term, but many post-ganglionic neurones have inhibitory actions on the structures they innervate. The term *effector* would therefore seem to be more suitable than 'excitor', so we will substitute this for the older term 'post-ganglionic' for the axons that arise from the autonomic ganglion cells.

The easily recognised two neurone pathway for outgoing impulses has naturally focused all attention on the efferent side of the ANS, but we must remember that this is less than half the story. The ANS operates reflexly, altering functions of the controlled organs only in response to receptor stimulation. Just as in the case of contractions of skeletal muscles, there are reflex responses, called into action by nerve impulses coming from receptors of one sort or another, so, in the organs and tissues innervated by the ANS, changes in function only occur as the result of receptor stimulation and the passage of nerve impulses from the receptors to the CNS and thence, via the ANS motor pathways, to the innervated structures. All the properties of reflex activity already described apply as exactly to activity in the ANS as they do to reflexes involving skeletal muscles. In fact, so long as the CNS is intact, most ANS reflexes operate through *centres* in the brain-stem, from hypothalamus to medulla oblongata. Some of the ANS centres can be identified as anatomically distinct groups of neurones, but mostly they are inextricably mixed together and with other cells, not themselves specifically concerned

125

with ANS reflexes. The word 'centre' is thus a functional rather than an anatomical term and most of the centres in the brain-stem cannot be recognised in histological sections as specific groups of cells (nuclei). The functions controlled through the ANS by such centres include the rate of the heart, arterial blood pressure and the body temperature. A survey of these follows later in this chapter, but the details are found in sections concerned with control of function in the description of each bodily system.

ANATOMY OF THE AUTONOMIC NERVOUS SYSTEM

The ANS has two components: the *sympathetic* and the *parasympathetic* systems, which are, to some extent, functionally antagonistic.

THE SYMPATHETIC NERVOUS SYSTEM

On each side of the vertebral column lies a chain of sympathetic ganglia (the *sympathetic trunk*) with approximately one ganglion for each spinal nerve (except in the cervical region where the original eight ganglia have become fused into three, the *superior, middle* and *inferior cervical ganglia*). From the first thoracic to the second lumbar spinal nerves, connector fibres run in *white rami communicantes*, to the corresponding ganglia of the chain. From all ganglia *grey rami communicantes* convey the non-myelinated effector fibres back to the spinal nerves, with which they are conveyed to their destinations (Figs. 6.1 and 6.2). In addition to these connections, effector fibres leave the upper pole of the superior cervical ganglion and run in a network along the carotid arteries and their branches to the head and neck. Effector fibres also leave the three cervical ganglia for the heart in cardiac nerves and connector fibres for the abdominal viscera run in the *splanchnic nerves* from the lower six thoracic ganglia and the first two lumbar ganglia to supplementary ganglia (Coeliac, superior and inferior mesenteric) in the abdomen. From these ganglia the effector nerve fibres run with the arteries to supply the various tissues they innervate, including the smooth muscle of the blood vessels themselves. These arrangements are shown pictorially in Fig. 6.2 while Table 6.1 shows the origin of connector fibres and the site of ganglionic synapses for the various organs innervated by the sympathetic system. Thus it may be said, in general, that the connector fibres are short, since the sympathetic trunk is not far from the spinal cord, while the effector fibres are long since they have to travel from a ganglion all the way to often remote parts of the body. The ganglia are visible to the naked eye.

THE PARASYMPATHETIC NERVOUS SYSTEM

Connector fibres leave the CNS in certain cranial nerves, particularly the vagus (tenth cranial nerve) and also in the second, third and fourth sacral nerves. For the most part the ganglia are visible only with the microscope and

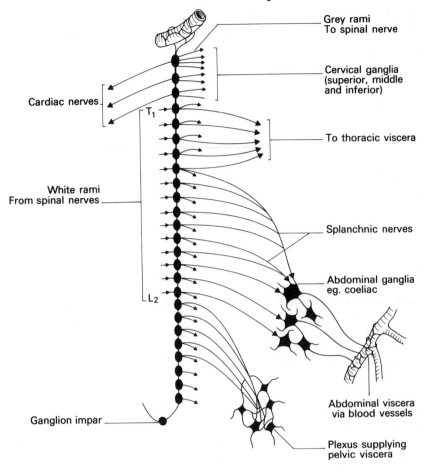

Fig. 6.1. The sympathetic nervous system. The sympathetic ouflow (white rami) extends from T1 to L2 only, but the distributing fibres (grey rami) pass to all the spinal nerves from the sympathetic ganglia along the trunk .There are also a number of outlying ganglia such as the coeliac ganglion.

are situated within or close to the organ supplied, the effector axons thus being very short and the connector fibres long. The ganglia are often dispersed and poorly defined. Figure 6.3 shows this diagrammatically, while Table 6.1 shows the organs supplied, the origin of the connector fibres and the sites or names of the ganglia wherein lie the effector cell bodies.

The *vagus* is by far the most important of the parasympathetic nerves. It contains a mixture of both motor and sensory fibres. On leaving the brain-stem it emerges through the base of the skull and runs through the neck into the thorax close to the carotid artery. In the thorax branches run off from both vagus nerves to the heart and to the lungs. Both vagus nerves then pass with the oesophagus into the abdomen where branches are given off to the stomach.

Table 6.1

SYMPATHETIC

Region	Origin of connector fibres	Site of synapse
Head and neck	T1–5	Cervical ganglia
Upper limb	T2–6	Inferior cervical and first thoracic ganglia
Lower Limb	T10–L2	Lumbar and sacral ganglia
Heart	T1–5	Cervical and upper thoracic ganglia
Lungs	T2–4	Upper thoracic ganglia
Abdominal and pelvic viscera	T6–L2	Coeliac and subsidiary ganglia

PARASYMPATHETIC

Region	Origin of connector fibres	Site of synapse
Head and neck	Cranial nerves 3, 7, 9, 10	Various parasympathetic macroscopic ganglia
Heart	Cranial nerve 10	Ganglia in vicinity of heart
Lungs	Cranial nerve 10	Ganglia in hila of lungs
Abdominal and pelvic viscera	Cranial nerve 10 (down to transverse colon)	Microscopic ganglia in walls of viscera
	S2, 3 and 4 (from transverse colon onwards and pelvic viscera)	Microscopic ganglia in walls of viscera

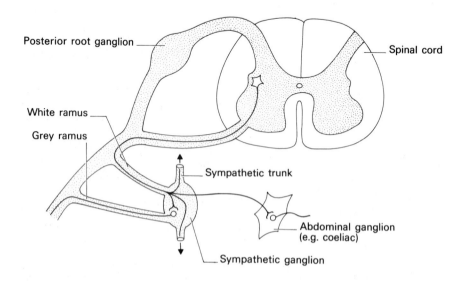

Fig. 6.2. Cross-section through the spinal cord to show the origin and distribution of sympathetic fibres.

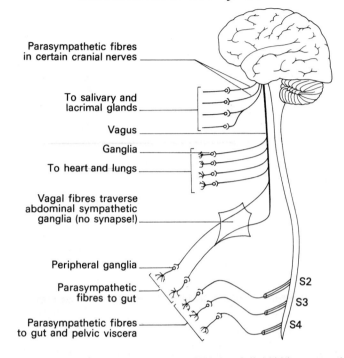

Parasympathetic fibres
in certain cranial nerves

To salivary and
lacrimal glands

Vagus

Ganglia

To heart and lungs

Vagal fibres traverse
abdominal sympathetic
ganglia (no synapse!)

Peripheral ganglia

Parasympathetic
fibres to gut

Parasympathetic fibres
to gut and pelvic viscera

S2

S3

S4

Fig. 6.3. The parasympathetic nervous system. This is subdivided into a cranial outflow (in 4 cranial nerves including the vagus) and a sacral outflow (S2, 3 and 4).

As recognisable nerves they end in the coeliac ganglion, but the vagal connector fibres go, together with the sympathetic effector nerve fibres, with the arteries to the gut and other abdominal organs, where they terminate in synapses on ganglionic cells (Auerbach's and Meissner's plexus cells in the wall of the gut). The widespread distribution of this nerve is responsible for its name, which has the same origin as the word 'vagrant'.

The other cranial parasympathetic connector fibres run to ganglia in or near the eyes and salivary glands. Effector fibres supply these glands, the muscles of the eyes, the lacrimal (tear) glands and the blood vessels of the nasal cavity. The sacral parasympathetic fibres supply the pelvic viscera (bladder, etc.), their ganglia being close to or incorporated with the viscera themselves.

FUNCTIONS OF THE AUTONOMIC NERVOUS SYSTEM

INTRODUCTION

As in all other parts of the nervous system, activity in and the effects produced by the ANS are part of reflex responses to stimuli affecting receptors. The

ANS acts generally as the efferent pathway for regulatory reflexes controlling the functions of many of the separate tissues and systems of the body, which are usually neither under voluntary control nor affect conscious awareness. However, in considering the functions of the ANS, one should never forget that these are but part of the reflex responses and cannot logically (or physiologically) be considered without reference both to the receptors, the stimulation of which initiates the reflexes, and the part played by the reflexes in regulation of bodily functions. In the accounts of the ANS and the various body systems which will be given later, the function of both receptors and the reflexes will be described.

SMOOTH MUSCLE

Since many ANS fibres act on smooth muscle, we must first consider some of the known features of this kind of muscle and how its activity is affected by the chemical transmitters released by impulses reaching ANS effector nerve terminals. A comparable study of the ANS effector fibres and their action on gland cells has not yet been made, so it cannot be described here.

'Smooth', 'plain', 'unstriated' or 'involuntary', these terms are interchangeable and are applied to the muscle of blood vessels, the gut, bronchi, ureters and bladder, uterine tubes and uterus, the ductus deferens from the testis, the internal muscles of the eyes and the muscle in gland ducts. Three of the terms refer to the absence of the cross-striations which are the striking microscopic feature of skeletal muscle (Chapter 7). However, smooth muscle contains the same actin and myosin threads, the basic materials producing contraction, as does skeletal muscle. Like nerve and skeletal muscle cells, smooth muscle cells have an electrical potential gradient across their cell membranes, usually of 50–55 mV with the positive charge on the outside. This is, in many cases, decreased by stretching or elongating the cell. Such decrease, or the decrease produced by a stimulating neural transmitter (this may be either noradrenaline or acetylcholine depending upon the site), will cause the generation of propagated action potentials. These, in turn, cause contraction of the muscle cell. It is thought that action potentials can be transmitted from cell to cell in smooth muscle, so that a nerve impulse, transmitted to one smooth muscle cell, can cause contraction in many neighbouring cells. As in skeletal muscle, the entry of calcium ions into the cells is probably the link between action potentials and shortening. Another distinctive feature of smooth muscle is that while one transmitter causes contraction, another may cause relaxation. For example, acetylcholine causes membrane depolarisation and contraction of the muscle of the small intestine. Noradrenaline, when applied to this same muscle, causes its membrane charge to rise, thus reducing the possibility of action potential development and of contraction. Thus the opposing actions of the two transmitters are balanced, and one finds that cholinergenic fibre (vagal) stimulation causes the gut to increase in motility, whereas noradrenergic fibre (sympathetic) stimulation causes it to relax. Perhaps this is the cause of the 'sinking feeling' so well known to oral examination candidates.

REFLEXES IN THE AUTONOMIC NERVOUS SYSTEM

The reflex activity of the ANS in the regulation of the different body systems will be described in the chapters dealing with the various systems. We shall only describe here the group of responses, mediated by the sympathetic component of the ANS that are known as either the *fight* or *flight* or the *alarm reaction* responses, following this with a brief survey of the actions of the ANS on all the body systems.

The initial receptor stimulation for these responses is in the *exteroceptors* (i.e. receptors of external stimuli) of the eyes, ears and nose. It should be remembered, however, that stimulation of pain receptors will evoke the same responses, as will stimulation of the receptors that are sensitive to a fall in the oxygen content of arterial blood (the carotid and aortic bodies; see Chapter 16 for a full description). Finally, in man, you must remember that thoughts, feelings or the remembering of past pain or other distress can also evoke the same responses as an actual threatening or painful state.

In the posterior hypothalamus and neighbouring parts of the midbrain, collections of neurones known as the *alerting reaction centres* are found to which impulses from the receptors listed above run and from which impulses go to mediate the various responses. These responses can be loosely summarised as increased activity of the sympathetic division of the ANS and decreased activity in the parasympathetic division. In the cardiovascular system the heart's rate and force of beating are increased (sympathetic division active). Blood vessels of skin, gut, liver and kidneys are contracted (sympathetic activity), while those in skeletal muscle are moderately dilated (sympathetic noradrenergic fibres inactive, but cholinergic fibres active). More blood flows to muscle and less to other regions. In the respiratory system, bronchial muscle relaxes (parasympathetic system inactive). In the digestive system secretion stops and motility decreases (parasympathetic inactive). Sphincter muscles are contracted (sympathetic active). The result is a dry mouth and a 'sinking feeling' in the abdomen. Sweat secretion, except on the palms, soles and axillae is reduced (sympathetic cholinergic fibre inactivity) and the hair is raised (sympathetic noradrenergic fibre activity). The concurrent vasoconstriction will make the skin feel cold and damp. In the eyes the pupils become dilated and the upper lids are raised (sympathetic activity). Sympathetic impulses also pass in this state to the medulla of the adrenal glands and by stimulation cause the production of adrenaline. Adrenaline circulating in the blood will mimic many of the actions already described, but it also increases the breakdown of muscle and liver glycogen and of fat in adipose tissue. The metabolic machinery of the muscles is thus provided with an increased amount of material for oxidative energy production.

The final results of the responses are that blood flow is diverted towards the important skeletal muscles and away from those organs that do not require a large supply continuously. The heart is able to increase its output as more blood is returned to it from the active muscles. Supplies of oxygen and nutrients for active muscles are available. We finally wish to emphasise strongly that the changes are to be regarded as *responses to receptor stimulation*, and, that, while they do serve a purpose (that of preparing the body for intense physical activity), the student physiotherapist should always regard them as effects of receptor stimulation, either immediate or in the remote past, and should remember that the ANS is so arranged that the various responses, whatever the route of impulses or transmitter employed, automatically all serve the same purpose.

OTHER SYSTEMS, ORGANS AND TISSUES

This detailed account of the fight or flight response of the ANS can serve as a model when considering the functions of the ANS generally. For a detailed study of its role in regulating function in the other bodily systems and tissues you should consult the separate chapters concerning these later in the book. We will here give only a brief outline of the regulatory functions of the ANS with respect to these systems, organs and tissues.

In the respiratory system the parasympathetic supply is the most important. The tissue lining the cavity of the nose is very richly supplied with blood vessels and secretory glands. Its function is to warm and moisten the inspired air as it enters the respiratory passages. Parasympathetic nerve impulses cause the vessels to dilate and the glands to secrete. Cyclic activity occurs in man, one nostril alternating with the other as preferred route for air flow over a period of 2–4 hours. The alterations in air flow are caused by variations in the swelling of the lining tissue due to the amount of blood in its capillaries, these variations in blood flow being caused by periodic variations in the parasympathetic nerve impulse supply to the two sides of the nose. The vagus (Xth cranial nerve) also supplies secretomotor nerve fibres to mucus secreting glands in the bronchi and motor fibres to the smooth muscle of the bronchial walls. Stimulation of vagal fibres causes contraction of bronchial muscle and secretion from bronchial glands. Both are protective when harmful dust or gas are inhaled into the lungs and are reflexly produced when such agents enter the nose, trachea or bronchi and stimulate receptors in the lining tissue.

In the digestive system parasympathetic nerve fibres from cranial nerves supply the salivary glands with secreto-motor impulses. The vagus similarly supplies gland tissue of the stomach and the pancreas. Nerve stimulation in these latter causes an output rich in digestive enzymes but in a small volume of fluid. Although the motility of the gut is largely controlled by a local network of nerve cells in its wall, these are further stimulated by the parasympathetic nerves and inhibited by the sympathetic. This latter also reduces the blood supply to the gut by causing vasoconstriction. The reflex control of defaecation and of vomiting are also mediated by the ANS, though the former (but rarely the latter) can usually be overruled, for a while at least, by conscious control. The last part of the hind gut, the anal canal, contains sphincter muscles which keep it closed. These are partly innervated by the sympathetic nervous system, impulses in which cause the sphincter muscle to contract. Parasympathetic fibres from the second and third sacral nerves also supply this muscle. Impulses in these cause the sphincter muscle to relax and the main muscle of the rectum to contract. The presence of material in the rectum stimulates receptors in its lining. Impulses from these reflexly initiate defaecation by stimulating the parasympathetic and simultaneously inhibiting the sympathetic outflow to the rectum and anal canal.

The voiding of urine from the bladder is reflexly controlled in a manner very similar to the control of defaecation, both components of the ANS being involved. It will be described in detail on p. 554.

Sympathetic nerves supply adipose tissues and are believed to stimulate fat release from these tissues (circulating noradrenaline certainly causes fat release). They supply skin with vasoconstrictor nerve fibres, also with cholinergic effector nerve fibres to sweat glands and noradrenergic fibres to the muscles that raise the hair (see Chapter 22).

The eyes receive both sympathetic and parasympathetic nerve fibres. The former cause the pupil to dilate by relaxation of the sphincter muscle in the iris. They also cause contraction in part of the muscle that keeps the upper eyelid in its normal raised position. Parasympathetic impulses cause both pupillary contraction and also contraction of the ciliary muscle, which makes the lens become more sharply curved and thus capable of focusing the images of nearby objects accurately on the retina. Parasympathetic fibres also supply and stimulate the lacrimal (tear) glands.

THE ADRENAL MEDULLA

This gland can be regarded as a collection of modified sympathetic ganglion cells, forming adrenaline and not noradrenaline, and then secreting this into the blood instead of employing it as an immediate neurochemical transmitter. Adrenaline has actions on a wide range of tissues and organs. It increases the heart rate and force of contraction. It promotes vasoconstriction in skin, gut and the kidneys' blood vessels, but dilates blood vessels in skeletal muscles. It causes contraction of the muscles that raise the hair and it dilates the pupils of the eyes. The muscular activity of the intestinal wall is reduced or stopped by adrenaline. It causes an increased breakdown of the carbohydrate storage compound, glycogen, normally present in both liver and muscles, and an increased release of fat (as non-esterified fatty acids) from adipose tissues. In most of these actions adrenaline is acting exactly as the various components of the alarm reaction. It is not unreasonable to state that the response to alarming stimuli acts, in part at least, through the release of adrenaline from the adrenal medulla gland, when stimulated by its connector neurones.

7 · The Microscopic Structure and the Function of Muscle

This chapter will not fulfil all that is promised in its title, for there are three types of muscle in the body and here we are concerned mainly with only one type, namely skeletal muscle whose function is, in general, to maintain posture and to carry out voluntary movements.

The three types of muscle mentioned above are *cardiac, smooth* and *skeletal*. Cardiac muscle is found only in the myocardium (the muscular component of the heart) and it has a number of extremely important properties which enable it to contract rhythmically, without rest or replacement, for a lifetime. Alterations in its rate or power of contraction can, along with other factors, control the cardiac output and blood pressure, and as all these factors are closely interrelated it will be convenient to defer a description of cardiac muscle to Chapter 14. This chapter will therefore deal, briefly, with smooth muscle and, in more detail, with skeletal muscle.

A number of alternative terms are used for these two types of muscle, the commonest being *involuntary* and *voluntary*. These terms are avoided here, since although smooth muscle is usually not controllable voluntarily, it can be controlled in certain places—for example, focusing (*accommodation*) of the eye is carried out by means of smooth muscle but this is entirely under the control of the will. On the other hand, the wall of the upper third or more of the oesophagus contains skeletal muscle but as soon as food or drink has passed over the back of the tongue and started on its way down, it is past recall, at least by contraction of the oesophageal muscle. For these reasons we shall here refer to smooth and skeletal muscle, although even these terms are not exact since skeletal muscle is not always attached directly to the bony skeleton—in the face, for instance, one or both ends of the facial muscles is inserted into the skin in order to produce various facial expressions.

SMOOTH MUSCLE

Smooth muscle is found wherever contraction and relaxation are controlled entirely by reflex action without any overriding voluntary control. It is therefore found, for example, in the walls of blood vessels, to control their diameter; in the gut, to squeeze food along the lumen by a process known as *peristalsis*, and in the bladder and uterus. All these organs proceed with their task quite automatically, and normally without our being aware of them. Like skeletal muscle, smooth muscle needs a nerve supply but this is always by the autonomic system, as might be expected (see Chapter 6).

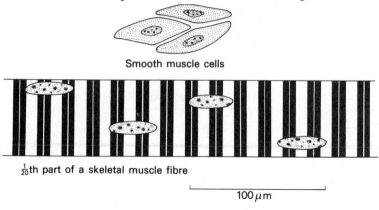

Smooth muscle cells

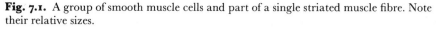

$\frac{1}{20}$th part of a skeletal muscle fibre

100 μm

Fig. 7.1. A group of smooth muscle cells and part of a single striated muscle fibre. Note their relative sizes.

The microscopic structure of smooth muscle

Smooth muscle, as a tissue, is made up of a number of single cells known as *smooth muscle fibres* (Fig. 7.1). They are elongated, spindle shaped cells, varying in length from about 15 mμ up to nearly 0.5 mm, the latter being found in the pregnant uterus. They contain a centrally placed elongated nucleus. Each cell is bounded by a cell membrane and the cytoplasm contains the usual organelles such as mitochondria, endoplasmic reticulum and a not very prominent Golgi apparatus. They also contain glycogen granules; but by far the most noticeable feature of these cells is their content of fine longitudinally running fibres or filaments. These resemble exactly the fine *actin* filaments which, as will be described later, are found in skeletal muscle fibres, associated with thicker *myosin* filaments in a highly organised fashion. In smooth muscle, however, no myosin filaments can be seen although biochemical analysis shows both actin and myosin to be present. This is difficult to explain but in spite of this, smooth muscle cells can contract very powerfully indeed, a force which is very evident during the birth of a baby. Figure 7.1 shows typical smooth muscle cells.

SKELETAL MUSCLE

As was explained in Chapter 2, a tissue does not consist of merely one type of cell and muscular tissue is not composed entirely of muscle fibres. A muscle, such as the biceps, contains a very large number of muscle fibres, but it also consists of connective tissue, dense collagen fibres in its tendon, nervous tissue (see Chapter 5) and blood vessels. The proportions of these components vary in different muscles and this, and the way in which the components are put together, will be described in considerable detail in Chapter 11. Here we shall be concerned only with the essential contractile element of skeletal muscle—the muscle fibre.

MICROSCOPIC APPEARANCE AND CHEMICAL NATURE
OF SKELETAL MUSCLE

By light microscopy a muscle is seen to be composed of fibres. Each fibre is a single multinucleate cell and it is bundles of many fibres that form the naked eye fibrous appearance of a muscle. Not only is the muscle fibre seen structurally to be single cell, bounded by a cell membrane like all other cells, (the *sarcolemma*), but it also behaves, in its contractions, as a single unit. Each fibre, along its length, is seen to consist of alternating light and dark bands. These refract light to differing degrees and are referred to as the *anisotropic* and *isotropic* bands (A- and I-bands, for short). In the centre of the A-band is a darker M-line, surrounding which is a somewhat paler H-zone. In the centre of the I-bands is the narrow, dark Z-line (Fig. 7.2). The A-bands are a constant 1.7 μm wide, but the I-bands vary in width with the state of contraction of the muscle (see below). Z-lines may be as much as 3.5 μm apart in fully stretched muscle. Muscle fibres may be from 1 mm to 5 cm long (though a fibre 34 cm long has been teased out of the sartorius muscle) and 10–100 μm in diameter.

SARCOPLASM

Immediately under the sarcolemma is a layer of cytoplasm, the *sarcoplasm*. This contains the nuclei of the cell, numerous mitochondria and granules of glycogen (which may make up as much as 1 % of the muscle and form an important source of energy—see below). The bulk of the fibre consists of myofibrils, separated by planes of sarcoplasm. Within the sarcoplasm run two systems of tubules. One is formed from extensions inwards of the sarcolemma; it is known as the *T-tube* system ('T' for transverse). T-tubes run, more or less straight and unbranched, from the outer surface of the fibre towards its interior. In human muscle they are related to the ends of the A-bands. Within the T-tubes are extensions of the extracellular space, with its characteristic high sodium ion content and electrical positivity. The second tube system is known as the *sarcoplasmic reticulum*. This consists of a truly intracellular system of branching tubules. Where these adjoin the T-tubes they expand into terminal cisterns and a T-tube with abutting sarcoplasmic reticulum on each side has been termed a *triad*. The sarcoplasmic reticulum tubes contain calcium ions and their bounding membranes contain an active, inwardly directed calcium transporting mechanism, which maintains the sarcoplasmic Ca concentration below 10^{-7}M.

Before we leave the sarcoplasm and study the myofibrils (which contain the muscle's contractile elements), we must consider the colour of muscles, as seen by the naked eye. While much muscle may be called pink or flesh coloured, some muscles are obviously darker than others. In extreme cases we have the red meat on a turkey's legs and its white breast muscle. In red muscles the sarcoplasm is more abundant; it contains a higher concentration of the enzymes that liberate energy and of myoglobin granules. It is the last named granules that give the colour, for they are red like the haemoglobin of blood and, like haemoglobin, myoglobin combines reversibly with oxygen. The blood supply of red muscle and the capillary density between the fibres is more rich than that of white muscle. A moment's thought about the use to which the wild turkey puts its wing and leg muscles should tell the student that white muscles undergo short powerful contractions, followed by complete relaxation producing the flapping to and fro movements of its wings, while the bird can stand still for considerable periods of time, perched upon its continuously contracted, red, leg muscles. White, then is for phasic movements and red for postural maintenance. This applies to man, also to other mammals and birds, though the division of muscles into red and white is by no means so complete. While the above may be true for the turkey, in many animals, including the human, muscles contain varying mixtures of both types of fibres. However, studies of the motoneurone axons that supply red and white muscle fibres respectively reveal how the two types of fibres are used. 'Red' motoneurones convey continuous streams of impulses at low frequencies, while 'white' axons convey impulses in irregular bursts at high frequencies. Furthermore, if reflex stimulation of a muscle be steadily increased, the red fibres are activated first and the white ones are only stimulated when a powerful contraction is required. Finally, if 'white' motoneurone axons are artificially stimulated at the normal frequency found in 'red' axons, their muscle fibres change to the red type, having an increased blood supply, an increase in oxidative enzyme content, and their contraction properties become the same as those of slow muscle fibres. It is suggested that these changes may be the basis of the alteration in muscle performance seen in endurance training regimes.

MYOFIBRIL PROTEINS AND THEIR ARRANGEMENT

A combination of chemical and electron microscopic analysis has clarified the essential nature of the contractile materials. Figure 7.3 shows a typical e.m. picture of the fibril cut along its length. The A-bands are made up of thick filaments and the I-bands of thin filaments which penetrate into the ends of the A-bands. The lighter H-zones within the A-bands are the regions beyond the penetrating thin filaments. Pictures taken at various stages of contraction show that in contraction the thin filaments penetrate farther into the spaces within the thick filaments until their ends meet from opposite sides of the A-band at the M-lines (thus obliterating the H-zones). Neither type of fibre alters in length. This is the basis of the *sliding filament* explanation of muscle contraction (Fig. 7.3). At the Z-lines the thin filaments are attached, from opposite sides, to a thin plate of electron-dense material. Since this conspicuous feature is unaffected by shortening and is repeated many times over down the length of a myofibril, the unit of tissue between two adjacent Z-lines is called a *sarcomere*. Under normal relaxed resting conditions a sarcomere might be 2.2–3.6μm long, with the thin filaments partly penetrating the thick ones. If, in contraction, sarcomere length falls below 2.0 μm, then thin filaments from each meet at the M-line and overlap. Further shortening down to a sarcomere length of 1.3 μm, can occur before the Z-lines abut onto the ends of the thick filaments of the A-band, and compress these, concertina-wise, from their normal length of 1.6 μm.

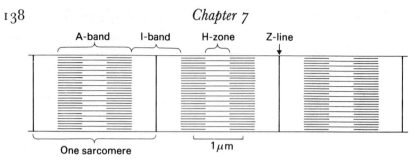

Fig. 7.2. The banded appearance of striated muscle that can be detected by the light microscope.

Chemical analysis, etc., shows that the thick filaments have the structure shown in Figure 7.4. The material, *myosin*, of which the units are made, is one of the family of fibrous proteins, like human hair or sheep wool proteins. Myosin molecules can be subdivided into the heavy globular heads and lighter, linear tails. The tails readily stick together in forming the thick filaments of the A-band. When the myosin units are packed together in a thick filament, the M-line is the central region devoid of the projecting heads of the units.

The thin filaments are made up of three separate proteins. First there is a coiled double chain of globular *actin* units. In the grooves of this double chain are short lengths of fibrous *tropomyosin* molecules and at the ends of these the rounded small *troponin* molecules. While the main action in muscle shortening is to be found in the interaction between the myosin heads and actin molecules, tropomyosin and troponin control the reaction between myosin and actin (see below).

THE MOTOR UNIT

All muscle fibres are wholly or partly controlled by the nervous system. The type being considered in this chapter, skeletal or striated muscle, will only contract after impulses reach it along its motor nerve supply. The nerve fibres or *axons* passing to an individual muscle usually leave a mixed spinal nerve as a single so-called motor nerve to a muscle (but see p. 99). At a well-defined point, the *motor point*, the nerve plunges into the substance of the muscle and then divides up. This point is important to physiotherapists for it is the point at which electrical stimulation of the motor nerve is most effective at causing the muscle to contract. Each axon from a single cell in the anterior horn of grey matter of the spinal cord (or equivalent part of the brain stem in the case of the cranial nerves—Chapter 5) splits up into 5–150 branches (the number is determined by the function of the muscle, fine movements requiring each axon to innervate only a few muscle fibres—Chapter 11), each of which runs to a single muscle fibre. The whole neurone (anterior horn cell, its axon and branches), together with the muscle fibres supplied by the axonal branches, is known as a *motor unit*. The muscle fibres of the motor unit are found scattered through the substance of the muscle, and are *not* bunched together.

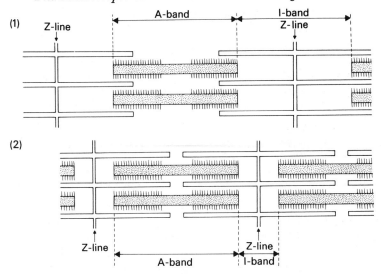

Fig. 7.3. The appearance of thick and thin rods found in the A- and I-bands respectively, when studied with the electron microscope. (1) Relaxed. (2) Contracted.

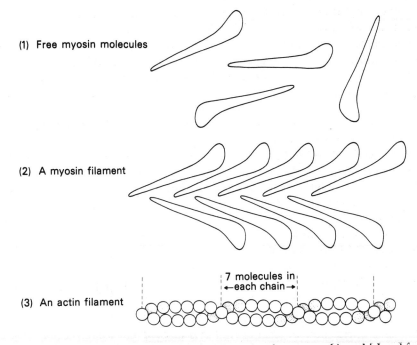

Fig. 7.4. (1) Free myosin molecules. (2) Myosin molecules arranged in a thick rod from the A-band. (3) The double and twisted chains of spherical actin molecules that form the thin rods of the I-band.

THE CONTRACTION PROCESS

RESTING MEMBRANE POTENTIAL

Skeletal muscle fibres, like nerve fibres, have a potential gradient across the sarcolemma, associated with chemical concentration gradients. In fact, the same three important ions, sodium, potassium and chloride ions are concerned and their extra- and intracellular concentrations are held so that the resting sarcolemma has a potential gradient of 90 mV, negative inside. The equilibrium potential (p. 55) for sodium is 65 mV, positive inside. That for potassium is 95 mV and for chloride 90 mV, both negative inside. Thus there is no tendency for chloride ions to move at all across the sarcolemma. The inward driving force caused by its high extracellular concentration exactly balances the repellant force of the negative charge of the interior. Potassium has a slight net outward drive, and sodium is strongly driven towards the interior by both concentration and electrical gradients. It is only the relative impermeability of the sarcolemma to sodium ions that keeps them out, together with an active sodium transport mechanism which extrudes any sodium ions that do penetrate the sarcolemma.

MOTOR NERVE FIBRES

Although the cells of origin of motor axons, and the physiology of nerve impulses in them have already been described in Chapter 4, some further aspects of their physiology must be mentioned here.

Usually these are large myelinated axons and they conduct nerve impulses rapidly, belonging to the Aα class of efferent nerve fibres. When an impulse is generated at the beginning of one of these axons in the spinal cord (or brainstem in the case of the cranial nerves) it passes down its trunk and into all of its terminal branches without any *attenuation* (weakening) of its strength as the impulse spreads through the terminations of the axon. Therefore the impulse arrives at each and every neuromuscular junction at the ends of the branches and the processes of transmission, etc. (to be described in the next section) result in a contraction in every one of the many muscle fibres which are innervated by the branches of the one motor nerve fibre.

Thus the motor unit, as already described, is not only an anatomical unit but also a functional unit. Once the anterior horn cell has generated its nerve impulse the whole motor unit is involved and all of its muscle fibres contract simultaneously in response to the impulses that reach them via the terminal branches of the motor axon coming from the anterior horn cell.

Anterior horn cells, according to the nature of the impulses reaching them, may produce impulses in slow or rapid succession, or they may be stopped (inhibited) from producing any impulses at all (but that is another story, see Chapter 4). The point we wish to emphasise here is that whatever may emerge from the anterior horn cell by way of nerve impulses in its axon will affect, equally, all the attached muscle fibres, the unit functioning as one single entity.

Each anterior horn cell is the end-point of many hundreds of reflex arcs (or chains of neurones) and the motor unit serves as the *final common path* for impules that may be part of as many different reflex responses. The anterior horn cell may not only be excited by the axons impinging upon it, for many of these are inhibitory. The final activity of any motor unit, acting as a final common path, will be the sum of all the excitatory and inhibitory influences reaching it.

MUSCLE ACTION POTENTIAL

When a nerve impulse reaches the neuromuscular junction, some of the vesicles in the axon terminals liberate, into the space between axolemma and sarcolemma, *acetylcholine* molecules. About 10^6 molecules may be released by each nerve impulse. When these reach the sarcolemma this becomes permeable to sodium ions, and these enter the sarcoplasm, causing the potential gradient to fall. In 0.2–0.3 ms the change in potential (the *end-plate potential*) has reached the threshold for action potential production, the fall in potential causing an increase in sodium permeability, causing a further fall in potential . . . and so on, until the sodium equilibrium potential is approached. As in nerve fibre this point is never reached, for potassium ions start to move out even as the sodium is entering the fibre, and this outward potassium movement causes the negative intracellular potential to be restored again. The action potential is propagated over the muscle fibre surface as in a non-myelinated nerve fibre. It takes 2–4 ms to pass any point and travels along the muscle fibre at about 5 metres/sec.

Unlike synaptic transmission at a neurone, neuromuscular transmission is invariably, in health, a 1:1 procedure. Each nerve impulse that arrives at the junction always causes just one impulse to spread over the muscle fibre. To put it another way, the end-plate potential caused by one pre-terminal impulse and the transmitter it releases, is always large enough to cause a propagated action potential to develop. Unless the neuromuscular junction is poisoned or diseased, nerve impulses invariably produce muscle impulses. The end-plate potential can only be separately recorded after poisoning or in other abnormal conditions.

ACTIVATION OF CONTRACTION AND THE MOVEMENT OF THE FILAMENTS

As the action potential passes over the surface of the muscle fibre, it also passes down the sarcolemmal T-tubes. From these it also passes to the terminal cisterns of the sarcoplasmic reticulum. These release some of their contained calcium ions into the sarcoplasmic fluid, through which the calcium diffuses into the myofibrillar substance and reaches the actin filaments. In sarcoplasm, Ca^{++} concentration is less than $10^{-7}M$ in the resting state. Release of Ca^{++} from the terminal cisterns raises it to above $10^{-5}M$, at which concentration it combines with tropomyosin in such a way as to cause troponin to move over the surface of the globular actin molecules. This movement of troponin now exposes a site on the actin molecule onto which a nearby myosin head can attach itself (Fig.

7.5). The processes of Ca^{++} leaving the sarcoplasmic reticulum, its passing to the thin filament and the resultant exposure of the myosin binding sites are collectively known as *activation*.

The myosin and actin combination, actomyosin, is an ATP-splitting enzyme, but in this process the myosin head rotates on its rod-like tail, dragging the actin molecule with it. The orientation of myosin molecules and the direction of this rotation pulls the actin molecule (and therefore the whole thin filament) further into the space between adjacent thick filaments. The bent myosin head releases its ADP molecule, combines with a fresh ATP and can then become detached from the actin and straighten out again. Provided Ca^{++} is still present in sufficient amount the myosin head finds a new actin molecule with exposed binding site, re-attaches itself, and begins the process again. The sequence, *myosin binding to actin*, ATP splitting, myosin head bending, ADP release, ATP binding, actomyosin splitting, myosin head extending, *myosin binding to actin* repeats over and over again so long as Ca^{++} remains in high ($> 10^{-5}M$) concentration (Fig. 7.6). In fact, within a few ms after the passage of the action potential, the sarcoplasmic reticulum is actively transporting Ca^{++} out of the sarcoplasm into the terminal cisterns and so eventually the Ca^{++} leaves tropomyosin and the active site on actin for binding myosin is no longer exposed. Thus Ca^{++} activation of the contractile process is brief, only about 10ms long.

In the average simple *twitch* contraction, produced by a single action potential, the Ca^{++} activation lasts long enough for the actin filaments to be dragged

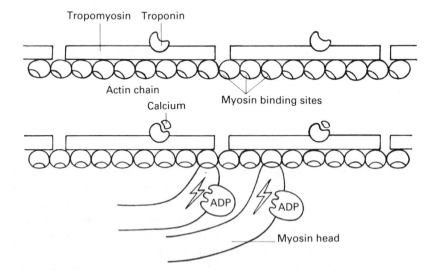

Fig. 7.5. (1) The thin rod of resting muscle. (2) When calcium binds to troponin, the actin molecules move so that myosin heads from an adjacent thick rod can bind onto actin molecules.

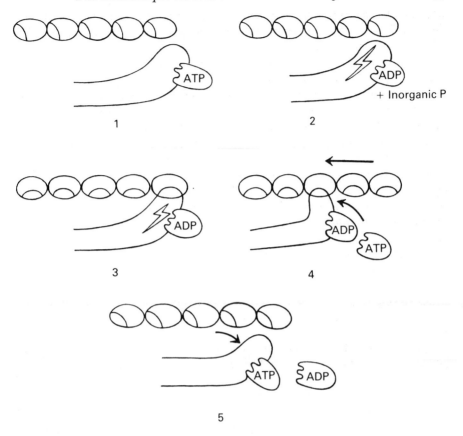

Fig. 7.6. (1) Actin activated by calcium binding to troponin. (2) A myosin head binds to actin X. (3) The ATP molecule is split, causing myosin head to distort. (4) A fresh ATP binds to the myosin and it is released from actin X and is then ready to combine with actin Y and repeat the process.

about one-third of the possible distance they can move into the spaces between adjacent myosin threads. In the normal situations, once the Ca^{++} has been removed the ATP-myosin complex, released from actin and with the myosin head once again extended, can find no binding site free on the actin filaments. These are now free to disengage and move out again from between the myosin filaments. Since the release of actin from myosin depends as much as does the flexing of the myosin heads upon ATP, this substance is required for the extension of the relaxed muscle as well as for the shortening of the contracting muscle.

The twitch response to a single action potential is divided into latent, contraction and relaxation stages. The latent period for the 'average' skeletal muscle fibre in man is about 2.0 ms. During this time the action potential is spreading over the sarcolemma and down the T-tubes. While the sarcolemma beneath the nerve terminals is depolarised (recovery takes about 2 ms) it cannot be stimulated again by a new release of acetylcholine, so the latent period is also the *refractory period*, during which the sarcolemma is inert to stimulation. The contraction phase of the 'average' skeletal muscle fibre in man lasts 10 ms, and the relaxation phase lasts 20 ms. Thus shortening is comparable in time with activation as one would expect. In some muscles the whole process may only take 7.5 ms and in other as long as 100 ms. White fibres have the shorter (faster) contraction and red fibres the longer contraction times.

Every time a muscle fibre undergoes a twitch contraction its response will be the same. Provided the stimulus to it (liberation of 10^6 acetylcholine molecules) is effective it will deliver a twitch contraction. Increasing the acetylcholine will not increase the twitch, for it cannot change the nature of the depolarisation wave nor the subsequent activation of the contractile apparatus. The muscle fibre thus obeys the *all or none* law, contracting either maximally or not at all.

SUMMATION IN MUSCULAR CONTRACTION

In such a twitch the period of activation is so short that only about one-third of the total possible movement of actin and myosin threads can occur. The refractory period following an action potential in skeletal muscle fibres lasts for about 2 ms, comparable to that of nerve fibres. This being the case, a fresh muscle action potential can pass over a muscle fibre before the previous period of activation (Ca^{++} falls to $< 10^{-5}M$) draws to a close. This new action potential produces a new wave of activation, and consequently a new period of interaction between actin and myosin is started. If a muscle is thus stimulated repetitively, activation and contraction periods following rapidly after one another, it will not be able to relax fully between successive contractions. If the impulses follow each other fast enough, then the contraction periods fuse completely and the muscle fibre will remain in a state of continuously sustained and steady contraction. This is known as a *complete tetanus*. (This state of sustained contraction of skeletal muscles in response to repetitive nervous stimulation, which we will all experience throughout our daily lives, must be distinguished from the disease of tetanus or lockjaw, which is an abnormal overstimulation of motor neurones due to toxins formed by an invading micro-organism, Clostridium tetani, always present in well-manured garden soil. Tetanus should also not be confused with *tetany*, a spasmodic muscular contraction seen when low extracellular Ca^{2+} levels allow an increased excitability of motor nerves, so that trivial stimuli, such as tapping a nerve trunk, cause contractions in the related muscle fibres.) The condition in which some relaxation occurs between the successive stimuli is called *partial tetanus*. Owing to the duration of the twitch contraction of the fibre, partial tetanus will occur when successive impulses reach it (assuming it is a fibre with shortening phase of 10 ms and a relaxation phase of 20 ms) at successive intervals of 10–30 ms, that is at 30–100 impulses/sec. Complete tetanus will occur when the interval between impulses is less than 10 ms, or the repetition rate is over 100 impulses/sec.

THE MECHANICAL EFFECTS OF CONTRACTION

We must turn now from the microscopic study of what happens when a single fibre shortens, to the effects on the whole muscle of which that fibre is but a small component.

A MECHANICAL MODEL

A model of a skeletal muscle can be built up as in Figure 7.7. The contractile element is, of course, made up of the myofilaments. These are embedded within the sticky sarcoplasm, the *viscous component*. The sarcolemma and connective tissue elements within the muscle form an elastic component which lies beside the contractile element. It is therefore known as the *parallel elastic component*. Finally the tendinous origin and insertion form a *series elastic element* for these are attached to the ends of the contractile fibres.

A muscle at rest may be under some tension, the *resting tension* or *muscle tone*. This is transmitted to all elements and is present in the series elastic element (tendon of insertion). It will be shared between the contractile and parallel elastic elements, the sum of these two tension being of course equal to that of the series element or resting tension. With the muscle at rest, its viscosity plays no part in its state.

The degree of resting tension experienced by a resting fibre determines its length (within the limits set by the non-elastic material in sarcolemma and other connective tissues). This is reflected in the sarcomere length, the distance between adjacent Z-lines or the degree of overlap between actin and myosin threads.

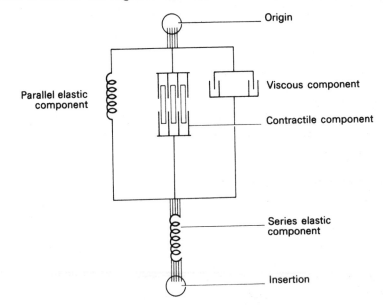

Fig. 7.7. A mechanical model of a skeletal muscle—see text.

The less these overlap at rest, the farther they can move during tetanic or sustained contraction. The result of this is that the initial resting length of the muscle, and thus the sarcomere length of all sarcomeres of all its fibres, determines the extent of its subsequent contraction. If the muscle is free to shorten, then the degree of shortening of the whole muscle is affected by its initial length. If, as is usually the case, the muscle is loaded, then the force that can be developed in its tendon will similarly be increased when the resting length of the muscle is increased.

ISOTONIC AND ISOMETRIC CONTRACTIONS

Isotonic contractions are contractions that occur with no increase in tone or tension developing in the muscle as its contractile element shortens (iso = the same). In the living body this, of course, never actually happens, but clearly, if the origin and insertion of a muscle can freely approach each other with no opposing resistance, the contraction would be isotonic and no tension would develop in the series elastic element.

In the ideal isometric contraction there is no overall change in length of the muscle, the origin and insertion remaining fixed. When the contractile element is activated, a shortening of the sarcomeres occurs as in any contraction, but this shortening causes a lengthening and development of tension in the series elastic element.

In the living body most muscular contractions are intermediate between isotonic and isometric. Scarcely any are truly isotonic, but the muscles that maintain posture contract with very little shortening, they are nearly completely isometric. It is the tension developed by their tendons, acting across joints between bones which maintains the stable position of the skeleton.

PRE- AND AFTER-LOADED CONTRACTIONS

All contractions thus involve a muscle in contracting against some load. Important differences in the performance of the muscle are found which depend on whether the muscle is loaded only when it begins to shorten (after-loading) or is already stretched by the load while at rest. Although these differences could, in theory, be studied in intact muscles in the living body, they are best investigated in an isolated muscle prepared as, for example, the frog gastrocnemius muscle shown in Figure 7.8. The lever used to record the muscle's contraction usually rests on a supporting stop which has been adjusted so that it takes only the lever and load weight. The tendon and thread are taut. The stop may be removed so that the resting muscle supports lever and load.

When the muscle contracts with the stop in position (after-loading), its contractile element has first to develop sufficient force to overcome the force of gravity acting on lever and load before it can begin to move them. The greater the load, the greater is the amount of the contraction required to overcome gravity. There is less left to raise the lever. With increasing load the lever is lifted by smaller amounts until the muscle cannot raise it at all, the force exerted

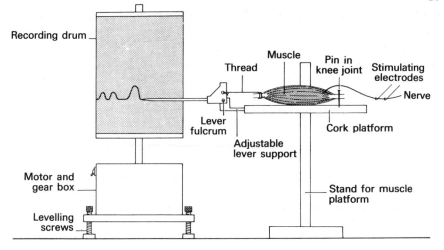

Fig. 7.8. The apparatus used for studying muscle contraction experimentally after a muscle has been removed from the body.

by gravity being greater than that developed by the contracting muscle. In a frog's gastrocnemius, this happens with a load of about 50 gm (*Rana temporaria* in Britain). The work done by the muscle (force exerted × distance load removed) similarly falls progressively to zero as the muscle lifts the lever by smaller amounts. When the load is so great that the muscle just fails to raise the lever the contraction is, of course, fully isometric.

When the supporting stop is removed, the lever and load, through the attaching thread, exert force upon the series and parallel elastic elements, and the contractile elements of the muscle, while the muscle is in the resting state. Though partially supported by the parallel elastic element, the sarcomeres will be partially elongated, actin filaments being pulled a small distance from between the myosin filaments, so the overlap between these filaments is reduced. When the muscle now contracts, the whole contraction is used to raise the lever, since the tension caused by gravity was present in the muscle before contraction began. Further than that, the increase in sarcomere length increases the amount of shortening possible in the sarcomere. The work that can be done by a free-loaded frog gastrocnemius muscle may increase until the load reaches 150–200 gm as a consequence, and such a muscle may, if free-loaded, lift 250 or even 300 gm. As the load is progressively raised, *whole muscle* shortening (rate of rise of lever) becomes slower, so the *power* of the muscle (= rate of doing external work) falls with increasing load. The same consideration would apply to reflex contractions in man, though in voluntary contractions one probably compensates for this decline in power by making an increased effort.

SUMMATION IN VIVO

If these experiments are done in isolated nerve-muscle preparations, using artificial (electrical) stimulation of the nerve, all the motor units (nerve fibres and innervated muscle fibres) are stimulated at the same time. Therefore it is necessary to stimulate, as has just been described, at 100 impulses/sec in order to obtain a smooth contraction in all motor units, and the resultant sustained contraction will be the maximum that the muscle can produce. Such a contraction can only be sustained for a very short time. Yet we know that steady sustained contractions are maintained by muscles at much less than maximum force, in everyday life, all the time. How is this achieved?

By recording simultaneously from several motor nerve fibres, it is found that they are desynchronised, impulses passing down them at different times. Moreover, some motor nerve fibres are resting while others are conducting impulses and they regularly alternate between the resting and conducting states. The frequency of impulses in any one nerve fibre would maintain its muscle fibres in a state of partial tetanus. It is the lack of synchronisation between the motor nerve fibres that ensures that some muscle fibres are contracting while others are relaxing, so the sum of the tensions produced by these out of phase partial tetanic contractions is a smoothly maintained tension. The overall tension is determined by the number of motor units in partial tetanus at any one moment, and fatigue is prevented because while some units are active, others are resting and they can alternate between the resting and the active state (see also p. 110).

THE ENERGY SOURCES FOR CONTRACTION

The flexion of myosin heads while they are attached to actin molecules, that follows the splitting of ATP to ADP and inorganic PO_4 ions, is the actual mechanism that causes sarcomere shortening. The energy required for myosin head flexion comes from ATP molecule itself, so this is the immediate source of energy for muscular contraction. But there are only 2–3 μmoles of ATP present in each gram of muscle, sufficient to sustain 100 twitch contractions, or 1 second of a full (100 stimuli/sec) tetanus. When muscles are stimulated to contract, whether they be isolated frog muscles or those of a man pedalling on a cycle ergometer, even at exhaustion there is very little reduction in the ATP content. It is restored, as fast as it is used, by a variety of means.

CREATINE PHOSPHATE (CrP)

Creatine is an organic base that forms high-energy bonds with phosphoric acid, similar to the high-energy bond of ATP. Skeletal muscle contains an enzyme that catalyses the transfer of the phosphate, and its energy, between CrP and ADP. As fast as ATP is broken down by the actin–myosin complex, the ADP that breaks away from the complex reacts with CrP to form fresh ATP and free creatine. Under equilibrium (resting) conditions there are about 3 μmoles/gramme of ATP, only 0.03 μmoles of ADP and about 22 μmoles of CrP. This reserve store of energy is thus about seven times the amount available in the ATP present, still sufficient for only short bursts of activity.

In further analysis of the provision of energy, it is usual in physiology texts to proceed from the most severe to progressively milder forms of exercise, or from emergency to normal maintenance systems of energy production. We think that it is more logical (perhaps more physiological) to proceed in the reverse manner, from the resting state, through mild and moderate to severe exercise, studying how the muscle obtains energy at each successive stage and the more general bodily changes that are evoked by the muscles' need for energy producing materials.

THE RESTING MUSCLE

In man, resting quietly, the body's skeletal muscles are responsible for about one-quarter of the total energy turnover of the whole body, using about 62.5 ml oxygen/min; each gram of muscle uses about 0.003 ml/min. This O_2 is used to maintain Na^+ and K^+ concentrations inside the fibres and to sustain its resting muscle tone. Energy production of muscle at rest is largely dependent upon oxidation of foodstuffs, or materials derived from them. Traditionally, and as a result of experimental work in the 1920s and 1930s, it has always been assumed that carbohydrates (starch and sugars) are the fuel of skeletal muscle. In the 1960s, though, direct studies of muscle metabolism in man all showed the principal fuel of resting muscle to be, in fact, fatty acids. These are found at all times present in the muscles and the blood (coming, as required, from the fat depots of the body), and can readily be taken up from the blood by muscle, broken down and oxidised and half the potential energy of the fat molecules transferred, in the oxidation process, to forming ATP from ADP and inorganic phosphate. Even under conditions after a meal, when there is plenty of glucose in the blood, and this glucose is being actively extracted from the blood by the muscles, they do not use it as fuel, but put it away in store, as glycogen. Skeletal muscle can also de-aminate some amino acids and use the residue for oxidative energy production, both while at rest and during exercise.

MILD EXERCISE

In some experimental studies muscles have been voluntarily exercised so that their oxygen uptake is doubled, or, in other studies, raised by six times their resting uptake. In both cases the muscles were still found to be almost entirely dependent upon the oxidation processes performed by this raised oxygen uptake for their re-formation of ATP, and in both cases the fuel used was still fatty acids. Both these sets of results were performed on human forearm muscles *in situ*. Similar results have been obtained in dogs' muscles with artificial stimulation of motor nerves, and in humans walking on a treadmill, where it was further seen that successive waves of mobilisation of fatty acids from the fat stores (adipose tissue) of the body were seen to occur during the 8-hour duration of these experiments.

MODERATE AND SEVERE EXERCISE

As exercise is progressively increased in severity, there is a switch in the source of the energy. In the dog studies on oxygen uptake 20 times the resting uptake is associated with oxidation of carbohydrate materials, mostly glucose from the circulating blood, but when the energy requirements are multiplied fifty-fold, energy is produced by non-oxidative means.

These involve the breakdown of the stored muscle glycogen to lactic acid. During this process about one-tenth of the potential energy of the glycogen becomes available for the rebuilding of ATP, the remainder residing in the lactic acid molecule. The glycogen content of human muscles is sufficient to maintain them in vigorous exercise for a few minutes (though in practice we never rely solely upon this source of energy, oxidation continuing during glycogen break-down). But the lactic acid formed in vigorous exercise must subsequently be disposed of. Its formation represents an oxygen deficit, and the debt must be repaid when the exercise ceases. Thus, in prolonged vigorous exercise, although oxygen uptake may increase sixteen-fold during the exercise, much lactic acid is formed and escapes into the circulating blood. Most of this can be re-stored as glycogen in the liver and in the muscles, but to convert lactic acid to glycogen requires energy (the energy that glycogen liberated when it became lactic acid) The energy, now that the body is again resting, comes from oxidation processes in muscle or liver, wherever lactic acid is being changed back again to glycogen. Thus the *oxygen deficit* or *debt* that was built up by vigorous exercise is finally repaid, and an athlete, performing a 100 metre sprint or running 1,500 metres in less than 4 minutes can obtain a considerable part of the energy needed to maintain ATP from the muscle glycogen store. He can for these short periods forgo his full need for delivering oxygen to the muscles via lungs and blood, but, during the ensuing 1–2 hours of rest must repay the oxygen debt and remove the excess lactic acid from his blood and tissue fluids.

Actin, its associated troponin and tropomyosin, and myosin, which are the contractile machinery of muscle, form the largest part of its solid material. They are proteins and so a significant part of the living body's protein is found in these substances. Proteins are of course also present as the myoglobin and the oxidative enzyme materials and as the Na^+–K^+ and Ca^{2+} pumps, so essential for muscular function. In fact, during periods of starvation and sickness all this muscle protein is treated by the body as a deposit account and is used for more essential purposes. Disuse of muscles is also followed by a withdrawal of their proteins. Wasting and loss of power occur therefore in all these conditions. Feeding, convalescence and athletic training are all accompanied by a build-up of muscle protein. Even in healthy normal life the proteins of muscle are constantly being destroyed and renewed, and some of their constituent amino acids are used in many other essential metabolic activities.

HEAT PRODUCTION AND ENERGY EFFICIENCY

When the splitting of ATP and bending of myosin heads occur, only about 40 % of the potential energy of the terminal phosphate bond is used to shorten the sarcomere. The remainder appears as heat. Sensitive methods of detecting temperature in a muscle have shown that this heat appears before the muscle as a whole shortens or develops tension. Further heat is evolved later on, when metabolic processes restore ATP and CrP. Again, in these, only 50 % of the potential energy of the fuel is trapped as high-energy phosphate-bond energy. The rest appears as heat. Thus only 20 % of the original energy of the foodstuffs can be converted into sarcomere shortening. The remaining 80 % appears as heat. If sarcomere shortening is opposed by viscous and elastic forces, then some of the mechanical energy is converted to heat. When you stretch an elastic band and then allow it to relax again, the energy you transferred from your muscles to the stretched band appears as heat in the relaxed band as it returns to its initial length. Exactly the same happens in the series elastic component in a contracting muscle. If tension has developed or increased in this during contraction (and virtually always it does) then, on relaxation, this energy appears as heat and is not available for performing external work (load lifting, etc.). In purely isometric work, of course, all the energy is converted into heat. So far as external work is concerned, skeletal muscle thus approaches, but never reaches, 20 % in efficiency, the remainder, > 80 % appearing as heat. In this it compares well with the steam engine (12 % of coal energy being converted to external work) and about equals that of the best internal combustion (Diesel) engine.

In conclusion, may we note that we have deliberately deferred to Chapter 23 a description of the general body changes that occur in muscular exercise, including the dissipation of the large amount of heat produced, since these depend on a knowledge of the circulatory, respiratory and other systems.

8 · Effects of Peripheral Nerve Damage

Nerves can become damaged in a variety of ways. Accidents and injuries can tear or cut through nerves, and sites where nerves are at risk are well recognised. Thus the junction of the 5th and 6th cervical roots of the brachial plexus is the point in the plexus that takes the strain when the shoulder is forcibly depressed. The radial nerve in the radial groove of the humerus can become torn if the shaft of the bone be broken. Pressure from a neighbouring object, although it does not immediately tear the nerve asunder, can have the same effect functionally, as is seen in a prolapsed intervertebral disc or when the intervertebral foramen becomes narrowed by shrinkage of the disc. Impulse conduction, the function of nerves, is interrupted when disease processes, 'neuropathies', attack the nerves or when they suffer from an insufficient blood supply. Initially, similar effects may be experienced by the patient, and similar treatment required of the therapist, when the site of the disease is in the cell bodies (in posterior root ganglion or anterior horn of the spinal cord) whose processes form the peripheral nerves.

Inevitably, then, there will be some overlap between matters discussed in this chapter and those already described in the last parts of Chapter 5.

ANATOMICAL EFFECT OF NERVE DAMAGE

From the patient's point of view, the most important feature of injuries to the nervous system is the prospect of recovery and this will depend to a large extent on the anatomical nature of the injury and whether the central or the peripheral nervous system is involved. As will be seen later, the essential requirement for nerve regeneration after injury (assuming that the blood supply and other local conditions are satisfactory) is that the regenerating axons can be guided to their original pathways so that they can link up with the appropriate nerve endings; and in order to do this it is important that they do not have to bridge too big a gap and that Schwann cells and endoneurial tubes are present to guide them on their way. In the central nervous system, Schwann cells and endoneurium are not present and no regeneration can therefore take place in the brain or spinal cord. This is why spinal cord injuries demand such a high standard of treatment from the physiotherapist. In the peripheral nervous system the chances of complete recovery will depend on the ability of the axons to grow out from the damaged central end of the nerve and to follow the appropriate course to the nerve endings. If the axons are damaged locally so that conduction is interrupted but the axons are still in place, the condition is known as *neurapraxia*. The axon peripheral to the site of injury will remain.

normal and, if stimulated, is still able to conduct. In such cases there is no gap to bridge; the axons, although damaged, are still in continuity, and recovery will occur in days or weeks (or even minutes in the case of mild compression of the nerve). The axons distal to the lesion do not degenerate but we have little information about the changes in the nerve at the site of the lesion. This is because, since recovery always occurs, pathological examination of the lesion (at least in humans) is not normally carried out.

Minor degrees of neurapraxia occur in everyday life when a nerve such as the ulnar, or common peroneal, became compressed against the underlying bone by an awkward posture. 'Saturday night paralysis' is a classical example in which a person who has imbibed over-enthusiastically, goes to sleep with his arm hanging over the back of a chair. When he finally regains consciousness, he has wrist-drop because of pressure on the radial nerve in the axilla (see Chapter 12).

A more severe type of injury to nerves occurs when the axons are completely severed but the connective tissue framework of the nerve remains intact. This is known as *axonotmesis*. Since the axons are interrupted, the portion of the axon distal to the lesion will degenerate and the ends of the axons on the proximal stump have to grow down the degenerated nerve to re-establish connections with the original nerve endings. Recovery is therefore relatively slow but the end results are good because the endoneurium is available to guide the axons along their original course. Axonotmesis can occur as a result of a severe stretching of the nerve or, more commonly, of a crush injury.

Finally, the most severe type of injury is that in which the whole or part of the nerve is completely severed, as may occur in wounds. This condition is known as *neurotmesis* and is less likely than the other two grades of injury to give rise to a good recovery. The nerve may be wholly or partly divided by a clean cut as may perhaps happen to the ulnar nerve at the wrist when a person puts his hand through a window pane, or there may be a gap between the ends of the nerve if severe damage has occurred. In this case it is obviously important for the surgeon to bring the ends of the nerve together and suture them as accurately as possible so as to increase the chances for the regenerating axons to find their way to the peripheral segment of the nerve.

The following account of nerve injury will deal firstly with the structural effects of interruption of a nerve by injury, then with the anatomy of the regeneration process and finally with the functional effects of nerve damage.

WALLERIAN DEGENERATION

The histological effects of nerve damage were first described by Waller in 1850, and the degenerative process which results has ever since been called *Wallerian degeneration* (Fig. 8.1). The process begins within 24 hours of the lesion. Distal to the site of injury, the axon first swells, then becomes irregular in outline and finally breaks up into fragments. Conduction in this part of the axon naturally ceases. A little later (1–4 days) the myelin sheath breaks up into fatty droplets which are later broken down to simple fats and finally, along with the axon

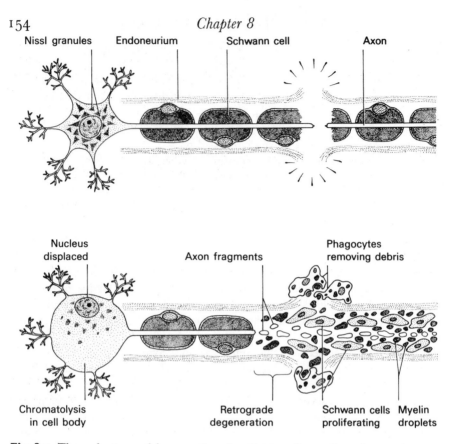

Fig. 8.1. The early stages of degeneration after division of a myelinated nerve (see text for description).

debris, removed by phagocytosis. The cells which remove the debris in this way are not only connective tissue phagocytic cells but also Schwann cells. Complete breakdown of the myelin sheath has occurred by the tenth day after the injury, while its chemical degeneration takes up to 1 month. Similar changes occur in the nerve proximal to the site of injury but here the changes occur over a varying distance, depending on the severity of the nerve lesion. These changes which occur *proximal* to the lesion are known as *retrograde degeneration*, and this process may also involve the cell body (see below). The Schwann cells, meanwhile, are dividing by mitosis and forming long strands of cells which grow out and enter the endoneurial tubes which were originally occupied by the intact nerve fibres (this is assuming that there is not a very large gap between the ends of the damaged nerve). The most active outgrowth of Schwann cells occurs from the *distal* stump, and the cells thus have to grow *proximally* to reach the central end of the nerve. This proliferation of the Schwann cells marks the first sign of regeneration and it begins even before the degenerative process is complete. Meanwhile, back at the cell, the Nissl granules break up and become dispersed and difficult to stain. This process is known as *chromatolysis* and the histological appearance is due to the loss of RNA in the cytoplasm. The cell becomes swollen, and the nucleus is displaced to one side of the cell. These changes commence during the first 24 hours after the injury and are another manifestation of retrograde degeneration. Later, the cell shrinks and may eventually die, although complete recovery can also occur. Degenerative changes are not necessarily restricted to the neurone which has actually been damaged. It may also affect adjacent neurones with which synaptic contact is made. This is known as *transneuronal* or *transynaptic* degeneration. For example, if the optic nerve is cut, retrograde degeneration will affect the cell bodies, which are in the retina, but changes will also occur in the cells of the lateral geniculate body of the brain, with which the ends of the optic nerve fibres make synaptic contact.

REGENERATION

As has been mentioned already, regenerative changes begin even before the degenerative process is completed and it is initiated by the division of the Schwann cells in the cut ends of the nerve (mainly the distal stump) and their growing proximally in long strands to enter the endoneurial tubes of the proximal stump. It is this outgrowth of Schwann cells which initially bridges the gap between the ends of the nerve and the smaller the gap, the more easily can the cell cords locate the tubes into which they must grow. If the gap is too wide, regeneration will not occur, so that, in severe nerve injuries, it is important that the cut ends of the nerve are exposed surgically and sutured together as accurately as possible.

The function of the cords of Schwann cells is to guide the axons into the endoneurial tubes of the distal part of the nerve. The axons, as a result of retrograde degeneration, terminate some distance proximal to the site of the lesion and they begin to grow down their own endoneurial tubes until they reach the gap (Fig. 8.2). They now put out large numbers of separate processes, or fibrils, which follow the cords of Schwann cells towards the distal stump, where they enter the endoneurial tubes. Although the Schwann cells guide the axon fibrils to the tubes of the distal stump, they do not guide individual fibres to their own particular endoneurial tube, so that many fibrils may enter the same tube and each tube may contain fibrils from several different axons. Moreover, many (perhaps most) fibrils do not follow the Schwann cell guides but grow out into the surrounding tissues, get lost, and ultimately degenerate. If the endoneurial tubes are intact (i.e. in axonotmesis) relatively few fibrils are formed and all except one for each axon rapidly degenerate. Each axon is thus enabled to grow down its own endoneurial tube and will eventually innervate its own original nerve ending. Complete functional recovery is therefore the rule. In neurotmesis, however, axons do not necessarily grow down their original tubes and many tubes remain empty while others contain many fibrils. Eventually all except one of the fibrils degenerates so that only one remains to form an axon and to continue to grow down the tube; but many tubes remain empty and may axons enter the 'wrong' tubes and so innervate

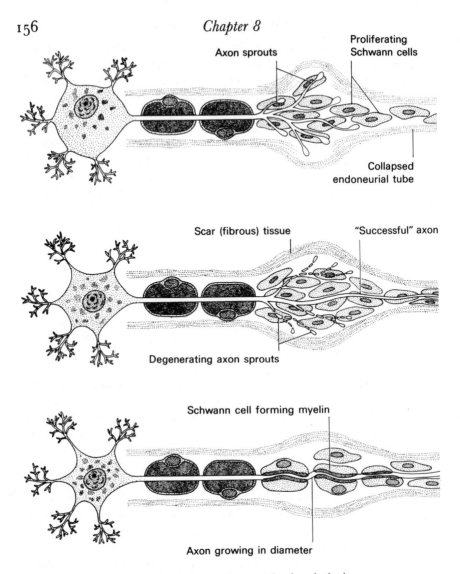

Fig. 8.2. Regeneration of a divided nerve (see text for description).

endings other than their own. Recovery, therefore, is much more variable than in axonotmesis. It is important to remember that the severed axons in the central cut end of the nerve retain their ability to sprout out into fibrils for a long time, perhaps for several years, so that a delayed suture of a nerve may be, at least, partially successful.

When a fibril has successfully entered an endoneurial tube, has enlarged to form an axon and has grown peripherally down the tube, it gradually becomes transformed into a complete nerve fibre. The diameter increases, but rather slowly and possibly not to the full size of the original axon. The Schwann cells begin to produce myelin and eventually form a myelin sheath complete with nodes of Ranvier. Functional recovery lags a little behind anatomical recovery.

The rate of regeneration is of great important to physiotherapists, who have to continue treatment until recovery is as complete as possible. The initial bridging of the gap by Schwann cells, in a neurotmesis, occurs at a rate of about 1 mm per day. The rate of axon growth is very much slower in the region of the injury than further distally where the axon simply has to grow along an endoneurial tube. The rate of growth in the injured region will therefore depend upon the extent of the damage to the surrounding tissues and to the density of the developing scar tissue through which the axons have to penetrate. In the scar region the rate of growth is therefore a good deal less than 1 mm per day, but in the endoneurial tubes a rate of growth of 3–4 mm per day is attained, although this is very variable. Myelination proceeds probably a little more slowly and rates of 1 and 2 mm per day have been recorded in human subjects, although a more rapid rate has been found in animal experiments.

PHYSIOLOGICAL (OR FUNCTIONAL) EFFECTS OF PERIPHERAL NERVE DAMAGE

MOTOR EFFECTS

The power to initiate voluntary contractions of muscles supplied by a nerve ceases when the nerve is cut or torn apart, thus disrupting the lower motor neurones. Equally, muscle tone and reflex contractions are also abolished, i.e. a *flaccid paralysis* is present. If the development of the lesion is gradual, nerve fibres ceasing to conduct one by one over some days or weeks, then the maximal power of the affected muscles will be reduced and they will fatigue more readily when performing activities that previously could be carried out with no effort. This is simply because, there being fewer motor units to perform the contraction, each will be working nearer its limit and will thus fatigue earlier, and the maximal power will be similarly affected. Eventually muscle atrophy will occur which becomes severe a few weeks after injury.

As well as these effects, two types of spontaneous and involuntary contraction emerge. *Fasciculations* are relatively coarse contractions. They are due to the severed peripheral ends of axons developing their own action potentials. Each impulse arises quite independently in one axon and passes down it to the muscle fibres of the motor unit. In fasciculation, then, the contractions seen are of whole motor units, but there is no relationship between the spontaneous twitches in the different units that make up a whole muscle. *Fibrillation* is a much finer contraction. Individual muscle fibres can themselves generate their own action potentials when none reach the axon terminal of the neuromuscular junction. This phenomenon is an aspect of a general law of neuro-effector junctions. When impulses cease to arrive at the axon terminals due to axonal degeneration, then the effector (muscle) membrane becomes hypersensitive and spontaneously depolarises when exposed to trivial concentrations of the neurochemical transmitter substance. In this way, some days after a nerve has been cut, individual muscle fibres generate their own action potentials and contract repetitively, with no relation to one another. Fibrillation can often be observed if a muscle is stimulated locally by flicking the overlying skin and can easily be recognised by electromyography.

Fasciculation, then, being dependent upon intact peripheral portions of axons, will disappear as these axons degenerate, which happens within a few days after damage. Fibrillation will only then emerge. After about three months the muscle fibres themselves begin to shrink and split both longitudinally and transversely. Eventually they become so disorganised that no fibrillation takes place, though reinnervation as late as three years after nerve injury may produce some recovery of voluntary and reflex contraction in a muscle.

If only a few motor nerve axons (or their cell bodies in the anterior horn of the spinal cord) have been severed, then the intact surviving axons will form buds which grow into adjacent unoccupied endoneurial tubes and so form new terminals on denervated muscle fibres. In this way, all the muscle fibres may become reinnervated, but there will be fewer and larger functional units in the muscle. These are commonly called *Macro motor units* for they contain many more muscle fibres than is normal for the muscle. Power and endurance are restored, but the range of variation of the strength of contraction is reduced along with the reduction in total number of axons surviving the injury.

THE REACTION OF DEGENERATION

Artificial electrical stimulation of an intact motor nerve will cause the muscles supplied to contract. Electric current can be passed through the living body when voltages are applied to electrodes placed on the skin. Greasy and dry skin is a poor electrical conductor, so it has first to be cleaned with a grease solvent and then kept moist with, for example, gauze soaked in a 1 % common salt solution. AC (faradic) and DC (galvanic) currents are both capable of stimulating nerve fibres through the skin, the latter *only when the circuit is closed or opened*. Normally a small *stimulating* electrode is placed over the nerve being studied and a large *indifferent* electrode is placed somewhere else on the body. With DC currents either electrode may be the positive anode or the negative cathode.

The most effective site for stimulating the motor nerve fibres to a single muscle is the point at which the nerve enters the muscle, the *motor point* of the muscle. With DC stimulation, in a normally innervated muscle, if the strength of the stimulating current is slowly increased, contractions appear in a regular order with the different modes of applying the stimulus. The *cathodal closing contraction* appears with the weakest stimulus, then, as the stimulus is increased, the *anodal closing contraction*. This is followed by the *anodal opening contraction* and finally by the *cathodal opening contraction*.

For a day or two after a motor nerve injury, electrical stimulation of the distal part of the nerve will produce a muscle contraction but as nerve fibres degenerate

they become unresponsive to the short faradic or AC impulses while still responding to DC shocks. After degeneration is complete, muscle fibres can be stimulated
by direct electrical shocks, with the following characteristics:

1   Both the contraction and relaxation phases are prolonged and sluggish
2   Short faradic pulses will not stimulate muscle fibres
3   The muscle responds to smaller currents than did the motor nerve. This relates
to the way it becomes generally hypersensitive after denervation. It may be
noticed that the motor point is no longer the most excitable area since the whole
muscle is excitable.
4   The anodal closing contraction is often elicited by a smaller current than is the
cathodal closing contraction, although this is unusual.

These changes are together known as the *complete reaction of degeneration*. If the
nerve can still be excited by DC but not by faradic currents, and the muscle
responds weakly, if at all, to faradic stimulation, the state is called *partial
reaction of degeneration*.

Other methods for studying motor activity are the *strength-duration (SD) curve*
and *electromyography* (EMG). The former involves the stimulation of a muscle for
varying times and a graph is then constructed of the minimum current necessary to
produce a contraction for each stimulus duration. (from 0.04–300 ms). In normal
muscle the required current remains the same over a wide range of duration of
stimulation but in a denervated muscle a much longer period of stimulation is
necessary to produce a response and as the duration is decreased the current
required to produce a contraction increases steeply.

Electromyography involves picking up action potentials from a contracting
muscle by surface or needle electrodes. When a normal muscle is contracted voluntarily, motor unit action potentials increase in frequency with increasing effort as
more and more units are recruited, until they eventually summate to give a maximal tracing in which the potentials from individual units cannot be identified. In
the denervated muscle, fibrillation potentials can be identified 2–3 weeks after the
injury. These are of low amplitude and short duration. When a muscle is only
partly denervated, the potentials of individual functioning motor units can be
identified but they do not summate to produce the high amplitude confused
pattern characteristic of normal muscle.

More sophisticated studies are sometimes performed in which electromyography is combined with nerve stimulation so that conduction velocity and the
site of the lesion can be studied.

SENSORY EFFECTS

The sensory effects of peripheral nerve damage have already been adequately
described in Chapter 5, since they do not differ from the effects of damage to
sensory pathways within the CNS. However, a peripheral nerve conveys
impulses from *all* receptors and therefore *all* modalities of sensation are affected.
The same applies to posterior root ganglion lesions, but damage within the cord

may affect some types of sensation and leave others unimpaired, due to the different routes they take.

Complete section of a major nerve carrying sensory fibres will result in a graded loss of skin sensation in part of the area supplied by the nerve. The area will not be as great as anatomical diagrams would lead one to suppose since there is a large area of overlap between adjacent cutaneous nerves. Also, the area affected for touch will be greater than that for pain, (estimated by pin-prick). Deep touch and proprioception are also lost and the skin will be warm and dry because of loss of sympathetic innervation.

One special effect of sensory nerve section may be seen following an amputation. This is known as the *phantom limb*. Following the amputation the patient may be 'aware' of his missing limb and imagine it still to be present. Nerve impulses generated spontaneously from the cut ends of the sensory nerve fibres cause this sensation and so the patient believes the limb to be still present. It is said that the phantom limb begins at its normal length, but progressively shrinks until the foot enters the stump (which may be below or above the knee) and then disappears. It may be dismissed as a curiosity by the patient, or may disturb him. In the latter case, re-operation of the stump often reveals the cut end of a nerve trapped in a mass of scar tissue and growing fibrils looking in vain for somewhere to go. Removal of this *amputation neuroma* usually then leads to disappearance of the unpleasant phantom limb sensation.

FUNCTIONAL RECOVERY OF PERIPHERAL NERVES

Once nerve cell bodies die, there is no replacement and they and their axons are lost for ever. Functions dependent upon them can never recover. This is the sad outcome of an attack of poliomyelitis, when the anterior horn cells affected may be killed by the disease process. The best that can be hoped for is the survival of some cells and the new budding of their axons to supply the neuromuscular junctions that have lost their nerve fibres, as has already been described. In sensory fibres there is no evidence that such a process can ever occur.

It is only in the peripheral nerves, in which each axon and its myelin sheath are surrounded by an endoneurial tube, that meaningful axonal regeneration can ever occur in the way already described. Regenerating fibres grow at 3–4 mm/day over the greater part of the length of the nerve, so it may be many weeks or months before they reinnervate neuromuscular junctions or receptor terminals, thus allowing a restoration of motor or sensory function. In axonotmesis each developing axon re-enters its original motor or sensory terminal and full recovery can occur. In neurotmesis, if some axons enter the wrong endoneurial tube, then they are lost completely if sensory axons enter 'motor' tubes or vice versa. If a regenerating sensory axon eventually reaches a receptor different in kind from its original one, considerable 'relearning' can take place and the patient may interpret the impulses as coming from a different receptor type, perhaps mediating a different sensation, in a different site. Since, after an injury, regeneration of axons proceeds at varying rates, the recovery of sensation is gradual. Owing to the various possibilities of the wrong route being taken, the recovery may be only partially complete since the relearning process described above is only partial, and some sensory fibres may totally fail to reinnervate receptors. Regenerating sensory fibres are particularly sensitive to mechanical stimulation, and this may be used to estimate the progress of recovery. Light percussion over the regenerated part of the nerve will give rise to a tingling sensation which is felt in the cutaneous distribution of the nerve (Tinel's sign).

Recovery of motor fibres, after a neurotmesis type of injury, proceeds as does that of sensory fibres. Motor axons entering sensory neurilemmal tubes will be lost for good. Some will reinnervate the original muscle and some other muscles. Again a considerable degree of relearning can occur if the 'wrong' muscles become reinnervated.

Inevitably, in motor nerve regeneration, there will be some deficit of axons. Those that reach the isolated muscle will form new branches which approach those muscle fibres lacking innervation. Thus, as in partial denervation of a muscle, there is the eventual formation of macro motor units.

9 · Bone and Bones

As was mentioned in Chapter 3, bone is classified as a connective tissue since its cells are embedded in a relatively large quantity of extracellular tissue. Since its function is to provide support and protection as well as to provide a reservoir for calcium and associated substances, the extracellular tissue or matrix is impregnated with mineral salts consisting mostly of calcium and magnesium phosphates and carbonates. Bone is therefore a tissue that is composed of organic and inorganic components. If a bone is taken from a body and placed in a strong acid, the inorganic component will be removed by chemical reactions, leaving only the organic material. This process is known as *decalcification* and a bone which has been so treated retains its shape but is quite soft and elastic—a long bone such as a fibula can almost be tied in a knot. On the other hand, the organic component can be removed by heating and drying, leaving a dry 'bone' such as is used by physiotherapy students for study. It may be kept in a cupboard and produced as required whereas a decalcified bone, since it consists of organic matter, must be kept in preservative or it would soon become offensive.

It is very important, therefore, to understand that the 'bones' which you study resemble live bones only in shape and are, in fact, only 'skeletons of bones'. Live bones are springy and resilient, have a nerve supply and a rich blood supply and are in a state of continual and rapid turnover as far as their inorganic material is concerned so that some of the calcium in your milk at breakfast time forms part of your femur by lunchtime and may be lost in your urine tomorrow. Live bones, too, can adapt their shape to their function, particularly when they are young, so that faulty posture or unsuitable shoes may cause distortion of bones. The plasticity of bone is best demonstrated by the work of the orthodontist who, by means of suitable splints and braces applied over a long period of time, can drastically rearrange the position of the teeth in the jaws and thus correct faulty dentition.

Bone is of two main types: *compact* and *cancellous* or spongy. The former is a solid tissue resembling ivory and is extremely strong. It is found in the shafts of long bones and forms a thin covering layer over their ends. Cancellous bone consists of a network of interlacing fine *trabeculae* and is found, for example, forming the main mass of the bodies of the vertebrae and the expanded ends of long bones. In all cases it is covered by a thin layer of compact bone.

THE MICROSCOPIC STRUCTURE OF BONE

The structure of compact bone is more complex than that of the trabeculae of cancellous bone. In the former the basic unit of structure is the *Haversian system* which consists of a series of concentric layers (*lamellae*) of bony matrix which surround a central canal—the *Haversian canal* (Fig. 9.1). The matrix consists of an organic ground substance, composed mostly of collagen fibres, which is impregnated with the inorganic salts mentioned above. Although, in a cross-section of bone, the Haversian systems resemble the skins of an onion, the systems are not spherical, but cylindrical, and the Haversian canals branch and inter-communicate, mostly running along the long axis of the bone. Each canal contains one or two blood vessels and perhaps some nerve fibres. The vessels communicate here and there with the blood vessels in the *periosteum* (the outer membrane which covers bones) and with the blood vessels in the cavity (marrow or medullary cavity) within the bone. Between the lamellae of the Haversian systems are a number of tiny spaces called *lacunae*, each of which contains one bone cell or *osteocyte*. The lacunae are connected to each other by a number of minute *canaliculae* each of which contains a cytoplasmic process of an osteocyte. In between individual Haversian systems are found a few lamellae which, as it were, fill in the spaces between the closely packed cylindrical systems; they are known as *interstitial lamellae*. These are remnants of previous Haversian systems

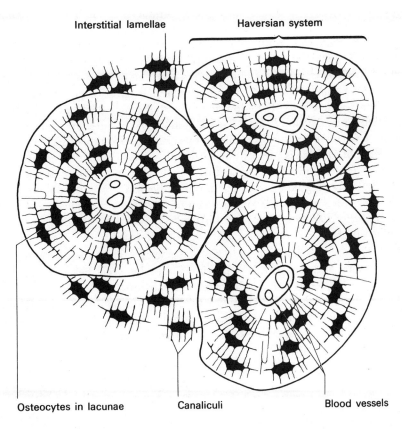

Fig. 9.1. The histology of bone as seen in a transverse section.

which have been partly replaced as new systems are laid down from time to time to replace the old. Bone is therefore able easily to adapt its structure to changed conditions and is just as liable as the soft tissues to weakness due to disuse (*disuse atrophy*).

Cancellous bone also consists of lamellae of bone with interposed osteocytes but organised Haversian systems are not found in this situation. The lamellae are arranged according to the lines of stress in any particular bone.

CALCIUM METABOLISM

The formation and maintenance of healthy bones (and teeth) demands an adequate supply of calcium, phosphates and other mineral salts in the diet and the ability to absorb these from the gut and to transfer them to the bones. Calcium is also an essential component of tissue fluid, where it is particularly important for controlling the excitability of nerve and muscle cells. Furthermore, calcium is a necessary part of the clotting mechanism of the blood. Calcium is found particularly in cheese, milk, beans, etc., and wheat, and for its absorption from the gut into the blood an adequate supply of vitamin D is required. This is a fat-soluble vitamin which is found in dairy products, in the liver (particularly fish livers) and the flesh of 'oily' fish, such as the herring and mackerel. It can also be manufactured in the skin by the action of sunlight on 7-dehydro-cholesterol. There is not an adequate supply of vitamin D in a normal British diet, but the ultraviolet content of sunlight makes good this deficiency (p. 561), always provided that skin colour and customary clothing habits allow. Dark-skinned immigrants to the United Kingdom may readily become deficient in vitamin D and consequently develop rickets. If extra vitamin D is required it can be provided in the form of cod or halibut liver oils, or by ensuring adequate exposure to ultra-violet light. Margarine is usually enriched with vitamin D and other fat-soluble vitamins. Supplementation of the diet with both vitamins and calcium should be provided during pregnancy since the mother's diet must contain enough of these substances to provide for the laying down of the baby's skeleton as well as the maintenance of her own and, after birth, the production of milk demands an adequate supply of calcium.

Having been absorbed, calcium circulates in the blood plasma, along with phosphate ions, and there is a very precise balance between the calcium levels in the blood, in the bones, and in the urine, for calcium phosphate is excreted by the kidney. The level in the blood is critical for health so that if it falls below a certain level, calcium will be withdrawn from the bones to bring the level back to normal, while if the level is too high, more calcium may be laid down in the bones or excreted in the urine.

The level of calcium in the blood is controlled by two hormones: *parathormone*, which is secreted by the parathyroid glands, and *calcitonin*, which is secreted by the clear cells of the thyroid. Parathormone helps to hold calcium in solution in the blood so that a deficiency will cause a fall in serum calcium level, while an excess of parathormone, which may be produced by a parathyroid tumour, will cause a rise, the extra calcium being withdrawn from the bones which become weakened and liable to fracture. Calcitocin stops the loss of calcium from bone so that an excess will cause a fall in the serum calcium level.

After the menopause, and in old age, calcium leaves the bones and they become fragile. It is not known exactly why this *osteoporosis* of the elderly occurs, but it causes many problems.

OTHER ROLES OF CALCIUM

Calcium is present in blood and tissue fluids at a concentration of 10 mg/100 ml, = 2.5 mM, of which about half is present as ionised calcium, Ca^{2+}. In other words, extracellular Ca^{2+} is about 1 mM. In this strength it is available for various important functions. It can be taken up by osteoblasts and used to form the hydroxy-apatite crystals of bone. It plays an indispensible role in five of the reactions leading to clot formation in shed blood. It stabilises nerve fibre membranes which would otherwise become permeable to Na^+ and spontaneously depolarise and so produce unwanted impulses.

Within cells, Ca^{2+} is present in far smaller amounts, around 0.1 mM. Even this value is probably an overestimate of free ionised calcium, since much is bound to intracellular organelles and membranes. Although the actual amounts are small, the concentration gradient across cell membranes is high and these must normally be impermeable to calcium and contain a pump mechanism to extrude any Ca^{2+} that crosses the membrane down the concentration gradient. A 100-fold rise within a muscle cell is the trigger (activation) for muscle contractions. In other cells, a comparable rise causes its attachment to a special protein, *calmodulin*. This then activates many other systems in different cells and causes:

1 Hormone and neurotransmitter release.
2 Secretion of enzymes by exocrine glands.
3 Breakdown of lysosomes, the intracellular organelles concerned with the cell's defence, immune and allergic reactions.
4 Ciliary movement.
5 Axonal transport in nerve fibres.
6 Phagocytosis in all cells, especially in leucocytes.

DEVELOPMENT OF BONE

The process of laying down bone in the fetus is known as *ossification* and is of two types. Ossification of the flat bones of the skull is said to be *in membrane* since the bone is laid down by special cells called *osteoblasts* in the membranous wall of the developing skull. Ossification of the other bones, which are of more interest to the physiotherapist, is unfortunately more complex and needs to be described in detail. The process is known as ossification *in cartilage* because the bone is preceded by a cartilaginous model which is a similar shape to the adult bone.

Ossification begins in a *primary centre* of ossification situated, in the case of long bones, somewhere near the centre of the shaft (Fig. 9.2). In most bones this develops between the 6th and 8th weeks of pregnancy. The matrix of the cartilage in this region becomes calcified (i.e. infiltrated with calcium salts); but this tissue is not true bone. Blood vessels grow in; the calcified cartilage is removed, and osteoblasts, which have grown in with the blood vessels, begin to lay down true bone. This process proceeds from the primary centre towards each end of the bone, but in the centre the newly formed bone becomes broken down to leave a cavity which will become the *medullary (marrow) cavity*. Further bone is laid down around the periphery of the shaft by osteoblasts at the surface; these cells are found in a vascular and fibrous layer which surrounds the shaft and is called the *periosteum*. Ossification in this region therefore resembles ossification in membrane.

Meanwhile one or more *secondary centres* of ossification begin to appear at each end of the cartilaginous model and these spread until the whole expanded end is ossified and is separated from the fully ossified shaft only by a plate of residual cartilage. The secondary centres of ossification all appear after birth (except for the centre at the lower end of the femur) and they form the *epiphyses*. The residual plate of cartilage is called the *epiphyseal plate*. It is extremely important because the cartilage cells divide continuously and would gradually increase the thickness of the plate were it not for the fact that, on the side of the epiphyseal plate adjacent to the shaft, the cartilage becomes calcified, breaks

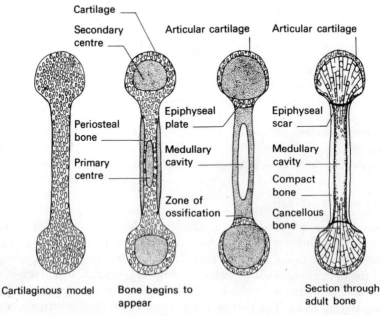

Fig. 9.2. Ossification of a long bone. Bone is represented by fine stippling (see text for description).

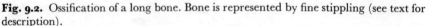

down and then becomes ossified, as was described above. Thus bone is continually being laid down at each end of the shaft and it is by this means that the bone grows in length. Growth in the diameter of the bone is brought about by the deposition of bone by the osteoblasts of the periosteum. Eventually between the ages of 18 and 25 (the exact age is different in different bones) the epiphyseal plate of cartilage becomes completely ossified so that further growth in length of the bone is impossible. This process is known as *fusion of the epiphyses* and when it is complete, the only sign of the original epiphyseal plate is a thin line of condensed bone known as the *epiphyseal scar*.

OSTEOBLASTS AND OSTEOCLASTS

As has been mentioned, the cells which lay down bone are called osteoblasts. During this process many of the osteoblasts produce bone so freely that they surround themselves with bone and wall themselves up in lacunae where they resign themselves to a lonely existence as osteocytes, coming into contact with their neighbours only by their fine cytoplasmic processes which occupy the canaliculi. The growth of a bone such as the femur, however, involves not only the laying down of bone but also its destruction. For example, as the bone grows the medullary cavity inside the bone must enlarge and this involves the destruction of the innermost layers of the bone. Similarly, growth of the shaft involves a good deal of remodelling at the ends. This is done by large multinucleate cells called *osteoclasts*.

The fusion of the epiphyses is final—once they have fused, no further growth in stature can occur and premature fusion will cause stunted growth. Similarly, damage to an epiphyseal plate as the result of disease or injury may interfere with its growth potential. This is particularly important in the lower limb since it may result in a child having one leg shorter than the other.

THE SHAPE OF BONES

Bones are usually classified into four types, according to their shape. *Long bones* are long in relation to their width; they are not necessarily long in the sense of measurement. They thus include not only the femur, humerus, tibia, etc., but also the metacarpals and phalanges. They usually have an epiphysis at each end during the period of growth but in the case of the metacarpals, metatarsals and phalanges, the epiphysis is at one end only. *Short bones* have approximately the same dimensions in all directions and thus include the carpal and tarsal bones and the patella. They consists of a core of cancellous bone enclosed in a thin shell of compact bone. *Flat bones* include the scapula, and the bones of the vault of the skull such as the parietal bones. They consist of a layer of cancellous bone sandwiched between two layers of compact bone. *Irregular bones* comprise all those which do not fit into the above classification and they include the vertebrae, hip bone, etc. Finally, mention must be made of a special type of bone known as a *sesamoid bone*. These are usually small and rounded and are found embedded in certain tendons, usually where the latter run in close proximity to another bone. Their main function is protective.

THE STRUCTURE OF LONG BONES

Since the long bones form the principal part of the skeleton, it will be necessary to describe their structure in some detail. This is shown in Figure 9.3. The shaft of the bone is formed of a thick layer of compact bone which thins out towards either end so that the mass of cancellous bone at the extremities is covered only by a thin shell of compact bone. Running through the cancellous bone one can usually see the epiphyseal scar, a condensed layer of bone which marks the position of the original epiphyseal plate of cartilage. The central part of the shaft is hollow—the cavity is called the *medullary (marrow) cavity*. In the normal adult it contains yellow marrow, which is simply fatty tissue. Red (haemopoietic) marrow may be found in the spaces between the lamellae of cancellous bone at the ends of long bones, particularly in children (see Chapter 17) but with increasing age this diminishes in amount as it is replaced by yellow marrow.

The surface of the bone is covered with a membrane known as the *periosteum* which is closely adherent to the bone and has two important functions. Firstly, it contains a plexus of fine blood vessels and from this plexus vessels enter and leave the bone to supply it with oxygen and nutrients. For this reason, stripping of periosteum from a large area of bone may cause its death. Secondly, the periosteum can lay down new bone on the surface since it can produce osteoblasts. It is therefore important for the growth in girth of the bone in children, and in the repair of fractures at all ages.

Somewhere along the shaft can usually be found a small hole—the *nutrient foramen*—through which pass the nutrient vessels on their way to and from the interior of the bone. The nutrient artery comes from one of the nearby major arteries of the limb. Other foramina are found at the ends of the bones—the arteries which enter these are derived from the arterial plexus around the joint. As well as a blood supply from periosteal and nutrient vessels, bones also have a nerve supply.

The ends of the bone are smooth, and shaped according to the type of joint in which they take part. The articular surfaces are covered, in life, by hyaline cartilage.

The surface of the bone is partly smooth and partly roughened and raised into ridges or tubercles (also known as tuberosities or trochanters). Wherever a mass of dense connective tissue is attached to the bone, its attachment is marked by a roughening of the surface. Such dense connective tissue may be represented by dense fascia, ligaments or tendons. Where muscle fibres are attached directly to the bone, with no intervening tendon, the bone surface is smooth. You can see this on the anterior (costal) surface of the scapula where the subscapularis muscle is attached. Most of its attachment is directly to the bone by muscle fibres, so that most of this surface of the scapula is smooth. This muscle, however, has a number of tendinous strands (intersections)

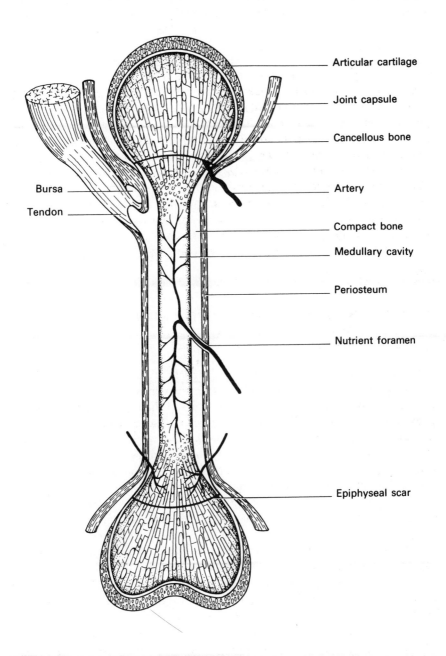

Articular cartilage

Joint capsule

Cancellous bone

Artery

Compact bone

Medullary cavity

Periosteum

Nutrient foramen

Epiphyseal scar

Bursa

Tendon

Fig. 9.3. Diagram of a typical long bone. Note the hollow tube of compact bone forming the shaft and the cancellous bone at each end.

running through it, and these raise several roughened ridges on the surface of the bone. The roughened ridges produced by tendinous or ligamentous attachments must be distinguished from the smooth ridges which represent reinforcing bars which are placed in areas of bone where extra strength is needed. A good example of such a bar is found along the lateral border of the scapula (see below).

THE INDIVIDUAL BONES

In the accounts of the individual bones which follow, the diagrams are meant only as an easy means of indicating the structures mentioned in the text, but ideally this chapter should be read with the bones themselves in your hand so that they can be articulated with each other, their shapes and surface texture felt and appreciated, and their positions in your own body recognised with particular reference to the bony promontories which can be palpated. Remember, however, that when articulating two bones to study the joint between them, the fit will be imperfect owing to the absence of articular cartilage (see Chapter 10).

THE BONES OF THE SKULL AND TRUNK

THE SKULL

The skull consists of a number of very complicated bones but you only need a knowledge of a few of them. Essentially, the skull consists of the cranium, which encloses and protects the brain, and the bones of the face which lie below and in front of the cranium (Fig. 9.4). The former bones are tough and rigid and are bound together very tightly, not only by fibrous tissue but, in the case of many of the bones, by their complex interlocking with one another. The bones of the face, however, are mostly thin and fragile, except for the mandible and parts of the upper jaw which obviously need to be powerful.

Look first at the dome or *vault* of the skull. It consists of six bones—a single *frontal*, a pair of *parietals*, an *occipital* and, at the sides, the flattened (squamous) parts of the *temporal* bone. These names are worth remembering because the lobes of the brain which lie deep to them are named similarly: the frontal lobes, parietal lobes, etc. The bones interlock by a series of complicated *sutures*, although in the baby's skull they are separated by membrane. At birth there are two frontal bones with a midline gap between them so that near the centre of the top of the skull there is a diamond shaped area between frontals and parietals which is free of bone (Fig. 9.5). This is called the *anterior fontanelle* and it can be felt as a soft pulsating area in a young baby's head. The temporal bone consists of a number of parts, the most important of which are the squamous part, which forms part of the side wall of the cranium, and the petrous part which lies inside the skull and houses the mechanism of the ear. A blunt

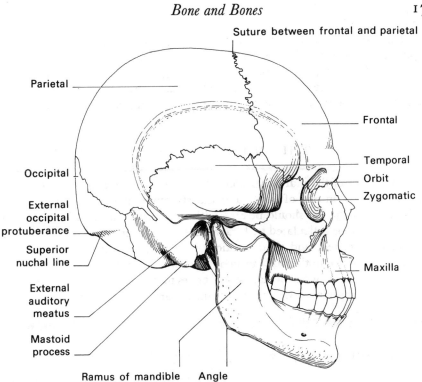

Suture between frontal and parietal

Parietal

Frontal

Occipital

Temporal

Orbit

Zygomatic

External
occipital
protuberance

Superior
nuchal line

External
auditory
meatus

Maxilla

Mastoid
process

Ramus of mandible   Angle

Fig. 9.4. Side view of the skull showing some of the palpable features.

projection downwards from the back of the temporal bone is called the *mastoid process* and serves for the attachment of several muscles, principally the sterno-mastoid. Just in front of the *external auditory meatus* (the external opening of the ear) a narrow process passes forwards to join a backwardly projecting process from the *zygomatic bone* to form the zygomatic arch. The occipital bone has a prominent projection in the midline called the *external occipital protuberance* from which a roughened line passes laterally on each side. This is the *superior nuchal line* and important muscles are attached to and below the line. The most noticeable feature of the occipital bone is the 'big hole'—called more elegantly in Latin the *foramen magnum* (Fig. 9.6). Through this the lowermost part of the brain (the medulla oblongata) is continuous with the spinal cord.

The base of the skull is thick and tough and is pierced by numerous holes which transmit various cranial nerves and important blood vessels. This part of the skull is, of course, inaccessible and cannot be palpated in life. The upper jaw is formed by the two *maxillae* (sing., maxilla) while the bone of the lower jaw is called the *mandible*. The rounded *head* of the mandible fits into a socket on the under surface of the squamous part of the temporal bone so that the joint is called the temporomandibular joint. The mandible itself consists of the *body*, which bears the teeth, and a vertical part the *ramus*, which ends at the *head*. The two are continuous at the *angle of the mandible* which is easily palpable.

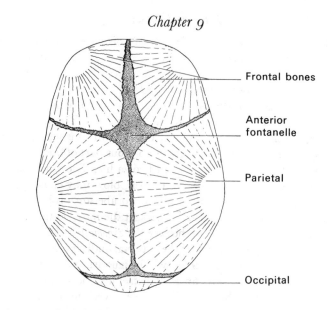

Fig. 9.5. The top of the skull of a newborn baby. The fontanelles represent areas where ossification has still not occurred.

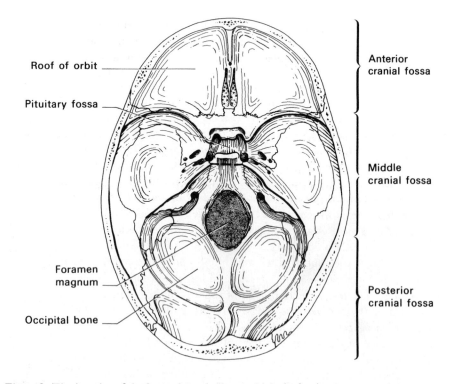

Fig. 9.6. The interior of the base of the skull, on which the brain rests.

The inside of the base of the skull (Fig. 9.6) is divided into three hollowed out spaces: the *anterior, middle* and *posterior cranial fossae*. The anterior cranial fossa lies above the eye sockets (*orbits*) and is occupied by part of the frontal lobe of the brain. The middle fossa is at a lower level and houses the temporal lobes. In the midline of this part of the skull is a saddle-shaped depression called the *pituitary fossa* in which the important pituitary gland lies. The posterior cranial fossa is at a lower level still and contains the cerebellum. It is separated from the middle fossa by the sharp upper border of the petrous part of the temporal bone.

Many features of the skull can be identified in life. The whole vault of the skull lies directly under the scalp so that a severe blow may cause a fracture in this region. Posteriorly, the external occipital protuberance is best palpated by running the fingers up to it from below—it is not the most posterior part of the skull; this lies above it. On the side of the face the *zygomatic arch* may be palpated along its length, starting just in front of the external auditory meatus. It is not as clear as might be thought because very powerful muscles and fascia are attached to it, flush with the surface. At its anterior extremity the margins of the orbit may be palpated just under the skin. The whole lower border of the mandible is clearly palpable, as is the angle. A finger's breadth behind the angle is the mastoid process.

THE VERTEBRAE

The vertebrae are a series of bones which collectively form the *vertebral column*. They are bound together by extremely powerful ligaments but although the movement between any two vertebrae is slight, the spine as a whole is a very mobile structure. There are 7 *cervical vertebrae* in the neck of almost all mammals (including the giraffe and the whale). There are 12 *thoracic vertebrae* (each of which carries a rib), 5 *lumbar vertebrae* in the lower part of the back, 5 *sacral vertebrae*, which are fused together to form the sacrum, and 1–4 *coccygeal vertebrae* which are fused together to form the *coccyx*. Although the structure of the vertebrae varies in the different regions of the spine, they all adhere to a basic plan so that it will be most convenient to describe first a vertebra from the region which most nearly conforms to the basic structure, namely the thoracic region.

THORACIC VERTEBRAE

Each vertebra consists essentially of a *body* and a *neural arch*. The body is the weight-bearing portion and the neural arch surrounds and protects the spinal cord. In addition, a number of processes are present which are for the attachment of muscles, for limitation of movement and for articulation with the ribs. A typical thoracic vertebra is shown in Figure 9.7. The body is rather heart-shaped (a playing-card heart rather than a human heart) and consists of a mass of cancellous bone surrounded by a thin shell of compact bone. The lower in the thoracic region a vertebra is situated, the larger the body, which is not surprising since the lower a vertebra is placed the more weight it has to bear. The neural arch surrounds a more or less circular *vertebral canal*. Project-

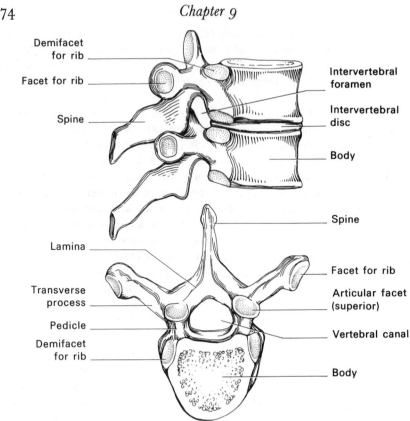

Fig. 9.7. Thoracic vertebrae. Notice how two demifacets become one complete facet when the vertebrae are articulated.

ing backwards and downwards from it is a *spine* or *spinous process* which is long compared to those of other vertebrae. Its downward inclination is enough to produce considerable overlapping of adjacent spines. Laterally are a pair of transverse processes which, again, are relatively long in the thoracic region. This is because one of their functions is to help to support the ribs, and on the anterior surface of each process (except for the 11th and 12th) is a roughly circular facet for articulation with the tubercle of the corresponding rib. The portion of neural arch between the spine and the transverse process is called the *lamina* and it is obvious that to expose the spinal cord at operation, this portion of the neural arch must be removed, an operation known as laminectomy. The other named part of the neural arch is the *pedicle* which lies between the body and the transverse process. The upper and lower borders of the pedicles are notched, particularly the lower borders; and if you articulate two adjacent vertebrae you will see that the two adjacent notches are converted into an *intervertebral foramen* which, in life, transmits the spinal nerve of that side. On the upper and lower edge of the body in the vicinity of the pedicle is a semicircular facet or rather *demifacet* for articulation with part of the head of a rib. The head of the rib articulates with the bodies of two adjacent vertebrae, its own and the one above, so on two adjacent vertebrae the two demifacets are converted into one complete facet, although this is not entirely complete since part of the head articulates also with the intervertebral disc which lies between each pair of bodies.

Finally, the upper and lower aspects of the neural arch carry a pair of *articular facets,* each of which articulates with the corresponding pair on the adjacent vertebra. These facets are, in life, coated with hyaline cartilage and form a plane synovial joint between the vertebrae. The joints do not carry any appreciable load but are responsible for controlling the type of movement between adjacent vertebrae. In the thoracic region the upper facets face backwards and slightly outwards and the lower facets forwards and slightly inwards. The line of the joint cavity thus lies in the plane of a circle whose centre is near the centre of the body. They will thus allow rotation of one vertebra on another, also flexion and extension and side flexion to some extent.

The first thoracic vertebra is in shape a little like the cervical vertebrae, although without the main distinguishing features of the latter, and it has whole facets rather than demifacets on its body since the first rib articulates only with the first thoracic vertebra. The lower thoracic vertebrae are a little like the lumbar vertebrae in shape and the 11th and 12th vertebrae (and possibly the 10th also) have whole facets rather than demifacets on their bodies. There are, however, no facets on the transverse processes of these vertebrae. An interesting point is that the 12th (last) thoracic vertebra has upper articular facets which are thoracic in type but inferior facets which resemble those of the lumbar vertebrae.

THE CERVICAL VERTEBRAE

A typical cervical vertebra (Fig. 9.8) has the principal features of a thoracic vertebra but there are certain modifications. The body is smaller, more oval and has overlapping lips anteriorly. The vertebral canal is larger and more triangular than in most of the thoracic region because the spinal cord is larger here owing to the presence of the cervical enlargement. The spine is smaller, not overlapping, and is bifid. The transverse process is perforated by a vertical foramen called the *foramen transversarium* which, in all vertebrae but the 7th, transmits an important vessel, the *vertebral artery*—important because it supplies the hindbrain with blood. The transverse process also has anterior and posterior *tubercles.* In fact, the posterior tubercle is the true transverse process and the anterior tubercle represents a rib or *costal element* which, like those in the thoracic region, articulates with both body and transverse process (Fig. 9.9). Between anterior and posterior tubercles is a deep groove which is occupied by the corresponding spinal nerve. Occasionally, the costal element of the 7th cervical vertebra is enlarged and may be connected to the sternum by a ligament or may even articulate directly with the sternum like a true rib. Such an enlarged costal element is called a *cervical rib* and is important clinically because it may

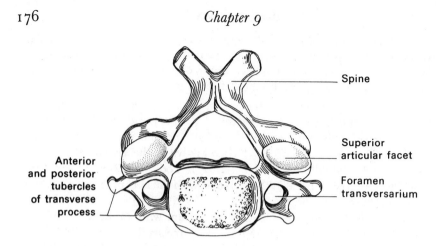

Spine

Superior
articular facet

Foramen
transversarium

Anterior
and posterior
tubercles
of transverse
process

Fig. 9.8. A cervical vertebra. Note the bifid spine and the foramen transversarium.

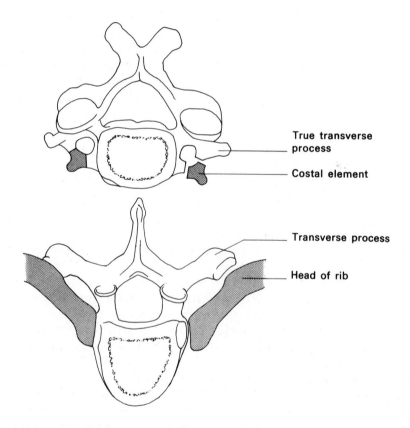

True transverse
process

Costal element

Transverse process

Head of rib

Fig. 9.9. To show how the foramen transversarium corresponds to the space between the neck of a rib and the vertebra.

produce symptoms by pressure upon the subclavian artery or the brachial plexus which are closely related to the cervical rib on their way to the upper limb. The upper articular facets are directed upwards and backwards and the lower facets downwards and forwards.

The first and second cervical vertebrae are quite different but the difference is easily understood if it is remembered that the embryonic body (the *centrum*) of the first vertebra is fused with that of the second and is detached from its own vertebra. The second vertebra, which is known as the *axis* therefore has a peg-like process or *dens* projecting upwards from its body while the first vertebra, the *atlas*, consists essentially of a ring of bone with various processes attached (Figs. 9.10 and 9.11).

The atlas is so called after the mythological giant who was believed to hold the world on his shoulders, and the atlas supports the skull in a similar manner. To do this, it is provided with a pair of kidney-shaped facets which articulate with a pair of reciprocally curved *occipital condyles* on either side of the foramen magnum. The transverse processes are very long so as to provide leverage for the muscles which turn the atlas and the skull together on the dens of the axis. A foramen transversarium is present at the root of the transverse process and near this there is a deep groove on the upper surface of the neural arch. This is occupied by the vertebral artery as it leaves the foramen transversarium and turns medially to enter the foramen magnum. The spine, on the other hand, is almost non-existent, being represented only by a small *posterior tubercle* on the back of the neural arch. A long spine would hinder extension of the head. The *anterior arch* of the atlas completes the neural arch in front, instead of a body. On the back of the neural arch is a facet which forms a synovial joint with a similar facet on the front of the dens. A pair of circular articular facets on the under surface of the atlas articulate with the upper facets of the axis. Note that the upper facets are kidney shaped, the lower are circular—this will enable you to hold the bone the right way up.

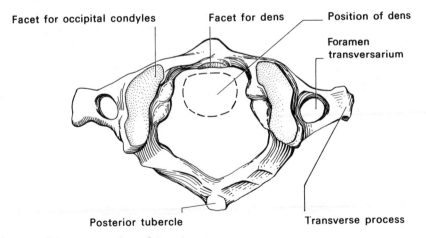

Facet for occipital condyles   Facet for dens   Position of dens

Foramen transversarium

Posterior tubercle   Transverse process

Fig. 9.10. The upper surface of the atlas.

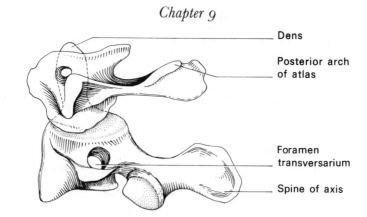

Dens

Posterior arch
of atlas

Foramen
transversarium

Spine of axis

Fig. 9.11. The atlas and axis, side view.

The axis is very different. Its transverse processes are much shorter than those of the atlas but its spine is large and bifid. It has a foramen transversarium on each side and upper and lower articular processes. The dens projects upwards and has a facet on its anterior surface. It has a slightly constricted neck—its old name, in Latin, meant the 'tortoise bone' and if you look at the dens end-on you will see why.

THE LUMBAR VERTEBRAE

There are five lumbar vertebrae and since they are in the lower part of the spine they are large and sturdy in appearance. The bodies, in particular, are massive in the lower lumbar region. They are wider than the thoracic vertebrae and their greatest diameter lies transversely (Fig. 9.12). The spines are large and quadrangular and do not overlap one another like the thoracic spines. It is therefore possible to introduce a needle, between the spines, into the vertebral canal in order to obtain a sample of cerebrospinal fluid. This minor operation is known as *lumbar puncture* and may be done without fear of damage to the spinal cord since this ends at the level of the lower border of the first lumbar vertebra.

The transverse processes are large, for the attachment of powerful back muscles, but they do not carry facets for ribs like those of the thoracic vertebrae. The superior articular facets face medially and the inferior, laterally and if you articulate two lumbar vertebrae you will see why only slight rotation can occur in this region. In fact, the vertebrae do not articulate very accurately in the dried state because in life the bodies are separated by intervertebral discs which are extremely thick in the lumbar region.

THE SACRUM

There are five sacral vertebrae but they are fused together to form a single mass of bone, the *sacrum*. As has been mentioned already, the size of the bodies increases towards the lower end of the spine since they have to carry an increased

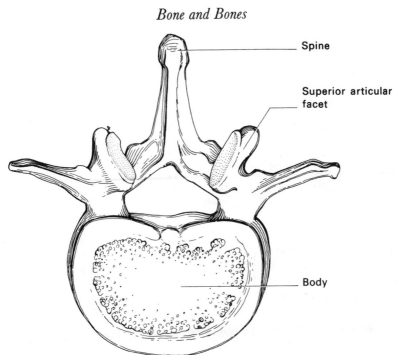

Spine

Superior articular facet

Body

Fig. 9.12. A lumbar vertebra.

load, but below the sacro-iliac joint the sacrum is no longer weight bearing and therefore the sacral vertebrae diminish in size from above downwards. Since the bone is formed by the fusion of five vertebrae, it will have four intervertebral foramina on each side but the spinal nerves which arise from the sacrum have already divided into anterior and posterior primary rami (see p. 90) before leaving the bone and therefore on each side there are four *anterior (or pelvic) sacral foramina* and four *posterior sacral* foramina (Figs. 9.13 and 9.14). The upper border of the first sacral vertebra projects forwards and is called the *promontory*. The anterior surface is markedly concave and is smooth. The posterior surface, on the other hand, is extremely rough and if you examine it while the sacrum is articulated with a fifth lumbar vertebrae, you will be able to identify elevations corresponding to the spines, articular processes and transverse processes of a normal vertebra. They are therefore known as *spinous, articular* and *transverse tubercles*. On the lateral side of the upper part of the sacrum, corresponding to the first $2\frac{1}{2}$ or 3 sacral vertebrae, is an *auricular surface* which articulates with the similarly named surface on the ilium to form the sacro-iliac joint. The apex of the sacrum articulates with a triangular piece of bone called the *coccyx* which is formed by the fusion of 3–5 small coccygeal vertebrae.

THE SPINE AS A WHOLE

In an articulated skeleton, of the type which is so often seen in lecture theatres,

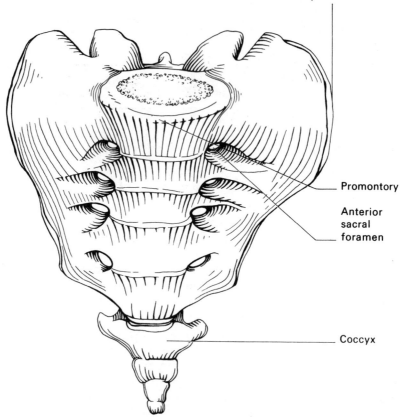

Position of auricular surface on lateral aspect

Promontory

Anterior
sacral
foramen

Coccyx

Fig. 9.13. The sacrum and coccyx, anterior view.

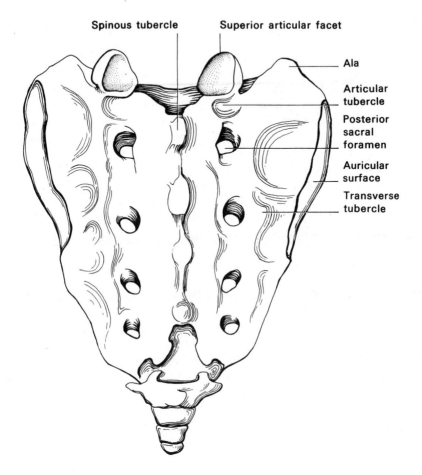

Fig. 9.14. The sacrum and coccyx, posterior view.

the vertebrae are separated by thick felt discs which are meant to represent intervertebral discs. You will see that the discs make up a considerable proportion of the total length of the spine so that if the vertebrae are simply strung together, the spine as a whole seems too short and moreover, it does not show the complex series of curves which can be seen in life. These curves are not present in the baby but develop during the first year of life as will be described in Chapter 10.

The range and direction of possible movements varies in different parts of the spine and depends on the thickness of the intervertebral discs and upon the direction of the articular facets. This will be discussed in detail in the next chapter, but it should be mentioned here that movement is most free in the cervical and lumbar regions. It is in these two regions that the vertebral canal is largest so as to allow free movement and also in order to accommodate the cervical and lumbar enlargements of the spinal cord and the cauda equina. The spinal cord ends at the lower border of L1 in the adult (L3 in the baby) so below this level the vertebral canal contains only the cauda equina. The most dangerous aspect of fractures of the spine is that they may damage the spinal cord. The usual type of fracture which can cause such damage is a crush fracture of the body, which you will remember is the weight bearing part of the vertebra. Small fractures of the spines and transverse processes are usually the result of violent muscular action and are painful but relatively harmless.

The only parts of the vertebrae which can clearly be felt and seen during life are the spines, which may be made more visible by flexion of the trunk. The prominent elevation at the root of the neck posteriorly is the spine of the 7th cervical vertebra and often, below it, the spine of the first thoracic is equally prominent. The spines of the other cervical vertebrae are not easily palpable because of the backward concavity of the cervical spine. The scapula lies opposite the spines of the 2nd to 7th thoracic vertebrae. The highest part of the iliac crest lies at the level of the spine of the 4th lumbar vertebra while a line joining the two posterior superior iliac spines passes through the spinous tubercle of the second sacral vertebra, the level at which the subarachnoid space ends.

THE RIBS AND COSTAL CARTILAGES

There are 12 pairs of ribs, 7 of which articulate anteriorly with the sternum by means of their *costal cartilages*, which are cartilaginous forward extensions of the ribs. These upper 7 ribs are known as *true* ribs. The remaining 5 are called *false ribs*—the 8th, 9th and 10th articulate by means of their costal cartilages with the cartilages of the rib above, the 11th and 12th ending freely without articulating anteriorly with anything. These last two ribs are called the *floating ribs*. Like the vertebrae, the ribs are best understood by examining a typical example and then studying the modifications in different regions.

The sixth rib (Fig. 9.15) may be regarded as typical. It consists of a head, neck, tubercle, and shaft. The head has two *demifacets* which articulate with the bodies of two adjacent vertebrae (its own, and the one above) and with the intervertebral disc. The *neck* lies between the head and the tubercle. The latter has an *articular part* which bears a facet for the transverse process of the corresponding vertebra and a *non-articular part* which is roughened for the attachment of ligaments. The shaft is long and gradually curved throughout its length but posteriorly it has a well-marked angle. Along the lower border of the shaft is a shallow *subcostal groove*.

The first rib (Fig. 9.15) is short and flattened, its anterior end being considerably lower than its head. It is important because various structures leave the thorax to enter the upper limb by passing across its flat upper surface. The second rib is also short but from the third rib downwards the features of a typical rib become more and more obvious. The eleventh and twelfth ribs are rather featureless strips of bone, especially the twelfth. The articulations of the ribs

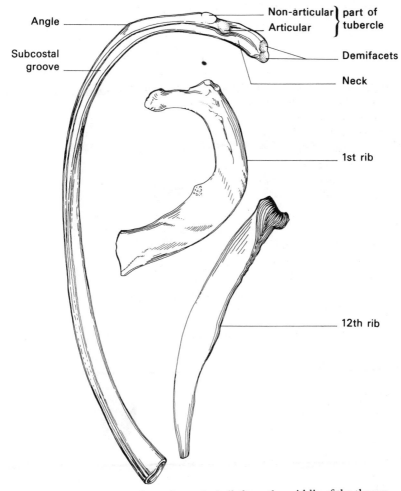

Fig. 9.15. The first and last ribs and a typical rib from the middle of the thorax.

with the vertebrae have already been described, and their mode of articulation with the sternum will be described in the next section.

THE STERNUM

This is a flattened bone consisting of an upper portion, the *manubrium sterni*, a body formed by the fusion of four segments or *sternebrae* and a lower pointed portion called the *xiphoid process* or *xiphisternum* which is cartilaginous in youth but later becomes ossified (Fig. 9.16). It is often deformed and may produce a prominent elevation on the surface. The sternum consists of an outer thin shell

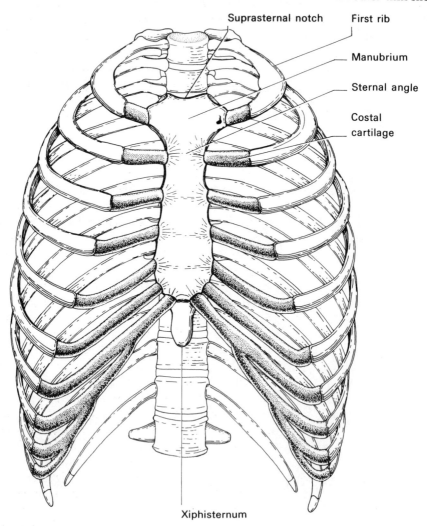

Suprasternal notch First rib

Manubrium

Sternal angle

Costal cartilage

Xiphisternum

Fig. 9.16. The skeleton of the thorax. Note that the sternal angle marks the level of the second costal cartilage. You may like to use this diagram to practise marking out the lungs and pleura (Fig. 19.5).

of compact bone and an inner mass of cancellous bone, the interstices of which contain red bone marrow. The sternum is often used to obtain a sample of red marrow in order to study certain blood diseases and since, in the midline, the sternum is subcutaneous, it is easy to puncture it with a needle to obtain such a sample.

The upper border of the sternum forms the *suprasternal notch* which is easily palpated at the base of the neck. Lateral to this is a deep indentation on each side which accommodates the rounded medial end of the clavicle to form the sternoclavicular joint. Just below this is a facet for the first costal cartilage. The manubrium is attached to the body of the sternum by a secondary cartilaginous joint (see Chapter 11) although in later life the two may become attached by bony union. There is a slight angle between the manubrium and body which is called the *sternal angle* or *angle of Louis*. This is easily palpable in life and is an important landmark. Laterally, at this level, there is a facet for the second costal cartilage. The 3rd–7th costal cartilages articulate with notches on the lateral border of the sternum, the 7th facet being in the angle between the body and the xiphoid process.

THE RIB CAGE

The skeleton of the thorax as a whole is made up of the thoracic vertebrae, the twelve pairs of ribs and the sternum (Fig. 9.16). The upper opening, bounded by the first thoracic vertebra, the first ribs and the manubrium is rather narrow for all the important structures which have to pass through it such as the trachea and oesophagus and the major vessels passing to and from the upper limb and the head and neck. This opening is usually called the *inlet* of the thorax, but sometimes, unfortunately, the outlet! The thorax is flattened anteroposteriorly but in infants it is almost circular in cross-section and in patients with certain long-standing lung diseases it may also tend to be barrel-shaped. Note particularly that, in general, the anterior ends of the ribs are lower than the posterior ends, while the lowest point on any rib is somewhere along the shaft. These facts are important when the mechanism of respiration is studied. The costal cartilages are elastic and can twist during respiration, but with increasing age they often become calcified and brittle. It is essential to be able to identify individual ribs because they are important landmarks used in mapping out the surface markings of the heart and lungs. The first rib is difficult to identify because it is tucked up under the clavicle, but the second rib and costal cartilages can be felt easily by moving the fingers laterally from the angle of Louis. From this point, therefore, the other ribs can be identified. The last costal cartilage to articulate directly with the sternum is the seventh but the tips of the 8th, 9th and 10th costal cartilages can be felt rather indistinctly along the *costal margin* which is the lower border of the rib cage. The angle between the right and left costal margins is known as the *subcostal angle*. It varies greatly in different individuals, being wide in short stocky people but narrow in long, thin individuals.

THE BONES OF THE UPPER LIMB

THE CLAVICLE

The clavicle or collar bone (Figs. 9.17 and 9.19) is a strut which helps to prop out the tip of the shoulder so that when it is fractured, the patient typically supports his arm with the uninjured arm and when, in a rare congenital condition, it is absent, the two shoulders can almost be folded together in front of the chest. Its shape can best be appreciated in the living body since it is placed just under the skin (i.e. it is *subcutaneous*) along its whole length, and a poorly aligned fracture produces an unsightly bulge. The medial end of the clavicle is expanded into an almost circular convex end which articulates with the sternum at the sternoclavicular joint, an intra-articular disc intervening. The under surface of the medial end is roughened for the attachment of the costo-clavicular ligament. The lateral end is flattened and ends in a small facet which articulates with the *medial* aspect of the acromion process of the scapula so that the line of the acromioclavicular joint is approximately in the sagittal plane (Fig. 9.19). This position of the joint is difficult to visualise and you should practise articulating the two bones and then palpating the line of the joint.

The lateral end of the clavicle is obviously much less strong than the medial end. This is because the acromioclavicular joint is relatively unimportant, the main bond between the bones being the *coracoclavicular* ligament which consists of two parts attached to the *conoid tubercle* and *trapezoid line* on the under surface of the bone (Fig. 9.17). The bone lateral to the conoid tubercle is therefore not in the direct line of weight transmission from the upper limb to the clavicle. If you remember, then, that the medial end is the larger and stronger, that the under surface is marked by the conoid tubercle and that, as you have found by palpa-tion, the medial half of the clavicle is convex forwards and the lateral half

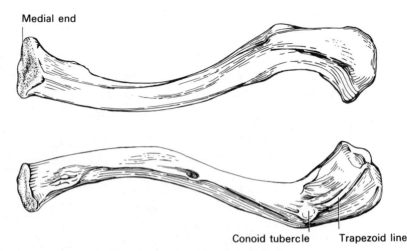

Medial end

Conoid tubercle   Trapezoid line

Fig. 9.17. The upper and lower surfaces of the left clavicle.

concave forwards, you should now be able to assign a bone to the correct side.

Four important muscles are attached to the upper and lower aspects of the medial and lateral aspects of the clavicle but their precise attachments are best appreciated by examining the muscles themselves (see Chapter 11). The muscles are the sternomastoid and trapezius above and the deltoid and pectoralis major below.

THE SCAPULA

The scapula is commonly known as the shoulder-blade and is much less easily palpable than the clavicle because of the many muscles which surround it and hold it in place. You should appreciate that the scapula is an extremely mobile bone and, with its attached muscles, more or less 'floats' over the chest wall on a layer of loose connective tissue. There is, in fact, only one rather small joint between the whole upper limb and the trunk: this is the *sternoclavicular* joint. The humerus, thus, articulates with the scapula at the shoulder joint; the scapula passes on the weight to the clavicle via the coracoclavicular ligament and to a small extent via the acromioclavicular joint and the clavicle passes it on to the trunk via the sternoclavicular joint. The main attachment of the upper limb to the trunk is therefore by means of the muscles attached to the clavicle and scapula which make up a system very like the independent suspension mechanism on a car. This is very different from the state of affairs in the lower limb, where the head of the femur fits firmly into its socket on the massive and rigid pelvis, which is attached almost immovably to the spine. You will come across this distinction between upper and lower limbs again when you study the joints and when you study the muscles. In the upper limb, mobility is achieved at the expense of strength and stability whereas the lower limb is built for support of the body weight and for maintaining the erect posture which involves a tricky balancing act, particularly when the whole body weight is taken on one leg.

The scapula consists essentially of a very thin, translucent, triangular plate of bone with two major appendages: the *coracoid process* and the *spine* which terminates in the *acromion process* (Figs. 9.18 and 9.19). The upper, outer angle of the triangle bears a shallow facet, the *glenoid fossa*, for articulation with the head of the humerus. Look first at the scapula from behind (Fig. 9.18). It is subdivided into two areas by the prominent spine. These are called, reasonably enough, the *supraspinous fossa* and the *infraspinous fossa*. They communicate between the lateral border of the spine and the glenoid fossa by means of the *spinoglenoid notch*. When the spine is followed laterally, it bends sharply forwards to form the *acromion process* which has a small oval facet on its medial surface to articulate with the flattened end of the clavicle (Fig. 9.19). The lateral border of the acromion process is usually rough and ridged which will indicate to you that the muscle which is attached here (the deltoid) has a partly tendinous origin. The upper border of the scapula is very thin and dips down laterally to

Chapter 9

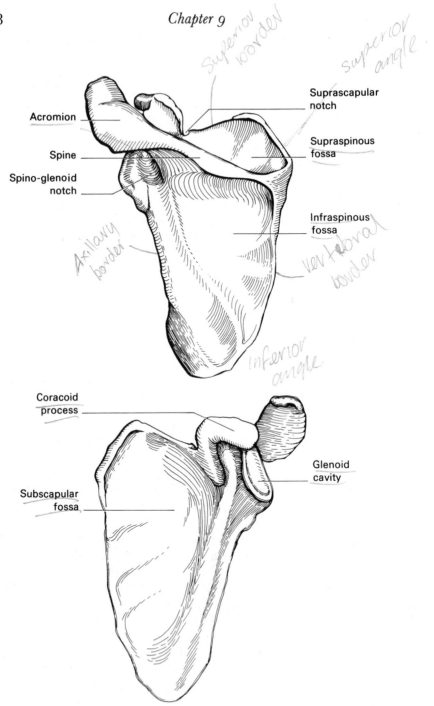

Acromion

Spine

Spino-glenoid
notch

Superior border

Axillary border

Suprascapular
notch

superior angle

Supraspinous
fossa

Infraspinous
fossa

vertebral border

inferior angle

Coracoid
process

Glenoid
cavity

Subscapular
fossa

Fig. 9.18. The back and the front of the left scapula.

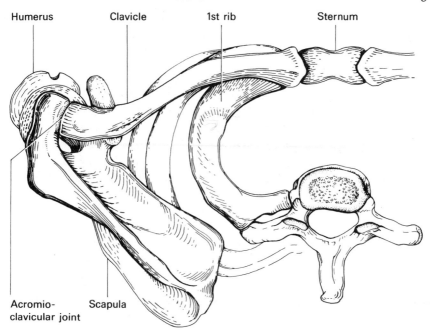

Humerus   Clavicle   1st rib   Sternum

Acromio-   Scapula
clavicular joint

Fig. 9.19. A view of the upper limb skeleton from above. Note the narrow triangular space between clavicle, scapula and first rib through which nerves and vessels have to pass into the axilla.

form a suprascapular notch which leads into the supraspinous fossa from in front.

Now examine the anterior or *costal* aspect. The whole scapula is concave from this point of view, the concavity being called the *subscapular fossa*. The fossa is occupied by a triangular muscle called *subscapularis* which therefore, along with another muscle, the *serratus anterior*, separates the scapula from the chest wall. The subscapularis takes origin partly by muscle fibres and partly by tendons so the subscapular fossa is mostly smooth but also has a number of roughened ridges. If you hold the scapula up to the light you will see clearly that the lateral edge is thickened to form a strengthening bar. This is because the serratus anterior muscle, mentioned above, exerts much of its pull on the inferior angle when it is used to rotate the scapula, so that reinforcement is necessary between the inferior angle and the main load-bearing area which is the glenoid fossa. Various roughened areas along the lateral border indicate the attachment of other muscles, while other rough protuberances above and below the glenoid fossa (the *supraglenoid* and *infraglenoid* tubercles) indicate the origins of the long heads of the biceps and triceps respectively.

At the upper outer angle of the subscapular fossa, and just lateral to the suprascapular notch, is the *coracoid process*—a smooth, anteriorly directed beak-like process to which are attached a number of muscles and ligaments including the main bond between clavicle and scapula—the *coracoclavicular ligaments*. The base of the coracoid process is closely related to the glenoid fossa.

At rest, with the arm by the side, the scapula extends from the second to the seventh ribs on the posterolateral (not posterior) aspect of the thorax. The glenoid cavity is therefore not directed laterally but laterally and forwards and the coracoid process points almost directly forwards. The direction of the glenoid fossa is important in considering movements of the shoulder joint (Chapter 10). Certain parts of the scapula are palpable and provide important landmarks. The inferior angle can easily be felt and it is instructive to compare its position when the arm is by the side and when the arm is lifted above the head. The whole medial border can also be palpated and seen and provides a useful indicator of the position of the oblique fissure of the lung when the arm is elevated. The other borders of the scapula are less easily felt owing to the muscles which cover them. The spine can be palpated along its length but not as easily as might be imagined because of muscle attachments and because the supraspinatus and infraspinatus muscles are covered by dense fascia which is attached to the spine. The acromion, however, is easily palpable. Remember that it is not the most lateral bony point in the shoulder region since the greater tuberosity of the humerus projects well beyond it (but not when the shoulder joint is dislocated). The tip of the coracoid process can be felt just below the lateral third of the clavicle, under cover of the medial edge of the deltoid muscle (Fig. 11.18).

Although the scapula is so thin, only a small part of it carries any great load and its position up against the chest wall protects it from trauma so that is is rarely fractured.

THE HUMERUS

The humerus is a typical long bone consisting of a sturdy shaft with expanded upper and lower ends (Fig. 9.20). The most prominent feature of the upper end of the humerus is the rounded head but you should notice the disparity between the extent of the smooth articular surface of the head and that of the glenoid fossa. There is thus a wide range of movement at the shoulder joint but it is rather unstable. The portion of the humerus that joins the head to the rest of the bone is naturally called the neck (anatomically speaking) but this is not particularly constricted as one would expect a neck to be so that it is a feature of academic interest only and is known as the *anatomical neck*. Rather more important is the *surgical neck*, which is just below the expanded upper end of the bone and a brief inspection will show that this is the area in which fractures of the upper end of the humerus are likely to occur, rather than at the true or anatomical neck. A 'fractured neck of humerus' therefore always refers to the surgical neck. The whole expanded upper end of the bone (above the surgical neck) consists not only of the head but also of two enlargements or tuberosities. The smaller of these is called the *lesser tuberosity (tubercle)* and is on the front of the bone medial to a shallow groove called the *intertubercular sulcus*, or sometimes, the *bicipital groove* since it is through this groove that the tendon of the long head of biceps travels on its way to the supraglenoid tubercle. The whole mass of bone lateral to the intertubercular sulcus is the *greater tuberosity* and an inspection of this part of the bone will show facets on the upper and posterior parts which are for the insertions of several muscles. If you place a scapula and a humerus together, you will see that the subscapularis muscle, which, as you know, arises from the subscapular fossa, can easily pass across the front of the shoulder joint to be inserted into the lesser tuberosity. Similarly, the muscle arising from the supraspinous fossa

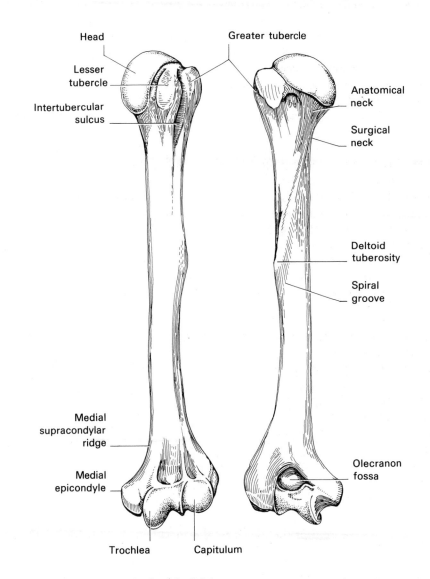

Head

Greater tubercle

Lesser
tubercle

Intertubercular
sulcus

Anatomical
neck

Surgical
neck

Deltoid
tuberosity

Spiral
groove

Medial
supracondylar
ridge

Medial
epicondyle

Olecranon
fossa

Trochlea   Capitulum

Fig. 9.20. The front and back of the left humerus.

(the *supraspinatus*) can pass across the top of the joint to reach the uppermost facet on the greater tuberosity, while the *infraspinatus*, which arises from the infraspinous fossa, will be inserted on the back of the greater tuberosity and another muscle, the *teres minor*, is inserted just below the infraspinatus. If you can picture these four short muscles, surrounding the upper end of the humerus, you will understand how they can act as controllable ligaments, contributing greatly to the stability of the shoulder joint. The four muscles are known collectively as the '*rotator cuff*'.

The lower part of the intertubercular sulcus is bounded by roughened lips to which the tendons of three powerful muscles are attached (Pectoralis major to the lateral lip, latissimus dorsi to the floor of the sulcus and teres major to the medial lip). On the lateral side of the upper part of the shaft is a very rugged V-shaped elevation called the *deltoid tubercle* to which is attached the tendon of the deltoid muscle. Posteriorly, the shaft is marked by a shallow and inconspicuous *spiral groove* (see Fig. 9.20) which has important relations to the radial nerve and the triceps muscle.

Look now, at the lower end of the bone. This is rather flattened anteroposteriorly and at the sides are two bony prominences: the *medial and lateral epicondyles*. The medial epicondyle is the more prominent of the two. Between the two are two facets for the radius and ulna. The facet for the radius is the more lateral of the two. It is called the *capitulum* and is more or less circular, and convex. Since the elbow cannot be extended much beyond 180°, the capitulum does not extend on to the posterior surface. The ulnar facet is called the *trochlea*, since it resembles a pulley. When you articulate an ulna with the trochlea you will see why the latter, unlike the capitulum, extends on to the posterior surface. At the same time, note that the medial lip of the trochlea extends farther downwards than the lateral lip. This throws the ulna laterally so that it lies at an angle to the humerus. This is known as the *carrying angle* because it is said to enable the forearm to be deflected laterally so that it clears the hips when carrying, for example, a bucket. It is also said to be more marked in the female than in the male because the former has wider hips but you should take this argument with a pinch of salt! The uppermost part of the ulna is called the *olecranon* and when the elbow is fully extended it fits into the *olecranon fossa* on the back of the humerus. Similarly, when the elbow is fully flexed the upper end (head) of the radius fits into the *radial fossa* and the beak-shaped coronoid process of the ulna fits into the *coronoid fossa*. Finally, above the medial and lateral epicondyles are a pair of sharp ridges, the medial and lateral *supracondylar ridges*. To these are attached intermuscular septa of dense connective tissue which separate the muscles of the front of the arm from those of the back. They also serve for the attachment of several muscles.

The greater part of the shaft of the humerus is surrounded by muscles, but parts of the upper and lower ends can be felt in life. The lesser tuberosity can be felt on the front of the humerus (through the deltoid muscle) and can be recognised when the humerus is rotated. The greater tuberosity is the most lateral bony structure of the shoulder region and can also be palpated through the deltoid muscle. The deltoid tuberosity may be palpable half-way down the lateral side of the shaft. At the lower end the medial and lateral epicondyles and the lower parts of the supracondylar ridges are easily felt. The epicondyles and the prominent olecranon process of the ulna form an equilateral triangle when the elbow is flexed.

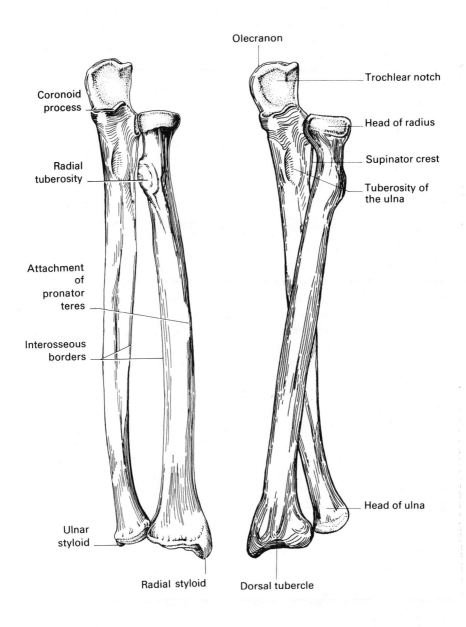

Fig. 9.21. The left radius and ulna seen from the front during (a) supination, and (b) pronation.

THE RADIUS

The upper end of the radius is almost perfectly circular when viewed from above. This is the *head of the radius* and it articulates with a suitably shaped notch on the lateral side of the ulna (Fig. 9.21). Below, on the medial aspect of the shaft, is a rounded elevation called the *radial tuberosity*. If you look closely at this you will see that its posterior part is rather rough but its anterior part is smooth. This is because the tough tendon of the biceps muscle is attached to the posterior part and since it has to pass over the anterior part to do so, a bursa is interposed between the tendon and the smooth anterior part of the tuberosity in order to prevent friction. Another small roughened area can be found half-way down the lateral side of the shaft. This is for the attachment of the pronator teres muscle—note that it is at the most lateral point of the rather bowed radial shaft. You will see later how this gives the muscle the maximum leverage. At the lower end, the radius is flattened antero-posteriorly and the posterior (or dorsal) surface can be recognised by the presence there of a *dorsal tubercle*. The anterior surface is smooth and slightly concave. The lateral side of the lower end of the radius is prolonged downwards to form a *styloid process*. The medial side is concave for it is here that the ulna articulates with the *ulnar notch* of the radius. The lower articular surface of the radius is concave for this surface forms part of the articular surface of the wrist joint. It articulates with two of the carpal bones: the *scaphoid* (most laterally) and the *lunate*.

The head of the radius can be palpated very easily by placing the index finger on the lateral epicondyle of the humerus with the middle finger immediately alongside and below it. The middle finger then lies over the radial head which can be recognised easily when it rotates during pronation and supination of the forearm. Its position is usually marked on the surface by a dimple. The shaft can be felt indistinctly, owing to the surrounding muscles. At the lower end the styloid process can be palpated quite easily at the upper end of a depression between some tendons at the base of the thumb. This is called the 'anatomical snuffbox' (Fig. 11.28), since, in theory at least, it would be a suitable receptacle in which to place snuff before sniffing it up the nose! The *dorsal tubercle* can be felt on the posterior surface of the wrist, more or less in line with the cleft between the index and middle fingers. On the front of the wrist the anterior surface of the radius can be palpated and the pulsation of the radial artery felt as the latter passes over the radius.

THE ULNA

Note the spelling! The bone is the *ulna*, the adjective derived from it is 'ulnar'; hence, for example, the ulnar notch of the radius or the ulnar artery. If you compare the radius and ulna you will see that the lower end of the radius is enlarged to articulate with the carpal bones while the upper end of the ulna is enlarged to articulate with the trochlea of the humerus (Fig. 9.21). Thus any force applied to the hand will be transmitted to the radius, thence to the ulna and then to the humerus. For some inexplicable reason the small lower end of the ulna is called the *head* so that the ulna and radius are placed head to tail like sardines in a tin. Examine first the upper end of the ulna. This has a deep hollow which fits around the trochlea and is therefore called the *trochlear notch*. The uppermost part of the notch is the *olecranon process* which, as you have seen, fits into the olecranon fossa of the humerus when the elbow is extended. The lower part is the *coronoid process* (don't confuse this with the coracoid process of the scapula) which fits into the coronoid fossa of the humerus during flexion of the elbow. Just below the coronoid process, a prominent roughened area is known as the *tuberosity* of the ulna. It is, of course, produced by the attachment of a tough tendon (of the brachialis muscle). Just below the trochlear notch, the lateral side of the ulna is hollowed out to form the *radial notch*, into which the head of the radius fits. Just behind this is a roughened, prominent vertical ridge to which part of the *supinator muscle* is attached. It is therefore called *supinator crest*. The main feature of the shaft of the ulna is its rather rounded posterior border which is an important landmark (see below). The lower end (head) of the ulna is small and rounded, to fit into the ulnar notch of the radius. It has a small and rather pointed styloid process with a small depression at its base. Inspection of the lower end of the ulna will tell you that it is not shaped to articulate with any of the carpal bones. Instead, it is in contact with a triangular plate of fibrocartilage whose apex is attached to the depression at the base of the styloid process and whose base is attached to the edge of the ulnar notch of the radius. It is this cartilage which, together with the radius, forms the articular surface for the carpal bones (Fig. 10.20).

Many parts of the ulna can be palpated easily in life. The olecranon has already been mentioned and from this point, which marks the tip of the elbow, the posterior border of the ulna can be followed right down to the lower end since it is subcutaneous throughout its entire extent. The groove which overlies it separates the flexor from the extensor groups of muscles. At the lower end of the ulna, the part which can be palpated depends on the position of the radius. If the forearm is supinated (palm facing forwards in the anatomical position), the head forms a prominent lump on the back of the wrist. When the hand is pronated, the styloid process can be palpated.

You should now examine the radius and the ulna together, properly articulated. Note how, during pronation, the lower end of the radius carrying the hand with it, moves around the head of the ulna while the upper end rotates 'on the spot' in the radial notch of the ulna. The axis of rotation is thus through the middle of the head of the radius above and through the base of the ulna styloid process below. This, however, is an approximation because the ulna itself moves a little during this movement. Note how the upper surface of the head of the radius and the olecranon notch of the ulna form an almost continuous surface for articulation with the humerus. The upper end of the radius therefore takes part in two joints: the elbow joint and the radio-ulnar joint. Still with the two bones together, look at the sharp edges of the bones where they face each other. These are the *interosseous borders* and they mark the attachment of a tough sheet of fibrous tissue, the *interosseous membrane*, which is interposed between the muscles of the front and those of the back of the forearm.

THE CARPAL BONES

There are eight small bones in the carpus, which is usually referred to loosely as the wrist region. However, as you will find when you palpate these bones, they are really situated in the proximal part of the hand. They are arranged in two rows of four bones each but the line of the joint between the two rows (the *mid carpal joint*) is not a straight transverse line but a rather complex line which is mostly concave downwards. The bones of the proximal row are from lateral to medial, the *scaphoid*, the *lunate* and the *triquetral*, with the small *pisiform* attached to the palmar surface of the triquetral. In the distal row are the *trapezium*, *trapezoid*, *capitate* and *hamate* (Fig. 9.22). The names of the bones describes their shape; for instance the scaphoid is boat-shaped, the lunate is crescent-shaped

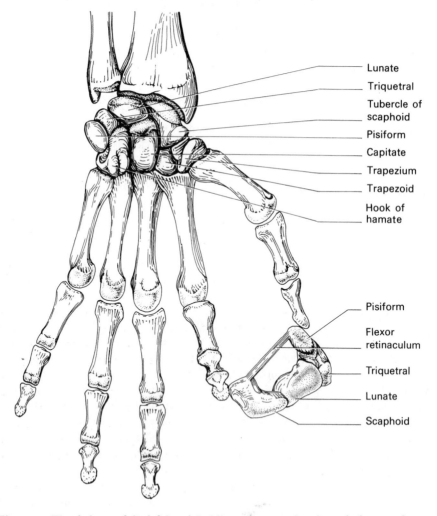

Lunate

Triquetral

Tubercle of scaphoid

Pisiform

Capitate

Trapezium

Trapezoid

Hook of hamate

Pisiform

Flexor retinaculum

Triquetral

Lunate

Scaphoid

Fig. 9.22. The skeleton of the left hand, holding a cross-section through the carpal tunnel.

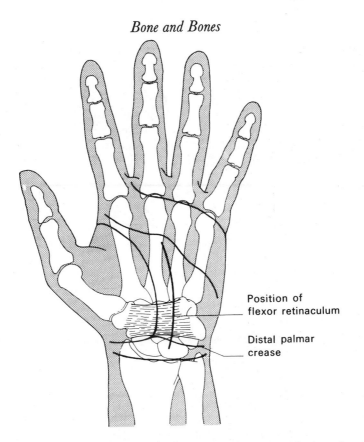

Position of
flexor retinaculum

Distal palmar
crease

Fig. 9.23. The position of the flexor retinaculum and of the creases in the palm of the hand. The distal palmar crease, not the webs of the fingers, marks the position of the metacarpophalangeal joints.

when seen from the side, the pisiform is like a pea and the hamate has a hook. If you get mixed up between the trapezoid and trapezium, remember that 'Trape*zium* supports the thumb'. If your bones are strung together by string or nylon, draw the string tight and you will find that the bones fit together so that they form a deep concavity facing anteriorly (Fig. 9.22). The bones forming the sides or edges of the concavity are the hook of the hamate and the pisiform medially and the trapezium and the scaphoid (actually the tubercle of the scaphoid) laterally. To these four bony points is attached a very strong sheet of fascia called the *flexor retinaculum* and this, together with the carpal bones, thus forms an osseofibrous tunnel: the *carpal tunnel*. It is through this tunnel that most of the tendons and one of the main nerves (the median nerve) enter the hand. It is a tight fit for so many structures and sometimes compression of the nerve occurs (see Chapter 12).

The proximal row of carpal bones, with the exception of the pisiform, form a smooth convex surface for articulation with the lower end of the radius and the fibrocartilaginous disc at the lower end of the ulna. The distal surfaces of the distal bones form an irregular castellated surface with which the bases of the metatarsals articulate. You do not need to examine these articular surfaces in detail but a few points are important and worth remembering. First of all, one of the most important joints of the hand is the joint between the trapezium and the first metacarpal. Examine the reciprocally curved surfaces of these two bones and note the extensive movement which can occur here. The base of the second metacarpal fits into an indentation between the trapezium, the trapezoid and the capitate. It is thus held rather firmly so that very little movement is possible here.

Owing to the concave shape of the carpus as a whole, only the four bones to which the flexor retinaculum is attached are palpable on the flexor surface (Fig. 9.23). The pisiform is at the medial 'corner' of the palm of the hand, usually just under the most distal of the skin creases of the wrist. The hook of the hamate is distal and lateral to this, in the palm of the hand in line with the cleft between the ring and little fingers. On the lateral side, the tuberosity of the scaphoid is prominent at the base of the elevation produced by the thumb muscles (the thenar eminence) but the trapezium can only be felt indistinctly. The scaphoid can also be palpated in the anatomical snuffbox (see above); in fact tenderness in the snuffbox is one of the diagnostic signs of a fractured scaphoid.

THE METACARPALS

The first metacarpal is the shortest but the most important since the thumb is the most important digit owing to its ability to rotate and thus to face the fingers (opposition) so that small objects can be held between them. Its joint with the trapezium has already been mentioned. The bases of the other metatarsals are rather flattened for articulation with the carpal bones and they also bear some small lateral facets for articulation with each other. The base of the third metacarpal has a small styloid process. The metacarpal heads form the knuckles. Look at your hand sideways and note that the prominences of the heads of the metacarpals lie opposite the distal crease in the palm of the hand (the 'line of head' of the palmists). The creases at the roots of the fingers therefore do *not* mark the level of the metacarpophalangeal joints (Fig. 9.23). Curiously, the metacarpals have only one epiphysis which is at the distal end except for the first metacarpal which has a proximal epiphysis.

THE PHALANGES

There are three phalanges in each digit except for the thumb which has only two. The terminal phalanges, of course, are not typical long bones for they only have a synovial joint at one end. The base of each phalanx is concave and the head convex. The only feature of real interest in the phalanges is the prominent bony ridge along each side of the palmar surface. This marks the attachment of a tough *fibrous flexor sheath* which encloses the flexor tendons (Fig. 11.33). Naturally, opposite the joints, such a tough sheath would hamper free movement, so you will see that the ridges only occur along the central portion of the shaft and fade out towards each end of the phalanx.

The heads of the metacarpals are easily palpable, as are their dorsal, or posterior, surfaces. The styloid process of the third metacarpal can be felt on the back of the hand in line with the cleft between the index and middle fingers. The phalanges are also palpable because there are no muscles, only tendons, in their vicinity.

THE BONES OF THE LOWER LIMB

THE HIP BONE (OS INNOMINATUM)

Although comparative anatomists can find many items of similarity between scapula and the hip bone (for example, the triangular blade of the scapula is obviously the counterpart of the large flat ilium component of the hip bone), for the clinical anatomists the scapula, clavicle and shoulder joint are very different indeed from the hip bone and hip joint. As has been pointed out before, the bones of the upper limb are built for mobility whereas those of the lower limb are responsible for weight transmission—free movement being much less important.

The hip bone consists of three components which can be recognised easily in the immature bone where they are separated by hyaline cartilage. In the adult bone, however, they are fused securely together to form the massive hip bone whose shape is so irregular that it is extremely difficult to describe. The three components are the *ilium* (not ileum, which is part of the gut), the *ischium* and the *pubis*. They are shown in Figures 9.24 and 9.25 and the original cartilaginous joints between them are indicated by dotted lines. The two hip bones are united in the midline in front by a secondary cartilaginous joint, the *symphysis pubis* (see Chapter 10); while posteriorly, each articulates with the sacrum at the *sacro-iliac* joint. This is a synovial joint like the hip or shoulder joints but is of a variety which is almost immovable, the bones also being bound together by massive ligaments. The whole unit of sacrum and the two hip bones forms the *pelvis* (Fig. 9.26) whose main function is to transmit the body weight from the spine in the midline to the lower limbs. The outer surface of the pelvis also gives attachment to many of the powerful muscles which act on the hip joint and also forms the socket for the head of the femur. The inner surface is related to, and gives protection to many of the pelvic viscera such as the bladder and the rectum.

If you bear these functions in mind and examine the hip bone, or better still, the whole pelvis (Fig. 9.26), you will be able to appreciate the way in which the structure of the bone is adapted to the function which it has to perform. Notice the thick strengthening bar of bone which runs from the *auricular surface* (for articulation with the sacrum) to the *acetabulum* and forms the main line of weight

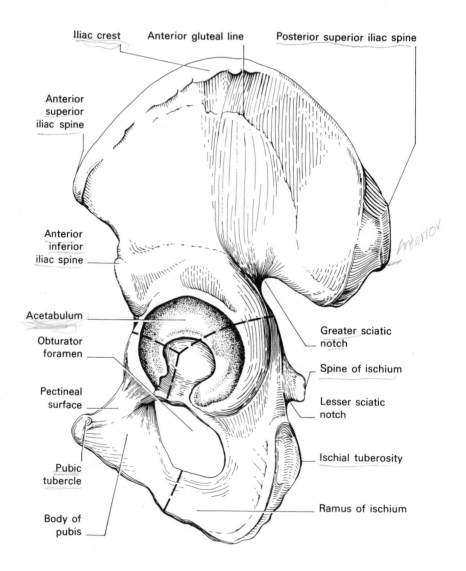

Fig. 9.24. The left hip bone, lateral aspect.

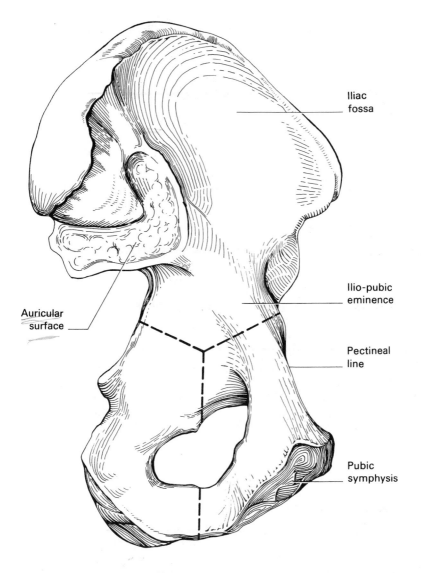

Fig. 9.25. The left hip bone, medial aspect.

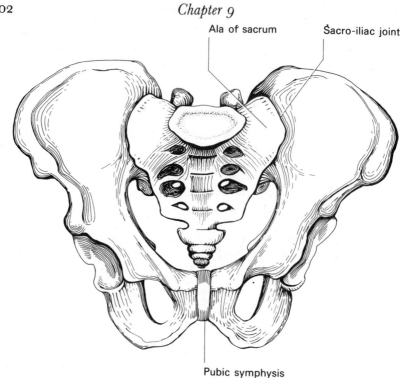

Ala of sacrum   Sacro-iliac joint

Pubic symphysis

Fig. 9.26. The bony pelvis consists of the two hip bones and the sacrum.

transmission. Were this the only component of the bone, the weight of the body would tend to splay the two hip bones apart so they are held together by tie beams in front, in the shape of the *ischium* and *pubis*. On the outer surface of the bone, particularly of the ilium, you will see roughened lines and tubercles which mark the attachments of powerful muscles, and you will also see the deep cup-shaped *acetabulum*, the socket for the head of the femur. (*Acetabulum* has the same origin as *acetic* acid—it means a vinegar cup!) Note the depth of the acetabulum and also that all three components of the hip bone help to form it. The inner surface of the hip bone is, naturally enough, smooth although a number of muscles are attached here by means of direct attachment of muscle fibres rather than by tendons. Finally, look at the whole pelvis and remember that in life the muscles which are attached to the inner surfaces of the bones will restrict the size of the cavity of the pelvis, and that in the female the fetus has to be expelled through this bony passage. If the pelvis is small or affected by disease, there may therefore be difficulties with birth.

In order to study the hip bone in detail, it will be convenient to deal with each of its components in turn.

The Ilium

The ilium forms the expanded upper part of the bone and gives attachment to a number of important muscles. The upper border of the bone (the *iliac crest*) is a important landmark since it forms a bony watershed between the muscles of the abdomen and those of the lower limb. Its anterior end is marked by a slight prominence called the *anterior superior iliac spine* while a little below this is a blunt elevation called the *anterior inferior iliac spine*. From these names you will deduce that there is also a *posterior superior iliac spine* and a *posterior inferior iliac spine*. The former is easy to identify but the latter is less definite. The iliac crest has an inner and an outer lip and an intermediate rough area; these are for the attachment of the abdominal muscles. Below the posterior inferior iliac spine, the border of the ilium has a large rounded indentation called the *greater sciatic notch*. The shape of this helps to distinguish a male from a female hip bone, since it is much more obtuse in the female whereas in the male the two sides of the notch are almost parallel. Below this, the ilium joins the ischium.

The outer surface of the ilium is called the *gluteal surface* for it is here that the three gluteal muscles (the main muscles of the buttock) are attached. The boundaries of the muscles attachments are marked out on the bone by the three *gluteal lines*: inferior, anterior and posterior. The inner surface of the ilium is smooth and hollowed out to form the *iliac fossa*, which gives attachment to the *iliac muscle*. In the lower part of the iliac fossa, the bone is thickened in the line of weight transmission, as has already been mentioned. The smooth ridge so produced is the *iliopubic eminence*. Posteriorly is the *auricular surface*, so called because its shape resembles the auricle of the ear. It is the surface with which the sacrum articulates and its appearance is misleading since this is a synovial joint and the auricular surface is, in life, covered with cartilage. Behind the auricular surface is the rugged *tuberosity of the ilium*, its rough appearance being due to the attachment here of the extremely powerful *sacro-iliac ligaments*.

The Ischium

The ischium forms the lower posterior part of the hip bone. The massive thickening, which bears the weight of the body in the sitting position, is the *ischial tuberosity*. It gives attachment to the tough *sacrotuberous ligament* and the tendinous origins of the *hamstring* group of muscles. It is therefore very rough and rugged in appearance. Above the ischial tuberosity is a sharp, pointed process called the *ischial spine* which is often broken off if the bone has been handled a great deal. Above the ischial spine is the *greater sciatic notch*, while below it is the *lesser sciatic notch*. The former has been mentioned already as providing an exit from the pelvis of various important nerves and vessels as well as a muscle (pyriformis). The smooth appearance of the lesser sciatic notch suggests that in life it is covered by cartilage and this is indeed the case—the tendon of the *obturator internus* muscle uses the notch as a pulley to enable it to turn through an angle of about 90° to reach its insertion. The ischium forms the lower posterior part of the acetabulum. Projecting forwards from the lower part of the ischial tuberosity is the *ramus* of the ischium. This unites with the inferior ramus of the pubis, although in the adult bone you will not be able to see the line of fusion.

The Pubis

The pubis consists of a body anteriorly, with superior and inferior *rami*. The superior ramus passes backwards to become continuous with the ilium and ischium and to help form the acetabulum. It has a rather sharp upper border— the *pectineal line*—inspection of which suggests that it is produced by a dense sheet of fascia. This fascia covers the *pectineus muscle* which arises from the smooth area below the pectineal line—the *pectineal surface*. Below the superior ramus is the *obturator foramen*, a large hole bounded by the pubis and ischium with the acetabulum above. This looks as though it ought to be provided for the passage of some large and important structures out of the pelvis but, in fact, most of it is filled in, in life, by an *obturator membrane*, leaving only a small gap superiorly through which the relatively small *obturator nerve and vessels* pass from the pelvis to the lower limb.

The body of the pubis takes part with its opposite partner in the midline joint called the *pubic symphysis*. The two bones are separated by a very thick disc of cartilage so they do not fit together accurately in the skeleton. On the upper surface of the body a poorly defined line, the *pubic crest*, runs from the symphysis laterally to end in the *pubic tubercle*, a small elevation at the medial end of the pectineal line. It is an important landmark and forms the medial attachment of the *inguinal ligament*.

The inferior ramus of the pubis is rough on its outer surface, for the attachments of lower limb muscles, and fairly smooth on its inner surface. It forms the lower margin of the obturator foramen, together with the ramus of the ischium.

Although many parts of the hip bone are inaccessible, especially its inner, pelvic surface, some of its features are easily palpable and are important landmarks in the abdomen, the lower limb and the back. The anterior superior iliac spine is very easily palpable and marks the outermost attachment of the inguinal ligament which is attached medially to the pubic tubercle. The latter can be palpated on the upper surface of the pubis in a thin person but may be more difficult to feel if it is covered by fat, particularly in the female. The tendon of a muscle, the *adductor longus*, is attached to the front of the pubis immediately below the tubercle and since this tendon is rather narrow and easily palpable (see Chapter 11) it serves a useful guide. From the anterior superior spine, the crest of the ilium may easily be followed backwards until it ends at the posterior superior iliac spine. This is easy to find since it is marked by a dimple on the surface. A line joining the posterior superior spines passes through the second sacral vertebra and marks, therefore, the lower limit of the subarachnoid space (see Chapter 5). It also marks the position of the centre of the sacro-iliac joint. The posterior inferior iliac spine can also be palpated but is less distinct and less important as a landmark, Finally, the tuberosity of the ischium may be palpated rather indistinctly in the gluteal region. When the hip joint is extended it is more difficult to feel since it is covered by the largest muscle in the body, the *gluteus maximus*. When the hip is flexed, as in the sitting position, the gluteus maximus slips upwards and the tuberosity is then covered by the very thick superficial fascia of this region. This is fortunate, since the fibro-fatty tissue which makes up the superficial fascia needs little blood supply, whereas gluteus maximus would never be able to withstand the insult of being sat upon, since its blood supply would be jeopardised.

Other parts of the hip bone are palpable in a thin person, but the bony landmarks mentioned above are those which should be known on account of their practical importance.

The Acetabulum

The acetabulum is the socket which receives the head of the femur. When the bones are covered with articular cartilage, the two components of the hip joint form a very good fit so that dislocation is uncommon. In normal posture, only the upper, anterior and posterior parts of the acetabulum are weight bearing so the articular cartilage is restricted to a horseshoe-shaped area which can easily be recognised on the dry bone. The remaining rougher part of the acetabulum is occupied by a pad of fat. Inferiorly the rim of the acetabulum is deficient to form the *acetabular notch*, through which blood vessels can enter the hip joint.

THE FEMUR

Like the hip bone, the *femur* (plural, *femora*) is also on the direct line of weight transmission so if you have already studied the humerus you will be impressed by the much tougher and stronger structure of the femur. You will also see that the femur has a very definite neck, unlike the humerus, which holds the shaft of the femur well away from the acetabulum and so helps in mobility (Fig. 9.27).

The upper end of the femur has a *head*, a *neck* and *greater* and *lesser* trochanters. The head is about the size of a rather large golf ball and fits firmly into the acetabulum in life, when both surfaces are covered by a layer of articular cartilage. In the centre of the head you will find a small pit which marks the attachment of a ligament—the *ligamentum teres*. This is not a very strong ligament but it is important because it transmits blood vessels to the head and you will easily be able to see the holes in the pit through which these vessels run. The other vessels of supply to the head enter the foramina around the base of the head.

The neck is set at an oblique angle (about 125°) to the shaft although developmentally it is part of the shaft to which separate epiphyses for the head and the greater and lesser trochanters are added. Since the line of weight transmission passes directly through the neck and down the shaft, it is obvious that it has to withstand a good deal of stress. A section through the upper end of the femur is interesting in this respect since it shows that the trabeculae of the cancellous bone, which forms the main mass of the upper end, are arranged specifically to transmit the weight. In the young bone the neck is thus more than strong enough to carry out its function, but in the elderly, fracture of the neck of the femur is very common, even as the result of a relatively slight stumble,

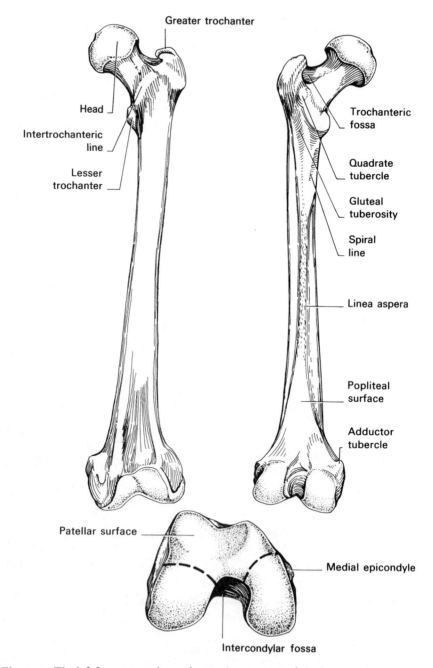

Fig. 9.27. The left femur, anterior and posterior aspects and the lower end.

and the long course of the blood vessels, which have to pass right up the neck to reach the head mean that a fracture of the neck may imperil the blood supply of the head and lead to difficulty with healing.

The *greater trochanter* is a large mass of bone which projects upwards at the base of the neck. Its roughened appearance denotes the strength of the tendons which are attached to it. They are the *gluteus medius* and *gluteus minimus*. Various other much smaller muscles are also attached to the greater trochanter and to the *trochanteric fossa*, a little depression, just large enough to take the tip of the little finger, that lies under the overhang of the greater trochanter. The *lesser trochanter* is, naturally, much smaller and is the elevation on the medial aspect of the upper end of the shaft. It is for the attachment of the *iliopsoas* muscle.

On the anterior surface of the femur a very broad roughened line runs between the greater and lesser trochanters; it is called the *intertrochanteric line* and marks the attachment of the *iliofemoral ligament,* which strengthens the front of the hip joint. The whole area of the femur above the intertrochanteric line, including the neck, is thus within the hip joint. Posteriorly, however, the ligaments of the hip joint are much weaker and are attached to the neck itself, so no markings can be seen. There is, however, an *intertrochanteric crest* on the posterior surface; but this is a smooth elevation and does not mark the attachment of ligaments. It has a smooth elevation half-way down called the *quadrate tubercle* to which is attached the *quadratus femoris* muscle.

The main feature of the posterior surface of the shaft of the femur is the 'rough line' which sounds much more impressive in Latin—the *linea aspera* is its official title. It marks the attachment of a number of muscles which are tendinous in this region. If you follow the linea aspera upwards, it diverges laterally and becomes more elvated to form the *gluteal tuberosity* to which is attached about one quarter of the *gluteus maximus* muscle. A less prominent offshoot of the linea aspera passes medially to form the *spiral line*.

If the linea aspera is followed downwards, it bifurcates into two rather faint lines with a smooth area between them. The smooth area is the *popliteal surface* of the femur and forms the floor of the *popliteal fossa* behind the knee. The lines are called the *supracondylar lines* because they lie above the two large *condyles* of the lower end of the femur. The medial supracondylar line has an elevation on the upper surface of the medial condyle. This is the *adductor tubercle* for the attachment of the tendon of the *adductor magnus* muscle.

The two condyles of the femur are very prominent when the bone is viewed from in front and they carry a very smooth articular area which forms the upper articular surface of the knee joint. On the lateral aspect of each condyle is an inconspicuous elevation called the *epicondyle*, which is easily palpable, and between the two condyles is a deep notch called the *intercondylar fossa or notch*. Now examine the articular surface. Its smoothness indicates that in life it is covered with articular cartilage. Careful examination of the surface will show faint shallow grooves which separate the two areas that articulate with the tibia from the upper patellar area. Note that the tibial area on the medial condyle is larger than that on the lateral condyle (Fig. 9.27) so that the area on the medial condyle is oval and that on the lateral, circular. These differing shapes have an important bearing on the movements of the knee joint and they are responsible for the 'locking' of the joint which occurs on full extension. If you examine the articular surface from the side you will be able to see one of the reasons why movements at the knee joint are a little complicated. The radius of curvature of the area that articulates with the tibia when the knee is extended is much greater than that of the surface in contact with the tibia when the knee is fully flexed (Fig. 10.31). One of the functions of the cartilages of the knee joint is to help adapt the femoral surface to that of the tibia in different positions of the joint. If you put the femur in place on the upper surface of the tibia you will see how obliquely the femur lies when the tibia is vertical. This, of course, is because in life the heads of the femora are separated by the whole width of the pelvis, whereas the lower ends are almost in contact. A knowledge of this obliquity is important for the proper understanding of the arrangement of the thigh musculature.

The shaft of femur is deeply buried in powerful muscles and so cannot be felt distinctly. Parts of the upper and lower ends, however, lie just below the surface and are important landmarks. The most prominent part of the lateral surface of the greater trochanter can be felt and is the most lateral bony point on the hip region. A line drawn from the anterior superior iliac spine to the ischial tuberosity passes through the greater trochanter (Nélaton's line). A tailor's 'hips' measurement is taken at the level of the greater trochanter in the male but in the female the greatest circumference is slightly below the trochanter because of a subtrochanteric collection of fat. At the lower end of the femur, the two epicondyles are clearly palpable on either side of the knee, while the adductor tubercle can also be palpated. It lies above the *medial* condyle of the femur and so can best be felt by running the finger down the medial side of, and slightly behind, the medial mass of muscles. Later, you will see how the tendon of the adductor magnus muscle is a guide to this tubercle.

The Patella

Patella is the Latin name for a limpet, a reasonably appropriate name for a bone which appears to be stuck on the front of the knee joint (Fig. 9.28). In fact, in life the patella is a bone formed in the tendon of the main muscle of the front of the thigh, the quadriceps femoris. It is, therefore, a sesamoid bone, although it is so large that the quadriceps muscle is said to be inserted into the patella, and the true tendon of the muscle, which is inserted into the tubercle of the tibia (see below), is called, inappropriately, the *ligamentum patellae*.

The patella has a lower, pointed *apex* to which the ligamentum patellae is attached. If you place a patella on the table with its apex pointing away from you, it will fall to one side and the side to which it falls is the side to which it belongs. You can see the reason for this if you look at the smooth posterior surface of the patella. This is divided into two by a vertical ridge. The area

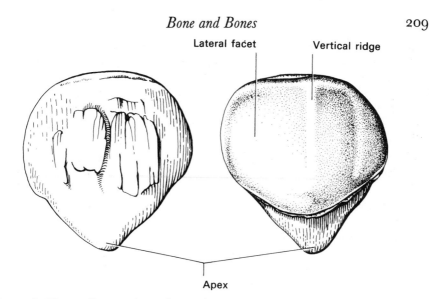

Lateral facet   Vertical ridge

Apex

Fig. 9.28. The patella, anterior and posterior aspects.

lateral to the ridge, which articulates with the lateral condyle of the femur, is a good deal larger than the area medial to the ridge. Thus the lateral side of the patella is heavier than the medial side so a right patella will fall to the right. Now articulate the patella with the femur and view the two bones end-on. Note how the lateral condyle of the femur extends farther forward than the medial condyle so that a prominent ridge of bone prevents lateral movement of the patella (Fig. 9.27). Owing to the oblique line of pull of the quadriceps muscles (see Chapter 11) which, in turn, is due to the obliquity of the femur, the patella has a tendancy to dislocate laterally and the forward projection of the lateral femoral condyle is one of the factors that helps to prevent this.

The patella is easily palpable over the front of the lower end of the femur. Its shape is best appreciated when the quadriceps muscle on the front of the thigh is relaxed (i.e. when the knee joint is extended passively) and in this position the whole outline of the patella is palpable and the bone can be held between the fingers and moved about.

The Tibia

The tibia is the principal bone of the leg (as opposed to the thigh), the fibula not being a weight-bearing bone. It is a typical long bone with a strong shaft and expanded upper and lower ends (Fig. 9.29). The shaft is triangular in cross section so that the tibia has anterior, medial and lateral (or interosseous) borders. The anterior border is very sharp and, to the layman, is called the 'shin'. The upper part of the rather smooth posterior surface has a roughened oblique ridge called the *soleal line*. To this is attached one head of the soleus muscle.

The upper end of the tibia has medial and lateral *condyles* which support the lower end of the femur in the standing position. In front there is a prominent *tibial tuberosity*. This has a lower, roughened area and an upper smooth area from which you might deduce that there is a powerful tendon or ligament attached to the roughened area, which is separated from the smooth area by a bursa. (Compare with the radial tuberosity (see above) and the back of the calcaneus (see below).) The tendon or ligament is the *ligamentum patellae*, which is in fact a tendon in spite of its name. On the under surface of the lateral condyle is a smooth circular facet about the size of a penny, which, in life, is covered by the articular cartilage of the *superior tibiofibular joint*. Note that the fibula does not articulate with the femur and is thus not involved in weight transmission. On

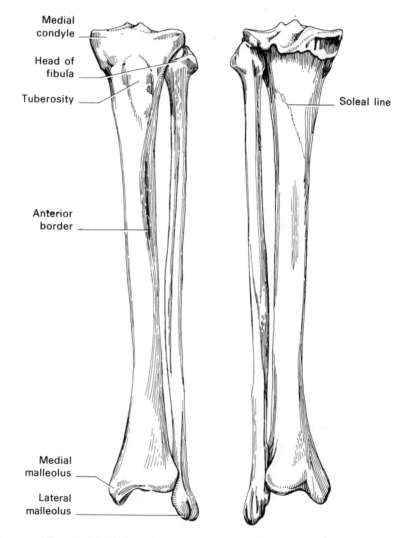

Medial condyle

Head of fibula

Tuberosity

Soleal line

Anterior border

Medial malleolus

Lateral malleolus

Fig. 9.29. The left tibia and fibula, anterior and posterior aspects.

the back of the medial condyle is a faint groove for the attachment of the tendon of the *semimembranosus* muscle.

The upper surface of the tibia is important because it takes part in the knee joint. You will remember that the lateral condylar surface of the femur that articulates with the tibia is approximately circular whereas the corresponding medial area is oval. Therefore, the lateral femoral facet on the tibia is circular and the medial is oval. Between the two facets is a roughened *intercondylar area* which is roughened for the attachment of ligaments. In the centre of this are the prominent *intercondylar eminences* (Fig. 9.30).

The posterior and anterolateral surfaces of the shaft have muscles attached to them, at least on their upper two-thirds, but since these muscles are attached directly rather than by means of tendons, the surfaces are smooth (except for the soleal line which has been mentioned already). No muscles are attached to the lower third of the tibia and if you remember that many of the small blood vessels which supply bone come from the attached muscles, you will understand why fractures of the lower tibia are often subject to delayed union.

The lower end of the tibia is very asymmetrical since on the medial side is a large process projecting downwards called the *medial malleolus*. On the lateral side there is no such projection but only a concave, rather roughened area for articulation with the fibula. The *lateral malleolus*, in fact, is formed by the lower end of the fibula so that you will not see the relative positions of the two malleoli unless you articulate the two bones when it will become apparent that the malleoli form an important part of the 'socket' into which the talus fits to form the ankle joint. The articulation between the tibia and fibula at their lower ends is thus very important and the two bones are held together by a very tough ligament—hence the roughened area on the bone. One other feature of the lower end is worthy of mention. Just above and behind the medial malleolus is a shallow groove which is used as a pulley by certain tendons (particularly that of tibialis posterior) so that they can pass forwards into the foot.

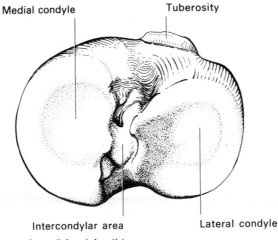

Medial condyle   Tuberosity

Intercondylar area   Lateral condyle

Fig. 9.30. The upper surface of the right tibia.

Chapter 9

A large part of the tibia can be palpated in life. At the upper end the medial and lateral condyles are close to the surface and their upper borders correspond to the line of the knee joint (see Chapter 10). The tibial tuberosity is subcutaneous, being separated from the skin only by a bursa, so that in the kneeling position the tibial tuberosity comes in contact with the ground. In the days when it was customary to scrub large areas of floor with a scrubbing brush, the bursa (and others in the vicinity) would sometimes be irritated and fill up with synovial fluid giving rise to 'housemaid's knee'. The tibial tuberosity is continuous below with the sharp anterior border of the bone which is subcutaneous along its entire extent, as is the anteromedial surface. Thus a very large area of the tibia can be palpated in life. At the lower end the medial malleolus forms a prominent projection on the medial side of the ankle; it is subcutaneous and continuous with the anteromedial surface.

The Fibula

The fibula is a long slender bone which is not weight bearing but which provides an attachment for numerous muscles and which provides the lateral part of the socket for the ankle joint by means of its lateral malleolus. Its lower end (Fig. 9.31) is easily recognised since it has a smooth triangular facet behind which is a small pit or fossa (remember that the Pit is Posterior—this will enable you to distinguish between a right and a left fibula). This is called the *malleolar fossa*. The upper end (or *head*) is distinguished by a circular facet which articulates with the corresponding facet on the tibia. The uppermost part of the bone is rather pointed since the cord-like lateral collateral ligament of the knee joint

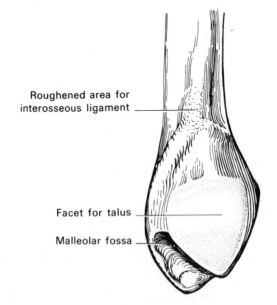

Roughened area for
interosseous ligament

Facet for talus

Malleolar fossa

Fig. 9.31. The lower end of the left fibula, medial aspect.

is attached here. The shaft has a number of well-marked ridges to which are attached various sheets of fascia, including an interosseous membrane which extends between tibia and fibula. The lower end presents a trap for the unwary student! The triangular facet is *not* for articulation with the tibia. It lies on the medial surface of the lateral malleolus and thus projects downwards below the tibia and is, in fact, for articulation with the talus. The tibial area lies above the facet and is roughened since it is to this area that the interosseous ligament is attached. The pit at the lower end is for the attachment of certain ligaments which will be described in Chapter 10. Behind the lateral malleolus is a faint groove which, like the rather similar groove above the medial malleolus of the tibia, is to provide a pulley around which two tendons (peroneus longus and brevis) can pass into the foot.

You should now articulate the tibia and fibula and note the relative positions of the malleoli—the lateral extends farther downwards than the medial.

Both upper and lower ends of the fibula are valuable landmarks. The upper end, or *head*, can be palpated a little below and behind the lateral condyle of the tibia. It is thus below the line of the knee joint. The lateral malleolus produces the prominent projection of the lateral side of the ankle.

THE BONES OF THE FOOT

These may be subdivided into those of the *tarsus* and those of the toes (*metatarsals* and *phalanges*) (Figs. 9.32 and 9.33). The tarsal bones are seven in number. The two largest (the *talus* and the *calcaneus*) lie posteriorly and are separated from the more anterior bones by a more or less transverse gap called the *transverse tarsal joint*. It is not a separate joint in its own right, however, but a combination of two joints, as will be seen later.

The *talus* is the uppermost bone and articulates above and at the sides with the tibia and fibula, below with the calcaneus and anteriorly with the *navicular*. The upper surface is curved antero-posteriorly, and is wider in front than behind (see Chapter 10 for the significance of this). It is continuous with medial and lateral facets on the sides. The medial facet is comma-shaped and articulates with the medial malleolus. It only extends over the upper part of the medial surface, the lower part being roughened for the attachment of the medial ligament of the ankle joint. On the lateral surface, however, the facet is triangular and larger, because the lateral malleolus extends farther downwards than the medial.

Anteriorly, the *head of the talus* is rounded and fits into the concavity of the navicular. It is connected to the main part of the bone by a slightly constricted *neck*. The facets on the under surface are complex and best understood by studying talus and calcaneus together. One large posterior facet is for a similar large facet on the calcaneus while anteriorly, one or two facets articulate mainly with the upper surface of the *sustentaculum tali* a prominent projection of the medial surface of the calcaneus. If you study the talus, navicular and calcaneus together, you will find, on the under surface, a smooth rather triangular part of

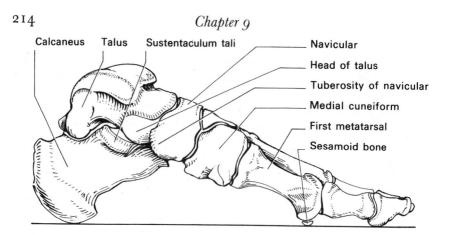

Calcaneus Talus Sustentaculum tali
Navicular
Head of talus
Tuberosity of navicular
Medial cuneiform
First metatarsal
Sesamoid bone

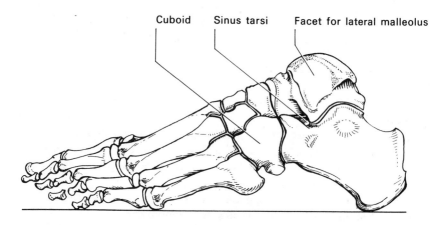

Cuboid Sinus tarsi Facet for lateral malleolus

Fig. 9.32. The skeleton of the left foot, medial and lateral aspects.

the head of the talus between the navicular and the sustentaculum tali (Fig. 9.33) which does not seem to articulate with any other bone but which is obviously part of a joint. This surface is in contact with a very tough ligament called the *spring ligament* which joins the sustentaculum to the navicular.

On the posterior surface of the talus there is a rather prominent posterior tubercle, alongside which is a groove, occupied in life by the tendon of *flexor hallucis longus* muscle. You may find it helpful to remember that although many tendons are related to the talus, no tendons or muscles are actually attached to it.

The *calcaneus* is the largest bone in the foot and forms the main bone of the heel. On its under surface there is a prominent *calcaneal tuberosity*, which in the standing position is separated from the ground by the thick skin and fascia of the heel. Anteriorly is a large facet for articulation with the *cuboid*. Medially, the *sustentaculum tali* forms a prominent projection; the groove below it is in line

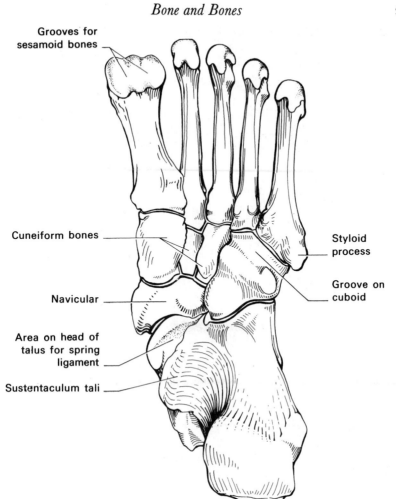

Grooves for sesamoid bones

Cuneiform bones

Navicular

Area on head of talus for spring ligament

Sustentaculum tali

Styloid process

Groove on cuboid

Fig. 9.33. The plantar surface of the skeleton of the left foot.

with the groove on the back of the talus and accommodates the tendon of *flexor hallucis longus*. The upper surface of the calcaneus (Fig. 9.34) has facets for articulation with the talus. Between the posterior large facet and the one or two more anterior facets is a deep transverse groove which, together with a corresponding groove on the talus, forms a tunnel called the *sinus tarsi* that accommodates several ligaments. The back of the calcaneus may be divided roughly into three zones. The lowermost part is covered by the fat and fibrous tissue of the heel; the central zone is roughened for the attachment of the main tendon on the heel—the *tendo calcaneus* or Achilles tendon; while the upper third is smooth because here lies a bursa which separates the tendo calcaneus from the bone. On the lateral surface there may be a distinct *peroneal tubercle* which lies between the tendons of the peroneus longus and peroneus brevis muscles.

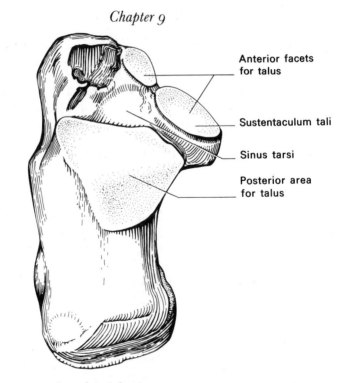

Anterior facets
for talus

Sustentaculum tali

Sinus tarsi

Posterior area
for talus

Fig. 9.34. The upper surface of the left calcaneus.

The *navicular* bone is so called because it is said to resemble a boat—it does resemble, to some extent, a coracle. It has a posterior concave surface into which fits the head of the talus and a convex anterior surface, which may be divided faintly into facets, for articulation with the three *cuneiform* bones. Below and medially is a rather prominent *tuberosity* to which is attached the tendon of *tibialis posterior*.

The *cuneiforms* are three in number: medial, intermediate and lateral. Cuneiform means wedge-shaped and the reason for this name becomes apparent if the bones are viewed end-on. The medial is the largest of the three. Each articulates with one (or more) metatarsals.

The cuboid bone is on the lateral side of the foot, articulating with the calcaneus behind and the 4th and 5th metatarsals in front. On its under surface is a deep groove in which lies the tendon of the *peroneus longus* muscle.

The *first metatarsal* is a massive bone compared to the other metatarsals which have very slender shafts. This is because it carries a great deal of the body weight during the 'take-off' stage of walking. Its base is kidney-shaped and articulates with the medial cuneiform. Its head is large and carries on its under surface two grooves in which lie two tendons, in each of which is embedded a sesamoid bone. The head of the metatarsal is rounded and articulates with the base of the proximal phalanx. The other metatarsals have facets at their bases for articulation with the tarsal bones and with each other. The fifth metatarsal has a prominent styloid process at its base which is a useful surface landmark. The tendon of the peroneus brevis is attached here. Note that in the articulated foot the second metatarsal extends farther forward than any of the others, so that if, for any reason such as muscular dysfunction, the first metatarsal does not bear its full load, the weight of the body may be too much for the second metatarsal, which responds by fracturing.

The *phalanges* are two in number for the great toe, but three for each of the others. They are smaller and less well shaped on the lateral side of the foot and in the little toe they may be fused into a small and ugly piece of bone.

You should now examine an articulated foot placed in contact with a flat surface. Owing to the arched arrangement of the bones, you will see that in the standing position the weight is taken on the medial tubercle of the calcaneus, the heads of the metatarsals, and to a variable extent, the lateral side of the foot. You may observe the same weight distribution in your wet footprint when you step out of the bath.

Many of the bones of the foot are palpable and are important landmarks. About an inch below and in front of the medial malleolus, the tubercle of the navicular may be palpated while immediately below is the sustentaculum tali. The neck of the talus may be felt between the prominent tendons of tibialis anterior and tibialis posterior on the upper medial surface of the foot. The main part of the calcaneus can best be felt when the heel is gripped between the fingers and thumb. The peroneal tubercle may often be felt about 2 cm below the tip of the lateral malleolus. The base of the first metatarsal and the joint line between it and the medial cuneiform can be felt on the medial side of the foot while on the lateral side the styloid process on the base of the fifth metatarsal forms a prominent projection. It may be more proximal than you expect. The heads of the metatarsals, especially the first, are, in lay terminology, the 'ball of the foot'. These can only be felt indistinctly because of the thick skin and dense fascia which lie over them for protection.

10 · Joints

It is important from a functional point of view that certain bones are held firmly together with little or no movement between them, for example, the bones of the vault of the skull. Such joints are, in general, of little interest to physiotherapists, whose work is much more closely involved with the freely movable joints since these are subject to various diseases in the treatment of which physiotherapists are closely concerned. Nevertheless, there are some exceptions to this general statement and it is important that the structure of all types of joint be understood. Joints are classified into three major groups: (a) Fibrous, (b) Cartilaginous and (c) Synovial.

FIBROUS JOINTS

In all fibrous joints the bones are held together by fibrous tissue consisting of fibroblasts and dense collagenous fibres (p. 38). When enough fibrous tissue is present to merit its being called an interosseous ligament, the joint is known as a *syndesmosis*. An example of such a joint is the inferior tibiofibular joint (Fig. 10.1). Other types of fibrous joint are dignified by various individual

Interosseous ligament

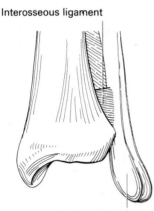

Lateral malleolus of fibula

Fig. 10.1. A syndesmosis—the inferior tibiofibular joint.

Periosteum (pericranium)

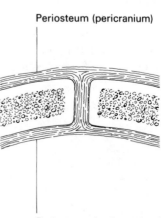

Periosteum (endocranium)

Fig. 10.2. A suture between the bones of the skull.

names, but the only one of these which need be remembered is that which unites the flat bones of the skull: a *suture* (Fig. 10.2).

CARTILAGINOUS JOINTS

These are subdivided into *primary* and *secondary cartilaginous joints* (Fig. 10.3). In the former, two bones are held together by hyaline cartilage only, as is the joint between the epiphysis and the diaphysis in a young bone (Fig. 9.2). Such a joint is sometimes known as a *synchondrosis* and with increasing age the cartilage becomes ossified so that bony union occurs (*synostosis*). Even before this stage is reached, however, there is little or no movement at primary cartilaginous joints, the most mobile of them being the joint between the first rib and the manubrium sterni. In secondary cartilaginous joints, (Fig. 10.3) the ends of the bones are covered by hyaline cartilage and the bones are also separated by a

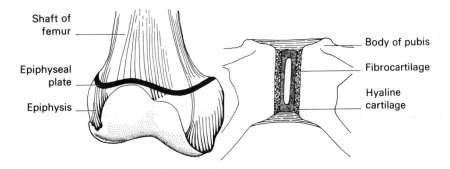

Fig. 10.3. Primary and secondary cartilaginous joints—the joint between the epiphysis and shaft of a young femur, and the pubic symphysis.

plate or disc of fibrocartilage. These joints are also known as *symphyses* and, by coincidence, they all happen to be in the midline sagittal plane. Examples are the manubriosternal joint and the symphysis pubis. The amount of movement depends on the thickness of the disc of fibrocartilage and the joints are usually described as 'slightly movable'. In the spine, however, there are a whole series of secondary cartilaginous joints between the vertebral bodies and although the movement at each joint is not very free, the sum of the movements at all the joints produces the normal free movement of the spine as a whole. Secondary cartilaginous joints do not have the same tendency to undergo fusion in later life as do the primary joints, but fusion, nevertheless, does often occur in the manubriosternal joint and in the symphysis pubis.

SYNOVIAL JOINTS

Synovial joints are the most complex and, in general, the most freely movable of all the joints of the body, and a good deal of the physiotherapist's work is concerned with the effects of pain and restricted movement in synovial joints. It is therefore most important that the basic structure of these joints be understood and that the direction and range of movement at all the synovial joints of the body (particularly those of the spine and the limbs) are known.

TYPES OF SYNOVIAL JOINTS

Synovial joints are classified according to the planes of movement which are possible and this, in turn, depends on the shape of the articular surfaces. The simplest type is the *plane*, or *gliding* joint, in which the articular surfaces are more or less flat or only slightly curved. Movements at this type of joint are restricted to a simple gliding or shifting movement; an example is the superior tibiofibular joint. In a *hinge joint*, movements occur in one plane only—i.e. only flexion and extension can take place, as is the case with the elbow joint or the interphalangeal joints. Another type of synovial joint in which only one type of movement is possible is the *pivot* joint. In these the axis of movement corresponds more or less to the long axis of the bone so that a rotatory movement occurs. Such a joint is the superior radio-ulnar joint in which the axis of rotation runs through the centre of the head of the radius.

On a slightly more complex level are *ellipsoid* (sometimes known as *condyloid*) joints in which movement can occur in two planes so that flexion and extension, abduction and adduction can take place, e.g. the wrist and the metacarpophalangeal joints. Finally, in a number of important joints the movements possible are a combination of those in ellipsoid and pivot joints, i.e. flexion and extension, abduction and adduction and rotation medially and laterally. These are *ball and socket* joints, and are typified by the hip and shoulder joints. There is, however, one other type of joint in which a similar variety of movements can occur, namely the saddle-shaped (or *sellar*) joints, of which the most important example is the carpometacarpal joint of the thumb.

The range and direction of the movements of a synovial joint are largely governed by the shape of the bones which form it, as will be understood when individual joints are described. The direction of the most important movements in a joint also governs the disposition of the ligaments so that in a hinge joint, for example, the main ligaments lie at the sides, the anterior and posterior ligaments being thin and pliable. The arrangement of the surrounding muscles also affects movements, especially when the muscles work on other joints as well. Thus, the hamstrings, which flex the knee and extend the hip, may limit full flexion of the hip when the knees are extended, although they may be stretched by exercise so that 'touching the toes' with straight knees then becomes quite easy.

In a typical synovial joint (Fig. 10.4) the ends of the bones are covered by hyaline *articular cartilage* and are held together by a *capsular ligament*. This ligament is lined by a *synovial membrane* which also covers all the structures within the joint which are not exposed to friction. Various other accessory structures may also be present and most of these are shown in Figure 10.4. The various tissues which make up synovial joints will now be described individually.

ARTICULAR CARTILAGE

The ends of the bones are reciprocally shaped; the exact shape determines the directions of movements which can occur at the joint. A study of the bones will often give a clue as to the relative stability of the joint, a comparison, for example, of the hip and shoulder joints in an articulated skeleton, will rapidly lead to the conclusion that the former is very much more stable than the latter. Attempts to articulate two bones on a skeleton are disappointing, however, since an accurate fit cannot be obtained. The 'fine adjustment' for congruence is produced by the layer of hyaline cartilage—the *articular cartilage*—which covers the ends of the bones and adapts the articular surfaces one to the other.

It is important to realise that the curvature of the articular surface is not constant all over the whole area of the surface. This is well shown at the lower end of the femur where, in extension of the knee, a surface with a relatively large radius of curvature is in contact with the tibia, whereas in flexion, a surface with a relatively small radius is involved (Fig. 10.31). In most joints there is one position in which the curvatures of the two articular surfaces correspond

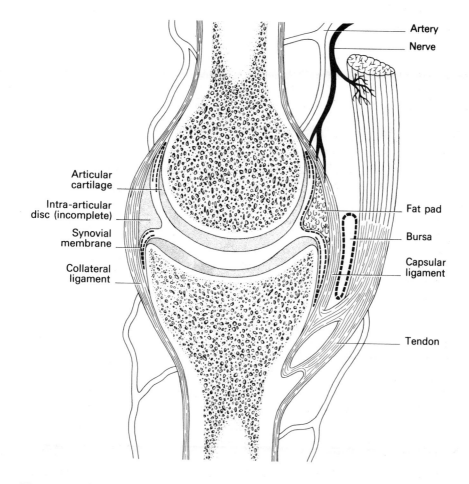

Fig. 10.4. A diagrammatic section through a typical synovial joint.

exactly and it is in this position that the ligaments which support the joints are most taut. The joint is then in its most stable conditions and this is known as the 'close-packed' position.

The articular cartilage is thick in the central part of a convex surface and thick at the periphery of a concave surface (Fig. 10.4) so that a rather accurate fit is obtained. Even so, the fit is not perfect (except in the close-packed position), as if it were so, the joint surfaces might become dry owing to the squeezing out of synovial fluid, which is a lubricant, from between the surfaces. The slight incongruences which are present help to distribute the synovial fluid between the articular surfaces during movement. The cartilage is also elastic and slightly compressible and is able to withstand the high shearing forces which occur in certain joints, particularly those of the lower limb. Cartilage is an avascular tissue (see Chapter 3) but its nutrition is extremely important for the proper functioning of a joint. The periphery of the cartilage receives its nourishment from the plentiful blood vessels in the adjacent synovial membrane, and the deepest layers of the cartilage adjacent to the bone is close to the vessels of the bone itself. The major part of the articular cartilage, however, is nourished entirely by the synovial fluid. That this is so can be deduced from the fact that in certain conditions, detached pieces of cartilage can lie freely within the joint cavity so that they cannot receive any nutrition from the blood vascular system at all. Such loose bodies are known as 'joint mice' since they can sometimes be felt but slip away from the examining finger to reappear in another part of the joint. It is significant that these loose pieces of cartilage not only remain alive but can actually grow in size when their only source of nutrition is synovial fluid. With increasing age, articular cartilage becomes thinner and less tough and its surface may show signs of damage and erosion.

THE CAPSULAR LIGAMENT

This forms a sleeve of dense fibrous tissue around the joint, being attached to the bone a varying distance beyond the articular surfaces. The thickness and the tightness of the capsular ligament will help to govern the mobility of the joint so that in the shoulder joint, for instance, the inferior part of the capsular ligament is lax so as to allow for the free abduction which takes place at the joint. The capsular ligament is strengthened here and there by other ligaments which are usually given specific names; such ligaments, however, are mostly only thickenings in the capsular ligament and diagrams which show them as completely separate bands are misleading. Some ligaments, however, are quite independent of the capsular ligament and these are usually known as *accessory ligaments*—see, for example, the lateral ligament of the knee joint. They may be intra- or extracapsular.

Ligaments are very inextensible, and when they are forcibly stretched or torn they give rise to severe pain, hence the popularity of the rack as a mediaeval instrument of persuasion.

SYNOVIAL MEMBRANE AND FLUID

All synovial joints, by definition, are lubricated by *synovial fluid* which is produced by a thin membrane: the *synovial membrane*. In the fetus the cells which will later form the synovial membrane line the whole of the inside of the joint but as soon as movement starts (during the fifth month of pregnancy) the membrane rapidly disappears from all the articulating surfaces, so that, from that time on, it is found lining the inside of the capsular ligament, the ends of the bone up to the edge of the articular cartilage and all the other intra-articular structures such as fat pads and ligaments which are not subjected to friction.

As well as producing synovial fluid, the membrane also has the important function of reabsorbing excess fluid from the joint cavity, and by virtue of its phagocytic cells, microscopic foreign bodies such as bacteria and broken-down blood cells can also be taken up. In most joint injuries an excess of synovial fluid is produced so that marked swelling occurs (this must be distinguished from the swelling caused by the accumulation of tissue fluid and blood when ligaments are torn) and the process of healing is accompanied by the reabsorption of this fluid by the synovial membrane, a process which may be speeded up by movement of the joint and by the application of external pressure.

Synovial fluid itself is a colourless, highly viscous fluid, rather resembling white of egg (hence the name syn-ovum meaning like an egg). It owes many of its properties to its content of complex carbohydrates known as *glycosaminoglycans* and its functions are to lubricate the joint and to nourish the intra-articular structures, particularly the avascular articular cartilage.

BURSAE

A bursa is a sac, lined by synovial membrane and containing synovial fluid which is found in regions whose friction is likely to occur. They are found, for example, where tendons cross over prominent bony areas and are situated between the tendon and the bone—such bursae are therefore very often found in the vicinity of synovial joints and may, indeed, communicate with the joint cavity. They also occur between closely apposed tendons and between superficial bones and the skin where friction is liable to occur, for example, over the tuberosity of the tibia.

For the sake of completeness, it should be mentioned here that synovial membrane is also found around certain tendons in the hands and feet in the form of *synovial sheaths*, and these are described in Chapter 11.

OTHER INTRA-ARTICULAR STRUCTURES

The majority of synovial joints contain pads of fat, covered with synovial membrane, which help to fill in unoccupied space in the joint and thus to aid the circulation of synovial fluid. Many joints contain an *intra-articular disc* of fibrocartilage which may divide the joint into two separate cavities; in some such joints the type of movement may be different in the two joints which are thus formed. The synovial membrane has the usual relationship to these discs, i.e. it covers all non-articulating areas so that it extends to the edges of the disc but the area subject to friction is devoid of membrane. As has been mentioned already, some joints contain intra-articular ligaments—see, for example, the cruciate ligaments of the knee joint.

BLOOD SUPPLY AND NERVE SUPPLY

All synovial joints are surrounded by a series of anastomosing blood vessels which are derived from major vessels in the vicinity. From this plexus branches are given off which supply the joint itself, particularly the capsular and other ligaments, and the synovial membrane. The nerve supply to joints is by means of branches of the nerves which supply certain of the muscles which act on the joint (Hilton's Law). A good example of this may be seen in the case of the femoral nerve. Branches to each of the vasti (which cross the knee joint) supply an articular branch to the knee joint, while the branch to rectus femoris (which crosses the hip joint) also supplies the hip joint. The vast majority of the articular nerves end in sensory receptors in the capsule and ligaments of the joint and in the peri-articular tissues. These are important for proprioception (Chapter 5). They also help to prevent overstretching of any part of the capsule since, when strongly stimulated, they initiate a reflex contraction of the particular group of muscles which can relax the stretched region of the joint.

Injuries to joints are intensely painful and the pain may induce reflex spasm of surrounding muscles. This is particularly so in dislocations. The pain from joints is often referred to skin or to another joint (pp. 105–109).

INDIVIDUAL JOINTS

JOINTS OF THE TRUNK

INTERVERTEBRAL JOINTS

As was pointed out in Chapter 9, the bodies of the vertebrae are responsible for weight transmission, while the articular facets only help to regulate the directions of the movements which take place between two adjacent vertebrae. The joints between the bodies are therefore extremely stable—they are secondary cartilaginous joints. The joints between articular processes need to provide a greater freedom of movement and are therefore synovial in type.

Each of the vertebral bodies is covered on its upper and lower surfaces with a thin plate of hyaline cartilage, while between two adjacent bodies is an *intervertebral disc*, which is composed mostly of fibrocartilage. The structure of the discs is extremely important from a clinical point of view. The peripheral zone consists of cartilage with a high proportion of fibres so that it forms a 'fibrous ring'; hence its name: the *annulus fibrosus*. The fibres are arranged in interweaving bundles which run obliquely through the thickness of the disc in various directions so that the annulus fibrosus is enabled to withstand torsional forces in any direction. The bundles of fibres are attached to the plates of hyaline cartilage and to anterior and posterior longitudinal ligaments. Trapped within the annulus fibrosus is the much softer and more elastic central material, the *nucleus pulposus*, which is situated nearer to the back than the front of the disc. It has a high water content, which diminishes with increasing age. As may be imagined, in the standing position the nucleus pulposus is under a great deal of pressure and it is not surprising that it may sometimes bulge or herniate, either through one of the plates of hyaline cartilage, to invade the vertebral body itself, or through the annulus fibrosus. The pressure may be increased still further during strong muscle contraction when the spine is flexed (Fig. 10.5), hence the importance of lifting heavy weights with the spine straight and vertical.

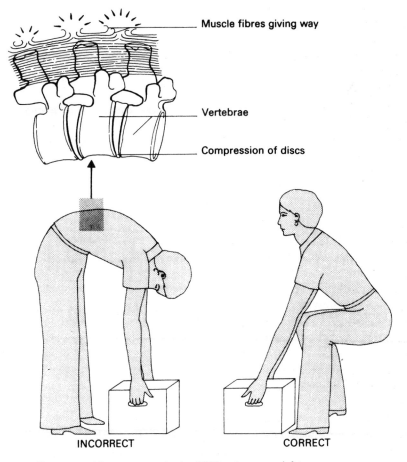

Muscle fibres giving way

Vertebrae

Compression of discs

INCORRECT   CORRECT

Fig. 10.5. Correct and incorrect methods of lifting heavy weights.

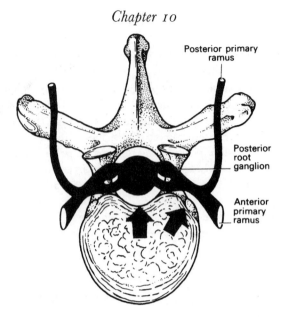

Posterior primary
ramus

Posterior
root
ganglion

Anterior
primary
ramus

Fig. 10.6. The arrows indicate how a disc herniation posteriorly may affect the spinal cord while a posterolateral herniation may affect a spinal nerve.

As is shown in Figure 10.6, herniation through the annulus fibrosus may compress a single spinal nerve, or the spinal cord itself. The commonly used term 'slipped disc' is thus a misnomer; the whole disc is never squeezed out from between two vertebrae.

The thickness of the intervertebral discs varies in different regions of the spine: they are relatively thin in the thoracic region, thicker in the cervical region and very thick indeed in the lumbar region. In the cervical and lumbar regions they are thicker in front than behind so that they contribute to the curvature of the spine in these regions. The movements between two adjacent vertebrae depend on the thickness of the disc so that movements are least free in the thoracic region (in the sacral region, of course, there are no discs and no movement).

The joints between the articular processes are plane synovial joints and they have the structure typical of such joints with hyaline articular cartilage, synovial membrane and a capsular ligament. You should study the direction in which the facets face in different regions of the spine, since this dictates the directions in which movement is possible (see Chapter 9).

The vertebrae are held together by powerful ligaments which connect all parts of the vertebrae except the pedicles—if these were joined by ligaments there would be no intervertebral foramina to allow the exit of the spinal nerves.

The most convenient way of remembering the intervertebral ligaments is to remember that each of the paired components (with the exception of the pedicles) has one ligament attached to it, while each of the unpaired components

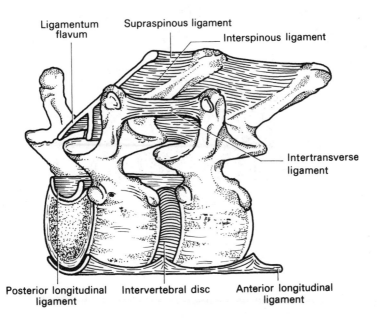

Ligamentum flavum

Supraspinous ligament

Interspinous ligament

Intertransverse ligament

Posterior longitudinal ligament

Intervertebral disc

Anterior longitudinal ligament

Fig. 10.7. The ligaments connecting two thoracic vertebrae.

has two (Fig. 10.7). The two ligaments joining adjacent bodies together are the *anterior* and *posterior longitudinal ligaments*. These run from the base of the skull to the upper part of the sacrum, the former covering the front of the bodies and the latter the back. The posterior longitudinal ligament is thus found inside the vertebral canal. Both ligaments are firmly attached to the intervertebral discs as well as to the bodies of the vertebrae. The spines are also joined by two ligaments. The *interspinous ligament* joins the main part of the spines and it is thickened along its posterior edge to form the *supraspinous* ligament which joins the tips of the spines. In the cervical region the posteriorly directed concavity formed by the tips of the spines is filled by a midline vertical sheet of fibrous and elastic tissue called the *ligamentum nuchae*. Each of the two laminae is joined to the lamina of the adjacent vertebrae by the *ligamentum flavum*. This contains a good deal of elastic tissue which is stretched during flexion of the spine. The transverse processes are joined by *intertransverse ligaments* and the articular processes by the *capsular ligaments of the synovial joints*.

Finally, mention must be made of the so-called *neurocentral joints*, or joints of Luschka, which are small cavities situated in the postero-lateral parts of the discs in the cervical region. Opinion is divided over whether these are true synovial cavities or simply splits in the annulus fibrosus caused by wear and tear, but their importance is due to their very close relationship to the emerging spinal nerves.

The joints between the atlas and axis and between the atlas and the occipital

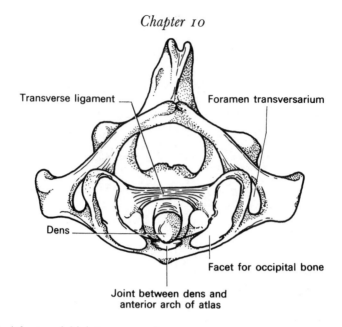

Fig. 10.8. Atlanto-axial joint, upper surface.

bone have a specialised structure. The median atlanto-axial joint is a pivot joint, the atlas rotating around the dens of the axis. There is a synovial joint cavity between the dens and the back of the anterior arch of the atlas, and the dens is held in place by a strong *transverse ligament* which bridges across it from one side of the articular facet to the other (Fig. 10.8). The lateral atlanto-axial joints are plane synovial joints and their capsular ligaments are thin and lax to allow rotatory movement.

The atlanto-occipital joint is between the rather kidney-shaped concave facets on the lateral masses of the atlas (Fig. 9.10) and the convex facets on the occipital bone. The bones are held together by the anterior and posterior longitudinal ligaments, the anterior and posterior atlanto-occipital membranes, which are attached between the corresponding arches of the atlas and the margins of the foramen magnum, and various other short ligaments. The principal movements at this joint are flexion and extension but a little lateral flexion can also occur.

THE SPINE AS A WHOLE

The functions of the spine are to protect the spinal cord and to transmit the body weight to the pelvis while at the same time to allow a wide range of movement. Thus the spinal cord is completely enclosed in a tunnel composed of alternating ligaments and bones, reminiscent of the cartilaginous rings and intervening tissue of the trachea. The only openings are the intervertebral foramina which allow the exit of the spinal nerves. The body weight is transmitted through the

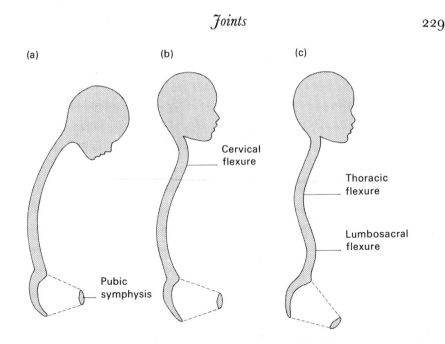

Fig. 10.9. Development of the curves of the spine (a) in the fetus; (b) development of the cervical flexure; (c) development of the lumbar flexure. From Beck, Moffat and Lloyd: *Human Embryology and Genetics*, Blackwell, Oxford.

bodies and intervertebral discs and it is obviously important that in order to minimise muscular effort the main axis of the spine should be perpendicular and any deviation from this line should be compensated for by another curve in the opposite direction. There are, in fact, a number of anteroposterior curves in the spine, whose disposition is best understood by following their development (p. 650). In the fetus and in the very young baby the whole spine forms a single curve, concave anteriorly (Fig. 10.9) although there are already early signs of the development of the cervical flexure.

Movements of the spine

In considering movements of the spine as a whole, it is necessary to think of both types of joints between two adjacent vertebrae, i.e. the secondary cartilaginous joints between the vertebral bodies and the synovial joints between the articular processes. The latter allow fairly free movement in certain directions but the movement between the vertebral bodies depends upon the integrity of the discs. During flexion, for instance, the anterior part of the disc will be compressed while the posterior region will be under tension. Thus movement here is limited to a tilting of one vertebral body upon another with the compression of the corresponding region of the disc. Although the amount of movement between any two vertebrae is limited, however, the sum of the

Flexion   Rotation   Lateral flexion   Extension

Fig. 10.10. The movements of the spine.

movements which take place at all the joints makes the spine, as a whole, a very mobile structure.

The most mobile regions of the spine are those corresponding to the secondary curves, and it is not surprising that these regions have the thickest intervertebral discs. Mobility in the cervical region is necessary in order to be able to turn the head to increase the field of vision, and anyone with a 'stiff neck' is well aware of the difficulty involved in, for instance, backing a car. Similarly, a mobile lumbar region is necessary for bending down, and for compensating for pelvic tilt. In the thoracic region, on the other hand, movements in a lateral direction are limited by the presence of the ribs.

The movements which are possible in the spine as a whole are illustrated in Figure 10.10. The extent and direction of the movements in any one region of the spine depend upon the thickness of the intervertebral discs and direction of the articular facets (see Chapter 9). In the cervical region. flexion and extension are free, particularly the latter. These movements are usually combined with flexion and extension at the atlanto-occipital joint. This is not invariably so, however, and movements of flexion of the cervical spine with extension at the atlanto-occipital joint will produce a forward movement of the face such as is seen when a short-sighted person peers forward to read. Lateral flexion in the cervical region is usually combined with rotation, owing to the slope of the articular facets, so that the ear is brought nearer to the tip of the shoulder (see the actions of sternomastoid, Ch. 11). In the thoracic region the principal movement is rotation, which occurs particularly in the lower thoracic region where the false ribs allow more free movement. Lateral flexion also occurs but is limited by the presence of the rib cage, while flexion and extension are similarly limited although some extension occurs in deep inspiration when it helps to increase the vital capacity. In the lumbar region, flexion and extension are particularly free; the latter may be seen particularly well in good gymnasts during a backward handspring. Lateral flexion can also occur but rotation is negligible.

Flexion of the spine is produced by longus cervicis, the scalene muscles, sternomastoid and rectus abdominis. In bending forward from the upright position, however, the movement is carried out by gravity and is controlled by the eccentric action of the extensor muscles. Extension is carried out by the various components of erector spinae, splenius capitis and semispinalis capitis. Lateral flexion is produced by sternomastoid, some of the short deep back muscles such as rotatores and by the oblique abdominal muscles—contraction of the right external oblique together with the left internal oblique will rotate the trunk to the left.

Flexion of the spine is limited by tension in the extensor muscles and in the ligaments of the posterior part of the spine and also by tension in the posterior parts of the intervertebral discs. In the cervical region it is also checked by contact between the overlapping anterior parts of the vertebral bodies (Fig. 9.8). Extension is less free than flexion and is limited by tension in the anterior longitudinal ligaments and by the approximation of the spines. Rotation and lateral flexion are limited by torsion and compression of the intervertebral discs and by tension in the ligaments of one side and in the antagonistic muscles.

COSTO-VERTEBRAL JOINTS

In general it may be said that each rib articulates posteriorly with two vertebrae

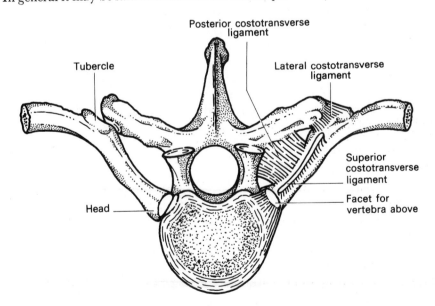

Fig. 10.11. The costo-vertebral joints.

—its own and the one above (Fig. 10.11). The head of the rib articulates with the two adjacent demifacets on the vertebral bodies and with the intervertebral disc between them, a factor which diminishes the amount of movement allowed in the thoracic spine. The articular facet on the tubercle articulates with the facet on its own transverse process. Exceptions to this are as follows: the first rib articulates only with the first thoracic body and transverse process; the tenth, eleventh and twelfth ribs also only articulate with one vertebra; the eleventh and twelfth have no articular tubercles and therefore do not articulate with transverse processes. The joints are all plane synovial joints and are surrounded by capsular ligaments and other accessory ligaments. There are, however, minor differences between the shape of the articular surfaces so that the lower ribs do not move during respiration in quite the same way as the upper ribs (see Chapter 19).

STERNOCOSTAL JOINTS

The first rib is joined to the manubrium by its costal cartilage and this is therefore a primary cartilaginous joint. The second to seventh costal cartilages unite with the corresponding notches on the sternum by plane synovial joints, although the cavities of these are sometimes missing. Small synovial cavities may also be found between the costal cartilages of the seventh to tenth ribs. The eleventh and twelfth ribs are floating—i.e. they do not articulate at their anterior ends with any other structure. Movements at all these joints are restricted to slight shifting but are reinforced by bending and twisting of the costal cartilage unless these are calcified, as happens with increasing age.

MANUBRIOSTERNAL JOINT

This is a secondary cartilaginous joint at which a slight hinging movement can take place. As a landmark, it is a useful guide to the position of the second rib (Fig. 9.16). In middle age and later the cartilage sometimes becomes ossified.

The movements at all the joints in the thoracic cage are involved in the mechanism of respiration and will be described in Chapter 19.

SACRO-ILIAC JOINT

Do not be misled by the rough appearance of the auricular surfaces of the bones—the sacro-iliac joint is a synovial joint of the plane variety and is therefore as subject to various types of arthritis as any other synovial joint. A capsular ligament is present which is thickened anteriorly and posteriorly (the *dorsal and ventral sacro-iliac ligaments*) and the articulating surfaces are covered with cartilage. The main bonds between the bones, however, are the very strong *interosseous sacro-iliac ligaments* which fill in the space between the back of the sacrum and the projecting posterior parts of the ilium (Fig. 10.12). Additional support is given by the *sacrotuberous* and *sacrospinous ligaments* (Fig.

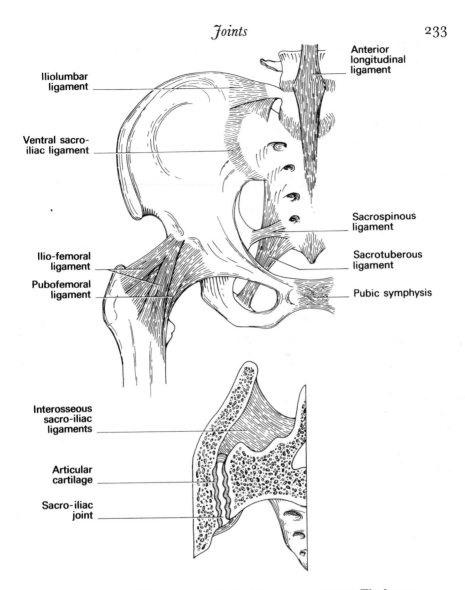

Iliolumbar ligament

Ventral sacro-iliac ligament

Ilio-femoral ligament

Pubofemoral ligament

Interosseous sacro-iliac ligaments

Articular cartilage

Sacro-iliac joint

Anterior longitudinal ligament

Sacrospinous ligament

Sacrotuberous ligament

Pubic symphysis

Fig. 10.12. The joints of the pelvis and the hip joint, anterior aspect. The lower diagram shows a section through the sacro-iliac joint.

10.13) which pass from the ischial tuberosity and spine respectively to the side of the sacrum and convert the greater and lesser sciatic notches into correspondingly named foramina. In addition, the *iliolumbar ligament* passes from the transverse process of the fifth lumbar vertebrae to the iliac crest and to the upper surface of the sacrum so that this also helps to bind the bones together. The position of the sacro-iliac joint is indicated on the surface by a line centred on the posterior superior iliac spines, since these lie at the level of the second sacral vertebra.

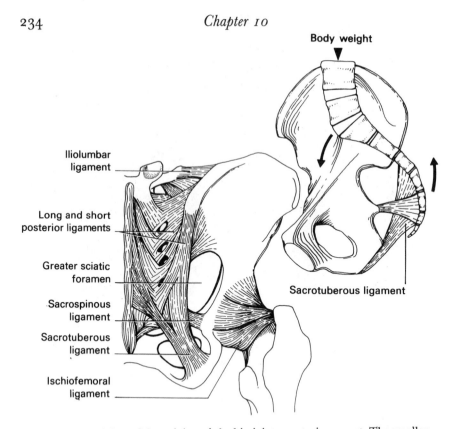

Body weight

Iliolumbar
ligament

Long and short
posterior ligaments

Greater sciatic
foramen

Sacrospinous
ligament

Sacrotuberous
ligament

Ischiofemoral
ligament

Sacrotuberous ligament

Fig. 10.13. The joints of the pelvis and the hip joints, posterior aspect. The smaller diagram shows how the sacrotuberous and sacrospinous ligaments resist rotation of the sacrum.

The sacro-iliac joints are subjected to great strain for they have to transmit the body weight from the spine to the pelvis and thence to the femoral heads. For this reason, only very little movement can occur at the joint although it is somewhat increased during pregnancy. The body weight tends to displace the sacrum downwards and forwards and this is resisted by a number of factors. The interosseous sacro-iliac ligaments pull the posterior parts of the ilia together so that the irregular but mutually congruent auricular surfaces are tightly interlocked. There is a tendency for the anterior parts of the joints to open out, but this is prevented by the pubic bones and pubic symphysis which act as a 'tie-beam'. Forward tilting of the sacrum is prevented by the sacrotuberous and sacrospinous ligaments (Fig. 10.13).

THE PUBIC SYMPHYSIS

This is a typical secondary cartilaginous joint and possesses a thick disc of fibro-cartilage. Although there is often a cavity in the centre of this it does not have a synovial membrane. The joint is reinforced by ligaments above and below.

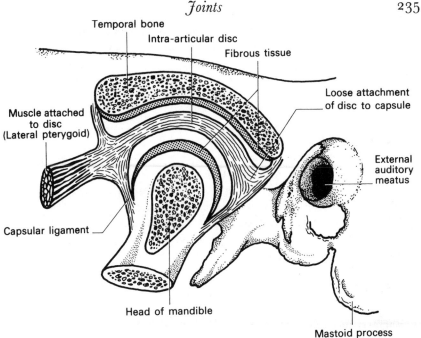

Fig. 10.14. A section through the left temporomandibular joint.

JOINTS OF THE SKULL

The only synovial joints of the skull are the minute joints between the ear ossicles and the joint between the temporal bone and the head of the mandible, so that only the latter (the *temporomandibular joint*) is of interest to physiotherapists. It is an atypical joint, perhaps because it is formed by two bones which ossify in membrane. The articular surfaces are the head of the mandible and the corresponding fossa on the temporal bone. The surfaces are covered by fibrous tissue rather than hyaline cartilage and the joint contains a fibrous disc which divides the joint cavity into two. There is a capsular ligament, thickened in places, and the attachment of the disc to the posterior part of the capsule is very lax (Fig. 10.14). When the mouth is opened, the head of the mandible together with the disc glide forward together on to the articular eminence which lies in front of the fossa. The hinge-type movement at the joint occurs between the head of the mandible and the disc. As a result of this complicated movement, the head of the mandible may be seen and felt as an eminence in front of the ear when the mouth is open, and the joint may be easily palpated. In this position the joint is unstable and a blow to the jaw may dislocate it very easily. The joint is so close to the ear that pain in the joint is often attributed to earache, and a clicking joint, which is common, causes some annoyance to the patient.

In addition to opening and closing the mouth (elevation and depression of the mandible) protraction and retraction can occur, together with some rotatory movements which are used in chewing. The muscles which open the mouth are weak and lie below the mandible but, of those which close the mouth, two are very powerful. They are the temporalis and the masseter which can be felt when the teeth are clenched in the temporal region and on the side of the face respectively.

JOINTS OF THE UPPER LIMB

The upper limb is built for mobility rather than strength and it comes as a surprise to many students to find that the only direct articulation between the upper limb and the trunk is at the relatively small *sternoclavicular joint*. The scapula articulates with the clavicle but is otherwise held in place only by its muscular attachments. Thus during extensive movements of the upper limb there are very free accompanying movements of the scapula and clavicle.

THE STERNOCLAVICULAR JOINT

This is a form of ball and socket joint between the enlarged medial end of the clavicle and the upper lateral angle of the manubrium sterni. The ends of the bones are covered by fibrocartilage and the joint contains an intra-articular disc which may be perforated. The disc is attached to the clavicle above and the first costal cartilage below and thus helps to stop medial displacement of the clavicle (Fig. 10.15). The capsular ligament is thickened in front and behind to form

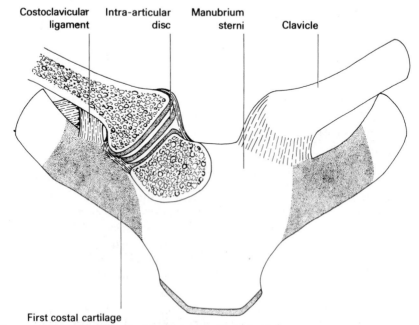

Costoclavicular ligament   Intra-articular disc   Manubrium sterni   Clavicle

First costal cartilage

Fig. 10.15 A section through the right sternoclavicular joint.

anterior and *posterior sternoclavicular ligaments* and there is also an *interclavicular ligament* which crosses the midline. The strongest ligament, however, is an accessory ligament—the *costoclavicular ligament*—which joins a rough area on the under surface of the clavicle to the first costal cartilage. This ligament forms a fulcrum so that when the lateral end of the clavicle is raised, as in shrugging the shoulders, the medial end is depressed. Movements at this joint accompany movements of the scapula and will be described later.

ACROMIOCLAVICULAR JOINT

This is a plane synovial joint between the flattened lateral end of the clavicle and the oval facet on the *medial surface* of the acromion (if you are in any doubt about the position of this joint, try to articulate a separate clavicle and scapula). The joint lies more or less in the sagittal plane and can easily be palpated. The joint cavity may be wholly or partly divided into two by an intra-articular disc. The capsular ligament is weak and the main bond between the bones is the *coracoclavicular ligament*. This is in two parts: a *conoid ligament* which is almost vertical and passes from the base of the coracoid process to the conoid tubercle on the under surface of the clavicle and a *trapezoid ligament* from the coracoid process to the trapezoid line on the clavicle (Fig. 10.16). These powerful ligaments

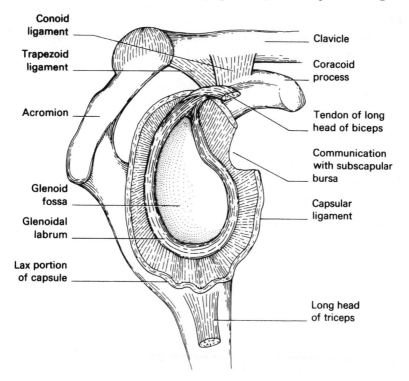

Fig. 10.16. Lateral view of the shoulder joint after division of the capsular ligament and removal of the humerus.

transmit the weight of the upper limb from the scapula to the clavicle and thence to the trunk. Therefore, the portion of clavicle lateral to the ligament is thin and fragile (and almost invisible on radiographs) and a fracture distal to the ligament does not produce severe effects.

MOVEMENTS OF THE SCAPULA AND CLAVICLE

The function of the clavicle is to act as a prop to the acromion and thus to brace the tip of the shoulder away from the midline. It performs this function through-out all movements of the scapula so that all movements at the sternoclavicular and acromioclavicular joints are secondary to movements of the scapula.

The scapula may be elevated as a whole, as in shrugging the shoulders, and from the elevated position it can be depressed again to its normal position, a movement which is usually carried out by gravity. It can be brought forward around the chest wall (*protraction*) and braced back so that medial border approaches the midline (*retraction*). The most important movement, however, is elevation of the lateral part of the scapula so that the glenoid points upwards. This is usually known as *lateral rotation* of the scapula and it plays an essential part in full abduction of the upper limb (Fig. 10.18). In all these movements there is a slight shifting or rotation at the acromio-clavicular joint and a more extensive movement at the *sternoclavicular* joint. Owing to the presence of the costoclavicular ligament, which acts as a fulcrum, elevation of the lateral end of the clavicle is accompanied by a downward movement of the medial end at the sternoclavicular joint, while a forward movement of the lateral end of the clavicle is accompanied by a backward movement of the medial end. All these movements take place more freely between the clavicle and the disc than between the disc and the sternum. Movements of the scapula are facilitated by the loose planes of connective tissue between the serratus anterior and subscapularis muscles and between serratus anterior and the chest wall.

THE SHOULDER JOINT

The shoulder joint is a good example of a ball and socket joint in which the very shallow glenoid fossa receives the head of the humerus. A comparison of the bones concerned will indicate that the shoulder joint is much less stable and much more mobile than the hip joint and this impression is borne out by examination of the ligaments around the joint. Both the articulating surfaces of the shoulder joint are covered with hyaline articular cartilage and the glenoid cavity is, in addition, deepened slightly by a ring of fibrocartilage, the *glenoidal labrum*, which is attached to its rim (Fig. 10. 16). The capsular ligament is attached to the margin of the glenoid cavity beyond the labrum and including the base of the coracoid process and the origin of the long head of biceps, and to the anatomical neck of the humerus except inferiorly, where its attachment is lower down (Fig. 10.17). The under surface of the capsular ligament is therefore lax so as to allow for abduction at the joint. Two or three thickenings

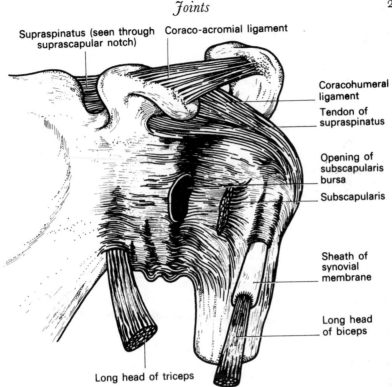

Supraspinatus (seen through suprascapular notch)

Coraco-acromial ligament

Coracohumeral ligament

Tendon of supraspinatus

Opening of subscapularis bursa

Subscapularis

Sheath of synovial membrane

Long head of biceps

Long head of triceps

Fig. 10.17. Shoulder joint, anterior aspect.

in the anterior part of the capsular ligament form the *glenohumeral ligaments*. In addition, the capsular ligament is reinforced above by a *coracohumeral ligament* which passes from the lateral border of the coracoid process to the greater tuberosity of the humerus. The joint is lined by synovial membrane, except for the cartilage-covered articulating surfaces. There are two openings in the capsule. Anteriorly, the subscapularis tendon crosses the joint on its way to the lesser tuberosity and it is separated from the joint by a large *subscapular bursa* which communicates widely with the joint through an opening in the anterior part of the capsule. The other, smaller opening is for the tendon of the long head of biceps. This arises from the supraglenoid tubercle and then traverses the joint above the head of the humerus before turning downwards to enter the intertubercular sulcus (bicipital groove). As the tendon pierces the capsule it takes with it a tubular prolongation of synovial membrane which helps to lubricate it in the sulcus.

Above the shoulder joint itself are a ligament and a bursa which are so important for the functioning of the joint that the bursa is sometimes called the secondary shoulder joint. The ligament is the *coraco-acromial ligament* which, together with the acromion and the bursa of the coracoid process, form the *coraco-acromial arch*. Below this is the bursa—the *subacromial bursa*—which is so large that it extends laterally to a subdeltoid position. The supraspinatus tendon

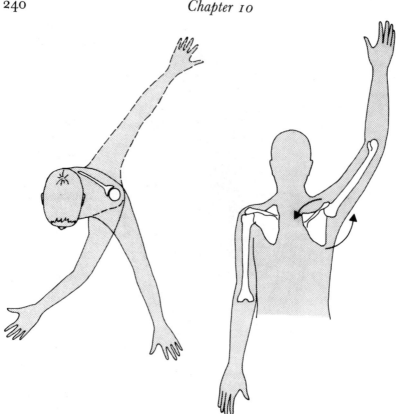

Fig. 10.18. Movements of the upper limb as shown are due not only to movement at the glenohumeral joint but also by movements of the scapula.

is intimately related to the floor of the bursa so that disease of the tendon readily affects the function of the bursa (Fig. 10.17). Below the supraspinatus tendon is the capsular ligament of the shoulder joint. Thus, the coraco-acromial arch forms a secondary socket for the head of the humerus and movements between the head of the humerus and the arch are facilitated by the bursa.

The edge of the glenoid cavity which forms the lower half of the whole joint may be represented on the surface by a line 3 cm long, drawn downwards from the lateral side of the coracoid process. The shoulder joint is relatively unstable compared to the hip joint owing to the shallowness of the glenoid cavity and the weakness of the ligaments. It is strengthened, however, by the four muscles of the rotator cuff (subscapularis, supraspinatus, infraspinatus and teres minor) which closely surround the joint, blend with the capsule, and act as mobile ligaments. The weakest spot in the capsule is the lax inferior part so that dislocation always occurs downwards in the first instance.

The movements of the joint take place about three axes, as is usual in a ball and socket joint. They are complicated by the fact that the glenoid cavity does not face directly laterally, but rather laterally and forwards (Fig. 10.18).

Hence, flexion, in which the humerus moves in the same plane as the glenoid, involves movement of the upper limb forwards and slightly medially. Similarly, extension carries the limb backwards and laterally. Abduction involves the arm in a movement away from the midline and slightly forwards from the coronal plane while in adduction the arm is brought back to the side in the same plane. Finally, medial and lateral rotation occur about an axis corresponding to the long axis of the humerus. The range of the latter movements may best be examined by first flexing the supinated forearm to a right angle since if rotation is attempted with the elbow extended it is difficult to distinguish from pronation and supination at the radio-ulnar joints.

The movement of full abduction of the upper limb needs to be described separately since it also involves movements of the shoulder girdle. The first 10°–20° of abduction takes place at the shoulder joint and is performed by the action of supraspinatus until deltoid takes over. Well before the humerus has reached the horizontal position, however, the scapula begins to rotate laterally so that the glenoid cavity begins to turn upwards. The combined movements of abduction at the shoulder joint and rotation of the scapula continue until full abduction of the upper limb is reached. In the final stages of movement at the shoulder joint, which brings it into the close-packed position, the humerus has to rotate laterally so that the greater tuberosity does not impinge upon the coraco-acromial arch. The scapular movement is, of course, accompanied by elevation of the lateral end of the clavicle and movement at the sterno-clavicular joint.

The movement of circumduction is performed by a smooth combination of the primary movements so that the humerus traces out a cone whose apex is the glenoid cavity or a circle whose centre is the glenoid. This type of movement is carried out when bowling at cricket.

Since the upper limb has such a wide range of movement, relatively long muscles are required so that an adequate range of contraction can be obtained. This statement does not, however, apply to some of the rotators. The principal abductors are supraspinatus (in the initial stages of the movement) deltoid (to abduct the shoulder joint), and the lower fibres of serratus anterior and both upper and lower fibres of trapezius (to rotate the scapula). The long head of biceps may assist when the humerus is laterally rotated. The chief adductors are the pectoralis major and latissimus dorsi, aided by the teres major and minor and the long head of the triceps. A simple lowering of the arm to the side, however, is performed by gravity with the deltoid contracting eccentrically. Flexion is carried out by the anterior fibres of deltoid and the clavicular head of pectoralis major, aided by coracobrachialis and the long head of biceps. Extension is the function of the posterior fibres of deltoid, teres major and latissimus dorsi with the sternocostal head of pectoralis major assisting during extension from the flexed position. The chief medial rotators are the pectoralis major, latissimus dorsi, subscapularis and the anterior fibres of deltoid while lateral rotation is carried out by the teres minor, infraspinatus and the posterior fibres of deltoid. All the major muscles which act on the shoulder joint are attached to the humerus at some distance from the joint so they act with a good mechanical advantage. When they contract strongly, any tendency of the head of the humerus to slip on the glenoid fossa is counteracted by contraction of the rotator cuff muscles which therefore act as stabilisers. Holding the upper limb in any position, especially if a weight is being carried, involves the contraction of various muscles around the scapula which act as fixation muscles for this bone.

Accessory movements at the shoulder joint are rather free because of the relatively lax capsule and the shallowness of the glenoid fossa. The head of the humerus can slide in various directions when the muscles are relaxed, and if the arm is abducted to a right-angle the two joint surfaces can be separated by traction.

The elbow joint and the superior radio-ulnar joint

These two joints share a single synovial cavity and a number of ligaments so that it is convenient to describe them together. The trochlear notch of the ulna articulates with the trochlea (trochlea = pulley) and the head of the radius with the capitulum (Fig. 10.19). In addition, the side of the head of the radius articulates with the radial notch of the ulna. The elbow joint is a hinge joint and the superior radio-ulnar a pivot joint. The concave head of the radius thus slides over the convex capitulum during flexion and extension and rotates upon it during pronation and supination. Since the elbow cannot normally be extended much beyond 180°, the capitulum does not extend on to the back of the humerus but the trochlea does, because of the shape of the trochlear notch of the ulna. All surfaces are covered by hyaline articular cartilage and the non-articular surfaces within the joint are lined with synovial membrane. The olecranon, coronoid and radial fossae are occupied by subsynovial fat pads which are displaced by the corresponding bony processes in full flexion or extension.

As is usual in hinge joints, the anterior and posterior parts of the capsular ligaments are relatively thin and lax while the medial and lateral parts are strong. The attachments of the capsular ligament are shown in Fig. 10.19. A special *annular ligament* (annulus = ring), lined by a thin layer of cartilage, is provided for the head of the radius. This is attached to the anterior and posterior edges of the radial notch of the ulna so that the bone and ligament together form a strong complete circle within which the head of the radius rotates. The capsular and radial collateral ligaments blend with the annular ligament laterally. The *ulnar collateral ligament* is attached above to the medial epicondyle of the humerus. Its fibres fan out to be attached by strong bands to the coronoid process and the olecranon process of the ulna and between these, to a band of fibrous tissue which joins these two bony points. The *radial collateral ligament* extends from the lateral epicondyle to the annular ligament. Many of the muscles which take origin from the region of the elbow have some of their fibres arising from the ligaments, for example, the common flexor and extensor origins. The ulnar nerve is closely related to the ulnar collateral ligament.

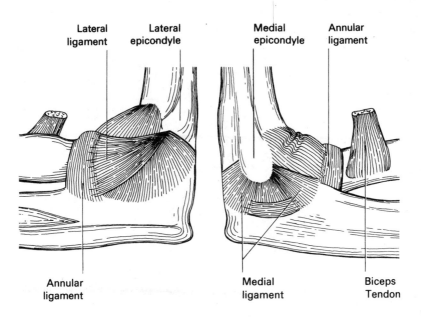

Fig. 10.19. Elbow, joint, lateral and medial aspects.

The movements which take place at the elbow joint are flexion and extension only. Extension is limited by contact between the olecranon and the olecranon fossa together with the tension in the flexor muscles and the anterior part of the capsular ligament. Flexion is limited by contact between the arm and the forearm. In full extension, the carrying angle can be seen, but owing to the shape of the trochlea this disappears during full flexion. The articular surface of the ulna does not fit the trochlea perfectly, so that flexion and extension are accompanied by a slight degree of rotation at the ulna. Movements at the superior radio-ulnar joint consist of a pure rotation—a fuller description of pronation and supination will be given in the next section. Accessory movements at the elbow joint are slight and consist only of a slight rocking of ulna on humerus and radius on capitulum.

The line of the elbow joint lies about 2 cm below the epicondyles. Rotation of the head of the radius may be felt by placing the index finger on the lateral epicondyle and the middle finger alongside and below it. The latter will then be lying over the radial head.

Flexion at the elbow joint is carried out by biceps, brachialis, brachioradialis (in the mid-prone position) and, in strong movements, the muscles arising from the common flexor origin on the medial epicondyle. The chief extensor is triceps but when the forearm is lowered from a flexed position, the movement is performed by gravity and controlled by an eccentric contraction of the flexor muscles.

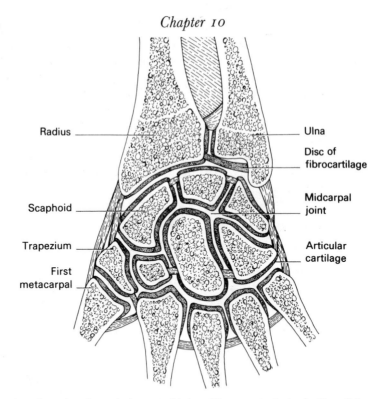

Radius —

Ulna

Disc of
fibrocartilage

Scaphoid —

Midcarpal
joint

Trapezium —

Articular
cartilage

First
metacarpal —

Fig. 10.20. A section through the carpal joints. Note particularly the line of the wrist joint, the midcarpal joint between the proximal and distal rows of bones, and the first carpometacarpal joint.

THE INFERIOR RADIO-ULNAR JOINT

This is a pivot joint although its movements are rather different from those of the superior radio-ulnar joints. The small ulnar head articulates with the ulnar notch on the radius and with a *triangular disc* of fibrocartilage whose apex is attached to the base of the ulnar styloid process and whose base is attached to the edge of the ulnar notch of the radius. These surfaces are included in a single joint cavity (Fig. 10.20). The ulna is thus excluded from the wrist joint so that during pronation and supination the radius rotates around the (more or less) stationary ulna, carrying the hand with it. In fact, the ulna does move a little during pronation and supination as may be easily felt with a finger on the ulnar styloid (the head of the ulna may only be felt clearly when the forearm is pronated).

Thus, the true axis around which pronation and supination occur is a line which passes through the centre of the radial head and approximately through the styloid process of the ulna. In supination, the shafts of the radius and ulna are parallel but in pronation they cross one another.

The capsular ligament of the joint is rather weak and the bones are held together principally by the triangular disc and the pronator quadratus muscle.

In addition, the integrity of the joint is ensured by the ligaments of the wrist joint (see below) and by the *interosseous membrane*, a sheet of dense fibrous tissue which extends between the interosseous borders of the shafts of the radius and ulna and whose fibres run downwards and medially. Several of the deep muscles take partial origin from the interosseous membrane. Pronation is carried out by pronator teres and pronator quadratus and supination by biceps and supinator. Brachioradialis can supinate from the fully prone position and vice versa so that its flexor effect can only occur when the arm is in the mid-position. The supinators are, in general, more powerful than the pronators, hence screws are made with a right-hand thread.

THE WRIST JOINT (radiocarpal joint)

The wrist joint is an ellipsoid joint in which the upper articular surface is formed of the radius and the triangular disc at the lower end of the ulna while the opposing surface comprises the three main bones of the proximal row of the carpus, i.e. the scaphoid, lunate and triquetral. These three bones are united by *interosseous ligaments* which fill up the spaces between them where they border on the joint cavity so that the bones and ligaments present a continuous surface. The line of the wrist joint is indicated approximately by a line joining the styloid processes of the radius and the ulna. The synovial membrane is uncomplicated and the capsular ligament is attached to the bones in the vicinity of the joint. It is reinforced in front and behind by *radiocarpal ligaments* and at the sides by *radial and ulnar collateral ligaments*. The capsular ligament is continued distally to form the capsular ligament of the midcarpal joint. The movements of this joint will be dealt with in the next section.

THE MIDCARPAL JOINT

This is a rather complex joint between the proximal and distal rows of carpal bones. The joint may be subdivided into two parts: firstly, a complex form of condyloid joint between distally, the hamate and capitate and, proximally, the concavity formed by the scaphoid and lunate. Secondly, there is a plane or gliding joint between the trapezoid and trapezium distally, and the scaphoid proximally (Fig. 10.20). The synovial cavity is continuous from one side of the carpus to the other, between proximal and distal rows of bones, and it sends extensions proximally and distally between individual bones. It may communicate with some of the carpo-metacarpal joints. There are strong *palmar* and *dorsal ligaments* and the bones are also held together by *interosseous ligaments*. The *flexor retinaculum* is also important for the integrity of the intercarpal joints since it helps to bind the whole structure of the carpus together to form the carpal tunnel.

The movements which can be carried out at the radiocarpal and midcarpal joints are flexion and extension, abduction (radial deviation) and adduction (ulnar deviation). In addition, a certain amount of passive rotation can be produced (not to be confused with pronation and supination). These movements take place at both joints together so that the back of the wrist makes a smooth curve during flexion. Flexion occurs partly at the midcarpal joint and, to a slightly less extent, at the radiocarpal joint. Extension is slightly greater at the radiocarpal joint. Abduction is more limited than adduction because the styloid process of the radius extends farther distally than that of the ulna. During this movement the scaphoid and lunate are in contact with the radius and the disc, while in adduction, the lunate and triquetral are the bones involved. Adduction takes place mainly at the radio-carpal joint but abduction occurs almost entirely at the midcarpal joint. Flexion is produced mainly by the flexor carpi radialis and flexor carpi ulnaris acting together, with the long flexors taking part during a powerful movement. Extension is produced by extensor carpi ulnaris and the extensors carpi radialis longus and brevis. In abduction the radial flexors and extensors act together while in adduction the ulnar carpal flexors and extensors are involved. Movements are limited by tension in the opposing muscles and ligaments. In flexion and extension the length of the muscles is particularly important; note, for example, how clenching the fist limits the range of flexion.

THE PISOTRIQUETRAL JOINT

This is a small circular synovial joint between the pisiform and triquetral bones. It has two accessory ligaments: the *pisohamate* and *pisometacarpal ligaments* which join the pisiform to the hook of the hamate and the base of the 5th metacarpal respectively. The latter probably represents the tendon of the flexor carpi ulnaris.

CARPOMETACARPAL JOINTS

Apart from the first, these are plane joints between the bases of the metacarpals

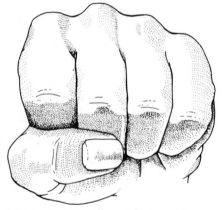

Fig. 10.21. A clenched fist to show the effect of flexion at the carpometacarpal joints of the ring and little fingers.

and the corresponding carpal bones and they are usually combined to form a single joint cavity which has offshoots passing between the bases of the meta-carpals. The capsular ligaments of these joints are strengthened by *dorsal, palmar* and *interosseous ligaments.* Movements are very restricted in the case of the 2nd and 3rd joints but a certain amount of flexion and extension occurs at the 4th and a good deal at the 5th. (These movements may be confirmed by holding the head of each metacarpal in turn and trying to flex and extend it.) It is because of this that the knuckles (metacarpal heads) in a clenched fist are not in a straight line (Fig. 10.21). When spherical objects are tightly gripped, the palm of the hand becomes cupped as a result of this movement and the medial side of the palm closes in around the object. No abduction can occur at these joints because of the deep transverse metacarpal ligament (see below).

THE FIRST CARPOMETACARPAL JOINT

This is one of the most important joints in the hand because it is here that the main movements of the thumb occur, particularly that of opposition. It is a saddle-shaped joint, the two surfaces being reciprocally curved, and the capsular ligament surrounds it completely so that the joint does not communicate with any other joint. Ligaments are formed by slight thickenings in the capsule and the anterior and posterior ligaments, which are oblique, are so arranged that when flexion occurs at the joint, medial rotation also takes place, while full extension is associated with lateral rotation.

The movements which take place at this joint are flexion and extension, abduction and adduction and medial and lateral rotation. Opposition results from a combination of flexion and medial rotation. Since the thumb, in the position of rest, lies in a plane at right angles to the fingers, its movements are not at first easy to understand. They are shown in Figure 10.22. Flexion involves bending the thumb across the palm of the hand while extension is the opposite movement. In abduction, the thumb is carried away from the plane of the palm and in rotation, the metacarpal rotates around its long axis. The most important movement is opposition since this is the means by which the pad of the thumb can be brought to lie against the pads of the fingers and small objects may thus be held delicately and fine manipulations carried out. You will appreciate the importance of this if you try to pick up a pen and write using only the fingers, which cannot be opposed to each other. This 'precision grip' will be analysed in more detail in Chapter 11.

The muscles which carry out the movements are as follows: Abduction—the abductors pollicis longus and brevis, and to some extent, extensor pollicis brevis. Adduction—the adductor pollicis and the first palmar interosseous (the latter muscle, however, acts mainly on the metacarpo-phalangeal joint). Flexion—flexor pollicis brevis, opponens pollicis and flexor pollicis longus. Extension—extensor pollicis longus and brevis and abductor pollicis longus. Medial rotation is always associated with flexion (= opposition) and is carried out by opponens pollicis and flexor pollicis brevis; it is assisted by muscles attached to

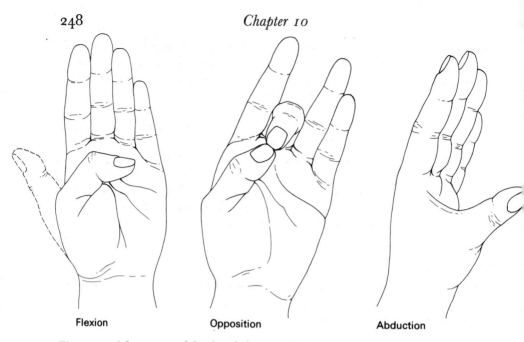

Flexion   Opposition   Abduction

Fig. 10.22. Movements of the thumb (see text for description).

the medial side of the proximal phalanx, while similar muscles on the lateral side are responsible for lateral rotation. Powerful opposition of the thumb, as in squeezing or pinching involves in addition, contraction of other muscles such as flexor pollicis longus and adductor pollicis. Movements at the first carpo-metacarpal joint are usually associated with movements at the metacarpo-phalangeal joint and the same muscles may work on both joints so that movements of the thumb and the muscles which produce them are very difficult to analyse. In addition, some of the short muscles have a dual nerve supply so that paralysis of individual muscles may be difficult to interpret.

METACARPOPHALANGEAL JOINTS

These are ellipsoid joints at which flexion, extension, abduction and adduction take place. The latter two movements occur in relation to the middle finger so that abduction involves the spreading out of the fingers, and adduction, bringing them together. Each joint has a capsular ligament, the palmar aspect of which is greatly thickened and infiltrated with fibrocartilage to form a *palmar ligament*. The palmar ligaments (except that of the thumb) are joined to each other by the strong *deep transverse metacarpal ligaments* (Fig. 11.31), which holds the heads of the metacarpals together. In addition, each joint has a *medial and lateral collateral* ligament. Muscles producing the movements of these joints are as follows: Flexion—flexor digitorum superficialis and profundus, flexor pollicis longus and brevis, interossei and lumbricals. Extension—extensor digitorum longus, extensor pollicis longus and brevis, extensor indicis and extensor digiti minimi. Abduction—dorsal interossei, abductor pollicis brevis, abductor digiti minimi. Adduction—palmar interossei, adductor pollicis. Flexion of the joints by the long flexors is associated with some degree of adduction and extension with abduction.

INTERPHALANGEAL JOINTS

These are simple hinge joints, the movements which occur being only flexion and extension. The joint surfaces can be separated by strong traction when the abrupt breaking of the film of synovial fluid between them is signalled by a sharp crack and the appearance within the joint (on X-ray) of a bubble of gas. The arrangement of the ligaments of these joints is similar to that in the metacarpophalangeal joints.

Flexion at these joints is brought about by flexor digitorum superficialis (for the proximal joint only), flexor digitorum profundus, flexor pollicis longus. Extension is by the interossei and lumbricals (via the insertions into the dorsal extensor expansion), extensor digitorum, extensor pollicis longus, extensor indicis, and extensor digiti minimi.

JOINTS OF THE LOWER LIMB

HIP JOINT

The hip joint is a very stable ball and socket joint between the head of the femur and the acetabulum. The head of the femur is covered with hyaline articular cartilage but is not completely smooth owing to the presence of the fovea, to which one of the ligaments is attached and which is pierced by one or more nutrient arteries. Cartilage also lines the acetabulum but is restricted to the weight-bearing area so that the articulur cartilage is horsehoe-shaped (Fig. 10.23). The remaining area of the acetabulum is occupied by a fat pad covered with synovial membrane. The acetabular notch is bridged by the transverse ligament which completes the circular rim and allows vessels to enter the acetabulum deep to it. The rim itself is reinforced by a fibrocartilaginous ring, triangular in cross section, which deepens the acetabulum and helps to keep the head of the femur in place. It is called the *labrum acetabulare*.

The capsular ligament is very thick, in marked contrast to that of the shoulder joint, and it is attached around the rim of the acetabulum beyond the labrum. Below, it is attached anteriorly to the intertrochanteric line of the femur but posteriorly it has a less strong attachment to the neck of the femur. It is reinforced around the region of the neck of the femur by deeply placed circular fibres (the *zona orbicularis*). Fibres from the femoral attachment of the capsule are reflected up the neck of the femur to form longitudinal ridges known as *retinacula*. Vessels from the capsule use the retinacula, which are covered by synovial membrane, as pathways to the head of the femur. Thickenings in the

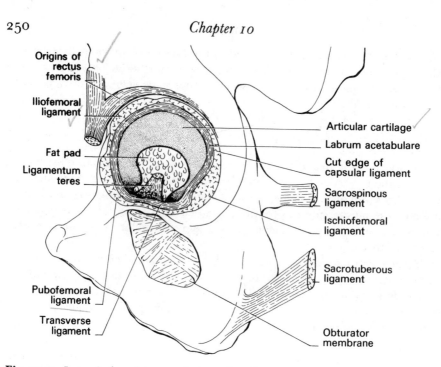

Origins of
rectus
femoris

Iliofemoral
ligament

Fat pad

Ligamentum
teres

Pubofemoral
ligament

Transverse
ligament

Articular cartilage

Labrum acetabulare

Cut edge of
capsular ligament

Sacrospinous
ligament

Ischiofemoral
ligament

Sacrotuberous
ligament

Obturator
membrane

Fig. 10.23. Lateral view of the acetabulum after divisions of the capsular ligament and
removal of the femur.

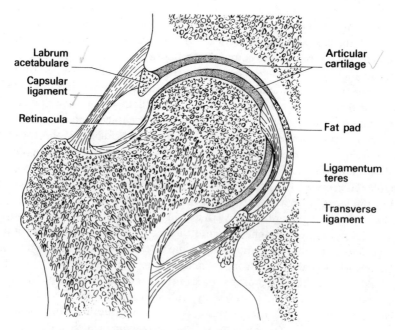

Labrum
acetabulare

Capsular
ligament

Retinacula

Articular
cartilage

Fat pad

Ligamentum
teres

Transverse
ligament

Fig. 10.24. A coronal section through the hip joint.

capsule form the three main ligaments of the joint: the *iliofemoral, ischiofemo*
and *pubofemoral* (Fig. 10.12). The most important of these is the iliofemoral
ligament, which is extremely strong, so that together with the flexor muscles of
the hip (especially iliopsoas), it counteracts the effects of gravity since the line
of gravity passes behind the joint. It is attached above to the anterior inferior
iliac spine and as it passes down it splits into two main bands which are attached
to the intertrochanteric line (its old name was the 'Y-shaped ligament'). The
capsular ligament between these bands is relatively thin. The pubofemoral
ligament is attached to the pubic part of the acetabular rim and the adjacent
pubic bone and passes across to the lower part of the intertrochanteric line,
extending also slightly upwards on to the back of the capsule. The ischiofemoral
ligament is attached to the ischial part of the acetabular rim and spirals over
the upper aspect of the joint to blend with the rest of the capsule at the base of
the greater trochanter.

The one intracapsular ligament is the *ligamentum teres* (ligament of the head),
which is attached below to the transverse ligament and the edges of the acetabular
notch (Fig. 10.24). It then winds over the surface of the femoral head to be
attached to the pit or fovea on the head itself. It is not a strong ligament and
does nothing to hold the bones together. Its main importance is that it is
clothed in synovial membrane which thus forms a tube around it through
which blood vessels are conveyed from the acetabular notch to the small foramina in the fovea.

The synovial membrane of the hip joint has the usual distribution. It lines
the inside of the capsular ligament and is reflected up the neck of the femur,
where it is raised into ridges by the retinacula, as far as the edge of the articular
cartilage. It is also reflected up around the ligamentum teres. Occasionally, the
synovial cavity of the hip joint communicates with the bursa which lies between
the tendon of psoas and the joint.

The position of the hip joint relative to the surface is not difficult to determine. Passing from lateral to medial, the femoral nerve is separated from the
hip joint by iliacus, the femoral artery by psoas and the femoral vein by
pectineus (Fig. 25.6). Since the tendon of psoas crosses the centre of the joint,
this point on the surface corresponds to the femoral point where the pulsations
of the femoral artery may be felt, i.e. half way between the anterior superior
iliac spine and the symphysis pubis.

The blood supply of the joint is derived from the plexus of vessels around
the joint in the usual way. Vessels reach the head of the femur via the retinacula
and via the ligamentum teres. There is little or no anastomosis between the two
sets of vessels, however, and a severe fracture of the neck may deprive the head
of most of its blood supply. The nerve supply is according to Hilton's Law.
Since the femoral and obturator nerves give branches to both knee joint and
hip joint, disease of the hip joint sometimes gives pain which is felt in the knee.

The movements which may be carried out at the joint are flexion and
extension, adduction and abduction, and medial and lateral rotation. Circumduction can also occur. In flexion, the anterior surface of the thigh is brought

towards the abdomen and, in fact, it is the contact between these two surfaces which limits flexion. Full flexion can only normally occur when the hamstrings are relaxed so that in examining the range of flexion at the hip, the knee must be fully flexed. Extension is a more restricted movement and is limited by tension in the flexor muscles and in the iliofemoral ligament. Abduction is much more limited than the corresponding movement in the upper limb owing to the length of the neck of the femur and the relatively little ancillary movement of the pelvic girdle compared to that of the scapula (but see below). It is limited by tension in the inferior part of the capsular ligament, in the pubofemoral ligament and in the adductor muscles. Adduction is limited by the presence of the other leg but the full range may be examined by abducting the opposite leg to make room for the leg being examined, or by a little flexion of the latter. As can be deduced from its attachments, the ligamentum teres is taut in adduction. Rotation at the joint occurs about an axis which, because of the length of the femoral neck, passes through the head of the femur above and (approximately) the lateral femoral condyle below, when the foot is on the ground; when the foot is free, however, the axis of rotation is variable.

It is extremely important to remember that movements at the hip joint are usually accompanied by movement of the pelvis and of the spine, and that limitation of movement at the joint may be disguised by compensatory move-

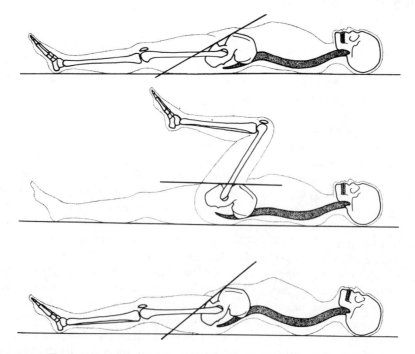

Fig. 10.25. Flexion of the hip joint can tilt the pelvis backwards and flatten the lumbar spine. The lowest diagram shows how a flexion deformity of the hip joint can be compensated by a forward tilt of the pelvis and an increase in the lumbar curvature.

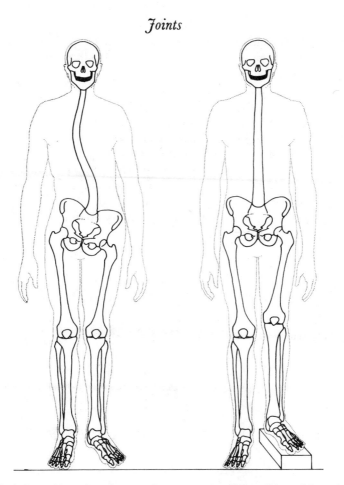

Fig. 10.26. A shortening of one leg may be compensated by a tilting of the pelvis. It may thus not be apparent unless the examiner investigates the relative levels of the anterior superior iliac spines.

ments of the whole pelvis. Thus, full flexion of the hip joint is usually accompanied by a backward tilt of the pelvis and a flattening out of the lumbar curve (Fig. 10.25). A permanent flexion of the hip joint as a result of disease may be disguised by a forward tilt of the pelvis and an increased lumbar curve (*lordosis*). The deformity may, however, be revealed if the lumbar curve is flattened out by fully flexing the opposite hip joint when the affected leg will be raised off the bed. Similarly, shortening of one leg may be disguised, when the patient is standing, by a sideways tilting of the pelvis (Fig. 10.26) and may only be revealed when the level of the two anterior superior iliac spines is compared.

Thus, when investigating the range of movement at the hip joint, it is essential to control movements of the pelvis by placing one hand on a convenient bony point such as the anterior superior iliac spine or by fixing the pelvis in some way.

The muscles responsible for flexion at the hip joint are the iliopsoas, the tensor fasciae latae, rectus femoris, sartorius, pectineus and the upper fibres of adductor longus. Extension is carried out by the gluteus maximus and the hamstrings—the former is particularly used for powerful movements, as in walking upstairs. The abductors are the gluteus medius and minimus and the tensor fasciae latae. The adductors are the adductors longus, brevis and magnus, the pectineus and the gracilis. Medial rotation is produced by the action of the anterior fibres of gluteus medius and minimus, the tensor fasciae latae and according to some authorities, psoas major, but this is uncertain. Lateral rotation is a function of the short muscles behind the hip joint and the gluteus maximus. In the normal standing position, the line of gravity passes behind the hip joint, which, in extension, is in its close packed position. Hence the posture may be maintained by only a minimal contraction of iliopsoas.

THE KNEE JOINT

This is essentially a hinge joint in which a small amount of rotation can also take place. The bones involved are the femur and tibia but there is also a large

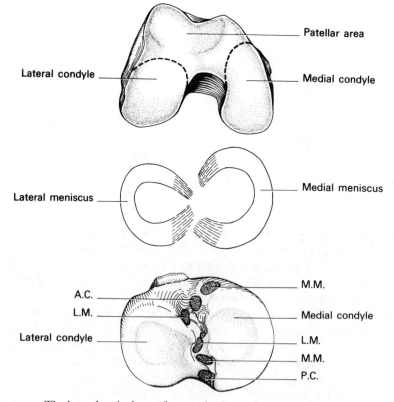

Fig. 10.27 The lateral articular surface on the femoral condyle, the lateral meniscus and the lateral articular surface of the tibia are all approximately circular. The corresponding surfaces on the medial side are all approximately oval.

sesamoid bone, the patella, which articulates with the femur. The fibula is not a weight-bearing bone and plays no part in the formation of the knee joint except to serve as an attachment for ligaments.

The medial and lateral articular surfaces of the tibia are separate and are each covered by articular cartilage. In the case of the femur, however, there is a single, extensive cartilage-covered surface for both the tibial condyles and for the patella. The areas which articulate with the tibia are, however, demarcated by faint grooves in such a way that the lateral area is shorter than the medial area (Fig. 10.27). This has an important bearing on the function of the joint. The deep (articular) surface of the patella is also covered by articular cartilage.

The joint is surrounded by a *capsular ligament* which is replaced anteriorly by the patella and the ligamentum patellae. There are various other deficiencies in the capsular ligament which allow the synovial membrane of the joint to become continuous with the lining of certain important bursae. These will be mentioned later. The capsular ligament is strengthened by wide expansions from the vastus medialis and lateralis which are called the *patellar retinacula*. They extend around the joint on either side to be attached to the edges of the patella and the ligamentum patellae and to the margins of the tibia as far back

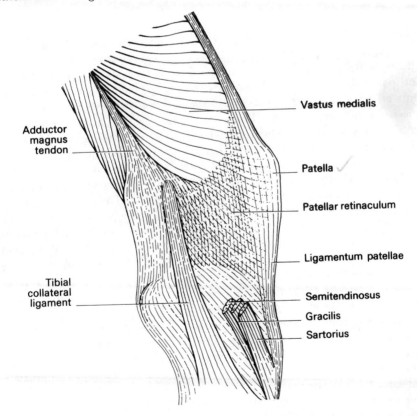

Fig. 10.28 Knee joint, medial aspect.

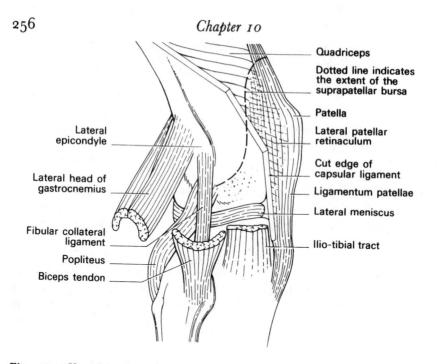

Quadriceps

Dotted line indicates the extent of the suprapatellar bursa

Patella

Lateral patellar retinaculum

Cut edge of capsular ligament

Ligamentum patellae

Lateral meniscus

Ilio-tibial tract

Lateral epicondyle

Lateral head of gastrocnemius

Fibular collateral ligament

Popliteus

Biceps tendon

Fig. 10.29. Knee joint, lateral aspect after removing part of the capsular ligament.

as the collateral ligaments. They thus have an important stabilising effect on the joint, particularly when quadriceps is contracting. The most important ligaments of the joint are the *tibial* and *fibular collateral ligaments* and the *cruciate ligaments*, which are internal (Figs. 10.28 and 10.29). The tibial collateral ligament is a wide strap-like ligament which blends with the capsular ligament. It extends from the medial epicondyle of the femur down to the upper one quarter or so of the shaft of the tibia although some of its posterior and deeper fibres are also attached to the upper part of the groove for the semimembranosus tendon. It is attached also to the medial meniscus. The fibular collateral ligament is very different. It is a strong rounded cord which extends between the lateral epicondyle and the upper end of the fibula where it is closely enfolded by the insertion of the biceps tendon. In contrast to the tibial collateral ligament it stands clear of the joint and is separated from the lateral meniscus by the tendon of popliteus. Posteriorly, a thickening in the capsule is produced by an expansion from the tendon of semimembranosus which passes upwards and laterally to form the *oblique popliteal ligament*.

The two cruciate ligaments are so called because they form a cross lying approximately in the sagittal plane. They are entirely within the capsular ligament but are not within the synovial cavity, since the synovial membrane covers them at the sides and anteriorly only (Fig. 10.30). The anterior cruciate is attached to the anterior part of the intercondylar area of the tibia and passes upwards, backwards and laterally to be attached to the medial side of the lateral femoral condyle (Fig. 10.32). The posterior cruciate is attached to the

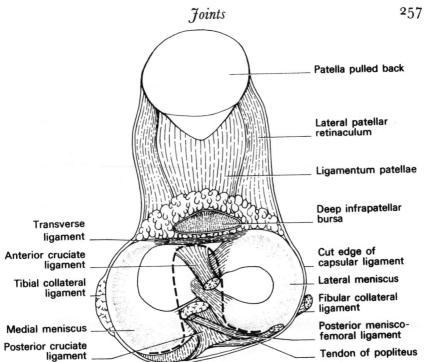

Fig. 10.30. The articular surface of the right tibia after disarticulartion at the knee joint and retraction of the patella. The dotted line indicates the synovial membrane in the vicinity of the cruciate ligaments.

most posterior part of the tibial intercondylar area and passes upwards, forwards and medially to the lateral aspect of the medial femoral condyle. These ligaments are extremely strong (Fig. 10.30). They are taut in all positions of the joint and are thus important in keeping the joint surfaces apposed. They also help to prevent anterior and posterior displacement of the tibia on the femur (Fig. 10.34). The anterior cruciate prevents forward displacement and the posterior, backward displacement.

The *medial* and *lateral* menisci are two crescentic pieces of fibrocartilage (but are more fibrous than cartilaginous) which are attached to the two articular surfaces of the tibia by means of their horns (Fig. 10.30) and by means of rather lax *coronary ligaments* around their edges. They are connected anteriorly by a *transverse ligament*. The shape of the menisci corresponds to the shapes of the articular surfaces of the tibia and of the femur (Fig. 10.27) but their degree of curvature can alter according to the curvature of the femoral surface, which itself is different in flexion or extension of the joint (Fig. 10.31). The lateral meniscus is attached by a strong *posterior meniscofemoral ligament* to the intercondylar area of the medial condyle near the posterior cruciate ligament, while a smaller *anterior meniscofemoral ligament* passes in front of the posterior cruciate ligament to the femur. The lateral meniscus is grooved by the popliteus tendon, which is attached to it by a few fibres. The menisci deepen the very shallow

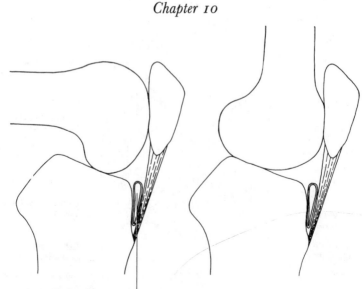

Deep infrapatellar bursa

Fig. 10.31. During flexion of the knee joint, the patella retains its relation to the tibia but changes its relation to the femur. There is also a change in the curvature of the part of the femur in contact with the tibia.

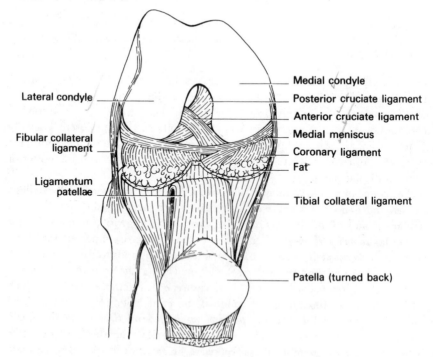

Lateral condyle

Fibular collateral ligament

Ligamentum patellae

Medial condyle

Posterior cruciate ligament

Anterior cruciate ligament

Medial meniscus

Coronary ligament

Fat

Tibial collateral ligament

Patella (turned back)

Fig. 10.32. Anterior view of the flexed knee joint after division of the quadriceps and retraction of the patella.

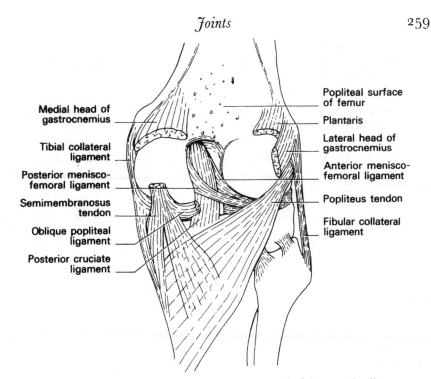

Medial head of gastrocnemius

Tibial collateral ligament

Posterior menisco-femoral ligament

Semimembranosus tendon

Oblique popliteal ligament

Posterior cruciate ligament

Popliteal surface of femur

Plantaris

Lateral head of gastrocnemius

Anterior menisco-femoral ligament

Popliteus tendon

Fibular collateral ligament

Fig. 10.33. Posterior aspect of the knee joint after removal of the capsular ligament.

tibial articular surfaces, help to fill in 'dead space' in the joint and thus help in the distribution of synovial fluid, act to a limited extent as 'shock absorbers' and, finally, adapt the tibial surfaces to the changing curvature of the femoral condyle. The medial meniscus is less mobile than the lateral, on account of its attachment to the tibial collateral ligament, and hence is more frequently injured than the lateral. Menisci can be removed surgically without greatly affecting the function of the joint; they later regenerate as fibrous structures.

The synovial membrane lines the inside of the capsular ligament and is reflected over the ends of the bones but, as is usual, stops short at the edge of the articular cartilage and at the edges of the articulating areas of the menisci. Anteriorly, where the capsule is deficient as a result of the presence of the patella, a large pouch of synovial membrane extends upwards between the deep surface of the quadriceps and the femur to form the *suprapatellar bursa* (Fig. 11.37). Its upper limit is about three fingers' breadths above the patella. When the knee joint contains an excess of synovial fluid, as result of disease or injury, fluid in this bursa produces a swelling or 'fullness' on either side of the patella and the patella may be lifted off its femoral articular area. Another outpouching of synovial membrane occurs in relation to the tendon of popliteus, which is thus lubricated as it passes over the back of the tibia (Fig. 10.33). Over the deep surface of the lower part of the patella and the ligamentum patellae, the synovial membrane covers an extensive fat pad and is thrown up into a pair of *alar folds* from which a band of synovial membrane (the *infrapatellar fold*) passes up to the intercondylar area of the femur (Fig. 10.32).

A number of bursae are found in the region of the knee joint. In front, subcutaneous bursae lie over the patella and the tibial tuberosity, while at the sides, most of the tendons which cross the joint have related bursae. One of these, which lies deep to the medial head of gastrocnemius and the semimembranosus tendon often communicates with the joint and may form a swelling when there is an excessive amount of fluid in the joint.

The position of the joint surfaces lies about a finger's breadth above the level of the head of the fibula and a finger's breadth below the level of the lower border of the patella. The gap between the tibia and the femur can be clearly felt on the medial and lateral sides of the joint.

It is a common error to imagine that the patella lies over the joint line. It does, in fact, articulate with the upper part of the femoral articular surface (Fig. 10.31) so that it is well above the joint in all positions of the knee. The articular surface of the patella is divided into two by a rounded vertical ridge (Fig. 9.28), the lateral part of the surface being more extensive than the medial. Only a part of the patella is in contact with the femur in any one position of the joint. The ligamentum patellae maintains a constant distance between the patella and the tibial tubercle and as the joint passes from full extension to full flexion, progressively higher parts of the patella surface make contact with the femur. In extreme flexion a narrow strip along the most medial part of the patella surface articulates with the femur. These different parts of the articular surface of the patella can often be distinguished on the dried bone as they may be marked out by faint lines.

The most extensive movements at the knee joint are flexion and extension, the former being limited by contact of the overlying soft parts, and the latter by tension in the ligaments, fascia and muscles behind the joint. Both cruciate ligaments are taut in extension and the anterior parts of the menisci are compressed between the femur and the tibia. Medial and lateral rotation can also occur, the range of passive rotation being greater than that of active. Rotation can best be appreciated by placing the foot on the ground while in the sitting position, when movement through about 45° can be obtained. Rotation also occurs during the movements of flexion and extension which will be described below. A small amount of abduction and adduction and of anterior and posterior gliding can also occur when the knee is semiflexed.

Flexion and extension are complex movements involving a rolling of the femur over the top of the tibia (or vice versa) combined with a gliding movement in which the femur moves forwards over the tibia and the menisci during flexion and backwards during extension (this may best be understood by simulating the movements with the dried bones—if a simple rolling were involved, the femur would roll backwards off the top of the tibia during flexion!). During these movements the curvature of the femoral surfaces is changing (Fig. 10.31), the radius of curvature diminishing as flexion occurs; this change is accommodated by the closing down of the menisci due mainly to shifting backwards of their anterior horns. In the final stages of extension (the last 30° or so) the femur rotates medially on the tibia. This is the result of the action of the quadriceps muscle rather than the rotators and is caused by the shape of the femoral articular surfaces. The lateral condylar articular surface is shorter than the medial (Fig. 10.27) so that during the rolling action which takes place during extension, the lateral condyle comes to the end of its range of travel before the medial condyle. As extension continues, the medial condylar surface continues to roll, and the result is a medial rotation of the femur (this, again, is best appreciated by an examination of the bones). This final 'screwing home' or locking mechanism brings the knee joint into its most stable position in which the articular surfaces are in the close packed position and all the ligaments are taut. In the first stage of flexion from this position, the femur has to rotate laterally and since popliteus can do this and is also a flexor, it is often described as the 'key' which unlocks the joint. In the comfortable erect standing position, the knee is not braced in this way by quadriceps. The line of gravity passes just in front of the transverse axis of the knee joint so that balance is maintained by the tension in the hamstrings, quadriceps being relaxed. Although the movements have been described above as though the feet were on the ground and the tibia fixed, it will be understood that the reverse is often the case, the femur being fixed and the tibia moving over it so that, for example, full extension of the knee is accompanied by a *lateral* rotation of the tibia on the femur.

Accessory movements are rather free at the knee joint and when the knee is semiflexed, a greater degree of passive rotation can be produced than is the case with active movements. The tibia can also be displaced slightly on the femur in a backward and forward direction and a limited degree of abduction and adduction can occur.

It would appear, from an examination of the bones alone, that the knee joint would be very unstable but, in fact, it is very rarely dislocated. Its stability depends less on the shape of the bones than on the very powerful ligaments and

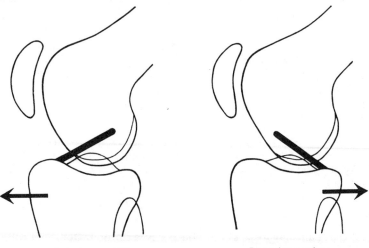

Anterior cruciate   Posterior cruciate

Fig. 10.34. Diagram to illustrate the functions of the cruciate ligaments (p. 257).

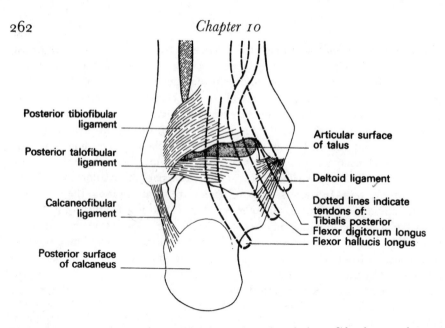

Posterior tibiofibular ligament

Posterior talofibular ligament

Calcaneofibular ligament

Posterior surface of calcaneus

Articular surface of talus

Deltoid ligament

Dotted lines indicate tendons of:
Tibialis posterior
Flexor digitorum longus
Flexor hallucis longus

Fig. 10.35. Posterior view of the ankle joint to show the relations of the three tendons on the medial side.

muscles which surround it. Thus lateral movements are prevented by the two collateral ligaments and distraction of the joint surfaces by all ligaments including the cruciates, which are taut in all positions of the joint. Anterior displacement of the tibia on the femur is prevented by the anterior cruciate ligament and posterior displacement by the posterior.

The principal muscles producing movement at the knee joint are as follows. Flexion is produced by the hamstrings and, to some extent, by the sartorius, gracilis, popliteus and gastrocnemius (the last named muscle coming into action when the foot is on the ground). Extension is mainly due to quadriceps but tensor fasciae latae also helps via the ilio-tibial tract. Semimembranosus, semitendinosus and popliteus are the principal medial rotators and biceps is the only lateral rotator if the rotating action of quadriceps is excluded.

THE TIBIO-FIBULAR JOINTS

The superior tibio-fibular joint is a plane synovial joint surrounded by a capsular ligament. The inferior tibio-fibular joint is important because the stability of the ankle joint depends upon the gripping of the talus between the two malleoli. It is a syndesmosis, the two bones being held together by means of a tough *interosseous ligament* (Fig. 10.1) as well as by *anterior* and *posterior tibiofibular ligaments*. The lowermost part of the latter, the *inferior transverse ligament* runs from the malleolar fossa of the fibula across to the medial malleolus. It is important because it deepens the posterior part of the tibial articular surface (Fig. 10.35) and so deepens the socket for the talus. The shafts of the bones are

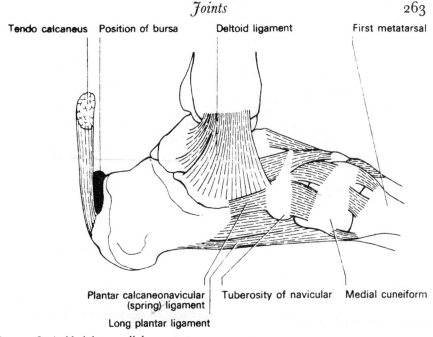

Tendo calcaneus   Position of bursa   Deltoid ligament   First metatarsal

Plantar calcaneonavicular | Tuberosity of navicular   Medial cuneiform
(spring) ligament

Long plantar ligament

Fig. 10.36. Ankle joint, medial aspect.

also held together along their length by the interosseous membrane. This membrane has a gap near the top to allow the anterior tibial artery to pass from the popliteal region to the front of the leg. Movements at the tibio-fibular joints are restricted to slight shifting; in particular, there is a slight lateral displacement of the lateral malleolus during dorsiflexion of the ankle joint because of the wider anterior part of the talus (see below).

THE ANKLE JOINT

This is a hinge joint between the lower ends of the tibia and fibula and the upper surface and sides of the talus. Since the lateral malleolus extends farther down than the medial, the articular surface on the lateral surface of the talus extends down to the lower edge of the bone but on the medial side the surface is comma-shaped, the lower part of the talus being roughened by ligamentous attachment. Since the upper articular surface of the talus is wider in front than behind, the ankle is most stable when the foot is in the dorsiflexed position. All articular surfaces are covered with hyaline cartilage.

The joint is enclosed by a *capsular ligament* which is relatively thin in front and behind but is strengthened by powerful ligaments at the sides. It is attached to the edges of the articular surfaces but extends forwards on to the neck of the talus. The medial ligament is usually called the *deltoid ligament* since it is, superficially at least, triangular in shape (Fig. 10.36). The tibial attachment is to the medial malleolus and the deep fibres are attached to the rough medial surface of the talus below the comma-shaped facet. The superficial fibres

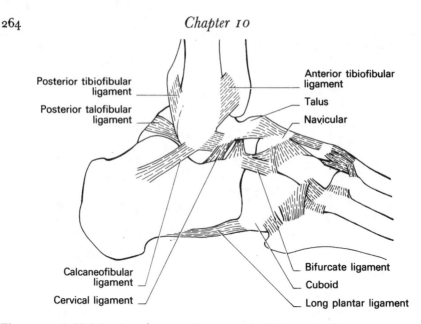

Fig. 10.37. Ankle joint, lateral aspect after removal of the capsular ligament.

spread out and are attached to the tuberosity of the navicular, the sustentaculum tali, and the posterior part of the talus. Between the navicular and the sustentaculum tali, the ligament is attached to, and helps to support, the *spring ligament* (see below). The medial ligament is so strong that during a violent eversion of the foot the medial malleolus may fracture before the ligament gives way. The *lateral ligament* is divided into three parts, all of which are attached above to the lateral malleolus. The *anterior* and *posterior talofibular* ligaments are attached to the anterior and posterior parts of the talus respectively. (The fibular attachment of the latter is to the malleolar fossa.) The strongest part of the ligament is the *calcaneofibular ligament*, which runs downwards and backwards to be attached to a tubercle on the lateral surface of the calcaneus (Fig. 10.37).

The synovial membrane has the usual distribution, lining the inside of the capsular ligament and covering the intra-articular parts of the bones, but stopping short at the edge of the articular cartilage. Anteriorly, the joint cavity extends forwards on to the neck of the talus so that a sharp knife, if dropped point down can enter the joint in front of the lower end of the tibia. At the sides and behind, however, the joint is well protected by the malleoli and the tendo calcaneus. The upper limit of the joint lies about 1 cm above the tip of the medial malleolus and corresponds to the anterior border of the articular surface of the tibia which can be felt anteriorly behind the tendons of the extensor muscles.

The principal movements at the ankle joint are dorsiflexion and plantarflexion (extension and flexion). The axis about which these movements occurs is slightly oblique so that the foot tends to turn outwards in full dorsiflexion and inwards in plantarflexion. Dorsiflexion is limited by tension in the muscles

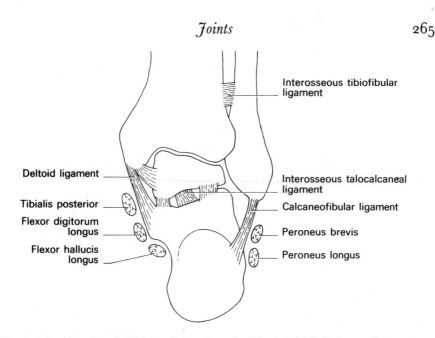

Fig. 10.38. Ankle joint; diagram to show how the talus is held by ligaments between tibia and fibula above and calcaneus below.

behind the joint and in the posterior parts of the medial, lateral and capsular ligaments. It is also limited by the shape of the talus, since during this movement the wider anterior part of the talar articular surface is wedged between the malleoli which 'give' slightly at the inferior tibiofibular joint to accommodate it. This is thus the close packed position of the joint. In plantar flexion the narrower part of the talus is between the malleoli and a very small amount of side-to-side movement can take place, which increases the total range of inversion and eversion of the foot. Plantar flexion is limited by tension in the extensor group of muscles and in the anterior parts of the medial and lateral ligaments.

The most powerful movement of plantar flexion occurs during the take-off stage of walking, when the whole weight is taken on one foot. The body weight, in this position, tends to cause the tibia and fibula to slide forwards on the talus and this is resisted by the shape of the articular surface of the talus (wider anteriorly), by the posterior lip of the tibial surface (sometimes called the *third malleolus*) which is deepened by the inferior transverse tibiofibular ligament, and by the posterior parts of the medial and lateral ligaments (Fig. 10.35). Nevertheless, if the malleoli are fractured or the ligaments severely torn, the leg will slip forward on the foot, i.e. the foot will be displaced backwards.

Plantar flexion is produced by gastrocnemius and soleus, helped by the deeper muscles, viz. flexors digitorum and hallucis longus and tibialis posterior. Dorsiflexion is by means of tibialis anterior, aided by extensors hallucis and digitorum longus and peroneus tertius.

THE SUBTALAR AND TALO-CALCANEO-NAVICULAR JOINTS

These joints are best discussed together since it is at the combined joints that the important movements of inversion and eversion occur. The subtalar joint is a synovial joint between the concave surface of the talus and the posterior convex facet on the upper surface of the calcaneus. The joint does not communicate with any other joints and it is surrounded by a *capsular ligament*. In addition, the bones are held together by *medial* and *lateral talo-calcaneal ligaments*, by an *interosseous talo-calcaneal ligament* which occupies the sinus tarsi, separating the joint from the talo-calcaneo-navicular joint and by a *cervical ligament* which attaches the neck of the talus to the upper surface of the calcaneus. Parts of the medial and lateral ligaments of the ankle joint also reinforce the subtalar joint and it is helpful to regard the talus as being held in position between the bones of the leg and the bones of the foot by the ankle joint ligaments (Fig. 10.38). The *talo-calcaneo-navicular joint* is between the rounded head of the talus and a complex joint surface comprising the concave surface of the navicular, the spring ligament and facets on the upper surface of the body of the calcaneus and the sustentaculum tali (Fig. 10.39). It is surrounded by a weak capsular ligament which blends posteriorly with the interosseous talo-calcaneal ligament. The *spring ligament* is a tough band of fibrous tissue stretching between the sustentaculum tali and the navicular, its official title being the *plantar calcaneo-navicular ligament*. It forms an integral part of the joint and therefore its upper surface is reinforced with cartilage. It supports the head of the talus and is supposed to give some 'spring' to the medial longitudinal arch of the foot (see below).

It is at these two joints, the subtalar and the talo-calcaneo-navicular, that the movements of inversion and eversion principally occur (occasionally the two joints are collectively known as the 'subtalar joint'). The movements are complex and the simplest way of thinking about movements in this region is to regard the talus as being part of the foot during flexion and extension at the ankle joint but as part of the leg during inversion and eversion, the remainder of the foot moving around the relatively fixed talus. The movement at these joints is supplemented by slight shifting at other tarsal joints which will be mentioned below, particularly at the transverse tarsal joint. The principal muscles involved in inversion are the tibialis anterior and posterior while the main evertors are the peroneus longus and brevis. Inversion is limited by tension in the peronei and in the lateral part of the interosseous talo-calcaneal ligament. Eversion is limited by tension in the invertor muscles and by the deltoid ligament.

THE CALCANEO-CUBOID JOINT

This joint is a saddle-shaped synovial joint between the anterior surface of the calcaneus and the cuboid. It is surrounded by a capsular ligament and is strengthened by dorsal and plantar ligaments. The plantar ligaments are particularly strong and are known as the *long* and *short plantar ligaments* (Fig. 10.40). The former arises from a large area on the under surface of the calcaneus and passes forwards to be attached to the ridge on the cuboid and the bases of

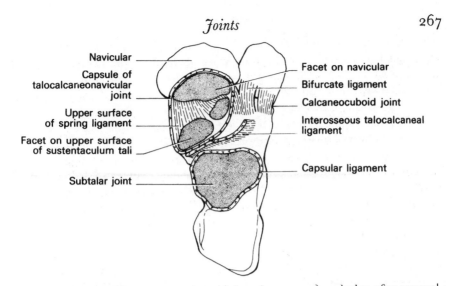

Navicular

Capsule of
talocalcaneonavicular
joint

Upper surface
of spring ligament

Facet on upper surface
of sustentaculum tali

Subtalar joint

Facet on navicular

Bifurcate ligament

Calcaneocuboid joint

Interosseous talocalcaneal
ligament

Capsular ligament

Fig. 10.39. The joints between the talus and the calcaneus and navicular, after removal of the talus.

the middle three metatarsals. It converts the groove on the under surface of the cuboid into a tunnel through which runs the tendon of peroneus longus. Deep to this ligament lies the short plantar ligament which simply bridges across the calcaneo-cuboid joint itself. In addition to these ligaments, there is, on the dorsal surface of the joint, a *bifurcate ligament* which is attached posteriorly to the calcaneus. Its anterior part divides into two bands which are attached to the cuboid and navicular respectively.

The calcaneo-cuboid and talo-navicular joints together are called the

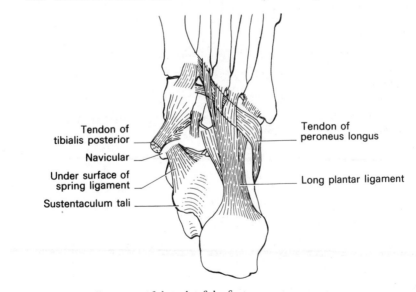

Tendon of
tibialis posterior

Navicular

Under surface of
spring ligament

Sustentaculum tali

Tendon of
peroneus longus

Long plantar ligament

Fig. 10.40. The deep ligaments of the sole of the foot.

transverse tarsal joint. They are almost in the same plane and separate the talus and calcaneus behind from the remainder of the foot (the forefoot) in front. At this compound joint the movements are restricted to slight shifting and some rotation. A lateral or outward rotation of the forefoot is often called supination by analogy with a somewhat similar movement in the forearm, and a rotation medially is pronation. Thus, when the foot is off the ground, supination adds to the total effect of inversion and pronation to that of eversion. When the foot is on the ground, however, inversion will result in passive supination, since the heads of the medial metatarsals, particularly the first, will be pressed against the ground and will rotate the forefoot into a more supinated position. This may be seen when standing with the feet wide apart. These complex movements will be analysed in more detail in Chapter 11.

THE JOINTS OF THE FOREFOOT

There are a number of plane synovial joints between the bones of the forefoot (Fig. 10.41). The navicular articulates distally with the three cuneiform bones and this synovial cavity is continuous with that of the joints between individual cuneiforms, between the lateral cuneiform and the cuboid, between the lateral two cuneiforms and the 2nd and 3rd metatarsals, and between the bases of the 2nd, 3rd and 4th metatarsals. Another cavity is formed by the joint between the lateral cuneiform and the 4th and 5th metatarsals, and, finally, there is a

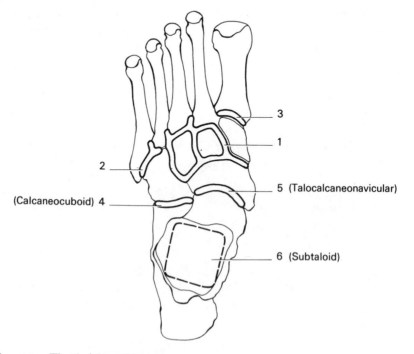

Fig. 10.41. The six joints of the carpus.

separate cavity for the joint between the medial cuneiform and the first metatarsal, which is thus similar to the joint between the trapezium and the first metacarpal. There are, therefore, 6 synovial cavities in the foot. The ligaments in the forefoot lie on the dorsal and plantar surfaces, the latter, as might be expected, being the stronger because of the arch of the foot (see Chapter 11). There are also a number of *interosseous ligaments*, particularly between the cuneiforms and these, too, tend to unite particularly the plantar areas of the bones. Movements at these joints are restricted to slight shifting which helps to adapt the foot to unevenness in the ground.

THE METATARSO-PHALANGEAL JOINTS

These are ellipsoid joints between the rounded metatarsal heads and the concave bases of the proximal phalanges; they are surrounded by a capsular ligament which is strengthened by *collateral* and by *plantar ligaments*. The latter are thick fibrous plates which are firmly attached to the bases of the phalanges but only loosely to the metatarsals. On each side they are continuous with the deep transverse metatarsal ligaments which, in the case of the foot, extend between all metatarsals (in the hand the first metacarpal is free) and help to hold them together. The first metatarso-phalangeal joint is a vulnerable joint and is much larger than the others. Its ventral aspect is reinforced by two sesamoid bones which are embedded in the capsular ligament and in the tendons of the short muscles of the big toe. The sesamoids articulate with two grooves on the undersurface of the first metatarsal. The movements possible at these joints are flexion and extension, abduction and adduction. These are similar to those at the corresponding joints in the hand but extension is more marked than flexion in the foot, particularly in the big toe. Accessory movements are restricted to slight shifting and some rotation.

Flexion is produced by flexors digitorum longus and brevis, flexors hallucis longus and brevis, flexor digit minimi brevis, the interossei and the lumbricals. Flexor digitorum accessorius also assists. Extension is by extensor hallucis longus and extensors digitorum longus and brevis. Adductor muscles are the plantar interossei and adductor hallucis while the abductors are the dorsal interossei and abductors hallucis and digiti minimi.

THE INTERPHALANGEAL JOINTS

These are synovial joints with *capsular ligaments* and thick *plantar ligaments* similar to those of the metatarsophalangeal joints. Movements are restricted to flexion and extension and are produced by the flexor digitorum longus and brevis and flexor hallucis longus for flexion and extensor hallucis longus and extensors digitorum longus and brevis for extension.

11 · The Muscular System

This chapter deals with one of the most important systems of the body to the physiotherapist, namely the muscles and tendons which produce movement and are responsible for the maintenance of normal posture. The basic tissue concerned is, of course, striated (or voluntary) muscle and the microscopic structure of the muscle fibre and its mode of functioning were described in Chapter 7. This chapter is concerned with the way in which the muscle fibres are combined together with connective tissue, nerves and blood vessels to form the units which are known as 'muscles', and also with the attachments and functions of the many individual muscles whose training and rehabilitation is so much part of the daily work of a physiotherapy department.

Although striated muscle fibres are often referred to as 'voluntary' muscle they are not necessarily entirely under the control of the will any more than all smooth muscle is entirely 'involuntary' in action. The muscle which controls the focusing of the lens of the eye (the *ciliaris*) is smooth muscle and is supplied by the autonomic nervous system but 'one can look onto glass or through it' quite voluntarily. On the other hand, the upper part of the oesophagus contains striated muscle in its wall, as does the peculiar cremaster muscle which helps to support the testis in the male, yet neither of these can be controlled voluntarily. Similarly, the intercostal muscles and the diaphragm, which are responsible for the movements of breathing, will (fortunately) continue to function even when you are asleep or deeply anaesthetised, although it is possible voluntarily to hold your breath for a limited time or, when you wish, to take deep breaths. The termination of the alimentary canal and the outlet of the bladder are guarded by *sphincters* of striated muscle which surround the lumen of the urethra and anal canal and are, without any conscious effort, kept in a state of contraction to prevent any social catastrophe, so that John Hilton, in the 19th century, was constrained to refer to 'that gallant and indomitable little sphincter'. However, the vast majority of the muscles of the body can be contracted at will, albeit usually in involuntary co-ordination with the contraction and relaxation of other muscle groups, as will be seen later in this chapter.

Each individual muscle fibre is between 10 and 100 μm in diameter, but its length is extremely variable, some being only a few millimetres long while others may be 30–40 cm or perhaps even more in a muscle such as sartorius. Each fibre is surrounded by a delicate sheath of connective tissue called the *endomysium* (Fig. 11.1). The fibres are arranged in small bundles or *fasciculi*, each of which is enclosed in a rather more substantial sheath called the *perimysium*. Finally, the whole muscle is surrounded by a sheath of connective tissue, continuous internally with the perimysium, which is called the *epimysium*. In this way, parts of the muscle can contract in their own compartment, while other parts may

270

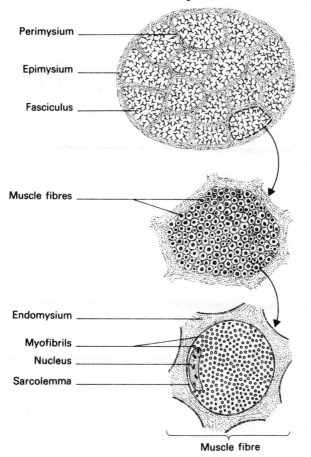

Perimysium

Epimysium

Fasciculus

Muscle fibres

Endomysium

Myofibrils

Nucleus

Sarcolemma

Muscle fibre

Fig. 11.1. The composition of a muscle. The top figure shows a cross-section through a muscle such as biceps, the next is a section through a single fasciculus and the lower figure is a single muscle fibre.

still be relaxed. It is important, therefore, to understand that different parts of individual muscles may have quite different, or even opposing, actions, while if a muscle is called upon to contract over long periods, collections of muscle fibres may work a 'shift system' whereby some are maintaining a state of contraction while others are 'resting' (see Chapter 7). The layers of connective tissue in muscle also form useful channels for the passage of nerves and blood vessels into the interior of muscles. Whole muscles are usually enclosed in compartments of deep fascia (see Chapter 3) so that they can easily slide over each other and are enabled to contract without interfering with the action of neighbouring muscles.

Muscle is a highly specialised tissue and is not easily able to regenerate. A small area of damage in a muscle may be repaired by the regeneration of the muscle fibres themselves, but an extensive lesion can only be repaired by means

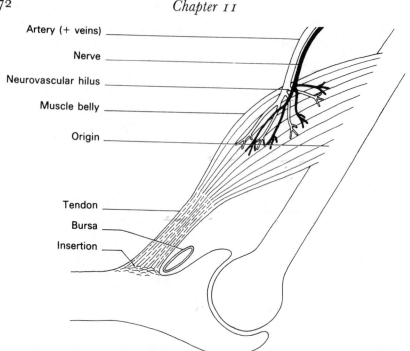

Artery (+ veins)

Nerve

Neurovascular hilus

Muscle belly

Origin

Tendon

Bursa

Insertion

Fig. 11.2. A typical muscle, showing its origin, its insertion by means of a tendon and its nerve and blood supply.

of a connective tissue 'scar'. If a whole muscle degenerates as a result of mechanical damage or deprivation of its nerve or blood supply, it can therefore never recover and it may then be the task of the physiotherapist to train other muscles to take over the function of those which have been destroyed.

Figure 11.2 shows a typical muscle in order to illustrate the various terms used in its description, although many muscles differ considerably from this plan; these will be mentioned later. The muscle in Figure 11.2 is attached at each end to a bone so that when it shortens it will produce movement at the joint between them. The muscle is attached directly to the upper bone but by means of a tendon to the lower bone. The tendon is separated from the bone by a *bursa*. Entering the muscle belly is a *neurovascular bundle* consisting of a nerve and an artery (one or more veins would accompany the artery but these are omitted for clarity). These various components of the whole muscle will now be considered individually.

MUSCLE ATTACHMENTS

In the muscle shown in Figure 11.2, the proximal attachment is directly to the bone but the distal attachment is by means of a tendon. As was described in the previous chapter, the bone to which the muscle fibres are attached directly is smooth whereas the attachment of the tendon to the bone is marked by a projection or tubercle. This has the effect of increasing the area of bone available for the attachment of the tendinous fibres. In the case of this muscle, one might intuitively expect that contraction of the muscle would cause the lower bone to move on the upper bone, the upper bone remaining stationary. The attachment of a muscle to the stationary bone is called its *origin* and that to the bone which moves is the *insertion*. These terms are not very satisfactory, however, since the majority of the muscles in the body are able sometimes to act from their insertion to their origin, that is to say, the bone which is usually fixed actually moves, while the bone which usually moves remains stable. For example, the muscles which pass from the chest wall to the humerus might be expected to pull the upper limb towards the trunk (adduction) and this they do. Supposing, however, that the upper limb is not free to move, as will occur if you hold tightly on to a wallbar above the level of your head. In this case, if the muscles contract powerfully enough they will pull the trunk towards the upper limb and raise the whole body. The muscles in question (e.g. pectoralis major and latissimus dorsi) are conventionally described as having their origin from the bones of the trunk and as being inserted into the humerus, because the latter is the bone which usually moves. In pulling oneself up on the wallbars, however, the humeral attachment acts as though it were the origin and the attachment to the trunk is really the insertion. It is extremely important that you should remember this reversed action of each muscle that you study. In some muscles, e.g. the rectus abdominis, it is very difficult to work out which attachment should be called the origin and which the insertion, and for this reason many anatomists prefer not to use these terms but to talk instead of the proximal and distal attachments of a muscle. This concept of reversed action is particularly important in the lower limb where muscles having their origin on the pelvis and their insertion on the femur may produce movement of the lower limb as in walking, but when standing they work from their insertions on their origins and are used to balance the body on the lower limbs.

Not all muscles are attached to bones directly. Some, such as the facial muscles, are inserted into the skin where they produce the changing expressions of the face. Others, the *sphincters*, form a complete circle of muscle which can occlude orifices such as the anus. Many muscles are attached partly to bone and partly to fascia. The largest muscle in the body, the *gluteus maximus*, is inserted directly into the femur by only about a quarter of its fibres, the remaining three-quarters being inserted into a wide band of fascia called the *ilio-tibial tract*. In general, one can say that in the upper limb, where movements are precise and well controlled, muscles tend to be inserted into precise bony points whereas in the lower limb, where movements are much less delicate and are often concerned with postural adjustments which need to move the greater part of the body over short distances, insertions are often poorly defined and tendons tend to spread out to become continuous with large tracts of deep fascia.

TENDONS AND TENDON SHEATHS

Tendons, which are composed of densely packed fibres of collagen, are immensely strong, much more so than muscle fibres. For this reason, muscles need to be attached over a large area of bone whereas tendons are attached to very limited bony points. Where a number of muscles have to be attached to a bone very close together, they are therefore provided with tendons so that, for example, in the lower limb, the attachments of numerous muscles to the narrow linea aspera are necessarily tendinous. Tendons are also useful to enable muscles to act from a distance without producing too much bulk—if the muscles of the forearm were not replaced by tendons at the wrist, it would not be possible for all the necessary muscles to pass into the hand without gross restriction of movement because of the large mass of muscle which would be necessary. As will be discussed below, muscles need a very good blood supply in order to work efficiently. Tendons, on the other hand, need only a network of extremely fine blood vessels and they are therefore much more able to resist pressure. They are also tougher and more able to withstand friction. Thus, muscular fibres are always replaced by tendons in places where a muscle has to pass over bone. Similarly, a tendon may be utilised to change the direction of pull of a muscle by hooking round a bony projection or running through a groove in a bone. Good examples of these structural adaptations may be seen in the gluteus maximus (Fig. 11.41), which is replaced by tendinous fibres where it plays over the greater trochanter of the femur, and the extensor pollicis longus (Fig. 11.28) whose tendon hooks around the dorsal tubercle of the radius before passing to the thumb. Very often a *bursa* is interposed between the tendon and the bone to minimise friction. A bursa is a small sac, lined by synovial membrane and containing a little synovial fluid. Tendons, because of their strength, are rarely ruptured; but a tear at the junction between a muscle and its tendon is not uncommon.

Tendons may be variously shaped in cross-section and, in general, the shape of the tendon is governed by the shape of its muscle. Thus a triangular or strap-shaped muscle usually has a flattened strap-like tendon, while a fusiform muscle has a tendon which is more or less circular in cross-section. Muscles which are very flat and thin may have a very thin tendon spread out in the shape of a broad sheet. This is called an *aponeurosis* (the 'neurosis' part of this word is a relic of the days when no distinction was made between tendons and nerves). The muscles of the abdominal wall fall into this category and a good deal of the anterior abdominal wall is aponeurotic rather than muscular. Many tendons, especially in the lower limb, besides being inserted into bones give off aponeurotic offshoots or *expansions* which blend in with the deep fascia or with the capsular ligaments of joints—the *oblique posterior ligament* of the knee joint is actually an expansion from the tendon of semimembranosus.

In the region of many joints, particularly at the wrist and ankle, many tendons are bound down to the bones by transversely running bands of deep fascia called *retinacula* (singular—*retinaculum*) and these are often subdivided

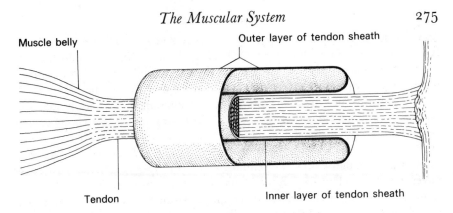

Fig. 11.3. A tendon sheath, partly opened up to show the inner and outer layers. The space between the two layers is exaggerated since normally it contains only a thin film of fluid.

into compartments for individual tendons. The function of the retinacula is to hold the tendons in place and, particularly, to prevent them 'bowstringing' or standing out excessively when the joint is bent.

In many situations, tendons are provided with a lubricating mechanism in the form of *tendon sheaths*. These are thin, double-walled sheaths of synovial membrane, one layer of which clothes the tendon while the other, outer, layer

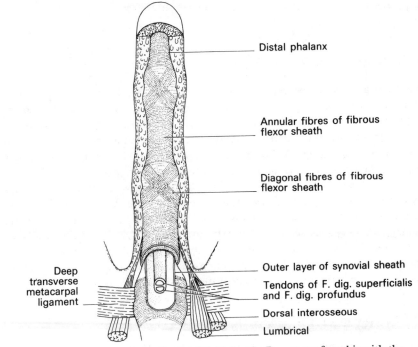

Fig. 11.4. A finger to show the fibrous flexor sheath. Do not confuse this with the synovial sheath, which is also shown.

lines the surrounding tissue (Fig. 11.3). The potential cavity between the two layers contains a thin film of synovial fluid which facilitates movement. The sites of these sheaths will be mentioned when the individual muscles are dealt with. They are of great importance to physiotherapists because they occasionally become infected and the aftermath of this may be the formation of adhesions between the inner and outer layers of the synovial sheath so that movement becomes severely restricted.

Synovial sheaths should not be confused with *fibrous flexor sheaths*. The latter are tunnels through which the tendons and their synovial sheaths pass in the fingers and toes. They are composed of dense fibrous tissue which arch over the flexor tendons and are attached to the phalanges at the sides (Fig. 11.4). The tunnels are thus described as *osseo-fibrous* channels. Opposite the joints the fibrous tissue is much thinner and is composed of criss-cross fibres so as not to restrict movement.

THE NEUROVASCULAR HILUS

As is shown in Figure 11.2, the nerves, arteries and veins tend to enter, a muscle together, bound around with connective tissue to form a *neurovascular bundle*. The site of entrance of the bundle into a muscle is called the *neurovascular hilus* and there may be more than one of these in a large muscle. The position of the neurovascular hilus in most muscles is fairly constant and this is helpful because this is the position on the muscle where electrical stimulation of the nerve may most effectively be applied.

BLOOD SUPPLY OF MUSCLES

Muscle, as a tissue, needs an extremely good blood supply for its proper functioning, so that numerous vessels are needed to supply a large muscle. Moreover, increased activity of a muscle requires a greatly increased blood supply and this is one of the causes of the increased venous return which accompanies physical activity (see Chapter 23). If the blood supply to a muscle falls below a critical level, the first symptom is pain in the muscle, which may be severe. This commonly occurs in arterial disease of the lower limb when the blood supply may be sufficient at rest, but when an increase is called for as a result of activity, severe muscular pain occurs. This is called *claudication* and the pain may be completely relieved as soon as the activity ceases. A similar pain occurs in cardiac muscle with an inadequate blood supply, when the condition is known as *angina*. Of course, if the blood supply to a muscle is severely and permanently restricted, the muscle fibres will degenerate and the muscle will become useless.

THE NERVE SUPPLY OF MUSCLES

The nerve enters the muscle at one or more hila and proceeds to break up into smaller branches. In histological terms, this means that bundles of axons leave the main trunk of the nerve and they eventually separate out into individual axons, each axon supplying a number of muscle fibres. In Chapter 5 the *motor unit* was described, each unit consisting of an anterior horn cell, its axon and all the muscle fibres which its axon supplies. The smaller the size of the motor units in a muscle (i.e. the fewer the number of muscle fibres supplied by each axon) the more precise and well controlled are the movements which the muscle can carry out. It is instructive to compare the relation between the size of a muscle and the size of its nerve in the case of such muscles as an interosseous muscle of the fingers and the gluteus maximus. The former is, perhaps, the size of a large almond with a nerve as thick as a piece of fine thread, while the latter may be the size of this book when well-developed but with a nerve only as thick as a shoelace. Obviously the motor units in the latter muscle are enormous compared to the former and this is in keeping with the functions of the muscles. The interossei are used for delicate movements such as threading a needle, while the gluteus maximus has to heave the whole weight of the body upwards when climbing stairs or getting up from a chair.

Many of the fibres in a nerve which enters a muscle are afferent rather than efferent, being responsible for the maintenance of muscle tone and, in the case of fibres coming from the tendon, for the sense of position (proprioception). However, when the nerve supply of a muscle is damaged the most obvious symptom is the loss of the motor activity.

John Hilton's comment on the anal sphincter has already been mentioned in this chapter and another of his observations is worth mentioning here. This is 'Hilton's Law' which states that the nerve supply of joints is derived from the nerve supply to the muscles which pass over the joint. (This does not mean that the nerve to *every* muscle passing over a joint gives an articular branch.) A good example of Hilton's law is seen in the quadriceps femoris muscle which has already been described in Chapter 10.

MUSCLE ARCHITECTURE

Muscles may be of many different shapes and to a considerable extent these differences are of functional importance. Many muscles have a very peculiar shape indeed when they have *specialised* functions to carry out, and it will not be possible here to describe the shape of all the muscles of the body. Nevertheless, it will probably be helpful to mention some of the common shapes of muscle and their functional significance. Basically, muscles may be subdivided into those whose fibres run in the direction of the line of pull of the muscle and those whose fibres run obliquely to the line of pull. In the former category (Fig. 11.5) are the strap-like and fusiform muscles such as sartorius in the lower limb and biceps brachii in the upper limb. In the latter category are the triangular muscles such as pectoralis major and also a very important group of muscles which are said to be of the *pennate* type. In these, the tendon is not at the extremity of the muscle but runs along a considerable part of the muscle's

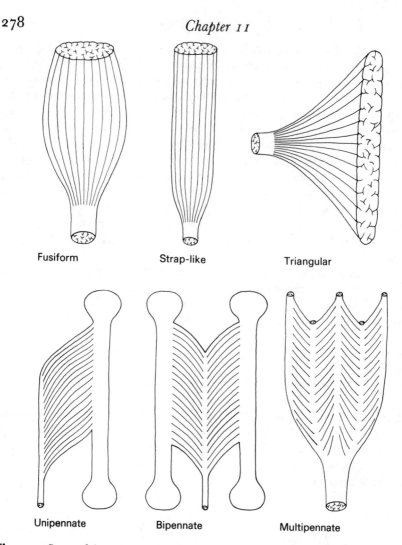

Fusiform   Strap-like   Triangular

Unipennate   Bipennate   Multipennate

Fig. 11.5. Some of the common arrangements of muscle fibres within a muscle.

length, the muscle fibres being attached along one or both sides of the tendon or, in some cases, all around it. 'Penna' is the Latin for a feather (the words for 'pen' and 'feather' are similar in many languages for obvious reasons: e.g. 'plume', 'Feder'); in *bipennate* muscles in which the muscle fibres are attached on both sides of the tendon the resemblance is obvious. Muscles in which the fibres are attached to only one side of the tendon are said to be *unipennate*. *Multipennate* muscles have a complicated structure consisting of a collection of bipennate subunits (Fig. 11.5).

There are important differences between the functions of the pennate muscles and those which have a simpler structure such as the fusiform and strap-like muscles. Firstly, as you know from Chapter 7, the strength of a muscle depends upon the number of muscle fibres, since each fibre, when it contracts, does so fully (the all-or-none law). Bulk for bulk, pennate type of muscles contain more muscle fibres than other types since the fibres are attached all the way down the sides of the tendon rather than simply into its end. Secondly, again because of the all-or-none law (p. 144), the distance through which a muscle can contract depends on the initial length of the individual fibres of the muscle or the sum of their lengths if they are arranged end-to-end. Pennate muscles obviously have relatively short fibres. In general, therefore, one can say that strap-like, fusiform and other similar muscles have a wide range of action but are not particularly strong in relation to their bulk but pennate muscles are powerful but act only through a rather restricted range. The latter are especially useful in postural activity when only small adjustments in the position of the body have to be made but considerable strength is called for. It is interesting, in this respect, to examine the soleus muscle whose deeper fibres are arranged in a bi- or multi-pennate manner but whose superficial fibres form a fusiform muscle. The deltoid muscle is an exception to this general rule. Although the central part of the muscle is multipennate, the fibres are relatively long and so the muscle has a rather wide range of movement.

MUSCLE ACTION

A knowledge of the origin and insertion of a muscle is very helpful in understanding the actions of a muscle since it is usually easy to work out the effect of approximating the insertion to the origin. The deltoid muscle, for example, passes from the shoulder girdle over the top of the shoulder joint to be inserted into the humerus so that when it contracts it will obviously abduct the upper limb by elevating the humerus. Muscle action is not nearly as simple as this, however. It was pointed out in Chapter 4 that whole movements, rather than individual muscles are represented on the cerebral cortex and to carry out such a simple movement as abduction of the upper limb involves the contraction of a large number of muscles, some of them a long way from the shoulder. You can demonstrate this for yourself by standing very close to a wall with your right arm by your side and touching the wall. Abduct the left arm. You will now find that you have to take emergency action with the feet in order to avoid falling. This is because when you abduct the left arm under normal circumstances, you change the position of your centre of gravity slightly and the muscles of the lower limb tilt the trunk slightly to the right so as to keep the centre of gravity over the feet. This movement is quite unconscious and you only become aware of it when the tilt of the trunk is prevented by the wall. Similarly, if you sit very close to a table you will find it impossible to stand up because in getting up from a sitting position, the first movement to occur is flexion of the trunk upon the lower limb so as to bring the centre of gravity over the feet. This again is quite unconscious until you analyse your movements very carefully. Apart from these postural adjustments, even local movements call for a number of other muscles to come into action. For example, when deltoid contracts, it does not necessarily elevate the humerus; it might just as easily depress the shoulder girdle. In order to abduct the upper limb by means of deltoid, therefore, it is necessary for muscles passing from the trunk to the shoulder girdle to stabilise the girdle and prevent it moving.

You will thus appreciate that in nearly all movements a whole group of muscles has to contract (and some to relax) in order to carry out the desired movement and no other. This is called the *group action* of muscles and it will now be necessary to analyse the various ways in which muscles can work when they are contracting as members of a group. You can study these actions very conveniently in your own upper limb.

PRIME MOVERS

A muscle is said to be acting as a *prime mover* when it is simply approximating its origin to its insertion. Thus, when you make a fist the muscles whose tendons run from the front of the forearm to the phalanges (flexor digitorum superficialis and flexor digitorum profundus) are acting as prime movers. Similarly, when you flex your elbow, biceps brachii is acting as a prime mover since its origin is from the shoulder girdle and it is inserted into the radius. Do not forget that, as has been mentioned already in this chapter, muscles can act with their origins and insertion reversed so that if you do a 'chin-up' on the wallbars, biceps is still acting as a prime mover.

ANTAGONISTS

If biceps were to contract on its own while all other muscles were relaxed, you might well find that flexion of the elbow would be so violent that you would inflict a nasty blow on your own face. In order to produce a controlled movement and avoid this unfortunate occurrence, the muscle which extends the elbow, namely the triceps, undergoes a controlled relaxation, the two muscles acting together in perfect co-ordination so that as biceps contracts, triceps relaxes. Triceps is then said to be acting as an *antagonist*. An antagonist is thus a muscle which is capable of carrying out the opposite movement to a prime mover but normally relaxes gradually (by eccentric action—see next section) as the prime mover contracts. If prime mover and antagonist contract simultaneously, very powerful contractions can be produced in both muscles without any movement being produced at all. Of course, if the arm is being extended, triceps becomes the prime mover and biceps is the antagonist.

An important result of this group action of prime movers and antagonists is seen when one or other of the opposing muscles is paralysed, usually by a lesion of its nerve supply. If biceps is paralysed, triceps acts unopposed and the arm will be held in a position of extension. Nerve lesions do not therefore result only in a loss of particular movements but they may also cause the affected part to be held in an unusual position (see, for example, the claw-hand produced by a lesion of the ulnar nerve).

FIXATORS

One example of a group of muscles acting as *fixators* or *fixation muscles* has already been mentioned. If deltoid is required to abduct the upper limb, it is first necessary that certain muscles which pass from the trunk to the upper limb girdle should contract in order to stabilise the girdle to allow deltoid to act from a fixed base. You can produce a similar fixation action on a smaller scale in your own wrist. Place your finger on the tendon of *flexor carpi ulnaris*. This is the thick tendon on the medial side of the front of the wrist (Fig. 11.30). Now abduct the little finger. You will find that the flexor carpi ulnaris contracts as you do so. The prime mover, in this case, is the *abductor digiti minimi* a small muscle which takes origin from the pisiform and is inserted into the base of the proximal phalanx. Theoretically, this muscle might bring the proximal phalanx nearer to the pisiform or might equally well approximate the pisiform to the proximal phalanx. In order to ensure that the former is the case, the flexor carpi ulnaris (which is inserted into the pisiform) contracts as a fixator, so that the abductor digiti minimi can act from a fixed base.

SYNERGISTS

Once more, make a fist. This is a movement of flexion but if you place your hand on the back of the forearm you will find that the extensor muscles are also contracting. They are acting as *synergists*. The muscles (flexor digitorum superficialis and flexor digitorum profundus) which produce flexion of the fingers also pass over the wrist joint and are capable of flexing this as well. This is an undesirable movement, as will be seen later, and in order to prevent it, the extensors of the wrist contract. Muscles acting as synergists, therefore, are contracting *to prevent an unwanted movement*, and they are liable to be called into play to assist any muscles which cross more than one joint.

CONCENTRIC AND ECCENTRIC ACTION

Think again about the deltoid muscle, which, passing from the upper limb girdle to the humerus over the shoulder joint, is obviously an abductor. Raise your arm to a right angle (using deltoid) and then lower it again to your side. The latter movement is one of adduction but if you palpate deltoid with the other hand you will find that it is in a state of contraction during the movement. Now raise your arm to a right angle again and let everything relax. Your arm will fall to the side quite limply because of the effect of gravity upon the limb. You will now understand that when the abducted arm is slowly adducted, deltoid is working by 'paying-out' to produce a controlled adduction, the prime mover in this case being gravity. The active lengthening of a muscle is known as its *eccentric action*; when a muscle is contracting as a prime mover (i.e. shortening in order to approximate its origin and insertion) it is said to be carrying out a *concentric action*. Eccentric action of muscles occurs very commonly, a major example being the action of bending forwards to pick something up. Here gravity is again the prime mover, and the action is controlled by the eccentric action of the extensors of the spine and the hip joint. Obviously if these were to suddenly relax you would fall flat on your face. In fact, it can be said that muscles will not act to produce a movement if the movement can effectively be carried out by gravity. Of course, if the movement is carried out against resistance, gravity is insufficient and, in this case, the prime movers will be acting concentrically. Thus, if you begin to lower your arm to the side but half way down your hand meets a table, continuing the action will produce a strong contraction of the *adductors*, i.e. pectoralis major and other similar muscles.

ISOTONIC AND ISOMETRIC ACTIONS

Isotonic and isometric contractions have already been mentioned in relation to single muscle fibres (p. 146), and the same principles can be applied to whole muscles. A muscle can therefore act isotonically to produce a movement but at times it may act isometrically to stabilise a joint or to maintain posture. A synchronous isometric contraction of prime movers and antagonists can produce powerful action in both groups of muscles with no movement occurring at the joint over which the muscles pass.

INSUFFICIENCY OF MUSCLES

Many muscles pass over more than one joint and have complex actions. In some cases the muscles may not be long enough to allow their antagonists (or gravity) to carry out their full actions on the joints. The hamstring muscles, at the back of the thigh, are extensors of the hip joint and flexors of the knee. If the hip joint is flexed and the knee joint extended, the hamstrings are often too short to allow the full range of movement so that many people are unable to touch their toes with their knees straight. If, however, they cheat by bending the knees slightly, full flexion of the hip becomes possible. When examining the hip joint, therefore, the knee is always flexed before testing for hip joint flexion. This effect is known as *passive insufficiency* of the muscle, which is too short to allow sufficient stretch. It can, of course, be overcome by exercises.

On the other hand, a muscle may be too long fully to carry out all the movement of which it is capable, even when fully contracted. This is known as *active insufficiency* and a good example can be seen in the action of the long flexors of the fingers which have already been mentioned. These muscles, flexor digitorum superficialis and flexor digitorum profundus, pass over the wrist joint, the carpal joints and the metacarpophalangeal and interphalangeal joints. Even when fully contracted they are not capable of fully flexing all these joints so that if you fully flex your wrist you will find that you are unable to give a tight grip.

By now you will have realised that muscle action is a great deal more complicated than a simple contraction producing a movement and when you are teaching a patient to exercise his muscles you will have to be able to understand many unexpected methods of producing *trick movements* for it is often possible, when a muscle is paralysed, for other muscles partly to take over its functions. For example, full flexion of the fingers is accompanied by their adduction and full extension by a tendency to abduct so that if the adductors of the fingers (the palmar interossei) are paralysed by a lesion of the ulnar nerve, it is possible for this action to be partly carried out by the flexors. When exercising the interossei, it is therefore important to keep the hand flat.

In the section which follows, individual muscles will be dealt with separately for the sake of convenience, but before reading this section you should be quite sure that you understand the principles of muscle action because there will be constant references to these in the descriptions of the actions of individual muscles.

THE MUSCLES OF THE HEAD AND NECK

There are numerous and complicated muscles in the head and neck, but from a clinical point of view the important ones may be classified into the following groups and the others ignored: (a) the muscles of mastication, (b) the muscles of facial expression, (c) the muscles which move the head and cervical spine.

THE MUSCLES OF MASTICATION

These may be subdivided into two groups: those which close the mouth or move the jaw forwards or sideways, and those which open the mouth. The latter are relatively weak since this movement is assisted by gravity and no great power is required. When attacked by an animal it is easier to hold its mouth closed than to try to escape once it has sunk its teeth into you! The powerful muscles which close the mouth are attached to the coronoid process and the body of the mandible. The most important are the *temporalis*, which occupies a fossa in the temporal region (the 'temple' in lay terms) and the *masseter* which passes down from the zygomatic arch to the region of the angle of the mandible (Fig. 11.6). Both muscles can easily be felt when the teeth are clenched, and in nervous people they can often be seen in intermittent contraction as a habit movement or *tic*. They are supplied by the *trigeminal nerve* (the fifth cranial nerve).

THE MUSCLES OF FACIAL EXPRESSION

These are a series of small but important muscles, most of which take origin from the bones of the skull and are inserted into the skin. Their names, most of which you need not remember, describe their functions, e.g. '*Levator anguli oris*'—the elevator of the angle of the mouth. The muscles which lie in the vicinity of the eyes and the mouth have their own names (such as levator anguli

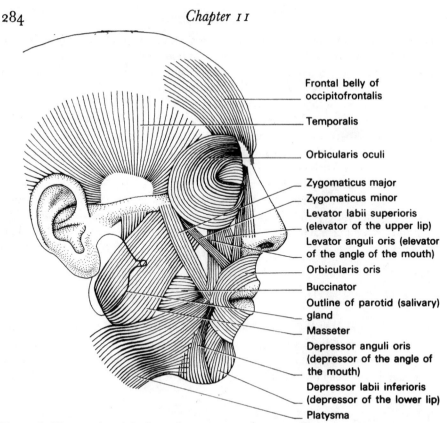

Frontal belly of
occipitofrontalis

Temporalis

Orbicularis oculi

Zygomaticus major
Zygomaticus minor
Levator labii superioris
(elevator of the upper lip)
Levator anguli oris (elevator
of the angle of the mouth)
Orbicularis oris
Buccinator
Outline of parotid (salivary)
gland
Masseter
Depressor anguli oris
(depressor of the angle of
the mouth)
Depressor labii inferioris
(depressor of the lower lip)
Platysma

Fig. 11.6. The muscles of the face: the masseter and temporalis are muscles of
mastication; the remainder are muscles of facial expression.

oris) but, together with muscle fibres which are not attached to bone at all,
form composite muscular rings around the corresponding orifices. The muscle
ring around each eye is called *orbicularis oculi* and that around the mouth, *orbicularis oris* (Fig. 11.6). The former closes the eyes either gently, as in sleep, or strongly
as when the eyes are screwed up in a bright light. The latter closes the lips or,
when in stronger contraction, can press them together menacingly, or, more
attractively, place them in position for a kiss. All the muscles of facial expression
are supplied by the facial nerve (the seventh cranial nerve). Their importance
to the physiotherapist is that this nerve is often affected by lesions within the
skull or by a peculiar condition which produces paralysis of one side of the face
known as *Bell's palsy*. Besides the disfigurement produced by such a lesion, the
loss of tone in orbicularis oculi allows gravity to pull down the lower eyelid
so that the tears spill down the cheek, while paralysis of the components of
orbicularis oris causes drooping of one side of the mouth with, sometimes,
dribbling of saliva, and inability to press the cheeks up against the gums so that
food accumulates between cheek and gums with unpleasant consequences.
These muscles can be tested by telling the patient to shut his eyes very tightly
(orbicularis oculi) and to whistle or smile (orbicularis oris).

MUSCLES WHICH MOVE THE HEAD AND CERVICAL SPINE

The former are those muscles which are attached to the skull, having taken origin on the trunk. They therefore include certain muscles which are described elsewhere, such as trapezius and erector spinae. Other muscles, however, are more local in action and will be described here.

STERNOCLEIDOMASTOID

This muscle is a prominent feature of the side of the neck since it is very superficial throughout its entire extent (Fig. 11.7). It is therefore a useful landmark and forms the anterior boundary of the posterior triangle of the neck (see Chapter 25). As is indicated by its name, it is attached to the sternum, clavicle and mastoid process but the name is more often than not shortened to *sternomastoid*. Its origin is by two heads—a narrow rounded tendon from the front of the sternum and a flattened muscular sheet attached to the upper medial one-

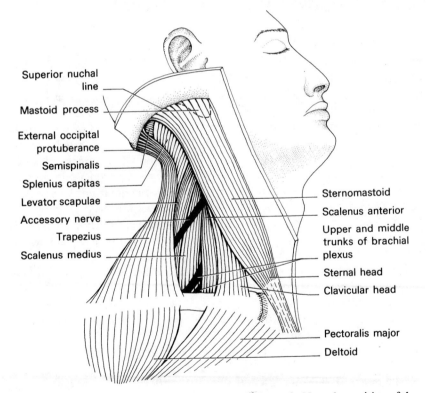

Fig. 11.7. The muscles of the posterior triangle of the neck. Note the position of the accessory nerve and the trunks of the brachial plexus.

third of the clavicle. It is inserted into the mastoid process and into the lateral two-thirds of the superior nuchal line where it meets trapezius (attached to the medial one-third of the superior nuchal line). The nerve supply is from the spinal root of the accessory nerve (the eleventh cranial nerve which has a spinal and a cranial root) and also from C2 and 3. The latter fibres are probably proprioceptive only. Sternomastoid rotates the head to the opposite side and produces lateral flexion to the same side—this composite action may best be described as bringing the ear nearer to the shoulder on the side of the contraction. This movement is sometimes seen as a tic, in which the head is repeatedly twitched into the position described above, a condition known as *spasmodic torticollis*. If both sternomastoids act together, they will flex the cervical spine against resistance, but normally, looking down at the feet does not require sternomastoid contraction but rather eccentric action of erector spinae. The muscle is used in turning the head to one side and is often seen in use by people who are deaf in one ear in order to turn the good ear towards the speaker. It often contracts strongly to resist the pull produced by combing long hair. Both muscles together are used in raising the head from a pillow. Acting from above, sternomastoid is an accessory muscle of respiration, elevating the upper part of the thorax in forced inspiration.

THE SCALENE MUSCLES

There are three scalene muscles: *scalenus anterior*, *scalenus medius* and *scalenus posterior*; but since their attachments and the direction of their fibres are very similar, they may be considered as a single unit—scalenus posterior is so small anyway that it may be ignored. The scalene muscles take origin from the anterior (scalenus anterior) and posterior (scalenus medius) tubercles of the transverse processes of the cervical vertebrae and they are inserted into the first rib (Fig. 11.8). They form the lower part of the floor of the posterior triangle of the neck and can be palpated in the angle between sternomastoid and the clavicle (Fig. 11.7). Since the cervical spinal nerves emerge from the vertebral canal between the anterior and posterior tubercles (see Chapter 9), the trunks of the brachial plexus emerge from between the scalene muscles, where the upper and middle trunks, at least, can be palpated. The scalene muscles are supplied by branches from the cervical spinal nerves.

The scalene muscles, acting from below, can flex the cervical spine, produce lateral flexion to the same side and rotate the cervical spine to the opposite side. Acting from above they can elevate the first rib which is important in energetic breathing, but even in quiet breathing the scalene muscles contract during inspiration (see Chapter 19).

There are many other small muscles in the head and neck which help to produce movements of the head and cervical spine and also of the pharynx, larynx, tongue, etc., but they are only of limited interest to the physiotherapist. One other large muscle is important in extension of the head and neck—this is the *erector spinae*, but this is more conveniently dealt with in the next section.

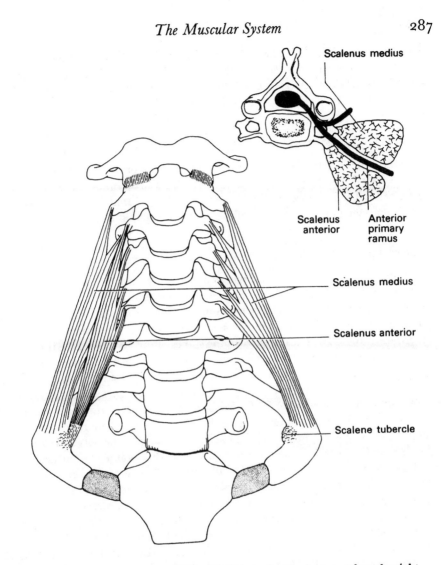

Fig. 11.8. Scalenus anterior and medius. The former has been removed on the right. The small diagram shows why the brachial plexus emerges between these two muscles because of their attachments to the vertebrae.

THE MUSCLES OF THE TRUNK

The muscles of the trunk may be conveniently subdivided into the muscles which move the spine, the muscles of respiration and the muscles of the abdominal wall. There is a good deal of overlap between these and other groups—for example, erector spinae extends the thoracic spine and so helps to enlarge the thorax during a deep inspiration, while rectus abdominis, a muscle of the abdominal wall, is an important flexor of the spine. The muscles of respiration comprise the intercostal muscles, the diaphragm and a number of other muscles such as sternomastoid that are called upon to help with forced respiration. The muscles of respiration will be described in Chapter 19.

THE MUSCLES WHICH MOVE THE SPINE

These are the erector spinae, the quadratus lumborum and certain other muscles which will be described elsewhere. The scalene muscles, for instance, have already been dealt with. For a complete list of all muscles that are responsible for carrying out movements of the spine you should consult Chapter 10.

ERECTOR SPINAE

This is a most complicated muscle whose detailed attachments you should make no attempt to learn or even to read. A very general account will suffice to explain the function of this compound muscle (Fig. 11.9). As its name implies, it is an extensor of the spine and it therefore fills up the vertical grooves on either side of the back that lie between the spines of the vertebrae, the transverse processes of the vertebrae and the angles of the ribs. (A lamb chop consists of half a vertebra with its attached rib, and the principal edible portion is the erector spinae.) The muscle is covered in its lower portion by a dense layer of fascia called the *thoraco-lumbar fascia*. In the sacral region the muscle is small but it increases rapidly in all dimensions in the lumbar region, where it is attached to all the nearby bones including the back of the sacrum, the vertebrae and the iliac bones. It is also attached to the many powerful ligaments that are found in this region. Higher up, the main bulk of the muscle divides into separate columns or slips and is attached to spines, transverse processes and ribs. The uppermost fibres reach the skull, where one of the most powerful parts of the muscle—the *semispinalis capitis*—is inserted into the area below the superior nuchal line and forms the main bulk of the thick rounded ridges which lie on either side of the midline at the back of the neck. Many of the deeper parts of this muscle mass are dignified by special names but these are the province of the specialist. It should be obvious to you that the longer parts of the muscle which span many vertebrae or ribs must be the most superficial while the shorter bands of muscle, that may only pass from one vertebra to the next, must be the deepest. The muscle is supplied, at many different levels, by posterior primary rami of the spinal nerves.

The various components of the erector spinae are extensors of the spine and of the head (in the case of those parts of the muscle that are attached to the skull). Contraction of one side only will result in lateral flexion. The deeper, short muscles, many of which are oblique in direction, can also produce rotation. The shorter muscles, particularly, are important for posture and for maintaining the position of one vertebra in relation to the next, the longer muscles being more concerned with carrying out movements. Erector spinae contracts

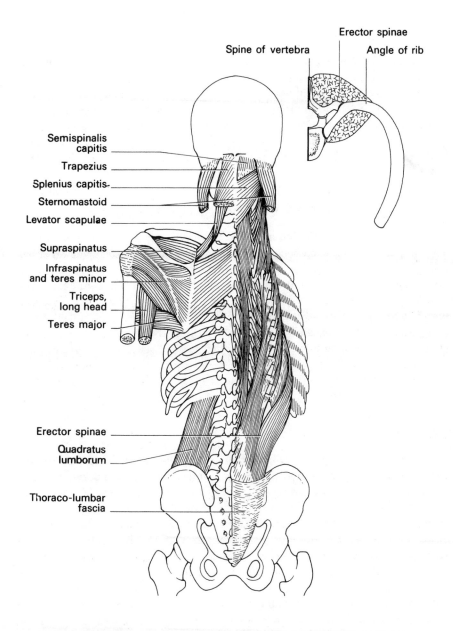

Fig. 11.9. The muscles of the back. The components of erector spinae have been separated. The small diagram shows the anatomy of a chop.

powerfully when raising the body to an upright position after stooping and, when contracting eccentrically, it is also important in bending forwards from an upright position.

THE QUADRATUS LUMBORUM

This muscle is attached below to the posterior part of the iliac crest and the ilio-lumbar ligament and above to the lower border of the twelfth rib (Fig. 11.9). Along with psoas major and the origin of the transversus abdominis, it forms part of the posterior abdominal wall and is therefore related anteriorly to the kidney, colon, etc. It is supplied by the lumbar nerves. Acting alone, one quadratus lumborum can produce lateral flexion of the spine to the same side, the two muscles acting together probably producing extension of the lumbar spine.

THE MUSCLES OF THE ABDOMINAL WALL

Three of these are flat muscles, arranged in layers, with extensive aponeuroses, while a fourth, the *rectus abdominis*, is a wide strap-shaped muscle lying on either side of the midline anteriorly. The rectus abdominis is enclosed in its own tough fascial *sheath*, formed from the aponeuroses of the other three muscles. The muscles of the abdominal wall lie immediately deep to the superficial fascia of the abdomen, deep fascia being absent. Since one of the functions of deep fascia is to form a rather firm and unyielding sheath for the underlying muscles, its absence is not surprising for if the abdomen were covered by a layer of deep fascia such as that in the thigh, a large meal would make one extremely uncomfortable. Similarly, in the thorax, the deep fascia is restricted to a very thin layer over the muscles.

THE EXTERNAL OBLIQUE

This is the most superficial of the abdominal muscles and its fibres, for the most part, run in the direction in which you put your hands in the pockets of a pair of jeans—i.e. downwards and forwards (Fig. 11.10). The muscle fibres arise from the outer surfaces of the lower 8 ribs, some of the slips of origin interdigitating with the slips of origin of the serratus anterior which arises from the upper 8 ribs (i.e. the middle 4 ribs give attachment to both muscles) and also with the latissimus dorsi. The most posterior fibres descend vertically to be inserted into the anterior half of the iliac crest. The muscle fibres farther forward give a rise to an extensive aponeurosis. This is attached to the anterior superior spine, the pubic tubercle and crest, and, above this, to a fascial thickening in the midline known as the *linea alba*. This extends from the xiphoid process to the pubic symphysis. Between the anterior superior spine and the pubic tubercle, the external oblique aponeurosis has a free lower border which is folded under to produce a slight thickening known as the *inguinal ligament*. This is unfortunate since it is not a ligament at all. You may hear surgeons refer to it by its alter-native name of *Poupart's ligament*, but do not be fooled by anyone who asks you the whereabouts of Poupart's junction; it is a signal box just outside Waterloo station in London! The anterior part of the external oblique aponeurosis forms part of the *rectus sheath* (see below).

THE INTERNAL OBLIQUE

This muscle is also a flat muscle with an extensive aponeurosis. Posteriorly, it is attached to the thoraco-lumbar fascia and below to the iliac crest, deep to the external oblique (Fig. 11.10). It is also attached to the inguinal ligament. Its fibres spread out and become aponeurotic and are inserted into the costal margin and into the linea alba. In general, its fibres run at right angles to those of the external oblique and it, too, helps to form the rectus sheath.

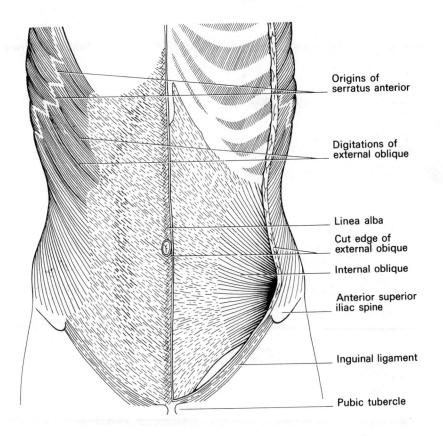

Origins of serratus anterior

Digitations of external oblique

Linea alba

Cut edge of external obique

Internal oblique

Anterior superior iliac spine

Inguinal ligament

Pubic tubercle

Fig. 11.10. The anterior abdominal wall. The external oblique has been removed on the right of the diagram.

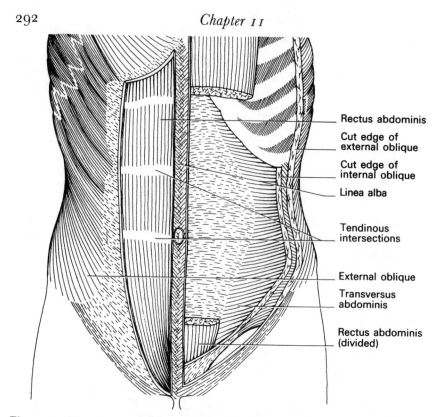

Fig. 11.11. The anterior abdominal wall. The anterior wall of the rectus sheath has been removed on the left and the rectus itself has been divided on the right.

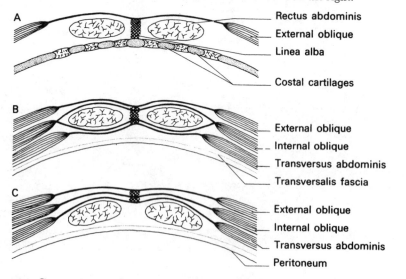

Fig. 11.12. Cross-sections through the rectus and its sheath; A. above the costal margin; B. above the umbilicus and C. above the pubic symphysis.

THE TRANSVERSUS ABDOMINIS

This muscle lies deep to the internal oblique and most of its fibres are transverse (Fig. 11.11). Its attachments are similar to those of the internal oblique except for its costal attachments. It takes origin from the thoracolumbar fascia, the iliac crest, deep to the other two muscles, and the inguinal ligament. Above, however, it passes *deep to* the costal margin and is attached to the *inner* surface of the lower 6 ribs, interdigitating with the origin of the diaphragm. Its aponeurosis is inserted into the linea alba and it takes part in the rectus sheath.

THE RECTUS ABDOMINIS

This is a wide, strap-shaped muscle on either side of the midline (Fig. 11.11). It is attached above to the outer anterior surface of the rib cage while below it is attached to the pubis and the front of the pubic symphysis. It has (usually) three tendinous intersections at and above the level of the umbilicus. These can be seen as transverse grooves on the surface in a muscular male (look at any Greek statue of Hercules), as can the slightly convex lateral border of rectus, which is called the *linea semilunaris*. The rectus muscle is enclosed in a sheath formed by the aponeuroses of the other abdominal muscles—the external oblique passes in front of the rectus, the internal oblique splits to enclose it and the transversus abdominis passes behind (Fig. 11.12). Above the costal margin, since the transversus passes deep to the ribs and the internal oblique is attached to the costal margin, only the anterior wall of the sheath is present consisting only of the external oblique aponeurosis. Posteriorly the rectus rests directly on the ribs and intercostal muscles. In the lower part of the abdomen all three aponeuroses pass in front of the rectus, the posterior wall of the sheath being formed only by the *transversalis fascia*, a sheet of fascia, dense in places, which lines the inside of the abdominal wall.

THE ABDOMINAL WALL IN GENERAL

You will now appreciate that the abdominal wall consists of three layers of muscles and aponeuroses that lie immediately deep to the superficial fascia. These consist of the external and internal oblique muscles and the transversus abdominis. Deep to these are various other thin layers before the peritoneal cavity is reached. In order to operate on the abdominal contents, the surgeon must therefore divide these three layers of muscles and must subsequently repair the layers by sewing them up. For this reason, patients after abdominal operations find movement painful and, in particular, any expulsive effort requiring contraction of the abdominal muscles such as micturition, defaecation and, especially, coughing can cause severe pain. It is, however, particularly important that patients should be able to cough after operations in order to clear the bronchi of secretions and it is often a vital part of the physiotherapist's duty to encourage this.

The nerve supply of the abdominal muscles is by means of the lower six thoracic nerves and the first lumbar nerve. You will see, in Chapter 12, that the 7th to 12th thoracic spinal nerves are not restricted to the thorax but continue beyond the costal margin to lie between the inner two layers of the abdominal wall musculature. They end by entering the rectus sheath. They are thus admirably placed to supply the abdominal muscles. Abdominal incisions are planned to avoid cutting these nerves which would cause paralysis of part of the abdominal wall and, perhaps, hernia (see below).

The actions of the abdominal muscles are to produce movements of the trunk; to contain, to protect and to compress the abdominal contents; and to help maintain normal posture. The external oblique, acting from below will pull the lower ribs of the same side forwards and thereby rotate the trunk to the opposite side. The internal oblique, whose fibres run at right angles to those of the external oblique, will rotate the trunk to the same side. This in rotating the trunk to the left, the left internal oblique and the right external oblique contract together. Contraction of the internal and external obliques of the same side will produce lateral flexion of the trunk to the same side. Contraction of the oblique muscles of both sides together will help rectus abdominis in producing flexion of the trunk. The oblique muscles are only used for rather forceful movements, other trunk muscles producing the more gentle movements. Rectus abdominis is a powerful flexor of the trunk and is used in sitting up from a lying position and similar movements. Flexing the trunk from a standing position is, of course, performed by gravity only, with the erector spinae working eccentrically. Rectus can also act as a fixation muscle in other flexor movements. If the legs are raised while the subject is lying down, rectus contracts powerfully to fix the pelvis. It can also be felt to contract if the head is raised from the pillow—a good method of making the muscle work in a feeble patient. In this case, the prime movers are the sternomastoid and the scalene muscles, rectus acting to fix the thoracic cage to allow these muscles to act from a stable base.

EFFECTS ON THE ABDOMINAL CONTENTS

The muscles of the abdominal wall help to keep the abdominal contents in place partly by their mechanical action and partly by maintaining the intra-abdominal pressure. Poor muscle tone and poor posture may cause the abdominal wall to sag with the production of a pot-belly; this must be distinguished from the protuberant abdomen produced by deposition of fat. When the abdominal muscles all contract together, they compress the abdominal contents, raise the abdominal pressure and help with all expulsive efforts such as expiration, defaecation, micturition, vomiting and childbirth. This effect is enhanced by the lowering of the diaphragm produced by the deep inspiration which occurs before such efforts. The transversus abdominis seems to be particularly important in compressing the abdominal contents and can be seen in action if the patient is told to 'pull your tummy in'. Increase in the intra-abdominal pressure may have the effect of causing a protrusion or *herniation* of some of the abdominal contents through any weak spots in the muscular wall of the abdomen. One such place is the *inguinal canal*. This lies just above the inguinal ligament and it consists of a valve-like opening which allows, in the male, the spermatic cord to pass out of the abdomen to reach the testis (the cord contains the *ductus deferens*, and the main blood supply and venous drainage of the testis). In the female the canal is also present but is much smaller and comparatively unimportant. Other weak places are the femoral canal (see later in this chapter), the umbilicus, and any surgical scars which have not healed to produce a strong mass of fibrous tissue. Protrusion of a sac of peritoneum containing some of the abdominal contents through one of these weak areas is called a *hernia* and the type of hernia is named according to the region in which it occurs—inguinal, femoral, umbilical, etc. Such hernias may be produced by repeated sudden rises in intra-abdominal pressure such as are produced by a chronic cough or lifting heavy weights.

Isometric contraction of the abdominal muscles is also important in protection of the abdominal contents, as anyone will know who has been caught unawares by a blow in the abdomen. Good development of these muscles is thus important in boxers. The muscles will also undergo protective contraction in intra-abdominal infections and in conditions causing peritonitis the abdominal wall may demonstrate 'board-like rigidity'.

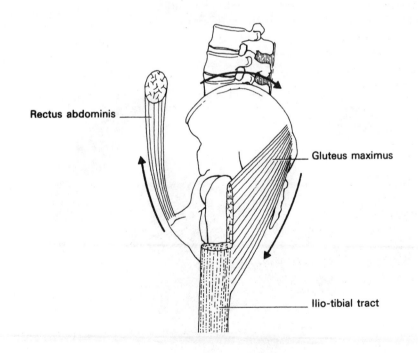

Fig. 11.13. To show the functions of rectus abdominis and gluteus maximus as postural muscles.

POSTURE

The role of the abdominal muscles in keeping the abdominal contents in place and preventing sagging of the abdomen has already been mentioned. In addition, rectus abdominis, because of its pubic attachment, can contribute to good posture by elevating the front of the pelvis (in co-operation with gluteus maximus), thus counteracting the forward tilting of the pelvis with lumbar lordosis which may be produced by poor posture. Patients may be shown how to use rectus in this way by the request to 'tuck your tail in' (Fig. 11.13).

THE MUSCLES OF THE PELVIS

The bony pelvis is lined with muscles, two of which (pyriformis and obturator internus) will be described with the lower limb muscles since they are attached to the femur. The third, and most important, is the *levator ani* which lies entirely within the pelvis. A similar muscle, in quadrupeds, is attached to the caudal vertebrae and is used to wag the tail but in man the tail disappears during development and the muscle is freed to form the floor of the pelvis. A firm pelvic floor is essential because of man's upright posture and the need to prevent the pelvic viscera from descending (prolapse) under the influence of both gravity and intra-abdominal pressure.

The levator ani arises from the back of the pubis, just short of the midline; from the spine of the ischium and in between these two origins, from the dense fascia that covers obturator internus (Fig. 11.14). Its most anterior fibres pass on either side of the prostate or vagina and are inserted into a central fibrous node—the *perineal*

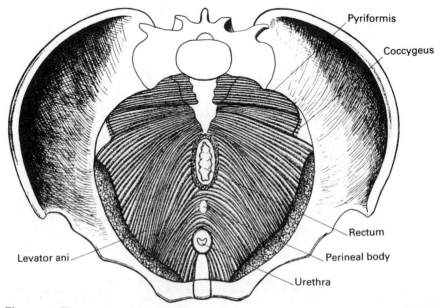

Fig. 11.14. The muscles of the pelvic floor.

body. Behind this, the muscle fibres form a sling round the angle between the rectum and anal canal (*puborectalis*) where it blends with the smooth and skeletal muscle of the sphincter ani to form the *anorectal ring*, the most important part of the sphincter in the maintenance of continence. Behind this again, the fibres pass into a fibrous band between the anus and the coccyx, and into the side of the coccyx itself. The nerve supply is derived from various sacral nerves, principally S4.

The muscle can be contracted voluntarily, the action being best described as a 'perineal shrug'. The pelvic floor is tightened and the pelvic viscera raised. The tone is responsible for preventing prolapse of the pelvic viscera such as the bladder, uterus and rectum while the puborectalis helps to prevent the rectal contents descending into the anal canal. As can readily be appreciated from Fig. 11.14, if the perineal body is split into two halves, contraction of the levator ani will pull the two muscles apart instead of tightening the pelvic floor. Such an injury may occur during a complicated labour, and the subsequent repair of the perineal body is essential if later prolapse is to be avoided.

THE MUSCLES OF THE UPPER LIMB

The muscles of the upper limb are characterised by a number of distinctive features. As has been mentioned before, muscles which act on the hand and fingers need to carry out extremely delicate and precise movements and they therefore tend to be inserted by means of tendons into well localised bony points instead of into broad bands of fascia which is the case in the lower limb musculature. On the other hand, the upper limb as a whole is extremely mobile, the shoulder joint being shallow and relatively unstable and the upper limb girdle articulating with the trunk only at the small sterno-clavicular joint. The muscles passing from the trunk to the upper limb therefore form the main attachment of upper limb to trunk and since they have to carry out movements through a very wide range, the muscle fibres must be extremely long. Thus the upper limb musculature spreads out over a large area of the trunk; the *latissimus dorsi*, for example, which is inserted into the humerus, extends down as far as the hip bone, a fact which has important consequences for paraplegics. Some of the muscles which pass from the scapula to the humerus, however, are very short. The shoulder joint is highly mobile and does not therefore depend on tough ligaments for its stability. Instead, a series of short muscles which form the *rotator cuff* act as mobile ligaments, being able to relax to allow free movement or to contract to produce stability.

THE MUSCLES OF THE SHOULDER AND SCAPULAR REGIONS

TRAPEZIUS

This is the most superficial muscle of the back. It is very extensive, being attached

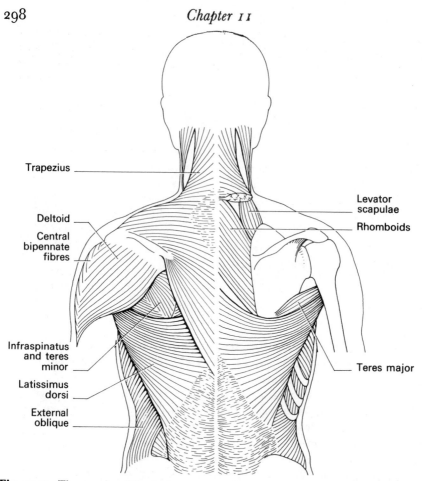

Fig. 11.15. The muscles of the back.

to the skull, the upper limb girdle and the spine down as far as the 12th thoracic vertebra. Its old Latin name meant the 'Monk's muscle', a very appropriate name in view of the shape of the two trapezius muscles together (Fig. 11.15). The origin of trapezius is from the superior nuchal line of the skull which brings it to the midline; its attachment then passes down the midline of the back so that in the cervical region it is attached to the ligamentum nuchae and below this, to the spines of all the thoracic vertebrae. From this long origin, the fibres converge on to the shoulder girdle. The uppermost fibres sweep downwards and forwards across the neck to be inserted into the lateral third of the clavicle, the anterior border of the muscle being the posterior border of the posterior triangle of the neck (Chapter 25). From here, the insertion forms a continuous line across the acromio-clavicular joint and along the whole upper border of the spine of the scapula. The lowermost fibres therefore ascend to the medial end of the spine of the scapula while the uppermost fibres descend to the lateral part of the shoulder girdle. These two sets of fibres form an important 'couple' which helps rotation of the scapula (see below). The central area of trapezius is tendinous so that a depressed diamond-shaped area can be seen between the scapulae of a muscular individual. The trapezius is supplied by the spinal accessory nerve which enters its anterior border after having supplied sternomastoid. It also receives some proprioceptive fibres from C3 and 4.

Since trapezius has such widespread attachments its actions are many. Acting from below, its uppermost fibres can work on the skull, extending it on the neck. Acting from above, these fibres will shrug the shoulders—the usual test for the proper functioning of the accessory nerve. Normal tone in the trapezius is responsible for the support of the shoulders from above, particularly when carrying heavy weights such as a shopping basket. Poor muscle tone leads to sloping shoulders— tailors are apt to build up the shoulders of male jackets to give the appearance of health and virility! The central fibres of trapezius help to brace back the shoulders. One of the most important actions of trapezius is to rotate the scapula laterally so that the glenoid points upwards. This movement is carried out by the 'couple' mentioned above and is vital to full abduction of the upper limb. As well as being a prime mover, trapezius is also very important as a fixation muscle for the very mobile scapula—one of the best ways of demonstrating trapezius in action is to tell the patient to hold his arms firmly out in front of him, parallel to the ground, while you try to push them downwards.

LATISSIMUS DORSI

This is another very extensive muscle which needs long fibres because it can move the humerus through a very wide range of movements. Its origin is from the spines of the vertebrae from T7 downwards; in the lumbar region the origin is via the thoraco-lumbar fascia which has already been described. It extends even further than this by travelling laterally from the lumbar region to be attached to the medial part of the iliac crest. The thoracic origin lies deep to the lower part of trapezius (Fig. 11.15). A small part of the muscle arises from the lower ribs as it crosses them and a little also arises from the lower angle of the scapula as the muscle overlies this. From this enormous origin, the muscle fibres converge towards the shoulder region, travel in the posterior wall of the axilla (Fig. 11.16), where the muscle can be palpated, along with *teres major*, between the fingers and thumb, and is finally inserted by means of a wide and very strong tendon into the floor of the intertubercular sulcus. The curved upper border of the muscle crosses the extreme lower angle of the scapula, holding it in a sort of waistcoat pocket (Fig. 11.15) and helping to keep the scapula in contact with the chest wall. As will be seen later, *teres major* arises from the lower part of the scapula and passes to the humerus. Latissimus dorsi therefore lies superficial to the origin of teres major (i.e. posterior to it) but its tendon turns under teres major to lie in front of the latter and, in doing so, reverses its surfaces, i.e. its anterior surface becomes posterior and its lowermost fibres become the upper fibres of the tendon. The two muscles are often partly fused in this region.

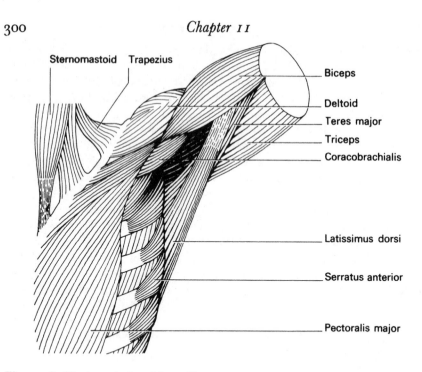

Fig. 11.16. The boundaries of the axilla.

Latissimus dorsi is supplied by the thoracodorsal nerve (C6,7,8) which comes from the posterior cord of the brachial plexus and has a long course down the medial wall of the axilla before entering the muscle.

Latissimus dorsi has a number of important actions. Acting as a prime mover, it is a powerful adductor and medial rotator of the shoulder joint. It can also extend the previously flexed upper limb. Acting with reversed origin and insertion it will bring the trunk nearer to the upper limb when the latter is fixed, i.e. it is used in raising the body to a trapeze or the upper wallbars or in any similar 'chin-up' situation. Because of its extensive coverage of the thorax and its attachments to the lower ribs it is an accessory muscle of respiration and one of the best ways of demonstrating its lateral border is by telling the patient to cough, when, for a second or less the lateral border stands out clearly as a ridge running from the back of the hip region to the axilla. One very important feature of this muscle is that, although a muscle of the upper limb, with a nerve supply from the brachial plexus, it can act on the lower limb via its attachment to the hip bone. In paraplegia, caused by a spinal lesion in the thoracic or lumbar regions, therefore, latissimus dorsi retains its nerve supply and will be the only surviving and functioning muscle below the level of the lesion. It can then be trained to elevate one side of the pelvis—'hip-hitching'. It is also important in elevating the pelvis when the humeri are fixed—this is seen when the hands are placed on the bed and the whole body lifted so that the bottom is raised from the bed and its position adjusted. Similarly, in crutch walking, the latissimus dorsi helps to support the weight of the body on the hands—the weight is *not* taken by the pads in the axillae.

SERRATUS ANTERIOR

This is the third of the large and extensive muscles which attach the upper limb and its girdle to the trunk. Its origin is by means of 8 bundles of fibres attached to the sides of the upper 8 ribs (Fig. 11.16). In a muscular person these are fairly substantial and give a serrated appearance to this region, hence the name. From the origin, the large flat muscle passes backwards around the chest wall, lying on the ribs and intercostal muscles. It is deep to the scapula and therefore separates the subscapularis muscle from the chest wall, and it is finally inserted into the whole length of the medial (vertebral) border of the scapula. It is not evenly spread out over the whole border however. The fibres from the first 2–3 ribs are inserted into the upper angle, the next few slips are spread thinly along the medial border and the fibres from the lower 3–4 costal origins converge on the inferior angle. The muscle forms a large part of the medial wall of the axilla. The spaces between the subscapularis and serratus anterior and between serratus anterior and the chest wall contain a great deal of very loose connective tissue and a little fat so that the scapula and its attached muscles can 'float' over the chest wall, allowing free mobility. The serratus anterior is supplied from the brachial plexus by the *long thoracic* nerve (C5, 6, 7). This nerve thus arises in the neck and has to have a very long course down the medial wall of the axilla in order to reach all the muscle fibres.

The actions of the muscle are extremely important. Acting as a prime mover, the whole muscle can obviously pull the whole scapula forwards around the

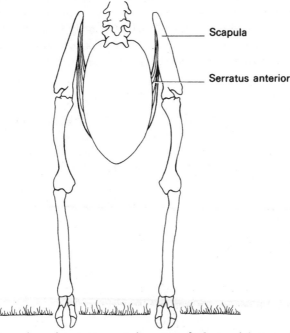

Scapula

Serratus anterior

Fig. 11.17. In quadrupeds, serratus anterior suspends the trunk between the forelimbs.

chest (*protraction*) which occurs when the arm reaches forwards, as happens powerfully when a boxer delivers a straight left. The lower fibres, which converge onto the inferior angle, will pull the angle forwards and therefore rotate the scapula laterally, i.e. will cause the glenoid to be directed upwards. This is extremely important in abduction of the upper limb (p. 241) so that a failure to be able to abduct fully is one of the signs of a lesion of the long thoracic nerve. Almost as important is the suspensory action of the muscle, which is best seen in quadrupeds (Fig. 11.17). A politician making a speech, resting his hands on the table, is thus using his serratus anterior to good effect, since were these muscles to give way suddenly, his body would fall forwards between his scapulae. In fact he would not, unfortunately, fall flat on his face on the table since other muscles would take up the strain, but some displacement would occur and the blades of the scapulae would project backwards. The standard test for the integrity of serratus anterior is therefore to tell the patient to lean up against the wall, taking his weight on both hands. If the muscle is damaged, the scapula on the affected side will project backwards (*winged scapula*). Curiously enough, although the muscle has such an extensive attachment to the ribs, it does not, apparently, function as an accessory muscle of respiration, although this is controversial.

THE RHOMBOIDS

There are two of these muscles on each side, *rhomboideus major* and *rhomboideus minor*, but since they lie edge to edge and have precisely similar functions they can be considered together. The rhomboids arise from the spines of the upper thoracic vertebrate and are inserted into the medial border of the scapula (Fig. 11.15). They are supplied by a special (dorsal scapular) nerve derived from C5. Their function is to brace back the scapula, aided by the trapezius.

LEVATOR SCAPULAE

This is a strap-shaped muscle which takes origin from the transverse processes (posterior tubercles) of the upper cervical vertebrae, and descends to be inserted into the medial border of the scapula between the spine and the upper angle (Fig. 11.15). Its function is to elevate the scapula, as in shrugging the shoulders, when it works in conjunction with trapezius. Just as important is its action in preventing depression of the scapula, so that its tone is important when carrying heavy weights.

PECTORALIS MAJOR

This is a triangular muscle which covers the upper part of the anterior surface of the chest. In the female a large portion of the breast overlies this muscle (the remainder lying on serratus anterior). Its origin is by two heads: a *clavicular* and a *sternocostal* head. The former arises from the lower border of the medial half of the clavicle. The sternocostal head arises from the sternum, lateral to the

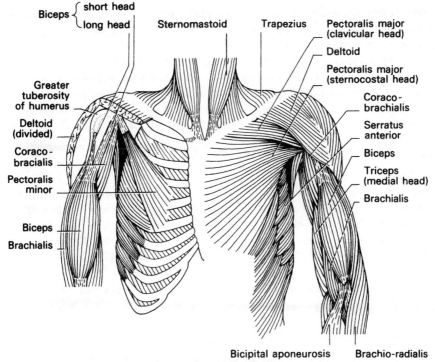

Biceps { short head, long head }
Sternomastoid
Trapezius
Pectoralis major (clavicular head)
Deltoid
Pectoralis major (sternocostal head)
Greater tuberosity of humerus
Coraco-brachialis
Deltoid (divided)
Serratus anterior
Coraco-brachialis
Biceps
Pectoralis minor
Triceps (medial head)
Brachialis
Biceps
Brachialis
Bicipital aponeurosis
Brachio-radialis

Fig. 11.18. The muscles of the shoulder region and arm.

midline, and the upper six costal cartilages (Figs. 11.16 and 11.18). The lowermost part of the muscle extends down as far as the anterior rectus sheath. The muscle fibres converge to a powerful tendon which is inserted into the rugged lateral lip of the intertubercular sulcus of the humerus. This tendon forms the main bulk of the anterior wall of the axilla, and can easily be palpated between the fingers and thumb. The nerve supply to the muscle is from the lateral and medial pectoral nerves.

The muscle is large and powerful and its actions are important. Acting as a prime mover, it is an adductor and medial rotator of the shoulder joint. It can thus be made to contract by placing the hands on the hips and pressing medially or, more powerfully, by placing the hands together in front of the body and pressing them together. In this case, the muscle is acting as a whole, but very often the two heads work independently. The clavicular head is a flexor of the shoulder joint and the sternocostal head is an extensor of the previously flexed joint, i.e. it will bring the arm back to the side. Acting from the insertion to the origin, the muscle as a whole, will raise the body when the hands are holding a support, such as a wallbar above the level of the head. The muscle as a whole can act as an accessory muscle of respiration.

PECTORALIS MINOR

This is a small muscle which lies immediately deep to pectoralis major. Its origin is from the 3rd, 4th and 5th ribs and it is inserted into the medial border of the coracoid process of the scapula (Fig. 11.18). It is supplied by the medial and lateral pectoral nerves.

The pectoralis minor can depress the tip of the shoulder or, more important, prevent it from being displaced upwards. It will thus assist serratus anterior as a 'politician's speech muscle'. It can also act as an accessory muscle of respiration.

Pectoralis minor is enclosed in a dense sheath of fascia, the whole being stituated deep to pectoralis major and forming, with the latter, the anterior wall of the axilla.

DELTOID

Delta is the fourth letter of the Greek alphabet and is written Δ. In fact the muscle is V-shaped. The origin is long, and is from the lower borders of the lateral one-third of the clavicle, the lateral border of the acromion, and the whole length of the spine of the scapula (Figs. 11.15 and 11.18). This origin thus lies directly opposite the insertion of trapezius. The muscle fibres converge, wrapping around the head and greater tuberosity of the humerus, and are inserted, by means of a powerful tendon, into the lateral surface of the humerus.

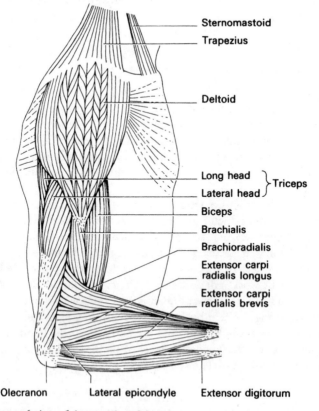

Fig. 11.19. Lateral view of the muscles of the arm.

As this attachment is tendinous, it produces an extensive roughened area on the bone which is known as the *deltoid tuberosity*. The central portion of the muscle, which arises from the lateral surface of the acromion, is multipennate, and the presence of a number of small but powerful tendons in this part of the muscle cause the roughness which is a feature of this edge of the acromion (Fig. 11.19). The anterior and posterior fibres, however, pass straight from the origin to their insertion. It is the deltoid muscle which gives the rounded contour to the shoulder region, and wasting of deltoid gives a characteristic 'square' appearance to this region. This should not be confused with the 'square' appearance of a dislocated shoulder, which is due to the displacement of the most lateral bony point in this region, namely the greater tuberosity. The deltoid receives its nerve supply from the axillary nerve (C5 and 6), which passes around the surgical neck of the humerus, giving a series of branches to the muscle as it does so.

Deltoid is the principal abductor of the upper limb. Its powerful multipennate central fibres, acting as a prime mover, will abduct the humerus to approximately 90°, provided that the initial stages (the first 15–20°) are carried out by supraspinatus. This is because, with the arm at the side, these fibres of deltoid are vertical so that their contraction will tend to cause the whole humerus to be displaced upwards. As soon as the supraspinatus has created an

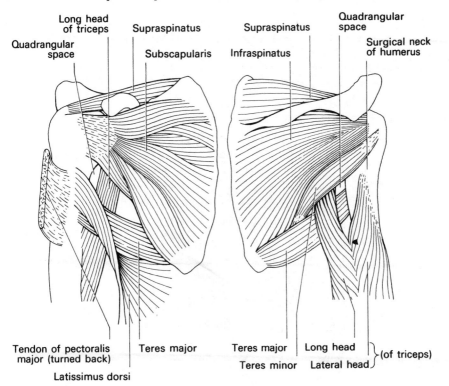

Fig. 11.20. The quadrangular space, from in front and behind.

angle, however, deltoid can act with a proper mechanical advantage and is then an abductor. When the arm is held out at right angles to the body by the central fibres, the anterior and posterior fibres act rather like the cables of a television mast and brace the humerus to prevent it swaying forwards or backwards. The anterior fibres, acting alone, will flex the shoulder joint and the posterior fibres will extend it. They also have a slight medial and lateral rotating action respectively. Deltoid is often used eccentrically, as in lowering the arm to the side.

TERES MAJOR

This is a strap-shaped but well-rounded muscle ('teres' means 'round'), which helps to form the posterior wall of the axilla along with latissimus dorsi (Figs. 11.16 and 11.20). Its origin is from the lower end of the lateral border of the scapula and it is inserted by a flattened tendon into the medial lip of the intertubercular sulcus. Its tendon is very closely related to that of latissimus dorsi as has already been described. It forms the lower boundary of the quadrangular space (see below) and it also indicates the lower boundary of the axilla so that, for example, the axillary artery changes its name to the brachial artery at the lower border of teres major. It is supplied by the lower subscapular nerve and it is an adductor and a medial rotator of the arm.

SUBSCAPULARIS

This muscle arises from the greater part of the subscapular fossa of the scapula by a mixed fleshy and tendinous origin, hence the ridges on this part of the

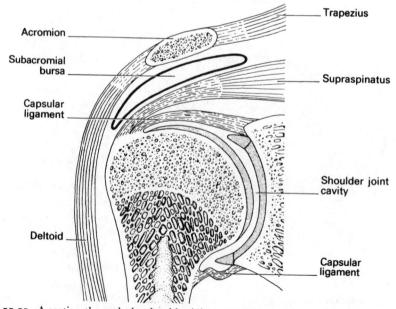

Fig. 11.21. A section through the shoulder joint to show the subacromial bursa.

scapula (see Chapter 9). It is inserted into the lesser tuberosity of the humerus. If you articulate the humerus and scapula you will see that the muscle will be an immediate anterior relation of the shoulder joint (Fig. 11.20), from which it is separated, as you would expect, by a large bursa which communicates with the joint (Fig. 10.17). (It is helpful to examine also the attachments of the next three muscles in the same way.) Subscapularis is supplied by the subscapular nerves. It is a medial rotator of the arm but its most important is as a member of the *rotator cuff* muscles which will be described below.

SUPRASPINATUS

This arises from a large part of the supraspinous fossa and is inserted into the upper part of the greater tuberosity. It is covered by dense fascia and is deep to trapezius and therefore cannot be palpated very easily. Examination of the articulated bones will show that it is related to, and is in fact fused with, the upper aspect of the capsule of the shoulder joint and that it has to pass under the coraco-acromial arch, i.e. the arch formed by the acromion and the coraco-acromial ligament. It is separated from this, and from the overlying deltoid, by a large *subacromial bursa* (Figs. 11.21 and 11.22). Conditions affecting the tendon can thus spread into the bursa and cause painful limitation of movement at the shoulder (see Chapter 10). The muscle is supplied by the suprascapular nerve. It is one of the rotator cuff muscles (see below) and is also partly responsible for the initial stages in abduction at the shoulder joint—the first 15°–20° according to some, but this is controversial. The important thing to understand is that when the arm is by the side, the fibres of deltoid are more or less vertical so that the contraction of this muscle alone would cause upward displacement of the head of the humerus. Supraspinatus is able to carry out a pure abduction, although with very little mechanical advantage, and this, along with the downward pull of teres major, enables deltoid to carry out abduction once a slight angle has been made. The tone of supraspinatus is largely responsible for preventing a vertically downward displacement of the head of the humerus when the arm is by the side. (It is often forgotten that the upper limb, on its own, is a *heavy* object which you will appreciate in the bath if you slowly raise the arm from a submerged position until it is clear of the water.)

INFRASPINATUS

Infraspinatus takes origin from the infraspinous fossa and crosses the back of the shoulder joint to be inserted into the greater tuberosity below supraspinatus. It is supplied by the suprascapular nerve and is a lateral rotator of the arm as well as being a muscle of the rotator cuff (Figs. 11.20 and 11.22).

TERES MINOR

This arises from the dorsal surface of the scapula along its lateral border and is often fused with infraspinatus. It is inserted into the greater tuberosity of the

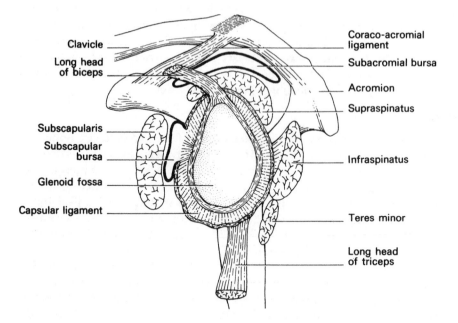

Fig. 11.22. Lateral view of the glenoid and the muscles of the rotator cuff.

humerus below infraspinatus. It is supplied by the axillary nerve and is a lateral rotator of the arm and a member of the rotator cuff group (Fig. 11.20).

THE ROTATOR CUFF

As you will have realised by now, the rotator cuff (Fig. 11.22) consists of the subscapularis, supraspinatus, infraspinatus and teres minor. These four muscles are arranged around the shoulder joint, lying in front, above and behind the joint (inferiorly, the joint is relatively unprotected which is one reason why the humerus normally dislocates downwards in the first instance). The muscles, as well as having the individual actions described above, also act as a set of 'relaxable ligaments'. You have seen in Chapter 9 how the shoulder joint is built for mobility rather than strength and is therefore not surrounded by tough inextensible ligaments. The capsular ligament, in fact, is relatively thin and lax (especially inferiorly) and has a large deficiency in front where the subscapular bursa communicates with the joint. The rotator cuff muscles thus have to play an important part in stabilising the joint by their contraction but they can, when the occasion demands, relax so as to allow the full mobility of the shoulder joint.

THE MUSCLES OF THE ARM

CORACOBRACHIALIS

This small and unimportant muscle passes from the coracoid process (where it is fused with the short head of biceps) to a point half-way down the medial side of the humerus (Fig. 11.18). It is supplied by the musculocutaneous nerve and moves the arm upwards and medially, as occurs when swinging the arm while walking.

BICEPS BRACHII

So called because it has two heads and must be distinguished from the muscle in the thigh known as biceps femoris. There is no need to describe its position since, to the layman, it is the best-known muscle in the body. It arises by two heads—the *short head* from the coracoid process in company with coracobrachialis, and the *long head* from the supraglenoid tubercle of the scapula (Fig. 11.18). To reach this, the tendon lies in the intertubercular sulcus and then passes through the shoulder joint itself. It is held in the sulcus by a ligamentous band (the *transverse ligament*) which bridges over it, and it is lubricated in this position by a prolongation of the synovial membrane of the shoulder joint, which surrounds it (Fig. 10.17). The two heads join to form a fusiform belly and it is inserted by a tough and easily palpable tendon into the posterior part of the radial tuberosity, being separated from the smooth anterior part by a bursa. An extension from the tendon called the *bicipital aponeurosis* passes medially and spreads out to blend with the deep fascia of the forearm. The aponeurosis has a prominent and very sharp edge which should not be confused with the biceps tendon. The muscle is supplied by the musculocutaneous nerve.

The biceps is an important supinator of the forearm and is also a powerful flexor when the forearm is in a supinated position. By means of its long head it is also a weak flexor of the shoulder joint. Like other muscles supplied by C5 and 6, it is thus important in putting food into the mouth by a combination of flexion and supination of the forearm and flexion of the shoulder. Acting eccentrically, it is used in lowering the forearm particularly when a weight is being carried. A good method of demonstrating the way in which increasing numbers of muscle fibres are brought into play when the load on a muscle is increased is to hold a bucket under a running tap while observing biceps. As the bucket fills, the muscle becomes more and more prominent. Biceps is used powerfully when lifting any heavy weight or in supinating movements such as occur when screwing in screws.

BRACHIALIS

The brachialis (Fig. 11.18) is a direct anterior relation of the elbow joint. It arises from the front of the shaft of the humerus by muscular fibres. It therefore leaves no mark and requires a wide area of attachment. It crosses the elbow joint and is inserted by a powerful tendon into the tuberosity of the ulna. It is a flexor of the elbow joint.

TRICEPS

As its name implies, triceps has three heads, which are called the *long*, *lateral* and *medial* heads. The names of the heads are not very informative and it is best to regard the triceps as consisting of two layers: a superficial layer comprising the long and lateral heads, and a deep layer formed by the medial head (Fig. 11.23). The medial head is attached to the lower part of the back of the humerus below the radial (spiral) groove. It does not leave a mark on the bone. The lateral head arises from a slightly roughened ridge above the radial groove, and the long head arises from the infraglenoid tubercle. These two heads unite to form the superficial layer of triceps which passes downwards to fuse with the medial (deep) head. Since the radial nerve descends in the radial groove, it lies between the two layers of the muscle. The insertion of triceps is into the olecranon by means of a strong tendon. It is supplied by the radial nerve, partly by branches which arise in the axilla (to the medial and long heads) and partly by branches which arise in the radial groove (to the medial and and lateral heads). A lesion of the radial nerve produced by a fracture of the shaft of the humerus will therefore not produce a complete paralysis of triceps.

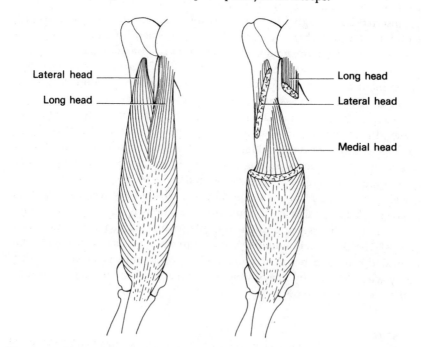

Lateral head

Long head

Long head

Lateral head

Medial head

Fig. 11.23. The three heads of triceps—the medial head is deep to the other two.

Triceps is a powerful extensor of the elbow joint, mainly by means of the medial head. This only occurs when the movement takes place against resistance or against gravity. When the arm is lowered to the side the movement is produced by gravity, with *biceps* contracting eccentrically. It contracts as a synergist, when biceps and supinator are supinating the forearm, in order to prevent flexion.

THE QUADRANGULAR SPACE

When the muscles of the posterior wall of the axilla are separated, a potential opening in it can be defined. The opening is roughly quadrangular and is bounded by the subscapularis and teres minor above, the teres major below, the surgical neck of the humerus laterally and the long head of triceps medially (Fig. 11.20). This *quadrangular space* is not a true space in the living body although a depression corresponding to the space can be seen and palpated in a muscular person. It is, however, an important landmark since it is through this space that the axillary nerve and its accompanying artery leave the axilla to reach the posterior aspect of the arm.

THE FLEXOR MUSCLES OF THE FOREARM

The muscles of the forearm are mostly slender muscles which have long and narrow tendons. This is necessary because the majority of the muscles act on the hand so that a large number of tendons have to be packed together in a relatively small space around the wrist in order to allow for mobility of the hand. The muscles of the front of the forearm (the flexors) are supplied by the median or the ulnar nerve while those of the back (the extensors) are supplied by the radial nerve. The brachioradialis however, forms an exception to this, as will be seen later. The flexor muscles arise from the medial side of the elbow region and the extensors from the lateral side. Do not be misled by Rembrandt's famous painting 'The Anatomy Lesson' in which the artist has slipped up and portrayed the flexors coming from the lateral side.

The muscles of the anterior surface of the forearm are best described in three layers, superficial (pronator teres, flexor carpi radialis, palmaris longus and flexor carpi ulnaris), intermediate (flexor digitorum superficialis) and deep (flexor pollicis longus, flexor digitorum profundus and pronator quadratus). At the wrist, however, where only tendons are found, the layers are less well marked.

RETINACULA AND THE CARPAL TUNNEL

An important thickening in the deep fascia called the *flexor retinaculum* is found in front of the wrist. As was described in Chapter 9, the carpal bones are arranged

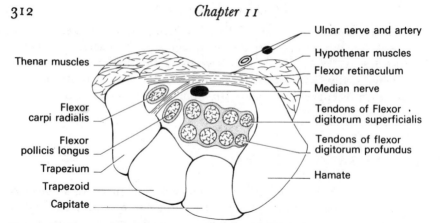

Fig. 11.24. Cross-section through the carpal tunnel. Note particularly that the ulnar nerve is superficial and the median nerve deep to the flexor retinaculum.

in such a way that they present a concavity directed forwards (Fig. 11.24). The flexor retinaculum is attached to the four most prominent bony points of the carpus, namely the tubercle of the scaphoid and the crest of the trapezium laterally and the hook of the hamate and the pisiform bone medially (Fig. 9.22). All these bones can be palpated so that the position of the retinaculum is easy to work out. Note particularly that its upper border is about the level of the distal skin crease of the wrist: the retinaculum does *not* lie in the position of a wrist-watch strap. The fibres of the retinaculum run transversely and are extremely strong, so that an almost rigid tunnel—the *carpal tunnel*—is formed between the bones and the retinaculum (Fig. 11.24). Through this a number of tendons with their synovial sheaths and also the median nerve have to pass. Compression of the nerve is not uncommon in this very restricted space.

On the extensor surface, a less well-defined *extensor retinaculum* is found. This is a simple strap-like thickening of deep fascia which helps to hold the extensor tendons in position (Fig. 11.28).

PRONATOR TERES

This muscle (Fig. 11.25) arises just above the medial epicondyle of the humerus and is inserted into the lateral side of the shaft of the radius about half-way down It also has a small and unimportant deep head. It is supplied by the median nerve and is a pronator of the forearm. Pronator teres form the medial boundary of a triangular depression in front of the elbow called the *cubital fossa*; the lateral boundary is formed by brachioradialis (Fig. 25.5).

FLEXOR CARPI RADIALIS

This muscle shares a common origin with the remainder of the superficial muscles from the medial epicondyle of the humerus (Fig. 11.25). Its long tendon passes through the carpal tunnel and it is inserted into the bases of the 2nd and

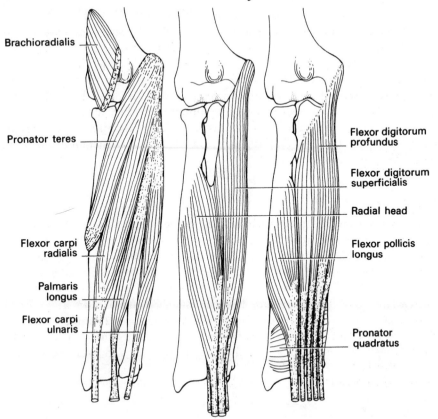

Fig. 11.25. The muscles of the superficial, intermediate and deep layers of the front of the forearm.

3rd metacarpals. It is supplied by the median nerve. Acting with other flexor muscles, especially flexor carpi ulnaris, it is a flexor of the wrist. Acting with the extensor carpi radialis muscles it is an abductor.

PALMARIS LONGUS

Small, and very often absent altogether, this therefore unimportant muscle arises from the common flexor origin (Fig. 11.25) and is inserted by means of a long, thin tendon into the front of the flexor retinaculum and the palmar aponeurosis (to be described later). It is supplied by the median nerve. When present, the tendon is prominent and gives an indication of the position of the median nerve at the wrist, the nerve being deep and slightly lateral to the tendon.

FLEXOR CARPI ULNARIS

Arising from the common flexor origin and by an aponeurosis from the posterior border of the ulnar, flexor carpi ulnaris is inserted into the pisiform bone (Fig.

11.30). A small ligament passes from the pisiform to the base of the 5th metacarpal and it is said that this represents the terminal part of the tendon, the
pisiform thus being a sesamoid bone. Whether this is true or not, it is useful to
remember the fact, since both the flexor and the extensor carpi ulnaris can then
be regarded as being inserted into the fifth metacarpal. The muscle is supplied
by the ulnar nerve. It is a flexor of the wrist (working with other flexors) and an
adductor (working with extensor carpi ulnaris). It often acts as a fixation
muscle, stabilising the pisiform so that abductor digiti minimi has a firm origin
from which to act. It also acts as a synergist to prevent abduction of the wrist
when the thumb is extended.

FLEXOR DIGITORUM SUPERFICIALIS

This muscle forms the intermediate layer, being deep to the preceding four
muscles. It arises from the common flexor origin and also has a head from the
front of the shaft of the radius (Fig. 11.25). The muscle belly cannot be palpated
since it is covered by the superficial layer of muscles but when the latter muscles
narrow down to tendons at the wrist, the four tendons of flexor digitorum
superficialis come to the surface between palmaris longus and flexor carpi
ulnaris. Here they are arranged in pairs with the tendons to the middle and ring
fingers in front and those to the index and little fingers behind. The tendons
pass through the carpal tunnel and then separate out and pass one to each
finger (Fig. 11.27).

In each finger is found a form of retinaculum called the *fibrous flexor sheath*

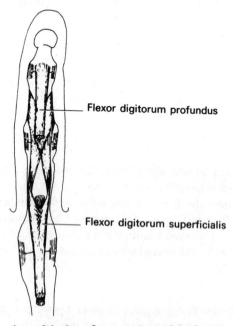

Flexor digitorum profundus

Flexor digitorum superficialis

Fig. 11.26. The insertions of the long flexor tendons of the fingers.

(Fig. 11.4). This takes the form of a tough fibrous arch which passes from one side of the phalanx to the other, forming, like the flexor retinaculum, an osseo-fibrous tunnel. Opposite the joints, however, the fibrous flexor sheaths are rather weak, with criss-crossing fibres, so as to allow free movement at the joints. In the fibrous flexor sheaths the tendons of flexor digitorum superficialis are accompanied by those of flexor digitorum profundus. As their name implies, the latter are deep to the former, at least in the hand and the proximal parts of the fingers. Just as they enter the fibrous flexor sheaths, each tendon divides into two to allow the passage of a tendon of flexor digitorum profundus (Fig. 11.26) The two slips then twist through 180° and reunite deep to the profundus tendon thus forming a groove in which the latter lies. The reunited tendon then divides yet again and is inserted into the sides of the middle phalanx. The rather complicated passage of the profundus tendon through the superficialis tendon is worth examining closely as it forms a beautiful mechanism by means of which the profundus tendon is perfectly free to move, no matter how much tension is developed in the superficialis tendon. A simple buttonhole would be useless, since contraction of superficialis would grip the profundus tendon and hamper its movement considerably.

Flexor digitorum superficialis is supplied by the median nerve and it flexes the middle and proximal phalanges and, in continued contraction, will flex the wrist.

FLEXOR POLLICIS LONGUS

This is the lateral muscle of the deep group which arises from the anterior surface of the shaft of the radius and the adjoining part of the interosseous membrane. Its fibres are arranged in a unipennate manner and join the long tendon along its medial side. The tendon runs through the carpal tunnel (Fig. 11.24) and is inserted into the base of the distal phalanx of the thumb. It is supplied by the median nerve via its anterior interosseous branch and it is a flexor of all the joints of the thumb.

FLEXOR DIGITORUM PROFUNDUS

This is the only muscle capable of producing flexion at the distal interphalangeal joint of the fingers, just as flexor pollicis longus is the only muscle able to flex the distal phalanx of the thumb. It arises from the shaft of the ulna, as far back as its posterior border, and from the adjoining part of the interosseous membrane (Fig. 11.25). It, like its tendons, thus lies alongside flexor pollicis longus. Its four tendons lie side by side as they pass through the carpal tunnel. In the palm of the hand they diverge and enter the fibrous sheaths of the four fingers. Each tendon passes through the tendon of flexor digitorum superficialis and is inserted into the base of the distal phalanx (Fig. 11.26). The medial half of the muscle is supplied by the ulnar nerve and the lateral half by the anterior interosseous branch of the median nerve. It is a flexor of the fingers and of the wrist joint. An interesting feature of this muscle is that the muscle belly does not separate out completely into its tendons until it has almost reached the wrist, although the portion of the muscle destined to move the index finger separates rather higher than this. The index finger thus is slightly more independent than the three medial fingers, which tend to move together. You can demonstrate this if you extend all your fingers (thus stretching the profundus) and then flex your ring finger on its own. The flexion is carried out by superficialis only, because profundus is unable to act on the ring finger alone and you will therefore find that the distal phalanx is quite uncontrollable.

PRONATOR QUADRATUS

A small muscle, easily identified because its fibres run transversely from the lower part of the shaft of the ulna to the corresponding part of the radius (Fig. 11.25). It is supplied by the median nerve via its anterior interosseous branch, and is a pronator of the forearm, being assisted by pronator teres.

SYNOVIAL SHEATHS OF THE FLEXOR TENDONS

In the wrist, hand and fingers, the flexor tendons are enclosed in a complicated arrangement of synovial sheaths which are shown in Figure 11.27. The tendons of flexor digitorum superficialis and profundus are enclosed in a common sheath which starts about 1 inch above the flexor retinaculum and ends, for the middle

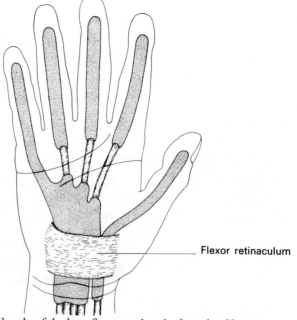

Flexor retinaculum

Fig. 11.27. The synovial sheaths of the long flexor tendons in the palm. Note the position of the main skin creases.

three digits, about half-way along the metacarpals. In the case of the little finger, the sheath continues as far as the distal phalanx but in the other three fingers the tendons are bare for a short distance and are then enveloped in synovial sheaths distal to the level of the metacarpophalangeal joints. The tendon sheath of flexor pollicis longus is separate from that of the fingers and extends from above the flexor retinaculum to the distal phalanx although there may be a small communication between the sheath for the thumb and that for the other tendons behind the flexor retinaculum. Infection of the synovial sheaths may cause adhesions between their parietal and visceral layers with consequent severe limitation of movement.

THE EXTENSOR MUSCLES OF THE FOREARM

Many of the extensor muscles arise from a common origin on the lateral epicondyle of the humerus (Fig. 11.28). They are supplied by the radial nerve either directly or via its posterior interosseous branch, and like the flexors, they mostly act on a number of joints.

BRACHIORADIALIS

This is an unusual and somewhat puzzling muscle since although it arises from the lateral side of the humerus, and is supplied by the radial nerve, it is actually a *flexor* of the elbow. It is sometimes said that it is really an extensor which, during the course of evolution, has migrated round the corner to lie on the anterior aspect of the forearm. The other peculiar thing about this muscle is that whereas most muscles cross the joint on which they act and are then immediately inserted into the bone (e.g. biceps), the brachioradialis crosses the elbow joint and then travels the whole length of the forearm before reaching its insertion (Fig. 11.19). The muscle arises from the lateral supracondylar ridge of the humerus and is inserted by means of a flattened tendon into the distal end of the radius on its lateral side. Brachioradialis is a flexor of the elbow joint especially when the forearm is in the midprone position (as when carrying a raincoat on the arm) and it is also said to act as a '*shunt*' muscle, i.e. a muscle which, by its proximal and distal attachments, keeps the bones which take part in a joint together and helps to prevent their distraction as might occur, for instance, when carrying a heavy weight. Such distractions may also result from centrifugal force when the arm is swinging freely and certainly brachioradialis contracts during both flexion and extension when these are rapid. When the forearm is semiflexed against resistance, brachioradialis stands out prominently, looking as if it needed a flexor retinaculum to hold it in place.

ANCONEUS

This is a small and unimportant triangular muscle which arises from the lateral epicondyle of the humerus and is inserted into the back of the upper part of the

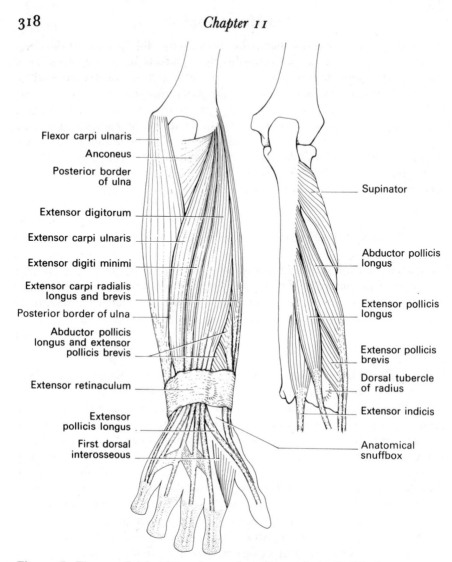

Flexor carpi ulnaris

Anconeus

Posterior border
of ulna

Extensor digitorum

Extensor carpi ulnaris

Extensor digiti minimi

Extensor carpi radialis
longus and brevis

Posterior border of ulna

Abductor pollicis
longus and extensor
pollicis brevis

Extensor retinaculum

Extensor
pollicis longus

First dorsal
interosseous

Supinator

Abductor pollicis
longus

Extensor pollicis
longus

Extensor pollicis
brevis

Dorsal tubercle
of radius

Extensor indicis

Anatomical
snuffbox

Fig. 11.28. The superficial and deep layers of muscles on the back of the forearm.

ulna, including the olecranon (Fig. 11.28). It is supplied by the radial nerve
and is an extensor of the elbow joint.

EXTENSOR CARPI RADIALIS LONGUS AND BREVIS

As was described above, there is only one radial flexor of the wrist (flexor carpi
radialis) but in the case of the extensors, two such muscles are present, although
they are very close together and cannot normally be identified separately during
life. Extensor carpi radialis longus arises from the lateral supracondylar ridge
of the humerus below brachioradialis and is inserted into the base of the

second metacarpal. The extensor carpi radialis brevis arises from the lateral epicondyle of the humerus and is inserted into the base of the third metacarpal (compare with the insertions of flexor carpi radialis). Both muscles are supplied by the radial nerve, the brevis via its posterior interosseous branch. Both can act with extensor carpi ulnaris to produce extension of the wrist or with flexor carpi radialis to produce abduction (or radial deviation). They are used more commonly as synergists to prevent flexion of the wrist occuring during flexor movements of the fingers.

EXTENSOR DIGITORUM

This is the main extensor muscle of the fingers although its action needs to be combined with that of the interossei and lumbricals under certain circumstances The muscle takes origin from the common extensor tendon on the lateral epicondyle. As it passes down the forearm it divides into four tendons which can easily be identified on the back of the hand (Fig. 11.28). They then travel on to the dorsal surface of each of the fingers. On the finger, each tendon spreads out into a triangular aponeurosis called the *dorsal digital expansion*. The base of the triangle (Fig. 11.29) receives the insertions of the appropriate interossei and

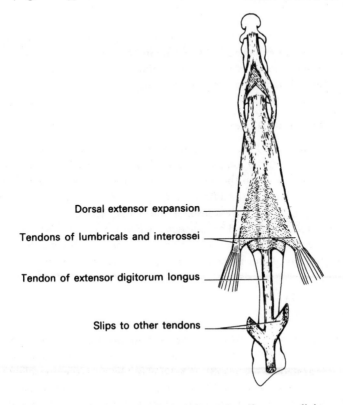

Dorsal extensor expansion

Tendons of lumbricals and interossei

Tendon of extensor digitorum longus

Slips to other tendons

Fig. 11.29. The insertions of the lumbricals, interossei and extensor digitorum (longus).

lumbricals (see below) and is bound down by transverse fibres to the deep transverse metacarpal ligament. Distally, the expansion splits into three slips, the middle of which is inserted into the base of the middle phalanx. The other two unite to be inserted into the base of the distal phalanx. On the back of the hand, tendinous slips pass between the tendons of extensor digitorum. These are variable in position and number, but usually the tendon to the ring finger is attached to the tendons of either side of it. This is why, if you make a fist, you are unable actively to extend the ring finger on its own, although it can be pushed into position passively by the other hand.

Extensor digitorum is supplied by the posterior interosseous branch of the radial nerve. It is an extensor of the wrist and fingers, during such actions as opening the hand to grasp an object. At the same time, owing to the divergence of tendons on the dorsum of the hand, it tends to spread the fingers apart. (Compare with the flexor muscles.) In using a *precision grip*, however, in which the fingers are used to manipulate small objects such as a needle, extension of the interphalangeal joints is combined with flexion of the metacarpophalangeal joints and in this case the interossei and lumbricals play an important role. This will be described later.

EXTENSOR DIGITI MINIMI

A small slip of muscle arising from the common extensor origin and inserting into the dorsal digital expansion of the little finger (Fig. 11.28). It is supplied by the posterior interosseous branch of the radial nerve and is an extensor of the little finger. Like all such small muscles, it is subject to a great deal of variation.

EXTENSOR CARPI ULNARIS

This muscle arises from the common extensor origin and also by an aponeurosis from the posterior border of the ulna (Fig. 11.28). Its tendon passes down in the groove between the head and the styloid process of the ulna, and it is inserted into the base of the fifth metacarpal (compare with the insertion of flexor carpi ulnaris). It is an extensor of the wrist, acting with extensor carpi radialis longus and brevis, and an adductor when it works with flexor carpi ulnaris. It also acts synergically with other extensors to prevent flexion of the wrist during flexion of the fingers. It is supplied by the posterior interosseous nerve.

SUPINATOR

This, and the main parts of the remaining four muscles form a deep stratum which cannot be seen until the other muscles that have been described above have been separated (Fig. 11.28). The lower parts of the muscle bellies of abductor pollicis longus and extensor pollicis brevis, however, can be seen since they come to the surface between extensor digitorum and extensor carpi radialis brevis; for this reason they are sometimes called the 'outcropping muscles'.

Supinator arises from the lateral epicondyle of the humerus, from the supinator crest of the ulna and from the elbow joint ligaments in between these bony points. It wraps itself around the radius from behind to be inserted into the upper third of the shaft of the radius. It is supplied by the posterior interosseous nerve, which actually passes through it, and it is a supinator of the forearm. It acts in conjunction with biceps to produce a very powerful supinating action which is used in tightening screws, provided they have the usual right-hand thread.

ABDUCTOR POLLICIS LONGUS

One of the 'outcropping muscles', this arises deeply from the upper parts of the posterior surfaces of the radius and ulna and from the interosseous membrane between. The muscle belly comes to the surface on the lateral side of the forearm and is inserted into the base of the first metacarpal (Figs. 11.28 and 11.30). It is supplied by the posterior interosseous branch of the radial nerve and is an abductor and an extensor of the thumb. Its tendon is easily palpable and forms a boundary of the anatomical snuffbox (see below).

EXTENSOR POLLICIS BREVIS

This muscle is very closely associated with the preceding muscle since its origin is from the radius and adjoining interosseous membrane immediately below the abductor and the muscle belly 'outcrops' between extensor carpi radialis brevis and extensor digitorum in the same way (Fig. 11.28). The tendon follows that of the abductor closely, forming, with the latter, the lateral boundary of the anatomical snuffbox. It is finally inserted into the base of the proximal phalanx of the thumb. It is supplied by the posterior interosseous nerve and it extend the proximal phalanx and the metacarpal of the thumb.

EXTENSOR POLLICIS LONGUS

This arises from the ulna just below abductor pollicis longus. Its slender tendon descends vertically to the lower end of the radius where it uses the dorsal tubercle as a pulley in order to change direction, lying in a well-marked groove in the bone as it does so (Fig. 11.28). It passes medial to the tubercle and then heads obliquely across to reach the thumb, where it is inserted into the base of the distal phalanx. As a result of its relation to the dorsal tubercle there is a wide gap between this tendon and those of the abductor pollicis longus and extensor pollicis brevis. This interval forms the base of the snuffbox, the tendon itself forming the medial boundary. The muscle is supplied by the posterior interosseous nerve and it extends all joints of the thumb.

EXTENSOR INDICIS

The last of the four deep muscles, this inconspicuous muscle arises from the ulna, below extensor pollicis longus and its tendon blends with the index finger tendon of extensor digitorum near the head of the second metatarsal (Fig. 11.28). It is supplied by the posterior interosseous nerve and it helps to extend the index finger.

THE ANATOMICAL SNUFFBOX

This is the depression which can be observed on the dorsum of the radial side of the hand when the thumb is extended. The depression lies between the tendons of abductor pollicis longus and extensor pollicis brevis laterally and extensor pollicis longus medially (Fig. 11.28). In its floor can be felt the styloid process of the radius and the scaphoid—tenderness in the snuffbox is an important sign of a fractured scaphoid. The pulsations of the radial artery can also be felt as the artery crosses the snuffbox before piercing the first dorsal interosseous muscle to reach the palm of the hand. It is helpful to remember that each of the three tendons which form the boundaries of the snuffbox is inserted into a different bone of the thumb—the abductor into the metacarpal, the extensor brevis into the proximal phalanx and the extensor longus into the distal phalanx.

THE SYNOVIAL SHEATHS OF THE EXTENSOR TENDONS

As might be expected, the extensor tendons are provided with synovial sheaths but only in the immediate vicinity of the extensor retinaculum (a transverse thickening in the deep fascia just above the wrist). There are six synovial sheaths here which are shared among the tendons as follows (from lateral to medial): (1) for abductor pollicis longus and extensor pollicis brevis, (2) for extensor carpi radialis longus and brevis, (3) for extensor pollicis longus, (4) for extensor digitorum and extensor indicis, (5) for extensor digiti minimi and (6) for extensor carpi ulnaris. There are thus no synovial sheaths on the backs of the fingers.

THE MUSCLES OF THE HAND

Inspection of the palm of the hand will show that it has a depressed central area with a swelling on either side, the larger of the two being, in lay terminology, the 'ball of the thumb'. These are known as the *thenar* (thumb-side) and *hypothenar* eminences, and the central depressed area lies over a dense band of fascia called the *palmar aponeurosis* (Fig. 11.30). You will find that over the thenar and hypothenar eminences the thick skin can be picked up with some difficulty but over the palmar aponeurosis it is almost impossible to pinch up the skin. This is because the skin of the palm is thickened (especially the stratum corneum) and is attached by dense fascia to the palmar aponeurosis. The palm of the hand shows numerous flexure lines, the 'line of life' of the palmist being the main flexure line for the thumb and the 'line of heart' for the metacarpophalangeal joints. In view of the anatomical interpretation of these lines, it seems doubtful if a particularly long flexure line for the first carpometacarpal joint indicates longevity, especially as very similar lines are found in the palms of apes. It would certainly be interesting to hear a palmist interpreting the flexure lines of a gorilla!

THE PALMAR APONEUROSIS

This dense triangular sheet of fascia extends from the insertion of palmaris longus (if present) and the central part of the flexor retinaculum to the bases

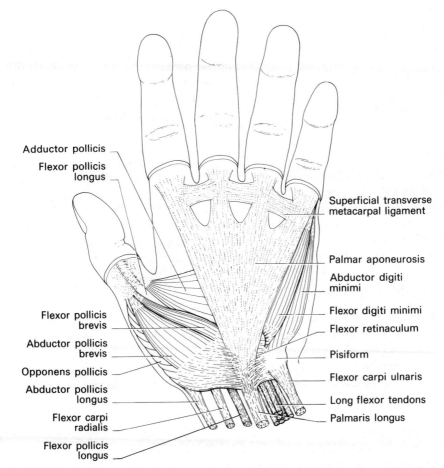

Fig. 11.30. The superficial muscles of the palm. Not all the tendons shown at the wrist are palpable.

of the fingers. Slips pass to each finger and are continuous with the fibrous flexor sheaths and to the deep transverse metacarpal ligaments. Much thinner extensions of the palmar aponeurosis cover the muscles of the thenar and hypothenar eminences. Fibrous septa pass from the deep surface of the palmar aponeurosis to some of the metacarpals and divide up the substance of the palm of the hand into potential *palmar spaces*. These are important in certain infections of the hand. Occasionally the fibrous tissue of the palmar aponeurosis undergoes shrinkage, for an unknown region, and the fingers may gradually become curled up into the palm of the hand (*Dupuytren's contracture*). The muscles of the thenar and hypothenar eminences comprise an abductor, a flexor and an opponens. In addition, a number of interosseous and lumbrical muscles assist movements of the fingers, particularly in adduction and abduction, while the thumb has an adductor of its own. Before reading the account of these muscles you should revise the rather complicated movements of the thumb.

ABDUCTOR POLLICIS BREVIS

This is the most lateral of the thenar muscles. It arises from the flexor retinaculum and, to some extent, from the lateral bones of the carpus to which the retinaculum is attached (Fig. 11.30). It is inserted into the lateral side of the base of the proximal phalanx of the thumb. It is supplied by the median nerve and it abducts the thumb and, to some extent, rotates it medially.

FLEXOR POLLICIS BREVIS

This muscle is medial to abductor pollicis. It arises from the flexor retinaculum and the adjacent carpal bones and is inserted, in common with abductor pollicis brevis, into the base of the proximal phalanx (Fig. 11.30). Its tendon contains a sesamoid bone. It is supplied by the median nerve (but usually by the ulnar as well) and is a flexor of the proximal phalanx of the thumb and also of the metacarpal. It also helps in medial rotation of the metacarpal.

OPPONENS POLLICIS

Opponens pollicis is the deepest of the muscles of the thenar eminence and is the only one to be inserted into the metacarpal. It arises from the flexor retinaculum and the adjacent bones and is inserted into the whole length of the shaft of the first metacarpal (Fig. 11.31). It thus acts only on the first carpometacarpal joint. It is supplied by the median nerve (and sometimes by the ulnar as well) and it produces opposition of the thumb.

ADDUCTOR POLLICIS

Although this is a short muscle of the thumb it contributes little to the thenar eminence since its main bulk lies deeply in the palm of the hand. It has two

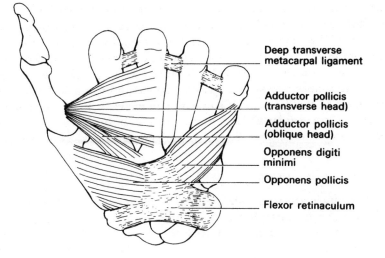

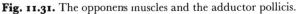

Fig. 11.31. The opponens muscles and the adductor pollicis.

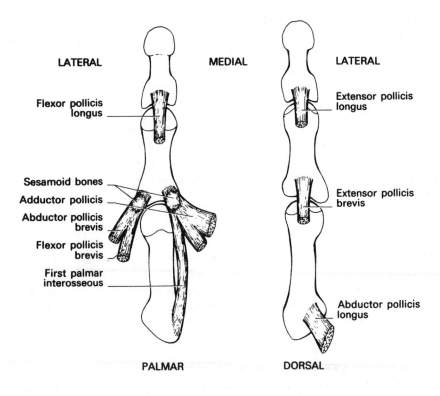

Fig. 11.32. The tendons inserted into the bones of the thumb.

heads of origin. The oblique head arises from the bases of the second and third metacarpals and is therefore blended with the insertions of flexor carpi radialis. The transverse head arises from the shaft of the third metacarpal (Fig. 11.31). The tendon of the muscle contains a sesamoid bone and is inserted into the medial (ulnar) side of the proximal phalanx (Fig. 11.32). The tendon of the flexor pollicis longus passes in the groove between this sesamoid bone and that in the tendon of the flexor pollicis brevis. The muscle is supplied by the deep branch of the ulnar nerve and it adducts the thumb. It acts strongly when the thumb and index finger are used in pinching movements.

ABDUCTOR DIGITI MINIMI

This small muscle arises from the pisiform and is inserted into the medial side of the base of the proximal phalanx of the little finger. It abducts the little finger, but when carrying out this action it needs the contraction of flexor carpi ulnaris, acting as a fixator, to fix the pisiform. It is supplied by the ulnar nerve (Fig. 11.30).

FLEXOR DIGITI MINIMI

Arising from the flexor retinaculum and adjacent bones, this muscle is inserted into the ulnar side of the base of the proximal phalanx of the little finger (Fig. 11.30). It is supplied by the ulnar nerve and it helps to flex the proximal phalanx of the little finger.

OPPONENS DIGITI MINIMI

This arises from the flexor retinaculum and adjoining bones and, like the opponens pollicis, is inserted into the whole length of the shaft of the metacarpal (Fig. 11.31). It is supplied by the ulnar nerve and helps to carry out the small degree of opposition which occurs at the carpometacarpal joint of the little finger, thereby 'cupping the palm'.

THE LUMBRICAL MUSCLES

The name of these muscles is derived from *Lumbricus terrestris* (the earthworm) and in size and shape there is a certain resemblance. There are four lumbricals and they arise from the tendons of flexor digitorum profundus. Each passes to the lateral (radial) side of the corresponding finger and is inserted into the dorsal digital expansion of the extensor digitorum tendon (Fig. 11.33). The first two, which move the index and middle fingers respectively, are supplied by the median nerve and the third and fourth by the ulnar nerve. Thus the whole of the flexor digitorum profundus, along with its attached lumbricals may be regarded as consisting of two parts—the lateral part and the corresponding lumbricals being supplied by the median nerve and the medial part, with its

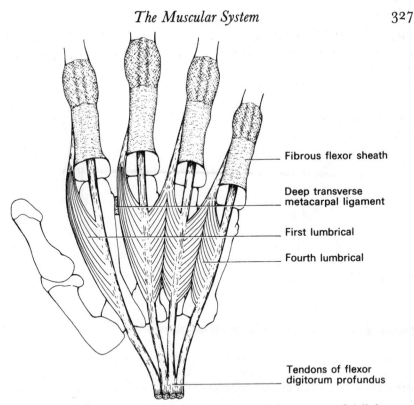

Fibrous flexor sheath

Deep transverse
metacarpal ligament

First lumbrical

Fourth lumbrical

Tendons of flexor
digitorum profundus

Fig. 11.33. The lumbrical muscles. The tendons of flexor digitorum superficialis have been omitted for clarity.

lumbricals, by the ulnar nerve. The lumbricals assist the interossei in producing flexion at the metacarpophalangeal joints and extension at the interphalangeal joints.

THE INTEROSSEI

The interosseous muscles are small but extremely important. As their name implies, they are 'between the bones', i.e. between the metacarpals and they are arranged in two layers of four muscles each, dorsal and palmar. Their attachments are difficult to memorise but if you want to do so you only have to remember two basic facts: (1) the palmar interossei arise from one metacarpal and the dorsal from two; (2) the dorsal interossei abduct the phalanges and the palmar adduct them. Some students remember D.Ab. and P.Ad. as an *aide memoire*. You will now be able to work out the attachments, shown in Figure 11.34. The four dorsal interossei arise from the adjacent sides of each pair of metacarpals. Their tendons pass on to the dorsum of the finger to join the extensor expansion and are also inserted into the side of the base of the proximal phalanx. The side depends on the finger, but since the dorsal interossei abduct you will deduce that the insertion will be on the side farthest away from the

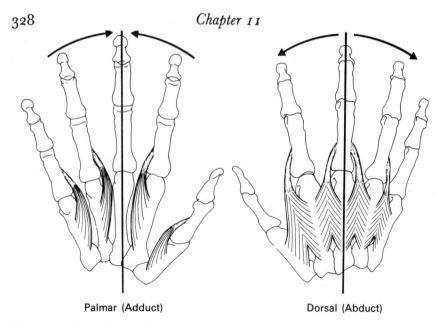

Palmar (Adduct)   Dorsal (Abduct)

Fig. 11.34. The palmar and dorsal interossei.

middle finger (remember that abduction and adduction of the fingers refers to movement away from and towards the line of the middle finger). The palmar interossei arise from only one metacarpal and are on the palmar side of the dorsal interossei. Each is inserted into the dorsal extensor expansion and into the side of the proximal phalanx nearest to the middle finger. All the interossei are supplied by the deep branch of the ulnar nerve. As has been mentioned, the dorsal abduct and the palmar adduct the fingers and since which ever way the middle finger moves, the movement is one of abduction, the proximal phalanx has two dorsal interossei attached. It is impossible to adduct the middle finger (unless it has first been abducted) so it has no palmar interossei attached to it. More important than these movements, however, are the movements which result from the insertions into the dorsal digital expansions. Since the tendons pass onto the dorsum of the proximal phalanges, both set of interossei will flex the metacarpophalangeal joints and by pulling on the dorsal expansion they can extend the interphalangeal joints. This combined movement is one which cannot be carried out by either the long flexors or by extensor digitorum working alone and is essential for the use of the *precision grip* which is described below.

MOVEMENTS OF THE HAND IN GENERAL

The essential and unique feature of the movements of the digits is the ability to carry out very fine and well-regulated manoeuvres such as picking up a pin and handling small tools. At the same time, the hand can be used for heavy work such as carrying a suitcase. A watchmaker may spend the week in carrying out the most meticulous and delicate operations on the minute components of a lady's wristwatch but in his garden at week-ends he may wield a sledgehammer during the manufacture of a rockery. These two basic types of activity of the hand are carried out respectively by the small muscles of the hand such as the interossei and lumbricals and by the long flexors of the fingers. It is convenient to think of two types of grip: the *precision grip* in which the fingers are used in conjunction with the opposed thumb and the *power grip* in which the whole hand is utilised as a vice or clamp to hold a hammer or pickaxe (Fig. 11.35). a third type of grip, the *hook grip*, is sometimes described which is used when the semi-flexed fingers are used to support a suitcase or handbag, but this is merely a variation of the power grip.

In the power grip, the flexors digitorum superficialis and profundus are used to flex all the joints of the fingers so that the fingers curl up into the palm of the hand. At the same time the extensors of the wrist (extensors carpi radialis longus and brevis and extensor carpi ulnaris) contract as synergists to prevent flexion of the wrist. The thumb is often opposed as well so that it comes into contact with the backs of the fingers. In this way an extremely tight grip can be produced, especially if the object being gripped is fairly large. If the wrist is allowed to flex at the same time, passive insufficiency of the extensors will cause the grip to be relaxed—a method of forcing a person to let go a weapon.

In the precision grip, the use of the long flexors is not required, at least for holding small and delicate objects, since it is essential that the pad of the thumb be opposed to the pads of the fingers, i.e. the interphalangeal joints must be held extended while the metacarpophalangeal joints are flexed. This is precisely the

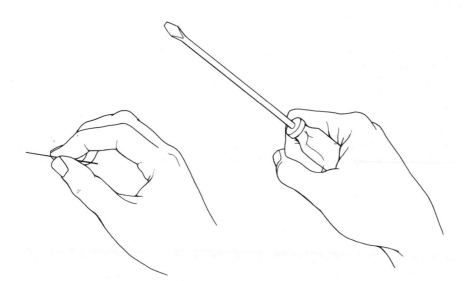

Fig. 11.35. The 'precision grip' and the 'power grip'.

action of the interossei which are supplied by the ulnar nerve. As soon as the fingers (usually the index finger) come into contact with the opposed thumb it is possible to apply more force to the grip by the contraction of the long flexors which then tend forcibly to flex the interphalangeal joints, although this is prevented by the thumb.

When gripping a rounded or spherical object such as a ball, the long flexors are operative and the medial border of the palm closes around the object by flexion at the metacarpophalangeal joints of the ring and little fingers. This is carried out partly by the action of the long flexors which 'mould' the palm around the object, and partly by the muscles of the hypothenar eminence, particularly opponens digiti minimi.

Abduction and adduction of the fingers is carried out in many skilled occupations such as typing and playing musical instruments. Adduction is used as a test for the integrity of the ulnar nerve—the patient is asked to grip a sheet of paper tightly between the fingers. Even if the interossei are paralysed, however, the paper may be gripped by a trick movement using the adducting action of the long flexors of the fingers, so the hand must be kept flat during this test.

The interossei, except for the first dorsal, are not prominent features of the hand. The first dorsal interosseous is responsible for producing the rounded bulge on the back of the hand between the metacarpals of the thumb and index fingers. It can be felt to contract when the index finger is abducted. When the interossei are wasted, perhaps as a result of a lesion of the ulnar nerve, their loss becomes noticeable because the skin sinks in between the metacarpals so that the back of the hand looks thin and very like a skeleton's hand.

It is very important that you should realise that the thumb is by far the most important digit in the hand because of its ability to carry out the movement of opposition. Any of the other digits may be removed without too severe an effect on the functioning of the hand but the loss of the thumb is a major catastrophe as may be appreciated if you try to do up a button or insert a hairgrip without it. Surgeons will also make every possible effort to preserve a damaged thumb and even if it is the only remaining digit, as long as there is a part of the hand to which it may be opposed, the hand can be used.

THE MUSCLES OF THE LOWER LIMB

The muscles of the lower limb are less complicated than those of the upper limb and they tend to be inserted into extensive sheets of fascia instead of into precise bony points as is usually the case in muscles of the arm. Many of the muscles play an important part in maintaining the stability of the joints as well as producing movements, and this is particularly true of the knee joint which has to support the body weight but which has rather flat articulating surfaces and a large gap in the anterior part of its capsule.

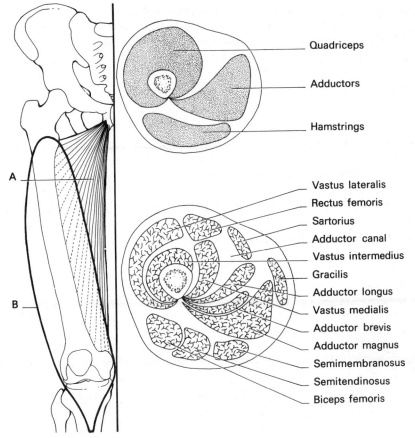

Quadriceps

Adductors

Hamstrings

Vastus lateralis
Rectus femoris
Sartorius
Adductor canal
Vastus intermedius
Gracilis
Adductor longus
Vastus medialis
Adductor brevis
Adductor magnus
Semimembranosus
Semitendinosus
Biceps femoris

A – Adductors

B – Outline of main quadriceps mass

Fig. 11.36. Diagrams to show the main groups of muscles in the thigh. Note the obliquity of the femur and hence of the quadriceps femoris.

THE MUSCLES OF THE FRONT OF THE THIGH

When considering the general structure of the thigh, it is important to remember the obliquity of the two femora which are separated by the whole width of the pelvis at their upper ends but may be almost in contact at their lower ends. The shaft of the femur does not therefore form a central axis down the middle of the thigh but is more laterally placed, with the quadriceps muscle wrapped around it. From the trunk, a series of muscles, especially the adductors, pass obliquely downwards to be inserted into the *back* of the femur along the linea aspera so that the medial muscle mass of the thigh is on a plane posterior to that of the femur and its surrounding muscles (Fig. 11.36).

The deep fascia of the thigh is thick and dense, and down the lateral side it is reinforced with longitudinal fibres to form a very strong band of fascia called the *ilio-tibial* tract. This extends from the crest of the ilium to the lateral condyle of the tibia. At its upper end, two muscles (the gluteus maximus and the tensor fasciae latae) are inserted into it. At its lower end it lies lateral to the knee joint and it is so tough and strong in this region that when under tension it stands out like a tendon.

ILIACUS

The iliacus muscle takes origin from the concave inner surface of the ilium by means of muscle fibres; the surface is therefore smooth. The bulky muscle passes out of the pelvis under the inguinal ligament and 'pours' over the edge of the pelvis to form the lateral part of the floor of the femoral triangle, which will be described later (Fig. 11.39). The muscle fibres are inserted into the tendon of psoas major (the next muscle to be described) or into the femur just below the lesser trochanter. It is supplied by branches of the femoral nerve.

PSOAS MAJOR

This muscle is, like iliacus, largely situated inside the abdomen although it is really a muscle of the lower limb. It arises from the transverse processes of all five lumbar vertebrae and also from the bodies and intervertebral discs. The muscle passes out of the pelvis under the inguinal ligament and is an anterior relation of the hip joint (Fig. 11.39). Its tendon receives most of the fibres of iliacus and is then inserted into the lesser trochanter. The psoas and iliacus are therefore often referred to as one, with the name '*iliopsoas*'. Psoas major is supplied by branches from the roots of the lumbar plexus (L1, 2, and 3). The psoas and iliacus muscles are both powerful flexors of the hip joint. They can thus bring the lower limb forward as in walking or kicking or, acting from below, they flex the trunk upon the lower limb as in sitting up from a lying position. They are active, too, when getting up from a sitting position when they bring the trunk forward so as to bring the centre of gravity over the feet before extension of the hip and knee joints raises the body off the seat. The muscles can also act as lateral rotators of the hip joint and psoas may have some effect on the vertebral column, probably in helping to balance the trunk in the sitting position.

The iliopsoas is separated from the hip joint by a bursa, which occasionally becomes distended with fluid. The muscle bellies, being intra-abdominal, are related to various viscera so that painful abdominal conditions may cause the patient to keep the thigh flexed so as to avoid stretching the muscles.

SARTORIUS

This is an extremely long strap-shaped muscle which runs obliquely across the thigh. It arises from the anterior superior iliac spine and is inserted by a very

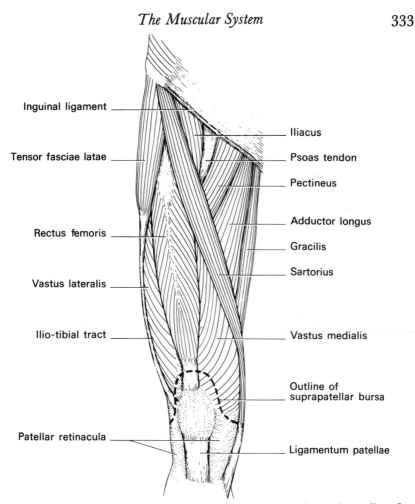

Fig. 11.37. The muscles of the front of the thigh. The dotted line shows the outline of the suprapatellar bursa.

thin flattened tendon into the medial surface of the upper end of the tibia, just in front of gracilis and semitendinosus (Fig. 11.37). As happens so often in the lower limb, the tendon also fans out to be attached to the capsule of the knee joint and the deep fascia of the leg. It is supplied by the femoral nerve and, as its name suggests, will move the lower limb into the position traditionally associated with tailors, i.e. the cross-legged position. In other words, sartorious flexes and abducts the hip joint and flexes the knee. In everyday life you use all the possible actions of sartorious when you inspect the sole of your foot.

QUADRICEPS FEMORIS

Quadriceps is a composite muscle with four components and should not be confused with *quadratus femoris*, whose name refers to its shape. The four parts of

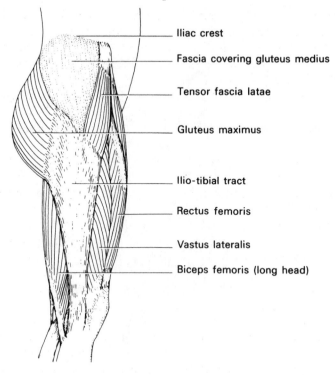

Iliac crest

Fascia covering gluteus medius

Tensor fascia latae

Gluteus maximus

Ilio-tibial tract

Rectus femoris

Vastus lateralis

Biceps femoris (long head)

Fig. 11.38. The lateral side of the thigh. Note the two muscles inserted into the iliotibial tract.

quadriceps (Fig. 11.37) are *rectus femoris, vastus lateralis, vastus intermedius* and *vastus medialis*. Their origins are different and must be considered separately. Rectus femoris is the only member of quadriceps which acts on the hip joint as well as the knee. Its origin is by two heads—a straight head from the anterior inferior iliac spine and a reflected head from the upper edge of the acetabulum and the adjoining part of the capsule of the hip joint. The former head is in line with the muscle belly when the hip is extended and the latter when it is flexed. The muscle belly is bipennate. The vastus lateralis (Fig. 11.38) arises by a thin aponeurosis from the *back* of the femur, principally from the lateral lip of the linea aspera, and the muscle is then wrapped around the lateral aspect of the femoral shaft. Vastus medialis arises from the medial lip of the linea aspera and adjoining parts of the back of the femoral shaft and it wraps around the medial side of the shaft. Vastus intermedius arises from the lateral and anterior surfaces of the shaft of the femur and is therefore deep to vastus lateralis and to the retus femoris. The quadriceps muscle is thus arranged around the femur (Fig. 11.36) and, owing to the obliquity of this bone, the muscle itself has an oblique pull in relation to the midline of the thigh. The four components fuse together in the lower part of the thigh.

Quadriceps is officially inserted into the upper, medial and lateral borders of the patella, with expansions spreading out on either side of the knee. The apex of the patella is attached to the tibial tubercle by the thick and strong *ligamentum patellae*. In fact, however, the ligamentum patellae is the real tendon of insertion of quadriceps and the patella is a sesamoid bone developed in the tendon. Some of the more superficial fibres of the quadriceps tendon, in fact, pass right over the patella to become continuous with the ligamentum patellae. As may be ascertained by examination of a muscular patient, the vastus medialis muscle fibres extend farther down the thigh than those of vastus lateralis. The lateral expansions of the tendons are called the *patellar retinacula* and they spread out on either side to blend with the capsule of the knee joint. The deep surface of quadriceps is separated from the knee joint and the lower end of the femur by the *suprapatellar bursa* which communicates with the knee joint widely and which extends for three fingers' breadths above the patella. A *deep infrapatellar bursa* lies between the ligamentum patella and the upper end of the tibia. Other bursae are present in the superficial tissues in front of the patella and its ligament. Quadriceps is supplied by the femoral nerve.

The actions of quadriceps are extremely important as far as the knee joint is concerned. Rectus femoris acts on both hip and knee joint but its weak flexing action on the hip joint is not important as there are many more powerful flexors. All four components of quadriceps are powerful extensors of the knee and are used in walking, kicking and in raising the body from a sitting or squatting position. The lowermost fibres of vastus medialis which extend right down to the medial border of the patella, are important in helping to prevent possible dislocation of the bone (Fig. 11.37). The oblique pull of quadriceps, as can be seen from this diagram, tends to cause a lateral dislocation of the patella, particularly when the muscle contracts suddenly and violently as may happen in trying to prevent a fall. This is resisted by the pull of the lower fibres of vastus medialis and also by the prominent lateral condyle of the femur (Fig. 10. 27). In the final stages of the extension of the knee, the joint is placed in the close-packed position by the lateral rotation of the tibia (or medial rotation of the femur). This has been described in Chapter 10 and is the result of the configuration of the joint surfaces and the extending action of quadriceps femoris. This action of quadriceps in the final stages of extension is most important and its loss is one of the first signs of quadriceps weakness. A useful test for the integrity of quadriceps is therefore to ask the patient, who is lying on the plinth, to press the back of the knee into the plinth (or on the physiotherapist's outstretched hand). This manoeuvre should result in the heel being raised slightly from the plinth and if the lower limb is then raised, the full extension of the knee should be maintained. If quadriceps is weak, however, the knee will flex slightly before the heel can be lifted from the plinth. This is known as *quadriceps lag*. The patellar retinacula also play an important part in maintaining the stability of the knee joint by means of their widespread attachment to the capsule of the knee joint, and a weak quadriceps can cause the knee joint to feel extremely unstable. Quadriceps exercises thus play an important part in the

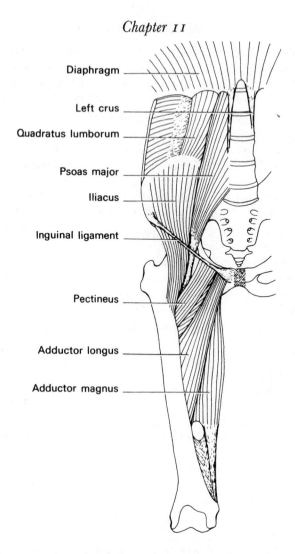

Diaphragm

Left crus

Quadratus lumborum

Psoas major

Iliacus

Inguinal ligament

Pectineus

Adductor longus

Adductor magnus

Fig. 11.39. Psoas, iliacus and the adductor group of muscles.

treatment of injuries to the knee, since the muscle wastes at a horrifying rate if the joint is immobilised for even a short time.

THE ADDUCTOR GROUP OF MUSCLES

In the introduction to this section the adductor group of muscles was described as lying in a plane posterior to the muscles (quadriceps) which are grouped around the femur. This is because the principal members of this group are inserted into the linea aspera on the posterior surface of the femur so that they form a well defined group of muscles which can be gripped in the hand and moved about when they are relaxed. Because they lie in a plane posterior to the femur, the medial part of the upper end of the thigh is rather concave in a muscular person who does not have too much superficial fascia and this depression marks the main part of the femoral triangle (see below).

PECTINEUS

This small quadrilateral muscle takes origin from the pecten pubis (pectineal line) and the area of bone in front of this and it runs downwards and laterally to be inserted into the back of the femur between the lesser trochanter and the linea aspera (Fig. 11.39). Iliopsoas is attached to the lesser trochanter and, as you will see later, adductor longus is attached to the linea aspera. These muscles are therefore arranged edge-to-edge and they form a stratum which forms the floor of the femoral triangle (see Chapter 25). Pectineus is supplied by the femoral nerve and sometimes by a branch of the obturator nerve. It is an adductor and a flexor of the hip joint.

ADDUCTOR LONGUS

The three adductor muscles which are now going to be described are arranged in three distinct layers, longus being the most anterior. It arises by a narrow and easily palpable tendon from the front of the pubis just below the pubic crest and then spreads out into a triangle, the base of which is inserted into the linea aspera over about the middle third of its extent (Fig. 11.39). It lies edge-to-edge with pectineus although there may be a gap between the two muscles through which adductor brevis can be seen. It is supplied by the obturator nerve and is an adductor of the thigh and can also help in flexion of thigh, on account of its anterior position. As you would expect, it is well developed in horse riders.

ADDUCTOR BREVIS

Smaller than adductor longus and lying posterior to its, this muscle arises from the inferior ramus of the pubis and is inserted into the upper part of the linea aspera (Fig. 11.40). Since its insertion arises higher than that of adductor longus the muscle can be seen from the front if there is a gap between adductor longus and pectineus. It is supplied by the obturator nerve and is an adductor of the thigh.

ADDUCTOR MAGNUS

This is a large and very bulky muscle with extensive attachments. It is the most posterior of the three adductors and forms the 'floor' of the back of the thigh. Its origin is from the inferior ramus of the pubis and from the ramus of the ischium (Fig. 11.40). Its most posterior fibres arise from the ischial tuberosity. It is inserted into the whole length of the back of the femur, i.e. to the gluteal

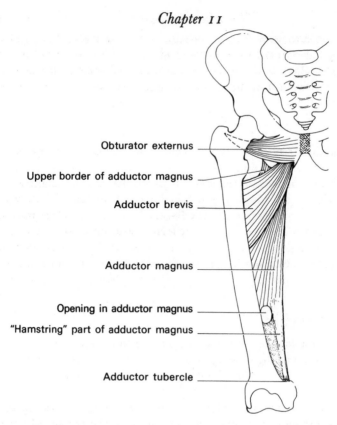

Obturator externus

Upper border of adductor magnus

Adductor brevis

Adductor magnus

Opening in adductor magnus

"Hamstring" part of adductor magnus

Adductor tubercle

Fig. 11.40. Adductor brevis and magnus (longus has been removed.)

tuberosity, the linea aspera and the medial supracondylar line. The most anterior fibres (from the pubis) have the highest attachment to the femur and are therefore almost horizontal. The most posterior fibres descend vertically from the ischial tuberosity to be inserted by means of a strong and palpable tendon into the adductor tubercle. With such a long insertion, it would be surprising if the whole muscle had a single unified action and, in fact, it is best considered as divided into two parts although there is no clear dividing line between the parts. The anterior and highest fibres are more or less parallel to the other adductors, have a similar adducting action and are supplied by the same nerve—the obturator. The posterior fibres are vertical, parallel to the hamstring muscles, arise from the ischial tuberosity (like the hamstrings) and are supplied by the same nerve as the hamstrings—the sciatic (tibial division).

GRACILIS

Gracilis is a long slender muscle that is found along the whole length of the medial side of the thigh (Fig. 11.37). It arises from the inferior ramus of the pubis and the ramus of the ischium and is inserted into the medial side of the upper end of the tibia between sartorius and semitendinosus. It is supplied by the obturator nerve and is an adductor, flexor and medial rotator of the thigh.

The adductor muscles as a whole form a powerful muscle mass but strong adduction of the thigh is not a commonly used activity and the muscles are more important in the maintenance of posture than in the production of movement.

THE MUSCLES OF THE GLUTEAL REGION

The muscles of the gluteal region form a massive muscular mass which is as important in maintaining posture as in producing movement. The muscles are covered by very thick superficial fascia with extensive fat deposits, especially in the female where the fat extends up smoothly into the loin as may be seen easily in any of Renoir's nudes. Over the ischial tuberosities, the fascia is especially dense and contains much fibrous tissue. This fibro-fatty tissue does not require a rich blood supply and can therefore withstand being compressed for long periods between the ischial tuberosities and a hard lecture theatre bench. It is, therefore, fortunate that gluteus maximus (see below) slips upwards off the ischial tuberosity when the thigh is flexed, since skeletal muscle would soon protest if it were thus insulted.

GLUTEUS MAXIMUS

Gluteus maximus is the largest muscle in the body and is the most superficial muscle in the gluteal region. It arises from the ilium behind the posterior gluteal line and from the adjoining portion of the sacrum and the associated ligaments, including the sacro-tuberous ligament. The coarse bundles of muscle fibres pass downwards and laterally to form a quadrilateral muscle (Fig. 11.41) which is inserted into the gluteal tuberosity of the femur (about 25%) and into the ilio-tibial tract (75%). The muscle fibres naturally become tendinous before they pass over the greater trochanter to reach the tract and the wide aponeurosis which is thus produced (Fig. 11.38) is separated from the greater trochanter by a bursa. The muscle is supplied by the inferior gluteal nerve and there is a striking discrepancy between the size of this relatively small nerve and that of the enormous muscle. The motor units must therefore be very large indeed so that although the muscle is extremely powerful, it cannot act with any finesse. The rather broad and generalised nature of the movements carried out by the muscle is also reflected in the fact that 75% of its fibres are inserted into an extensive tract of deep fascia rather than into a localised bony point.

Gluteus maximus is a powerful extensor of the hip and is particularly important when extension is used to raise the body from a sitting position or in walking upstairs. It also helps to keep the thigh extended and the trunk upright during walking. It is a lateral rotator of the thigh and its upper fibres can help with abduction. Its wide insertion into the ilio-tibial tract enables it to be useful in balancing the body on the tibia and femur. Its postural activity is also seen

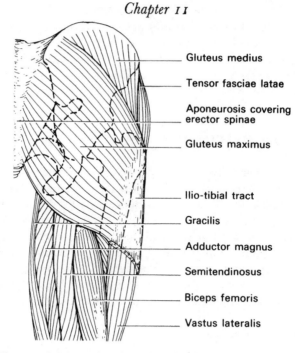

Gluteus medius

Tensor fasciae latae

Aponeurosis covering erector spinae

Gluteus maximus

Ilio-tibial tract

Gracilis

Adductor magnus

Semitendinosus

Biceps femoris

Vastus lateralis

Fig. 11.41. The superficial muscles of the gluteal region.

when, acting from below in conjunction with rectus abdominis, it helps to maintain the correct tilt of the pelvis (Fig. 11.13).

GLUTEUS MEDIUS

This muscle and those lying below it in the gluteal region cannot be seen fully in a dissection until gluteus maximus has been removed (Fig. 11.42). About two-thirds of it does extend forwards, however, in front of the anterior border of gluteus maximus so that it can be palpated below the iliac crest (Fig. 11.41). It arises from the outer surface of the ilium deep to and in front of gluteus maximus. It is a fan-shaped muscle and its fibres converge to be inserted into the outer surface of the greater trochanter.

GLUTEUS MINIMUS

Another fan-shaped muscle which lies deep to gluteus medius and covers the capsular ligament of the hip joint. It arises from the outer surface of the ilium deep to gluteus medius and it is inserted into the outer surface of the greater trochanter (Fig. 11.42).

The gluteus medius and minimus may be considered together. Both are supplied by the superior gluteal nerve and both are powerful abductors of the hip, while their anterior fibres rotate it medially. By far their most important

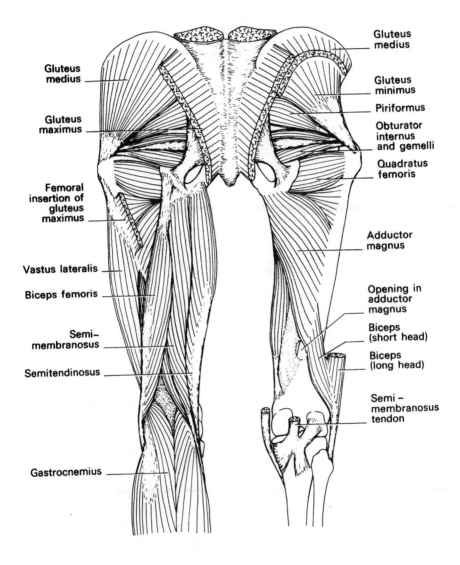

Gluteus medius

Gluteus medius

Gluteus maximus

Gluteus minimus

Piriformis

Obturator internus and gemelli

Quadratus femoris

Femoral insertion of gluteus maximus

Vastus lateralis

Biceps femoris

Semi-membranosus

Semitendinosus

Adductor magnus

Opening in adductor magnus

Biceps (short head)

Biceps (long head)

Semi-membranosus tendon

Gastrocnemius

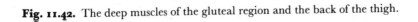

Fig. 11.42. The deep muscles of the gluteal region and the back of the thigh.

Fig. 11.43. When the weight is taken on the *right* leg, the *right* gluteus medius and minimus support the pelvis on that side.

function, however, is in balancing the pelvis on top of the femur when one foot is off the ground. Thus if the *right* foot is lifted off the ground, the *left* glutei, acting from below, will support the left side of the pelvis and even tilt it slightly so as to bring the centre of gravity over the supporting foot (Fig. 11.43). The lever on which the muscles act during this action is the neck of the femur and the hip joint. Thus any condition which affects the contraction of the muscles, the length or angulation of the femoral neck or the function of the hip joint will cause the pelvis to drop to the opposite side when the weight is taken on the affected leg. This is known as *Trendelenberg's sign*. The supporting action of the glutei is extremely important in normal walking when the opposite leg is carried forwards for the next step.

TENSOR FASCIAE LATAE

This muscle is often grouped with the anterior muscles of the thigh but since it is supplied by the superior gluteal nerve it is best described in this section. It arises from the anterior part of the iliac crest and descends vertically, for a short distance only, to be inserted into the ilio-tibial tract which it shares with gluteus maximus (Fig. 11.38). It extends the knee, via the ilio-tibial tract, and also laterally rotates the leg. It probably assists in abduction of the hip but it is more important as a postural muscle, helping to maintain the position of the two joints over which it passes. If the left leg is lifted from the ground, the right tensor fasciae latae, together with the ilio-tibial tract, form an elastic tie between the hip bone and the tibia, and help to prevent the pelvis tilting to the left.

PIRIFORMIS

Piriformis (Fig. 11.42) lies deep to gluteus maximus and just below gluteus medius. It arises from the *front* of the sacrum and leaves the pelvis by passing out through the greater sciatic foramen. It is inserted into the uppermost part of the trochanter. Piriformis is therefore a lateral rotator of the thigh and can help in abduction. It is supplied by branches from the sacral plexus.

OBTURATOR INTERNUS

This muscle arises from the inner surface of the hip bone and the obturator membrane which fills up most of the obturator foramen, where it has important relationships to various abdominal viscera. It leaves the pelvis via the lesser sciatic notch and is inserted by means of a tendon into the greater trochanter. It is supplied by a branch from the sacral plexus. The tendon is accompanied by two small and unimportant muscles called the *gemelli* (the twins) which lie above and below it (Fig. 11.42). The obturator internus is a lateral rotator and is supplied by a branch from the sacral plexus.

QUADRATUS FEMORIS

Arising from the lateral surface of the ischial tuberosity, this muscle is inserted into the quadrate tubercle of the femur. It is a lateral rotator of the thigh and is supplied by a branch from the sacral plexus (Fig. 11.42).

THE OBTURATOR EXTERNUS

This muscle arises from the obturator membrane and the surrounding parts of the hip bone. It passes spirally beneath the hip joint and its tendon is inserted into the trochanteric fossa of the femur. You will probably find it difficult to visualise this muscle. It is supplied by the obturator nerve and is a lateral rotator of the thigh.

THE FUNCTIONS OF THE LATERAL ROTATORS

All the short lateral rotators of the thigh which have been described above are inaccessible to palpation as they lie deep to gluteus maximus. With the exception of obturator externus, the muscles are arranged edge-to-edge below gluteus medius in the order in which they have been described (piriformis, obturator internus and the gemelli, quadratus femoris). Below the last-named muscle is the posterior surface of adductor magnus (Fig. 11.42). The muscles may be compared to those of the rotator cuff of the upper limb as they are probably most important as postural muscles and in helping to maintain the stability of the hip joint. Acting together they are lateral rotators of the thigh and may work in this way to keep the feet pointing forwards when the corresponding side of the pelvis swings forwards during walking.

THE HAMSTRINGS

The hamstrings (Fig. 11.42) comprise three muscles whose origin is from the ischial tuberosity and which cross both the hip joint and the knee joint. They are extensors of the former and flexors of the latter; they are supplied by the sciatic nerve. They lie superficial to (i.e. posterior to) the adductor magnus, which forms a 'floor' for the back of the thigh. In fact, the most posterior part of adductor magnus resembles the hamstrings in many ways and is often called the 'hamstring portion' of the muscle.

BICEPS FEMORIS

This is the lateral hamstring. Arising from the ischial tuberosity, its long head passes laterally, crossing the sciatic nerve, and is joined by the short head which is a broad sheet of muscle arising from the greater part of the length of the linea aspera. The narrow tendon is inserted into the upper end of the head of the fibula where it partially enfolds the fibular collateral ligament of the knee joint (Fig. 10.29). It is supplied by both parts of the sciatic nerve—the tibial part to the long head and the peroneal part to the short head. Its actions will be considered with the other hamstrings.

SEMIMEMBRANOSUS

This muscle arises by a flattened tendon from the ischial tuberosity and it remains tendinous for almost half its length. The tendon is replaced by a bulky mass of muscle which narrows down again to a thick tendon which is inserted into a groove on the back of the medial condyle of the tibia, part of the tendon being deep to the tibial collateral ligament of the knee joint. From its tendon, an expansion passes upwards to form the *oblique posterior ligament* of the knee joint (Fig. 10.33). It is supplied by the tibial component of the sciatic nerve.

SEMITENDINOSUS

The semitendinosus arises from the ischial tuberosity and its fusiform belly is replaced about half-way down the thigh by a long slender tendon which is inserted into the medial side of the upper end of the tibia behind the insertions of sartorius and gracilis. Semitendinosus and semimembranosus are thus arranged head to tail like sardines in a can, and this helps to reduce the bulk of the thigh. Semitendinosus is supplied by the tibial component of the sciatic nerve.

ACTIONS AND FUNCTIONS OF THE HAMSTRINGS

All the hamstrings, acting from above, flex the knee and so are important in walking during the carrying through of the foot when the weight is taken on the other leg. They are also extensors of the hip and are used especially, in conjunction with gluteus maximus, when the body is raised from a stooping position by extension at the hip joint. They also contract powerfully when climbing stairs or stepping up on to a platform. In many people the hamstrings are too short to produce full extension at the knee joint combined with full flexion at the hip so that they are unable to touch their toes with their knees straight. In investigating the range of movements at the hip joint, it is therefore important to flex the knee before attempting full flexion of the hip. When the knee is semiflexed, biceps is a lateral rotator and semimembranosus and semitendinosus are medial rotators of the leg. The hamstrings are also important as postural muscles, any forward sway of the body being immediately counteracted by a strong contraction of the hamstrings.

THE EXTENSOR GROUP OF MUSCLES OF THE LEG

The movements of flexion and extension have been explained in Chapter 10 and, if you are not fully familiar with these terms, you should read this explanation again. Flexion of the foot is also known as *plantar flexion* and is the position adopted by a ballet dancer 'on the points'. Extension (*dorsiflexion*) is the opposite movement and the extensor muscles are therefore found on the *front* of the leg. One of the most important functions of this group of muscles is to dorsiflex the foot during the 'carry through' phase of walking so as to keep the toes clear of the ground. If the muscles are paralysed, for example by a lesion of the deep peroneal nerve, the condition is known as *foot-drop* and the toes drag along the ground during walking. It is easy to diagnose an untreated case of foot-drop simply by examination of the shoes, the toes of which are scuffed and worn.

The deep fascia of the leg is thick and is strengthened by transverse fibres to form two extensor retinacula which are shown in Fig. 11.44 and which serve to prevent the tendons standing out too far during dorsiflexion.

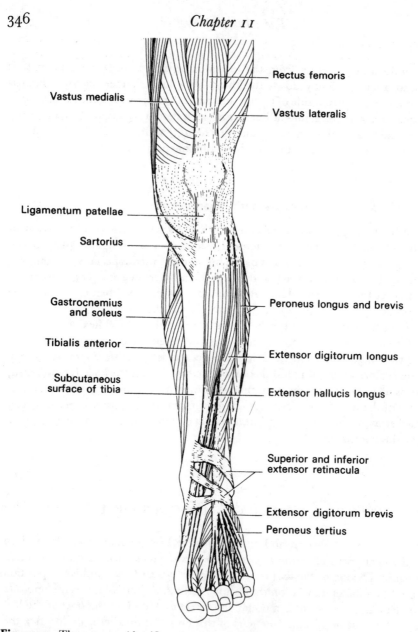

Vastus medialis

Rectus femoris

Vastus lateralis

Ligamentum patellae

Sartorius

Gastrocnemius
and soleus

Peroneus longus and brevis

Tibialis anterior

Extensor digitorum longus

Subcutaneous
surface of tibia

Extensor hallucis longus

Superior and inferior
extensor retinacula

Extensor digitorum brevis

Peroneus tertius

Fig. 11.44. The extensor (dorsiflexor) group of muscles.

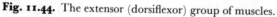

TIBIALIS ANTERIOR

This is the largest muscle on the front of the leg (Fig. 11.44) and its thick
tendon is easily palpable on the medial side of the ankle. It arises from the
lateral surface of the shaft of the tibia and, after passing under the extensor
retinacula, is inserted into the medial side of the base of the first metatarsal

and the medial cuneiform. It is supplied by the deep peroneal nerve. It is a dorsiflexor and an invertor of the foot and acts particularly when these two movements are combined. It plays an important part in postural activity and may be easily seen if its tendon is watched when standing on one leg although since the line of gravity of the body passes slightly in front of the ankle joint, it is not very active in a normal relaxed posture. It is also important in maintaining the medial longitudinal arch of the foot by elevating it during walking or running.

EXTENSOR HALLUCIS LONGUS

This muscle arises from the middle third of the shaft of the fibula and is at first buried between tibialis anterior and extensor digitorum longus. Its tendon comes to the surface in the lower part of the leg, passes deep to the extensor retinacula and traverses the foot to be inserted into the base of the distal phalanx of the big toe (Fig. 11.44). It is supplied by the deep peroneal nerve and is a dorsiflexor of the big toe, including its terminal phalanx, and, in continued action, a dorsiflexor of the foot.

EXTENSOR DIGITORUM LONGUS

Arising from the upper three-quarters of the shaft of the fibula, extensor digitorum longus passes deep to the extensor retinacula and splits into four tendons for each of the four lateral toes (Fig. 11.44). The arrangements of the tendons is very similar to that of the corresponding muscle in the upper limb. Each tendon spreads out to form a dorsal digital expansion into which the corresponding interossei and lumbricals are inserted. The tendon then divides into three, the central slip being inserted into the base of the middle phalanx and the other two slips uniting to be inserted into the base of the distal phalanx. Each tendon, except that to the little toe, also receives the insertion of extensor digitorum brevis. The muscle is supplied by the deep peroneal nerve and it is a dorsiflexor of the toes and of the foot.

PERONEUS TERTIUS

This is a small and unimportant muscle, sometimes missing, but it is included here because its tendon is sometimes quite prominent. It arises from the lower part of the shaft of the fibula, its origin being a continuation downwards of the origin of extensor digitorum longus. It is inserted into the base of the fifth metatarsal. It is supplied by the deep peroneal nerve and it is a weak dorsiflexor of the foot and also helps in eversion.

EXTENSOR DIGITORUM BREVIS

This muscle is responsible for the slight bulge on the dorsum of the foot just in front of the lateral mallolus (Fig. 11.44). It arises from the upper surface of the calcaneus just behind the calcaneo-cuboid joint. After passing obliquely across the dorsum of the foot it splits into four tendons. The most medial is inserted into the lateral side of the base of the proximal phalanx of the big toe while the other three join the tendons of the extensor digitorum longus to the 2nd, 3rd and 4th toes. It is supplied by the deep peroneal nerve and it helps in extension of the phalanges of the middle three toes and of the proximal phalanx only of the big toe.

THE LATERAL MUSCLES OF THE LEG

These comprise the peroneus longus and peroneus brevis (Fig. 11.45). As was mentioned in Chapter 10, if the lateral surface of the fibula is followed downwards, it passes posteriorly and leads to a groove behind the lateral malleolus. This is utilised by the tendons of the peronei as a pulley to enable them to turn through about 90° in order to pass into the foot.

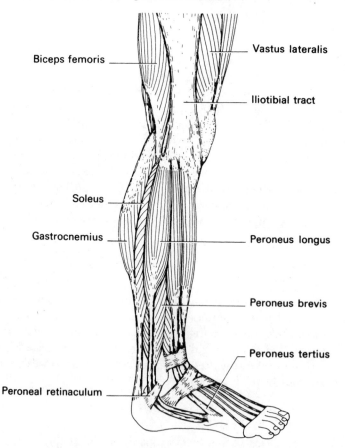

Fig. 11.45. The peronei.

PERONEUS LONGUS AND PERONEUS BREVIS

Peroneus longus arises from the upper two-thirds of the lateral surface of the fibula and peroneus brevis from the lower two-thirds (Figs. 11.45 and 11.47). Longus is pierced near its origin by the common peroneal nerve. The two tendons are very closely related at the lower end of the fibula, peroneus brevis being in front. Both tendons turn forwards as they reach the back of the lateral malleolus and they then pass forwards on the lateral side of the calcaneus. Peroneus brevis continues its course to reach the base of the fifth metatarsal. Peroneus longus turns medially, passes through the groove in the cuboid and is inserted into the lateral side of the base of the first metatarsal and the medial cuneiform. The insertions lie opposite the insertions of tibialis anterior on the same bones. The peroneal tendons are enclosed in a common synovial sheath in the ankle region and they are held down by a couple of *peroneal retinacula*. Both peronei are supplied by the superficial peroneal nerve. They are evertors of the foot and peroneus longus can act as a dorsiflexor. In the natural standing position, they act as postural muscles only and they are also important in helping to maintain the arches of the foot during walking, peroneus longus and tibialis anterior acting almost as a sling under the foot.

THE MUSCLES OF THE CALF

The main bulk of the calf is formed by two muscles only: the *gastrocnemius* and the *soleus* (Fig. 11.46). Associated with these is a very small muscle, the *plantaris*. Deep to these large superficial muscles there are four muscles which lie in direct relation to the tibia and/or fibula: the popliteus, tibialis posterior, flexor hallucis longus and flexor digitorum longus. The two superficial muscles play a very important part in walking and running since they are extremely powerful plantar flexors of the foot and come into play when the heel is lifted off the ground. In addition, these muscles, particularly soleus, are extremely vascular and contain a dense plexus of veins so that their contraction is particularly important for the functioning of the muscle pump which is described in Chapter 16. All the muscles on the back of the thigh are supplied by the tibial nerve.

GASTROCNEMIUS

The gastrocnemius arises by two heads from the upper posterior part of the medial condyle of the femur and from the lateral side of the lateral condyle (Fig. 25.9). The two heads unite to form the main mass of the muscle and this, rather more than half-way down the calf, gives way to an extremely powerful tendon which is common to gastrocnemius and soleus (Fig. 11.46). The medial part of the muscle extends rather farther down than the lateral and this can easily be seen in the living subject. The tendon is called the *tendo calcaneus* although, fortunately, its older, more picturesque name is still often used—the

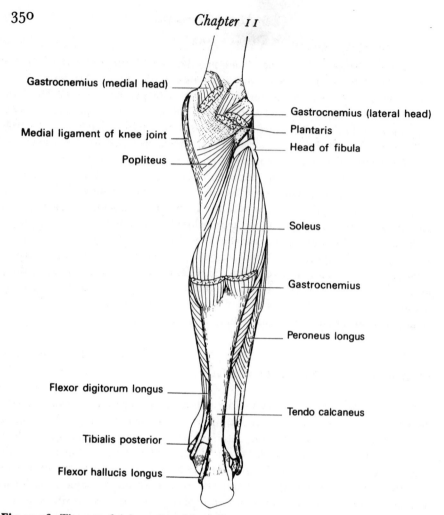

Fig. 11.46. The superficial muscles of the calf.

Achilles tendon. Achilles' mother dipped him in the river Styx as an infant, in order to render him invulnerable to wounds, but unfortunately held him by the heel to do so and forgot to wet the region of his tendon calcaneus so that an arrow wound here later had unfortunate consequences.

The tendo calcaneus is inserted into a roughened area occupying about the middle one-third of the posterior surface of the calcaneus. The lower third is covered by the fibro-fatty pad of the heel while the upper third is related to a bursa which lies between the tendon and the bone.

Gastrocnemius is supplied by the tibial nerve by means of branches which arise both in the popliteal fossa and deep in the calf of the leg. It is a powerful plantar flexor of the foot. When the foot is off the ground it is used in such actions as pedalling a bicycle, but its commonest, everyday action is in the take-off stage of walking when it lifts the heel, and with it the whole body weight, off the ground. It contracts even more powerfully when the body is actually propelled forwards as in running and its action is enhanced by the leverage produced by the backward prolongation of the calcaneus behind the ankle joint. It can also act as a postural muscle, although in this respect soleus is the more important, and since it crosses the knee joint it can act as a flexor of the knee. If the knee is fully flexed, therefore, plantar flexion can only be carried out by soleus and this fact is helpful when investigating the power of individual muscles.

SOLEUS

This is the deeper of the two large calf muscles. It arises by two heads, the medial from the soleal line and the upper part of the medial border of the tibia and the lateral from the upper quarter of the back of the fibula. Between these two origins is a fibrous arch, deep to which the popliteal vessels and the tibial nerve enter the leg from the popliteal fossa (Fig. 14.15). The muscle is inserted, via the tendo calcaneus, into the middle third of the back of the calcaneus. The superficial fibres of soleus run along the length of the muscle so that the muscle belly appears to be of the fusiform type. The deep fibres, however, are arranged in a bipennate fashion so that this part of the muscle is extremely powerful but does not have a very wide range of contraction. Soleus is supplied by the tibial nerve. It is a powerful plantar flexor of the foot and is able to carry out this movement when the knee is fully flexed. It is used in walking and running in the same way as gastrocnemius but since it is nearer to the ankle joint than the latter it has less mechanical advantage and is commonly used more as postural muscle than for active movement. This is particularly true of its deep bipennate fibres. The action of a strong postural muscle in this region is necessary because the line of gravity passes in front of the ankle joint. The capacious venous plexus in this muscle, as well as that in gastrocnemius, forms an important part of the muscle pump (p. 458). In the absence of muscle contraction as occurs in very ill patients who lie motionless in bed, thrombosis is liable to occur in these veins. This produces pain in the calf when the muscle is stretched (or squeezed) by passive dorsiflexion of the foot (Homans' sign).

PLANTARIS

A small muscle which arises from the lateral supracondylar line. Its belly terminates in a thin tendon just below the knee joint and this travels on its own down the length of the calf until its insertion into the tendo calcaneus. The tendon is sometimes called the 'Freshman's nerve', since students dissecting for the first time are rumoured sometimes to mistake it for a nerve. The nerve supply and actions of plantaris are similar to those of gastrocnemius.

POPLITEUS

Popliteus lies deeply on the back of the upper end of the tibia and so is inaccessible to palpation. It is attached to the triangular area above the soleal line of the tibia by muscular fibres (this attachment is usually referred to as its insertion). It narrows down to a tough tendon (Fig. 11.46) which passes upwards and laterally, grooving the posterior part of the lateral meniscus, to which it is attached. It then passes deep to the fibular collateral ligament of the knee joint (Fig. 10.29) and is attached to a depression on the lateral surface of the lateral condyle of the femur. It is supplied by the tibial nerve, which is closely related to it since the muscle forms the floor of the lower part of the popliteal fossa (Chapter 25). It is a flexor of the knee joint and is a medial rotator of the tibia on the femur (or a lateral rotator of the femur on the tibia). You will remember that quadriceps femoris, in the final stages of extension, 'locks' the knee by laterally rotating the tibia on the femur. Popliteus, therefore, as a flexor and medial rotator of the tibia, is the key which 'unlocks' the knee joint at the

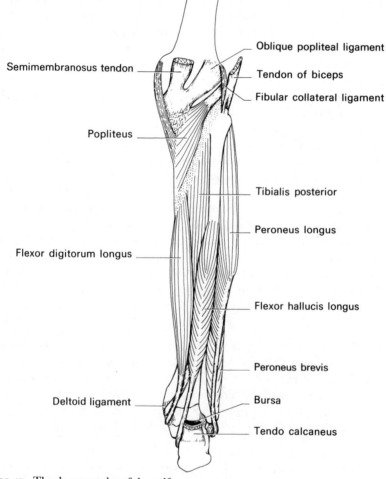

Fig. 11.47. The deep muscles of the calf.

beginning of flexion. It is likely that it also helps to avoid damage to the lateral meniscus by pulling it backwards during knee flexion so that it does not become trapped between tibia and femur.

TIBIALIS POSTERIOR

This is the most deeply placed muscle in the calf; it arises from the interosseous membrane and from both tibia and fibula where they adjoin the membrane. Its tendon is strong and grooves the back of the tibia behind the medial malleolus in a groove which it shares with extensor digitorum longus (Fig. 11.47). It then bends forwards into the foot, passing deep to the small flexor retinaculum and superficial to the medial (deltoid) ligament of the ankle joint. It lies just below the spring ligament, which it helps to support. It is inserted mainly into the tuberosity of the navicular but also sends slips to most of the tarsal bones and metatarsal bases (but not to the talus, to which no muscles or tendons are attached). When the foot is strongly inverted, the tendons of both tibialis anterior and posterior stand out prominently to form a sort of 'anatomical snuffbox of the foot', which has the head of the talus in its floor. Tibialis posterior is supplied by the tibial nerve and it is a strong invertor and plantar flexor of the foot. During active movements, it helps to support the medial longitudinal arch of the foot, a function which is helped by its widespread insertion to the tarsal bones.

FLEXOR HALLUCIS LONGUS

This is a long slender muscle which arises from the shaft of the fibula below the attachment of soleus. It crosses the lower end of the tibia more or less in the midline and it therefore lies deep to the wide tendo calcaneus and is inaccessible to palpation (Fig. 11.47). It passes through the groove on the back of the talus which leads it to the groove on the under surface of the sustentaculum tali. From here it enters the sole of the foot and lies in the second layer (see below), being crossed by the tendon of extensor digitorum (Fig. 11.50), to which it is attached by a slip. It passes between the two sesamoid bones at the head of the first metatarsal and is inserted into the base of the distal phalanx. It is supplied by the tibial nerve. It has a similar action on the big toe to that of flexor digitorum longus on the other four toes.

FLEXOR DIGITORUM LONGUS

This muscle arises from the back of the tibia below popliteus and medial to tibialis posterior. Its tendon passes superficial to that of tibialis posterior in the groove behind the medial malleolus and it then passes forwards lateral to the sustentaculum tali; it thus lies between tibialis posterior and flexor hallucis longus in this region (Fig. 11.47). In the sole of the foot it lies in the second layer and is connected to the tendon of flexor hallucis longus, which it crosses super-

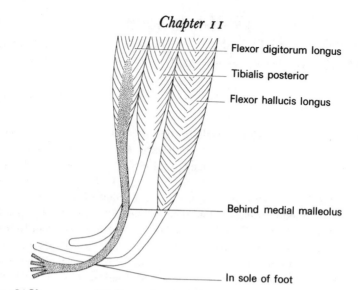

Flexor digitorum longus

Tibialis posterior

Flexor hallucis longus

Behind medial malleolus

In sole of foot

Fig. 11.48. If you are confused over the relations of the tendons of the flexors, remember that flexor digitorum longus tendon is superficial twice, once behind the medial malleolus and once in the sole of the foot.

ficially. (If you get confused over the varying relations of the tendons of this region, remember that the tendon of flexor digitorum longus is *superficial twice* (Fig. 11.48)—it crosses superficial to tibialis posterior behind the medial malleolus and superficial to flexor hallucis longus in the sole of the foot.) It then splits up into four tendons which enter the fibrous flexor sheaths of the lateral four toes. These sheaths and their contents are very similar to those of the fingers which have already been described. Thus, each tendon of flexor digitorum longus passes through an opening in the corresponding tendon of flexor digitorum brevis (which corresponds to flexor digitorum superficialis in the upper limb) and is inserted into the base of the distal phalanx. The muscle is supplied by the tibial nerve. This muscle, and the flexor hallucis longus, between them flex the toes, particularly the distal phalanges, and also act as plantar flexors of the foot. During standing and walking, the muscles hold the toes firmly in contact with the ground and they receive important assistance from the interossei and lumbricals. As in the hand, these muscles flex the metacarpophalangeal joints while, at the same time, they extend the interphalangeal joints. They thus tend to hold the toes flat against the ground and prevent any 'buckling', a phenomenon which will be described later.

THE SYNOVIAL SHEATHS OF THE ANKLE REGION

Each of the tendons on the front of the ankle joint has its own synovial sheath in this region, although extensor digitorum longus shares its sheath with peroneus tertius. Peroneus longus and peroneus brevis share a partly divided sheath in the region of the lateral malleolus and, near the medial malleolus, all three flexor tendons (including tibialis posterior) have individual sheaths. Various retinacula are found around the ankle region and these have been mentioned above where appropriate.

THE MUSCLES AND TENDONS OF THE SOLE OF THE FOOT

The sole of the foot has a complex system of muscles and tendons in its concavity. Weight is taken, in the standing position, on the posterior part of the calcaneus and on the heads of the metatarsals. No muscles cross these bony points, therefore, although some of the tendons are related to the metatarsal heads. Inspection of your own toes will rapidly convince you that the more lateral the toe, the less well developed it is, the little toe often being a deformed and useless appendage. The same is true of the structures in the sole of the foot, and dissection on the lateral side is difficult because many of the muscles are poorly defined, partly fibrous and difficult to separate from adjacent ligaments. Often one or more tendons to the little toe may be extremely thin or even missing altogether.

The muscles and tendons of the sole are usually described as being arranged in in four layers (Figs. 11.49–11.54). This is purely for the purposes of description and this convention certainly makes it easier to remember the relations of the muscles. In fact, however, the muscle layers are not very clearly defined. The muscles and the nerves which supply them are very similar to those in the hand and you will find it a time-saving ploy to study both together. Do not, however, try to use the terms 'medial' and 'lateral' because, owing to the different embryological development of the limbs, the thumb is the lateral digit of the hand but the big toe is the medial digit of the foot. If, however, you place your hand and foot together with the palm and sole facing medially (Fig. 5.18) you will be able to compare the big toe and the thumb and when you come to study the nerves, you will realise that the median nerve corresponds to the medial plantar and the ulnar nerve to the lateral plantar. You will also be able better to understand the use of the terms 'flexion' and 'extension' as applied to the foot.

Another trap you should avoid in describing the muscles of the foot is the use of the terms 'above' and 'below'. If you should be lucky enough to get the chance to dissect a foot, you will naturally start by removing the skin of the sole, and will then expose the muscle layers in turn. It is tempting to refer to the muscles of the second layer as being below those of the first layer, but remember that in the anatomical position, when the foot is on the ground, the 2nd layer muscles are actually *above* those of the first layer. It is best, therefore, to use the terms 'superficial' and 'deep' so that the second layer is said to be deep to the first layer. In this way you will avoid confusing errors.

When describing the muscles and tendons of the four layers of the sole of the foot, you may find it helpful to remember that the layers consist of: (1) three muscles, (2) two muscles, two tendons, (3) three muscles, (4) two muscles, two tendons. You may have to cheat a little in the descriptions, but nevertheless, this is a useful aide-memoire.

THE PLANTAR APONEUROSIS

This corresponds to the palmar aponeurosis in the hand and it consists of a central thick portion with much thinner extensions covering the muscles on either side (these muscles correspond to those of the thenar and hypothenar eminences). It is a tough fibrous structure, attached to the calcaneal tuberosity behind (Fig. 11.49) and splitting into five slips in front. These slips pass to each of the toes where they fuse with the fibrous flexor sheaths and with the deep transverse metatarsal ligaments. In the upper limb, palmaris longus is partly inserted into the palmar aponeurosis, but in the lower limb the backward projection of the heel has, as least partly, cut off the continuity between the flexor surfaces of the leg and the sole of the foot.

THE MUSCLES OF THE FIRST LAYER

This consists of three muscles, a flexor between two abductors. Since these are the most superficial of the short muscles of the foot, they are the longest and they span from the calcaneal tuberosity to the phalanges.

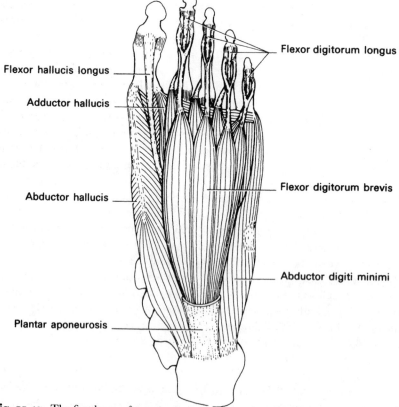

Flexor digitorum longus

Flexor hallucis longus

Adductor hallucis

Abductor hallucis

Flexor digitorum brevis

Abductor digiti minimi

Plantar aponeurosis

Fig. 11.49. The first layer of muscles in the sole.

ABDUCTOR HALLUCIS

Note that the big toe, unlike the thumb, has no long abductor so that 'brevis' is not necessary. It arises from the calcaneal tuberosity and passes up the medial side of the foot to be inserted into the medial side of the proximal phalanx of the big toe (Fig. 11.49). It is supplied by the medial plantar nerve. As its name implies, it can be used to abduct the big toe (remember that 'abduction' of the toes refers to movement away from the second toe) but many people find this difficult to perform without practice as this is not normally a common movement. It can also act as a supplementary flexor. The muscle is relatively large, however, so it seems more likely that its most important function is to help to support the medial longitudinal arch of the foot by acting as a tie-beam between the bases of the arch. This will be discussed at the end of this chapter.

ABDUCTOR DIGITI MINIMI

A small muscle which runs from the calcaneal tuberosity to the lateral side of the base of the proximal phalanx of the little toe (Fig. 11.49). It is often partly replaced by fibrous tissue. It is supplied by the lateral plantar nerve and it abducts and helps to flex the little toe, which can be done with practice. It may help to support the lateral arch of the foot, at least during walking or running.

FLEXOR DIGITORUM BREVIS

This muscle lies between the abductors and corresponds to the flexor digitorum superficialis in the upper limb, i.e. it has four tendons which are perforated by the tendons of flexor digitorum longus. It arises from the calcaneal tuberosity, extends forwards deep to the thick central part of the plantar aponeurosis and then divides into four tendons (Fig. 11.49). These pass to the lateral four toes, and are arranged in a precisely similar way to the tendons of flexor digitorum superficialis, being inserted into the sides of the middle phalanx. The muscle is supplied by the medial plantar nerve and is a flexor of the toes, except for the distal phalanges.

THE MUSCLES AND TENDONS OF THE SECOND LAYER

The second layer comprises two 'sets' of muscles—the lumbricals and the flexor digitorum accessorius. Running through the layer, and providing attachments for these muscles are the tendons of flexor hallucis longus and flexor digitorum longus, which have already been described (Fig. 11.50).

LUMBRICALS

There are four lumbrical muscles in the foot and they arise from the four tendons of flexor digitorum longus (Fig. 11.50). The first arises from the medial

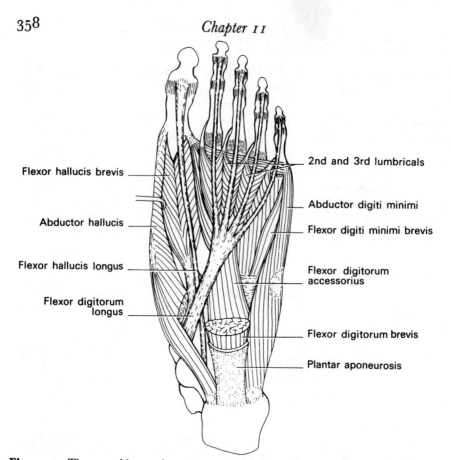

Flexor hallucis brevis

Abductor hallucis

Flexor hallucis longus

Flexor digitorum longus

2nd and 3rd lumbricals

Abductor digiti minimi

Flexor digiti minimi brevis

Flexor digitorum accessorius

Flexor digitorum brevis

Plantar aponeurosis

Fig. 11.50. The second layer of muscles in the sole after removal of flexor digitorum brevis

side of the tendon to the second toe and the other three from the adjacent sides of the four tendons. They are inserted into the dorsal digital expansions. The first lumbrical is supplied by the medial plantar nerve and the others by the lateral plantar. Their action will be discussed with the interossei.

FLEXOR DIGITORUM ACCESSORIUS

This muscle arises from the under surface of the calcaneus by two heads which lie on either side of the calcaneal attachment of the long plantar ligament. It is inserted into the lateral border of the undivided tendon of flexor digitorum longus or sometimes into the individual tendons (Fig. 11.50). It is supplied by the lateral plantar nerve. It is said to convert the oblique pull of the long flexor tendons into a straight fore-and-aft pull, but this has never been proved. It can also maintain flexion of the toes when flexor digitorum longus has relaxed, as, for example, when a simultaneous dorsiflexion of the foot is required. This, in the initial stage of walking, occurs when the foot is in contact with the ground and the leg is brought forwards on the foot by contraction of the dorsiflexors.

THE MUSCLES OF THE THIRD LAYER

These muscles lie relatively deep and they are therefore short and restricted to the anterior part of the foot. They comprise an adductor between two flexors.

FLEXOR HALLUCIS BREVIS

This arises from the under surface of the cuboid and adjacent structures. It splits into two portions which are inserted on either side of the base of the proximal phalanx of the big toe (Fig. 11.51). The lateral component is inserted in conjunction with adductor hallucis and the medial component with abductor hallucis. Each of these conjoined tendons contains a sesamoid bone and the tendon of the flexor hallucis longus runs between the two sesamoids. The muscle is supplied by the medial plantar nerve and it flexes the metacarpophalangeal joint.

FLEXOR DIGITI MINIMI BREVIS

This muscle has a mysterious name since there is no flexor digiti minimi longus. It arises from the base of the fifth metatarsal and is inserted into the lateral side of the base of the proximal phalanx of the little toe (Fig. 11.51). It is supplied by the lateral plantar nerve and is an unimportant flexor of the little toe.

ADDUCTOR HALLUCIS

Like the adductor pollicis, this muscle has two heads. The oblique head arises from the base of the second, third and fourth metatarsals and the transverse head from the ligaments related to the heads of the third, fourth and fifth metatarsals. The tendon is inserted into the lateral side of the base of the proximal phalanx of the big toe along with the insertion of flexor hallucis brevis (Fig. 11.51). Adductor hallucis is supplied by the lateral plantar nerve and it adducts the big toe and, by means of its transverse head, helps to maintain the transverse arch of the foot.

THE MUSCLES AND TENDONS OF THE FOURTH LAYER

The fourth layer has two 'sets' of muscles—the dorsal and plantar interossei —and two tendons, namely those of peroneus longus and tibialis posterior which have already been described.

THE INTEROSSEI

The plantar interossei, three in number, arise from the medial sides of the third, fourth and fifth metatarsals and are inserted into the medial sides of the proxi-

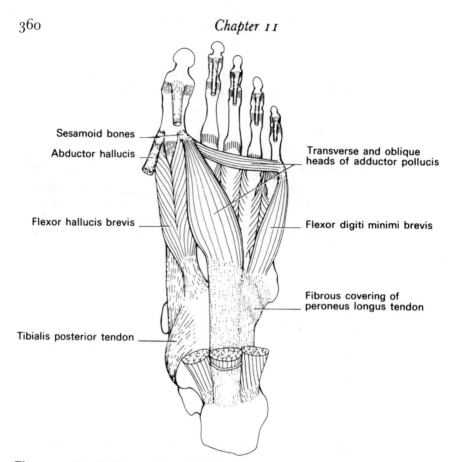

Sesamoid bones

Abductor hallucis

Transverse and oblique heads of adductor pollucis

Flexor hallucis brevis

Flexor digiti minimi brevis

Fibrous covering of peroneus longus tendon

Tibialis posterior tendon

Fig. 11.51. The third layer of muscles in the sole.

mal phalanges of the corresponding toes. Each of the four dorsal interossei arises from two adjacent sides of a pair of metatarsals. They are inserted into the bases of the proximal phalanges and the dorsal extensor expansions as shown in Fig. 11.52. It will thus be seen that the mnemonic which was mentioned for the fingers applies to the toes (D.Ab. and P.Ad) as long as you remember that the second toe is the central axis to and from which adduction and abduction occur. The interossei are all supplied by the lateral plantar nerve. Besides abducting and adducting the toes (not a common movement) the interossei and lumbricals, by pulling on the dorsal digital expansions, can extend the interphalangeal joints while flexing the metacarpophalangeal joints. This again is an uncommon movement and these muscles, are more important in acting as synergists, i.e. in preventing unwanted movements. When the foot is on the ground, the toes are held in contact with the ground by the action of the long flexors and this action continues until the foot leaves the ground on stepping off. This action is well seen in an exaggerated form when wearing sandals held in place by only a strap across the dorsum of the foot. During the take-off stage, when the heel leaves the ground, the long flexors contract vigorously and the sandal, pivoting on the

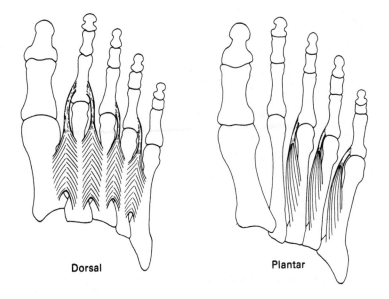

Fig. 11.52. The fourth layer of muscles in the sole.

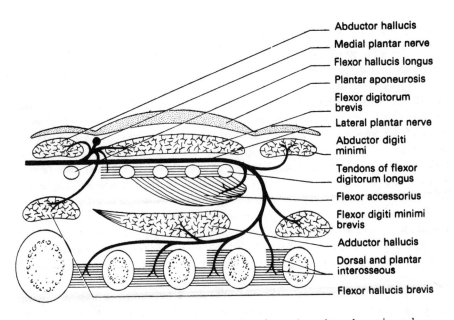

Abductor hallucis
Medial plantar nerve
Flexor hallucis longus
Plantar aponeurosis
Flexor digitorum brevis
Lateral plantar nerve
Abductor digiti minimi
Tendons of flexor digitorum longus
Flexor accessorius
Flexor digiti minimi brevis
Adductor hallucis
Dorsal and plantar interosseous
Flexor hallucis brevis

Fig. 11.53. A highly diagrammatic cross-section of the sole to show the various planes.

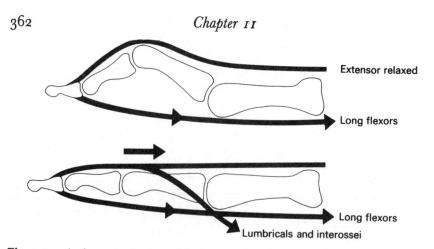

Fig. 11.54. An important action of the lumbricals and interossei in the foot is to prevent buckling of the toes.

strap, flies upwards to strike the heel, producing the characteristic slapping sound. The long flexors tend to buckle the toes in the same way as the links in a bicycle chain will buckle if the two ends of a short length of chain are pushed together. The interossei and lumbricals counteract this buckling movement, particularly at the proximal interphalangeal joint (Fig. 11.54). The interossei also help to maintain the transverse arch of the foot by bunching the metatarsals together.

THE NERVE SUPPLY OF THE MUSCLES OF THE FOOT AND HAND

You will have noticed many similarities between the foot and the hand (as well as considerable differences) and the purpose of this paragraph is to point out the advantages of learning the distribution of the ulnar and lateral plantar and the median and medial plantar nerves together. Apart from the fact that there are no opponens muscles in the foot (since the toes cannot be opposed, at least in the human foot), and that the second lumbrical needs a little care, you will find that the hand and foot are very similar. Thus the medial plantar nerve supplies the abductor and flexor hallucis brevis, the first lumbrical and the flexor digitorum brevis while the median nerve supplies the abductor and flexor pollicis brevis (*and opponens*), the first *and second* lumbricals and the flexor digitorum superficialis (which corresponds to the flexor digitorum brevis). Both nerves also provide a cutaneous supply to 3½ digits. The lateral plantar nerve supplies the abductor and flexor digiti minimi brevis the *2nd*, 3rd and 4th lumbricals, the adductor hallucis and all the interossei. It also supplies flexor digitorum accessorius which is not found in the hand. The ulnar nerve supplies the abductor and flexor digit minimi (*and opponens*), the 3rd and 4th lumbricals, the adductor pollicis and all the interossei. Both nerves also provide a cutaneous supply to 1½ digits.

THE ARCHES OF THE FOOT

The foot essentially serves two important functions. At rest, the two feet need to provide a firm and wide basis for support, like an elephant's foot. During activity, however, the foot is a lever which is used to propel the body along. It is quite astonishing that a heavy person, weighing perhaps 15 stone (95 kg) is able, every time he takes a step, to support the whole body weight on one foot. The weight is transferred from the leg to the talus and thence to the other bones of the foot but the only parts of the foot in contact with the ground are the metatarsal heads and the pads of the toes. It is necessary, therefore, that the numerous small bones of which the foot is composed, be held together extremely firmly and this is done both by ligaments and muscles and is helped by the arched form of the foot. The foot is usually described as having three arches—a *medial* and a *lateral longitudinal arch*, and a *transverse arch*. The usual description of these arches is a little rigid and it must not be thought that the arches are discrete and separate parts of the foot—in fact the greater part of the sole of the foot presents a general concavity as can be seen from a wet footprint. Nevertheless, the concept of separate arches is useful from a descriptive point of view. The medial longitudinal arch consists of the calcaneus, talus, navicular, the three cuneiforms and the medial three metatarsals, the weight being born on the tuberosity of the calcaneus and the metatarsal heads. The summit of the arch is in the region of the head of the talus, which rests upon the spring ligament between the sustentaculum tali and the navicular. The weight of the body transmitted to the talus tends to push the latter two bones apart and this is resisted by the spring ligament, which thus gives a certain resilience to this part of the foot. The lateral longitudinal arch is much lower than the medial arch and comprises the calcaneal tuberosity, the cuboid and the lateral two metatarsals. The transverse arch is extensive and is formed by the metatarsals and the distal tarsal bones.

The arches should not be thought of in terms of an architectural arch, which it is possible to support merely by the keystone shape of the stones without any mortar or ties between the components. In the foot, the articular surfaces of the bones are mostly reciprocally curved, the extremities of the arch are not (normally!) embedded in the ground and the bones are held together by ligaments which are not very extensible and muscles which are. Thus the arches of the foot are normally supple and capable of being flattened by the body weight or elevated by muscular action.

The factors which help to support the arches are not well understood and will only be discussed briefly here. In the first place, the shape of the bones is of some importance and, in particular, the cuneiform bones are wedge-shaped with the thin end downwards. The ligaments undoubtedly play an important part, particularly when the foot is at rest as in standing. When the body weight is tending to flatten the arches, it is obvious that the strong ligaments need to be on the plantar surface since this is the region where the bones tend to separate—the dorsal parts of adjacent bones are actually pressed together (Fig. 11.55).

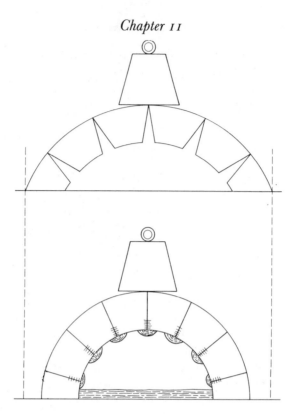

Fig. 11.55. The main ligaments in the foot are on the plantar surface, since this is where loading of the arches produces tension. The plantar aponeurosis and the muscles of the first layer help to support the longitudinal arches by preventing flattening of the whole foot.

Thus the dorsal ligaments of the intertarsal bones are very thin whereas on the plantar surface such ligaments as the long and short plantar ligaments provide a very powerful bond. The plantar aponeurosis also plays an important part in this respect since it prevents separation of the calcaneal tuberosity and the metatarsal heads and so preserves the longitudinal arches. Similarly, the muscles of the sole are important, particularly those of the first layer, since these have the greatest span. Tibialis anterior and posterior and peroneus longus help to support the arches from above by a sling action and peroneus longus also helps to maintain the transverse arch since it runs across the sole of the foot. It is probable that when the foot is at rest and the arches are relatively flattened, the ligaments are the most important factor in maintaining the arches and they are probably taut in this situation. During walking or other activity, however, the arches are accentuated and here the action of the muscles is extremely important. The reason for our lack of knowledge of the precise mechanism of the foot is that it is obviously extremely difficult to carry out electromyographical studies on the very small and closely packed muscles of the foot during walking or running and moreover, clinical observations are not always in accord with anatomical theories. For the moment, therefore, this discussion of the activities of the foot must remain vague and unsatisfactory.

STANDING, WALKING AND RUNNING

The movements at various joints and the actions of muscles during these activities has been described under various headings in this and in previous chapters, but it may be helpful here to sum up the activities involved.

In standing, the outstanding feature is the economy of effort required and only slight muscle contractions are necessary to keep the whole body balanced upon the heads of the femora and to keep the lower limbs in a suitable supporting position. In quadrupeds, a continued contraction of the extensor muscles of the head is necessary to keep the head up and the eyes pointing directly forwards but in man, the head is balanced upon the cervical spine so that a very minor degree of contraction of the muscles is required. Similarly, the four curves in the spine—two primary and two secondary (p. 650)—all help to keep the line of gravity in the right place so that the whole trunk is nicely balanced upon the femoral heads. Any description of posture must begin with a knowledge of the position of the centre of gravity of the body which is usually at a level of a little more than half the total height from the ground and more or less centrally placed in the body. The *line of gravity* is a perpendicular dropped from the centre of gravity.

The lower the centre of gravity, the more stable the position; it is higher in the infant than in the adult so the toddler walks with feet wide apart for the maximum stability. For a stable posture the line of gravity must lie within the supporting base, i.e. the feet, and the nearer to the centre of the base, the more stable the position. Any movement of any part of the body away from the centre of gravity will affect it, so that a simple abduction of the arm will move the centre of gravity towards that side and the position will become unstable unless the centre is returned to its original position by movement of the trunk to the opposite side. If you stand with your feet and right arm close to a wall and then abduct your left arm, you will fall to the left because of your inability to re-position the centre of gravity. This effect is enhanced if a heavy weight is carried in one hand, so that it is usually easier to carry two heavy suitcases than one (Fig. 11.56).

The line of gravity in the normal standing position passes a little behind the hip joint, in front of the knee joint and reaches the ground between the two feet and a little in front of the ankle joint. It is not absolutely stationary, however, and the body is constantly swaying slightly, mainly in the antroposterior plane. This is controlled by alternating contraction of the flexor and extensor groups of muscles, and the muscle activity involved aids venous return by means of the muscle pump (p. 458). Since the line of gravity passes in front of the ankle joints, the calf muscles, especially soleus, are particularly important, while tension in the hamstrings opposes the effect of gravity on the knee joint and that in iliopsoas and other flexor muscles the effect on the hip joint. A sudden unexpected blow in both popliteal fossae will flex the knee joints forcibly, bring the centre of gravity behind the joint and cause collapse of the victim unless quadriceps can contract in time.

Fig. 11.56.

With the feet in the normal position, the weight is transmitted through the arches and is taken on the heels, the outer borders of the feet and the metatarsal heads. If, however, the feet are turned outwards, or if they are widely separated, the body weight tends to flatten the medial arch and the feet roll over medially.

The position of the pelvis in the antero-posterior plane is regulated by the combined actions of rectus abdominis and gluteus maximus (p. 296), which tilt the pelvis backwards, and by iliopsoas and other hip flexors which can tilt it forwards. Since the line of gravity passes behind the hip joint, the muscle tone in the flexors is important in keeping the pelvis in position but if the line of gravity is displaced anteriorly, as can happen if poor muscle tone allows the head and shoulders to sag forwards, the pelvis may tilt forwards and the lumbar lordosis increase to keep the centre of gravity over the feet. The sideways tilt of the pelvis is controlled by gluteus medius and minimus and tensor fasciae latae (p. 341) and this also has its effects on the spine.

The anteroposterior and lateral curvatures of the spine are affected both by muscle tone and by gravity. In the former case, the tone in erector spinae, the retractors and elevators of the shoulder girdle and the extensors of the head are all important for preventing the trunk 'slumping' forwards while adjustment of the spinal curves may be necessary to keep the line of gravity over the feet in conditions affecting the lower limb and the pelvic girdle.

The role of muscle tone in posture is described on p. 118. It should be remembered that, normally, the muscle tone takes the strain in maintaining posture, rather than the ligaments. In faulty posture, or with poor muscle tone, the ligaments may become stretched and give rise to pain.

Finally, it may be worth mentioning the effect of wearing very high heels. This has the effect of shifting the centre of gravity forwards and in order to bring it back to the normal position the knees are slightly bent, and the pelvis tilted forwards. There may be a lordosis. In addition, with continual wear of such shoes, the calf muscles may become shortened so that they will be painful when flat shoes are worn.

In walking, it is helpful to subdivide the process into various stages. The first activity is to brace the arches of the foot for action by muscular contraction which draws the bones together, accentuates the arches and transforms the whole foot into a more or less rigid arched form. The body is brought forwards by a contraction of the dorsiflexors of the supporting foot and the line of gravity is transferred to a position well in front of the ankle joint. This results in the necessity to raise the heel, which is done by the powerful contraction of the plantar flexors, particularly gastrocnemius and soleus and these muscles also begin to propel the body forwards. This is known as the 'take-off' stage and propulsion is increased by gluteus maximus which extends the hip. Meanwhile, the opposite leg is swinging forwards and in order to enable this to happen, two important adjustments have to be made in addition to contraction of the flexors of the hip and extensors of the knee. Firstly, the gluteus medius and minimus must contract on the same side as the weight bearing leg so as to prevent the pelvis dropping on the unsupported side. Secondly, the foot which is being carried forwards must be dorsiflexed so that the toes will clear the ground—as was mentioned previously, dragging of the toes is one of the disabilities suffered by patients with foot-drop. As a result of this dorsiflexion, the heel of the foot meets the ground first and there is a momentary eccentric action of the dorsiflexors so as to lower the rest of the foot to the ground in a controlled fashion. This foot then becomes the supporting foot and the opposite foot is carried through. This is an extremely simplified description of a most complex and incompletely understood action. No mention has been made of the adjustments which have to be made in the trunk in order to maintain the rhythm of walking and to keep the centre of gravity of the body over the supporting foot. For example, the arms are swinging and the whole trunk rotates slightly to one side and then the other during normal walking, but consideration of these movements must be left to more specialised textbooks.

In walking up a slope, or upstairs, the most powerful actions are those of gluteus maximus and quadriceps femoris which are responsible for propelling the body upwards by their extensor actions. In running, the basic movements are similar to those in walking but the centre of gravity is placed more anteriorly by inclining the body forwards so that the 'take-off' of the weight bearing foot is much more powerful. The whole body is propelled forwards by the contraction of the plantarflexors and the extensor muscles of the knee and hip in an effort to keep up with the centre of gravity. All the other movements are correspondingly exaggerated and the vascular effects of violent exercise will be pronounced.

12 · The Peripheral Nerves

In Chapter 4 you saw how the nervous system can be subdivided into the central nervous system (CNS) and the peripheral nervous system. Chapter 5 has described the anatomy of the CNS in some detail and it is now necessary to give an account of the anatomy of the peripheral nervous system, or at least those parts of it which are relevant to a physiotherapist's work.

The peripheral nervous system consists of the 12 pairs of cranial and the 31 pairs of spinal nerves. Most of the former will be dealt with in other chapters as they are concerned with the special senses, with movements and sensations of the mouth and pharynx and with the autonomic nervous system. Only a very brief summary of the cranial nerves will therefore be given here, except for the 7th (facial) nerve which is important because it supplies the muscles of facial expression and will therefore be described in some detail.

THE CRANIAL NERVES

1 *The olfactory nerve*

This nerve supplies the upper part of the nose with nerve endings which subserve the sense of smell. This will be described in Chapter 13.

2 *The optic nerve*

Strictly speaking, this is an outgrowth from the brain rather than a true cranial nerve, and it is therefore surrounded by meninges, the subarachnoid space containing cerebrospinal fluid. It is possible to examine the end of the optic nerve directly by means of an ophthalmoscope so that examination of the eye may give valuable information not only about the eye itself but also about the optic nerve and about the pressure of the cerebrospinal fluid. Vision will be dealt with in Chapter 13.

3, 4 *and* 6

These three nerves, whose names need not be remembered, supply the muscles which move the eye. The 3rd nerve also supplies the muscle which elevates the upper eyelid (*levator palpebrae superioris*).

5 *The trigeminal nerve*

So called because it divides into three major branches—*ophthalmic, maxillary and mandibular*—which supply sensory fibres to the upper, central and lower regions of the face respectively. The ophthalmic division is important since this supplies the surface of the eye and if it is damaged (or anaesthetised), foreign bodies in the eye will cause no pain and may therefore produce serious damage before they are detected and removed. The mandibular division also contains motor fibres which supply the muscles of mastication, i.e. the muscles which close the jaw. Two of these can be easily felt when the teeth are clenched—*temporalis* in the temporal region (i.e. over the temporal bone) and *masseter* which lies between the zygomatic arch and the angle of the mandible (Fig. 11.6).

7 *The facial nerve*

This is the nerve which supplies the *muscles of facial expression*—this term includes all the muscles of the face and scalp with the exception of the muscles of mastication which are supplied by the trigeminal nerve. The facial nerve arises from the under surface of the pons, within which lies its nucleus— equivalent to the anterior horn cells of the spinal cord. Corticonuclear fibres descend from the lower part of the motor cortex, cross the midline and synapse with the cells of the facial nucleus, thus forming the upper motor neurones. The facial nerve, having left the pons passes laterally and enters a foramen on the internal aspect of the temporal bone. It then lies in close relation to the middle ear cavity where it may be damaged in operations in this area. It emerges from the base of the skull near the mastoid process and immediately enters the parotid gland. This is one of the salivary glands which lies just in front of the ear—it is this gland which becomes swollen in mumps. Within the gland, the facial nerve splits up into 5 main branches, although some other smaller branches are also given off to unimportant muscles. If you place your right hand over the right side of your face, with the fingers spread apart and the thenar eminence over the ear, you will get a good idea of the course of these five branches and you will understand their names: *temporal, zygomatic, buccal, mandibular* and *cervical*. These branches emerge from the anterior border of the parotid gland and spread out over the face, supplying the muscles in the vicinity. The cervical branch supplies platysma (Fig. 12.1).

Paralysis of the facial muscles can occur in many ways, since the nerve has such a long and complicated course. An upper motor neurone lesion is common as a result of haemorrhage or thrombosis affecting the arteries which supply the internal capsule, or it may be caused by a tumour anywhere between the cortex and the facial nucleus. A lesion of the lower motor neurone may be caused by intracranial tumours, by damage to the ear in disease of the middle ear, or by tumours of the parotid gland. Finally, paralysis of the lower motor neurone type may occur on its own, with no other apparent lesions being present. This condition is called *Bell's palsy* and its aetiology is uncertain.

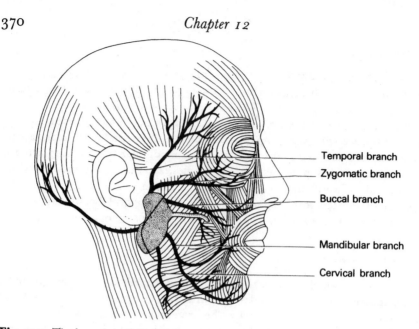

Fig. 12.1. The branches of the facial nerve on the face.

Facial nerve lesions cause an obvious asymmetry of the face, which is made more apparent by asking the patient to smile, to screw up the eyes or to purse the lips. Sagging of the musculature causes loss of the normal skin creases and it is important that the stretched muscles should be relieved of tension by means of a hook in the corner of the mouth or other supporting mechanism, otherwise they will become permanently lengthened. The muscle paralysis also has other effects: sagging of the lower eyelid and of the lower lip may cause tears and saliva to overflow with resultant skin damage. Paralysis of the buccinator will prevent the cheeks being held against the teeth and will allow food to accumulate in the space thus produced.

8 *The vestibulo-cochlear nerve*

Unfortunately, this nerve has suffered a number of unnecessary changes of nomenclature in recent years and you may find it referred to as the *auditory* or the *stato-acoustic* nerve. These clumsy hyphenated names are meant to remind you that the nerve is not only concerned with hearing but also with the perception of movement and position of the head and body. The nerve will therefore be discussed more fully in Chapter 13.

9 *The glossopharyngeal nerve*

This small nerve is mostly sensory to the back of the tongue and the pharynx. It is the nerve which carries the unpleasant sensations when a foreign body such as a finger is thrust to the back of the throat.

10   *The vagus*

This, along with the cranial root of the accessory (see below), supplies various muscles in the larynx and pharynx. In addition, the vagus is the largest nerve carrying parasympathetic fibres to a large part of the body (see Chapter 6). It runs down the neck near the carotid artery and internal jugular vein, enters the thorax to supply the lungs and heart, and finally passes through the diaphragm where it breaks up to form plexuses in common with sympathetic nerves. These supply the alimentary canal down as far as about two-thirds of the anterior border of the trapezius two fingers' breadth above the clavicle. It takes over.

11   *The accessory nerve*

This nerve has two roots of origin: *cranial* and *spinal*. The former arises from the medulla, joins the vagus and is distributed with it (it is thus *accessory* to the vagus). The spinal root arises from the upper part of the spinal cord, enters the skull through the foramen magnum and leaves it after joining the cranial root. It then leaves the cranial root again and enters sternomastoid which it supplies. Leaving sternomastoid half-way down its posterior border, it runs obliquely downwards and backwards across the posterior triangle of the neck and enters the anterior border of the trapezius two fingers' breadth above the clavicle. It forms the motor supply to this muscle.

12   *The hypoglossal nerve.*

Supplies the muscles of the tongue.

THE SPINAL NERVES

The course of a typical spinal nerve and the names of the branches have already been described in Chapter 4. It will now be necessary to consider the course and distribution of individual nerves. In the cervical region, the thorax and the abdomen, the nerves have a fairly typical course. The *cervical plexus* is hardly worthy of the name since there is no real plexus formation except for an interconnection between adjacent nerves. The first cervical nerve is peculiar in that its posterior primary ramus supplies only a few small muscles at the back of the neck while its anterior primary ramus joins the hypoglossal nerve and is distributed to some small muscles in the front of the neck. The cervical plexus is therefore limited to C2, 3 and 4, because C5 is the first nerve to enter the brachial plexus. These nerves, then, supply skin and muscles of the back of the neck by their posterior primary rami while their anterior primary rami supply a number of muscles in the vicinity and also give rise to 4 cutaneous nerves and the important *phrenic nerve*. The cutaneous branch from the posterior primary ramus of C2 is called the *greater occipital* which is a large nerve supplying the skin of the back of the scalp. The four nerves from the anterior primary rami are the *lesser occipital* (C2) which supplies an area behind the ear; the *greater auricular* (C2 and 3) which is the only nerve other than the trigeminal to supply the face. It supplies sensory fibres to an area corresponding roughly to the parotid gland. The *transverse cutaneous nerve* (C2 and 3) runs forwards across sternomastoid to supply the skin of the front of the neck. Finally, the *supraclavicular nerves* (C3 and 4), three in number, spread out to cross the clavicle at its medial and lateral ends and at its centre. They supply the skin of the supraclavicular region and the upper part of the chest.

The phrenic nerve

This nerve arises mainly from C4, but also receives contributions from C3 and C5 ('C3, 4, 5—keeps the diaphragm alive'). It runs in front of the scalenus anterior down into the thorax where it passes along the side of the great vessels and the heart, i.e. on the right side it is related to the superior vena cava and the right atrium and on the left to the arch of the aorta and the left ventricle. It finally reaches the diaphragm which it supplies with both motor and sensory fibres. It is the sole motor supply to the diaphragm, so that damage to one phrenic nerve will paralyse its skeletal muscle. The diaphragm will therefore rise on the affected side, although respiration by means of the other respiratory muscles can still maintain adequate oxygenation of the blood. The sensory fibres in the phrenic nerve only supply the central region—the periphery is supplied by branches from the lower six intercostal nerves. Pain from structures related to the diaphragm is often referred to the tip of the shoulder. For example, an inflamed gall bladder may affect the peritoneum on the under surface of the diaphragm causing pain impulses to travel up the phrenic nerve, the pain being referred to the corresponding dermatomes (C3, 4 and 5) which are shown in Figure 5.17.

THE INTERCOSTAL NERVES

The course of a typical spinal nerve has already been described, so it will now only be necessary to mention a few special features of the thoracic nerves. The first thoracic nerve (T1) joins the brachial plexus. The remaining thoracic nerves have the distribution of typical spinal nerves, their anterior primary rami forming the intercostal nerves and the subcostal nerve (T12). The posterior primary rami of the thoracic nerves pass backwards between the ribs to supply the skin and the muscles of the back. Their anterior primary rami (the intercostal nerves) send white rami communicantes to the corresponding sympathetic ganglia and receive grey rami from them. They then run along the intercostal space (i.e. the space between two adjacent ribs) together with the intercostal artery and vein. The three structures are arranged with the vein uppermost and the artery in the middle and they lie along the lower border of the corresponding rib between the internal intercostal muscle and the innermost layer of intercostals which is sometimes called *transversus thoracis*. They have lateral and anterior cutaneous branches and each has a *collateral branch* which arises posteriorly and runs along the lower part of the same intercostal space. The second intercostal nerve has a particularly large lateral branch which lies high in the axilla and supplies the skin of the medial side of the arm. It is called the *intercosto-brachial nerve*. The 7th-11th intercostal nerves, having followed their intercostal spaces, continue into the abdomen to supply skin and muscles of the abdominal wall, lying between those layers of the abdominal muscles that correspond to those of the intercostal spaces, i.e. between the internal oblique and transversus abdominis. The 7th intercostal lies just below and parallel to the costal margin (i.e. the lower border of the rib cage), the 10th lies at the level of the umbilicus (Fig. 5.17) and the subcostal nerve a little above the pubic region. Part of the first lumbar nerve (*ilio-lumbar*) has an anterior branch which lies just below the subcostal. These lower intercostal nerves are therefore extremely important because they are responsible for supplying the muscles of the abdominal wall. Abdominal incisions are carefully arranged to avoid them since if they are cut, a portion of the abdominal musculature will be paralysed and this may lead to the formation of a hernia.

THE BRACHIAL PLEXUS

This is formed by the anterior rami of C5, 6, 7, 8 and T1, which are known as the *roots* of the plexus. The 5th and 6th cervical nerves unite to form the *upper trunk*, the 7th cervical nerve continues unchanged to form the *middle trunk* and the 8th cervical and first thoracic nerves unite to form the *lower trunk*. Each trunk then divides into an anterior and posterior division. All three posterior divisions unite to form the *posterior cord*, the anterior divisions of the upper and middle trunks unite to form the *lateral cord* and the anterior division of the lower trunk continues unchanged to form the *medial cord*. This all sounds a little confusing and it is best to refer constantly to the diagram (Fig. 12.2) in order to understand the brachial plexus. It is also helpful to compare this arrangement with that in some animals in which there are only two cords: anterior and posterior. The branches of the anterior cord supply all the skin and muscles on the front of the limb and those of the posterior cord, the skin and muscles on the back. In the human brachial plexus, and anterior cord has become split into two (the lateral and medial cords) so that, with a very few exceptions which will be mentioned later, the front of the upper limb is supplied by these two cords: the back by the posterior cord.

The roots of the brachial plexus are situated between scalenus anterior and medius muscles. The trunks lie in the lower medial angle of the posterior triangle of the neck where they may be palpated in the angle between sternomastoid and the clavicle. They divide into anterior and posterior divisions behind the clavicle so that the cords are found in the axilla where they are related closely to the axillary artery. Branches are given off from the roots, from the trunks and from the cords. They are best subdivided into *supra-* and *infraclavicular* branches. The supraclavicular branches are the dorsal scapular nerve (C5), the long thoracic nerve (C5, 6 and 7) and the suprascapular nerve.

The dorsal scapular nerve arises from the uppermost root (C5) and descends along the medial border of the scapular to supply the rhomboid muscles. There are also branches to the local muscles such as the scalenes.

The long thoracic nerve arises by three roots from C5, 6 and 7. These usually penetrate scalenus medius before uniting. The long thoracic nerve then descends on the surface of this muscle behind the brachial plexus and the axillary artery and vein, and finally comes to lie on the surface of serratus anterior which it supplies on the medial wall of the axilla. The length of the nerve (it is aptly named) renders it liable to damage with resulting paralysis of serratus anterior. The characteristic sign of this condition is *winging of the scapula*, the medial border of which stands out from the back as a prominent vertical ridge. It may be accentuated by asking the patient to lean against the wall, supporting himself on his hands. One of the functions of serratus anterior is to prevent the body falling forwards between the scapulae (Fig. 11.16), but this will occur, at least partially, when the muscle is paralysed so that the scapula projects backwards to form a 'wing'. The patient will also find difficulty in fully elevating his arm because serratus anterior is important in lateral rotation of the scapula (Chapter 10).

The suprascapular nerve (C5 and 6). This nerve arises from the upper trunk of the brachial plexus and passes backwards through the suprascapular notch of the scapula into the supraspinous fossa and thence through the spinoglenoid notch into the infraspinous fossa. It supplies supraspinatus and infraspinatus. The nerve is not often damaged on its own but it is affected in lesions of the upper trunk of the brachial plexus (see below).

The infraclavicular branches of the brachial plexus arise from the three cords in the axilla.

BRANCHES OF THE LATERAL CORD

THE MUSCULO-CUTANEOUS NERVE (C5 AND 6)

This nerve arises from the lateral cord in the axilla, lying lateral to the axillary artery. It pierces coraco-brachialis, which it supplies, and then comes to lie sandwiched between biceps and brachialis. It supplies both these muscles and emerges at the lateral border of biceps to become subcutaneous near the elbow region. From here onwards it is called the *lateral cutaneous nerve of the forearm* and it supplies an area of skin on the lateral side of the forearm down to the base of the thenar eminence. (This nerve is sometimes known as the B.B.C. nerve on account of the muscles that it supplies.) The nerve may be represented

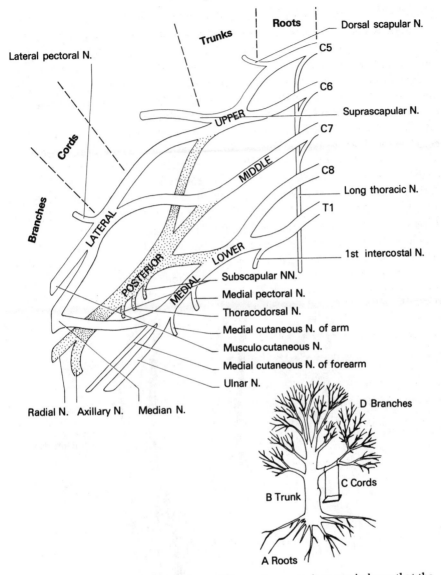

Fig. 12.2. Diagram of the brachial plexus. The garden scene is to remind you that the order is roots, trunks, cords, branches.

on the surface by a line from the axilla to the lateral side of the tendon of biceps at the elbow.

THE LATERAL PECTORAL NERVE (C5, 6 AND 7)

A small nerve, arising from the lateral cord which passes forwards to supply pectoralis major and minor.

THE MEDIAN NERVE (LATERAL ROOT, C5, 6 AND 7)

The median nerve (C5, 6, 7, 8 and T1) arises by two roots—lateral and medial—from the corresponding cords, but, for convenience, the course of the whole nerve will be described here. The medial cord lies medial to the axillary artery and the medial root of the median nerve crosses in front of the artery to join the lateral root so that the median nerve, at its origin, is lateral to the artery. It descends along the medial side of the arm, crossing the axillary artery from lateral to medial about half-way down, i.e. at the level of the insertion of coraco-brachialis. It can usually be palpated easily in this position, especially if it is put on the stretch by abducting and extending the arm. It continues its course on the medial side of the artery to the cubital fossa (Fig. 25.5), where both it and the brachial artery can be palpated as they lie on the brachialis muscle, medial to the tendon of biceps. Continuing through the cubital fossa, it passes between the two heads of pronator teres, under the arch formed by the union of the two heads of flexor digitorum superficialis and is then adherent to the deep surface of the latter muscle, being bound to it by dense connective tissue. When the muscle narrows down to form tendons at the wrist, the median nerve emerges and comes to lie just lateral to the tendon of palmaris longus,

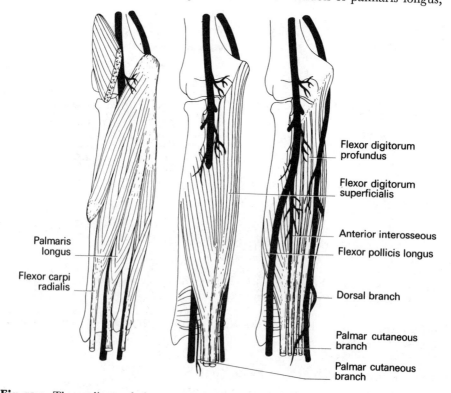

Palmaris longus

Flexor carpi radialis

Flexor digitorum profundus

Flexor digitorum superficialis

Anterior interosseous

Flexor pollicis longus

Dorsal branch

Palmar cutaneous branch

Palmar cutaneous branch

Fig. 12.3. The median and ulnar nerves on the front of the forearm. Compare with Figure 11.25.

if it is present (Fig. 12.3). In the forearm it thus runs down more or less in the midline— hence its name. It enters the palm of the hand by passing deep to the flexor retinaculum, i.e. through the carpal tunnel. It then divides into its terminal branches.

The median nerve usually gives no branches above the elbow joint but occasionally its first branch in the forearm (to pronator teres) may arise from it in this region. In the forearm itself, the median nerve supplies pronator teres, flexor carpi radialis, palmaris longus and flexor digitorum superficialis. In the upper part of the forearm it gives an anterior interosseous branch which descends on the interosseous membrane, giving branches to the muscles on either side of it—namely the flexor pollicis longus and the lateral half of flexor digitorum profundus. It ends by supplying the pronator quadratus.

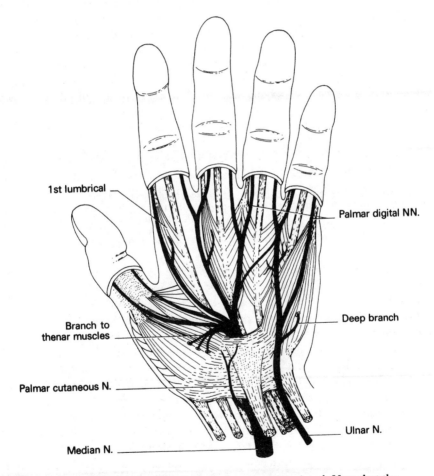

Fig. 12.4. The median and ulnar nerves in the palm of the hand. Note that the important branch of the median nerve to the thenar muscles is given off at the distal margin of the flexor retinaculum.

Just above the flexor retinaculum, a palmar cutaneous branch arises, crosses the retinaculum and supplies the skin of the palm of the hand. The largest branch in the hand is a branch to the muscles of the thenar eminence which arises at the distal edge of the retinaculum and turns abruptly laterally to supply these muscles (Fig. 12.4). Of these, the abductor is the only muscle which is supplied wholly and invariably by the median nerve. The opponens *often* receives an additional branch from the ulnar and the flexor brevis *usually* receives such a branch. The median nerve also supplies small branches to the first two lumbrical muscles (see note about the nerve supply of flexor digitorum profundus on p. 326), and then divides into palmar digital nerves whose branches supply both sides of the thumb, index and middle fingers and the lateral side of the ring finger. The last-named branch is usually connected to the corresponding digital branches of the ulnar nerve so that there is a good deal of overlapping in the territories supplied by each nerve and some variation in the number of fingers supplied by each. In the fingers the branches of the palmar digital nerves are called the (proper) *digital nerves* and they are closely related to the anterior part of the sides of the phalanges. They supply sensory fibres to the whole palmar surface of the fingers and also send dorsal branches which supply the nail bed and a varying amount of the skin over the distal one or two phalanges of each finger.

The course of the median nerve is easily marked on the surface. Beginning in the axilla, it runs down the faint groove on the medial side of the arm between biceps and brachialis and it can be palpated half-way down where it crosses the brachial artery. It then lies midway between the medial and lateral epicondyles of the humerus where it may again be palpated medial to the pulsating brachial artery. From here it passes down the midline of the forearm to a position just lateral to the palmaris longus tendon (or medial to flexor carpi radialis) and it then passes into the palm of the hand and breaks up into branches.

The median nerve is exposed to injury above the elbow but in the forearm it is deeply situated and injury is unlikely except in very severe wounds. At the wrist it becomes vulnerable again, and the large branch to the thenar muscles, distal to the flexor retinaculum, is also liable to accidental injury from wounds and surgical incisions in the palm of the hand. In the carpal tunnel the medial nerve is liable to be compressed because of the restricted space through which it has to pass. Indeed, the normal median nerve is rather flattened as it passes through this region. The motor and sensory disturbances produced by compression in this region are known as the *carpal tunnel syndrome*.

Interruption of the median nerve above the elbow leads to paralysis of most of the muscles of the front of the forearm with inability to flex the interphalangeal joints of the thumb, index and middle fingers. The little and ring fingers, however, can be flexed because the medial half of flexor digitorum profundus is still intact and the metacarpophalangeal joints of all the fingers can be flexed because the interossei are intact. In addition, wasting of the thenar muscles will be noticeable, giving the hand an appearance like that of an ape and there will be loss of sensation of the thumb, index and middle fingers (see below). The thumb cannot be abducted or opposed. Pronation of the forearm is weak and flexion of the wrist is accompanied by ulnar deviation. Interruption of the nerve at the wrist, such as occurs in the carpal tunnel syndrome, may lead to surprisingly little motor disturbance. Abductor pollicis brevis, however, is always affected so that true abduction of the thumb (i.e. bringing the thumb vertically away from the palm of the hand) becomes impossible and opposition is also usually affected. This is therefore the test to apply in suspected lesions of the median nerve. The sensory loss is severe, however, since it involves the most important digits—the thumb, index and middle fingers—along with a varying area of the ring finger. As has been mentioned, there is a considerable area of overlap with the ulnar nerve and the area of loss of crude sensation such as deep touch and pain may be much less than that of light touch and discrimination. In any case, anaesthesia of even the thumb, index and middle fingers makes fine manipulative work surprisingly difficult if not impossible. In partial lesions of the median nerve, such as is most likely to occur in the carpal tunnel syndrome, there will also be paraesthesia ('pins and needles'), pain, or other sensory disturbances in the affected areas of skin.

BRANCHES OF THE MEDIAL CORD

THE MEDIAL PECTORAL NERVE (c8, T1)

This nerve supplies the pectoralis major and pectoralis minor. It is joined by a loop of communication to the lateral pectoral nerve so that both nerves can supply both muscles.

THE MEDIAL ROOT OF THE MEDIAN NERVE (c8, T1)

This large branch of the medial cord crosses the axillary artery and its subsequent course has already been described.

THE MEDIAL CUTANEOUS NERVE OF THE ARM (c8, T1)

A small and unimportant nerve which supplies an insignificant area of skin on the medial side of the upper arm close to the area supplied by the intercostobrachial nerve.

THE MEDIAL CUTANEOUS NERVE OF THE FOREARM (c8, T1)

This is the smaller of the two terminal branches of the medial cord, the other being the ulnar nerve. It runs down the medial side of the arm, becoming superficial by piercing the deep fascia half-way down, near the place where the basilic vein goes deep, but passing in the opposite direction. It continues down the medial side of the forearm to the wrist, supplying a large area of skin in this region, and ends by breaking up into branches (Fig. 12.7).

THE ULNAR NERVE (c(7) 8, t1)

This nerve is the larger of the two terminal branches of the medial cord and it lies, at first, medial to the axillary artery and then to the brachial artery. It pierces the medial intermuscular septum half-way down the arm so that it becomes related to the front of the medial head of triceps (Fig. 3.6). It then runs behind the medial epicondyle where it may be palpated. This is the so-called 'funny-bone' (which therefore has nothing to do with the humorous) so that a blow on the nerve in this region will cause numbness and tingling on the medial side of the hand and in the little and ring fingers.

It enters the forearm by piercing from behind the aponeurotic origin of flexor carpi ulnaris which arises from the posterior border of the ulna. It thus comes to lie under cover of flexor carpi ulnaris and on the surface of flexor digitorum profundus. It descends in this plane to the wrist. It is joined about one-third of the way down the forearm by the ulnar artery which, since it comes from the brachial artery in the midline, lies on its lateral side (Fig. 12.3). At the wrist, the nerve is just lateral to the pisiform bone (because flexor carpi ulnaris is inserted here) and it then enters the palm of the hand by passing superficial to the flexor retinaculum (although it is covered by a superficial slip of the retinaculum). As it passes the pisiform it divides into superficial and deep branches (Fig. 12.4). The superficial branch continues its course, passing superficial to the hook of the hamate and dividing in this region into digital branches which supply the medial side of the little finger and the adjacent sides of the ring and little finger. These nerves also communicate with the digital branches of the median nerve. They can often be palpated as they pass over the hook of the hamate. The deep branch passes between abductor and flexor digiti minimi, supplying both of them and also opponens. It then reaches the deepest part of the palm of the hand, running laterally across the metacarpals and interossei, accompanied by the deep palmar (arterial) arch. This deep branch supplies all the interossei, the medial two lumbricals and ends by supplying adductor pollicis. It also usually gives a branch to flexor pollicis brevis and often to opponens pollicis.

The branches of the ulnar nerve along its course supply, in the forearm, the flexor carpi ulnaris and the medial half of flexor digitorum profundus. (It is helpful to remember that the median nerve supplies motor branches to all except 1½ muscles in the front of the forearm and supplies sensory branches to all except 1½ fingers.) About two thirds of the way down the forearm, the ulnar nerve gives a dorsal branch which winds round the medial side of the forearm, deep to flexor carpi ulnaris, and supplies the dorsal aspect of the little finger and the medial side of the ring finger. The trunk of the ulnar nerve also provides a small palmar cutaneous branch.

The course of the ulnar nerve may be marked on the surface by a line from the axilla to the back of the medial epicondyle and thence to the lateral side of the pisiform bone.

The effect of damage to the ulnar nerve will be different according to whether it is interrupted in the forearm or at the wrist. It is liable to injury in both situations since it is very near the surface as it passes behind the medial epicondyle of the humerus and also as it passes superficial to the flexor retinaculum. Damage in the forearm will cause paralysis of the flexor carpi ulnaris and the ulnar half of flexor digitorum profundus. This will have the effect of weakening ulnar deviation of the hand so that the hand is held in radial deviation. Also there will be inability to flex the terminal phalanges of the ring and little fingers. In addition, paralysis of the small muscles of the hand has severe effect on the manipulative ability of the hand and it also produces a characteristic deformity known as the '*main en griffe*' or *claw hand*. If you remember that the interossei and lumbricals produce flexion at the metacarpophalangeal joints and extension at the interphalangeal joints, you will understand that if these muscles are paralysed, the result will be extension at the metacarpophalangeal joints and flexion at the interphalangeal joints. This gives the hand a clawed appearance. The deformity is less marked in the index and middle fingers than in the ring and little fingers because the lumbricals which are inserted on the former two are supplied by the median nerve and are therefore able to correct the clawing to some extent. The claw-like appearance is accentuated by wasting of the interossei, which fill in the spaces between the metacarpals. This is particularly obvious on the back of the hand, where the skin sinks into hollows between the metacarpals. There is also wasting of the hypothenar eminence. It is interesting that in a lesion of the ulnar nerve at the level of the elbow joint, the visible claw-like deformity is less marked than in the low lesion which has just been described. This is because, in addition to the small muscles of the hand, the flexor carpi ulnaris and the ulnar half of flexor digitorum profundus are paralysed. Loss of the latter muscle means that the ring and little fingers will not be pulled into flexion at the interphalangeal joints. This is the so-called *ulnar paradox*. However, although the deformity is less, the function of the hand will be more severely affected than in the low lesion.

The most important functional effect of ulnar nerve damage is that the action of the interossei is lost, in particular their ability to flex the metacarpophalangeal joints and extend the interphalangeal joints. It is not therefore possible to oppose the pads of the fingers to the thumb so that fine work cannot be done. The sensory loss is relatively unimportant. The tests for ulnar nerve function have been described in Chapter 11.

BRANCHES OF THE POSTERIOR CORD

As was explained earlier in this chapter, the posterior cord supplies all the muscles and skin on the back of the upper limb, but you must remember here that brachioradialis is a muscle which really belongs to this region although it has become a flexor and shifted round to the front. It is therefore the only flexor supplied by the radial nerve. The posterior cord is very large and lies behind the axillary artery. It gives off three fairly small branches before dividing into its two terminal branches—the radial and the axillary nerves.

UPPER AND LOWER SUBSCAPULAR NERVES (C5, 6)

These two branches are given off high up from the posterior cord and, since they are already lying near the posterior wall of the axilla, it is easy for them to supply the subscapularis. The lower subscapular nerve continues its course over the lower border of subscapularis and supplies also the teres major.

THE THORACODORSAL NERVE (C6, 7, 8)

This arises between the two subscapular nerves and descends until it reaches the anterior border of latissimus dorsi. It continues downwards on the deep surface of latissimus dorsi, and supplies the muscle with all its motor fibres. The importance of this muscle and its nerve supply in paraplegia has been discussed in Chapter 10.

THE AXILLARY NERVE (C5, AND 6)

This is a large branch of the posterior cord which immediately passes downwards and backwards through the quadrilateral space. It thus reaches the back of the shoulder region. After giving off a small branch to the teres minor and an even smaller cutaneous branch called the *upper lateral cutaneous nerve of the arm* which emerges near the posterior border of deltoid (Fig. 12.7), it winds around the surgical neck of the humerus deep to deltoid and is expended upon the nerve supply of that muscle. The nerve is obviously exposed to injury along its course—by fractures of the surgical neck of the humerus and by dislocation downwards of the shoulder joint. A small area of anaesthesia over the posterior part of deltoid is a characteristic sign of these lesions, in which abduction obviously cannot be tested. Lesions of the nerve will limit, but not entirely abolish, abduction of the arm since this can also be carried out by supraspinatus and by scapular rotation produced by trapezius and serratus anterior.

THE RADIAL NERVE (C5, 6, 7, 8, T1)

The radial nerve is the second and largest terminal branch of the posterior cord in the axilla and passes medial to the shaft of the humerus. It then winds downwards, backwards and laterally along the radial groove behind the humerus, passing at first between the long and medial heads of triceps and then sandwiched between the lateral and medial heads (Fig. 12.5). For a part of its course here it is separated from the bone by the medial head but for a short distance it may be directly in contact with the bone. It then pierces the lateral intermuscular septum just below the insertion of deltoid and comes to lie an-

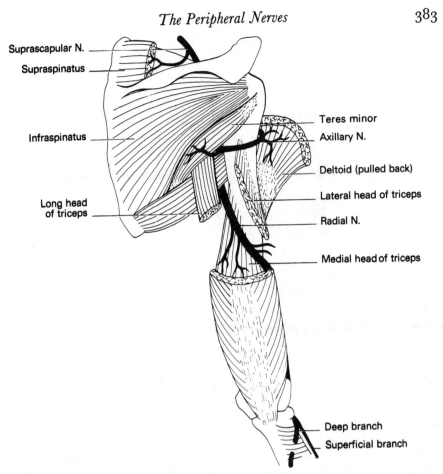

Fig. 12.5. The radial nerve in the spiral groove.

teriorly in the deep gutter between brachialis medially and brachioradialis and extensor carpi radialis longus laterally, In this position it crosses the elbow joint and divides into its large *deep* or *posterior interosseous branch* and a smaller *superficial branch* (Fig. 12.6). The latter is now lying just under cover of brachioradialis and it continues in this position until about two-thirds of the way down the forearm. Here, it inclines backwards, deep to brachioradialis, and finally divides into five terminal branches which cross the region of the anatomical snuffbox and supply the lateral and medial sides of the thumb, the lateral side of the index finger and the adjacent sides of the index and middle and the middle and ring fingers. These branches do not, however, supply the skin all the way to the fingertips as the dorsal branches of the digital branches of the median nerve supply the nail bed and part of the dorsum of the fingers.

The branches of the radial nerve arise in the axilla, in the arm, and just below the elbow joint. In the axilla, a branch descends to supply the long and medial heads of triceps. In the spiral groove both muscular and cutaneous branches arise. The muscular branches supply the lateral and medial heads of

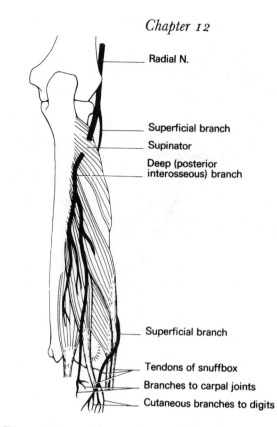

Radial N.

Superficial branch

Supinator

Deep (posterior
interosseous) branch

Superficial branch

Tendons of snuffbox

Branches to carpal joints

Cutaneous branches to digits

Fig. 12.6. The radial nerve in the back of the forearm.

triceps and the anconeus while the cutaneous branches are the *lower lateral cutaneous nerve of the arm*, the *posterior cutaneous nerve of the arm* and the *posterior cutaneous nerve of the forearm* (Fig. 12.7). The lower lateral cutaneous nerve of the arm runs laterally, pierces the deep fascia and supplies an area of skin over and below the region of the insertion of deltoid. The posterior cutaneous nerve of the arm supplies an area of skin on the back of the arm and is unimportant. The posterior cutaneous nerve of the forearm is large. It emerges from beneath the deep fascia on the lateral side of the arm just above the elbow and descends on the back of the forearm to the wrist supplying a large area of skin. There is, however, a great deal of overlap between the territories supplied by individual cutaneous nerves so that the sensory loss occasioned by radial nerve lesion is less than might be expected on purely anatomical grounds.

Just above the elbow, where the nerve is lying in the groove between brachialis and the closely apposed brachioradialis and extensor carpi radialis longus, the radial nerve supplies branches to all three muscles and also the extensor carpi radialis brevis. The brachialis thus has a dual nerve supply but that from the musculocutaneous nerve is the more important.

The most important branch of the radial nerve is the deep branch (posterior interosseous nerve) and this arises just below the elbow joint. This nerve winds

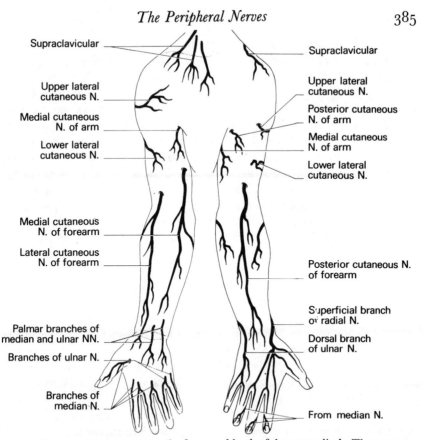

Fig. 12.7. The cutaneous nerves on the front and back of the upper limb. The nerves, in life, are considerably smaller than shown here.

round the upper end of the radius, passing through supinator and dividing it into two strata. It supplies this muscle and sometimes the extensor carpi radialis brevis and then descends on the back of the forearm lying at first on the deep muscle layer and then on the interosseous membrane. It supplies all the muscles on the back of the forearm.

The radial nerve is liable to be damaged in the axilla and upper arm as a result of pressure or of fracture of the humerus. In the axilla, pressure from an incorrectly adjusted crutch may cause 'crutch palsy', while another cause of similar pressure is the traditional 'Saturday night paralysis' in which a person falls into a drunken stupor with one arm over the back of a chair and awakes to find himself with not only a headache but also a radial nerve lesion. In the spiral groove the nerve is liable to damage from the broken ends of the bone in fracture of the shaft of the humerus, or it may be affected by callus formation when the fracture is healing. Lesions of the nerve below the axillary region do not necessarily give rise to complete paralysis of triceps owing to the early origin of the branch to the long and medial heads. Weakness of triceps will, however, occur, even though the medial and lateral heads retain part of their nerve supply.

In lesions lower than half-way down the humerus the triceps will be unaffected. The characteristic lesion in radial nerve palsy is *wrist drop* in which, when the forearm is pronated, the wrist is held in a flexed position by the flexors and by gravity. The lesion is not so evident when the forearm is supinated because the wrist may then be extended by gravity, the flexors acting eccentrically. The loss of the synergic action of the extensors becomes evident when the patient is asked to make a fist. Normally, the extensors of the wrist will contract in order to prevent the unwanted flexor effect at this joint but when the extensors are paralysed, the wrist will flex and the grip will be weakened. The sensory loss in radial nerve lesions is unimportant, partly because of overlap by the territories of neighbouring nerves and partly because sensory perception on the dorsal surface of the hand and fingers is so much less important than that on the palmar surface.

THE LUMBAR PLEXUS

The lumbar plexus is much simpler than the brachial plexus and it is also less exposed to injury since it is entirely within the abdomen. It is formed by the upper four lumbar nerves, although the contribution from L4 does not involve the whole nerve. The fourth lumbar nerve lies on the borderline between lumbar and sacral plexuses so that it divides into two large trunks, the upper of which joins the lumbar plexus and the lower, the sacral plexus (Fig. 12.8). When the lumbar nerves emerge from the intervertebral foramina they immediately enter the posterior layers of the psoas major muscle. The branches of the lumbar plexus are therefore formed within the psoas major and they emerge, with two exceptions (the obturator and genito-femoral nerves) from the lateral border of

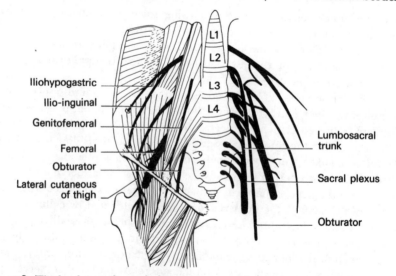

Fig. 12.8. The lumbar and sacral plexuses.

the muscle. There are no complexities such as the trunks and cords of the brachial plexus—each named branch of the plexus simply receives its contribution from one or more roots. It is only necessary, therefore, to describe the branches and give their root values. Before the named branches are given off, the roots of the plexus give branches to quadratus lumborum, psoas major and iliacus. (In the case of the last two muscles, the nerve supply may travel via the femoral nerve.)

THE ILIO-HYPOGASTRIC NERVE (L I)

This is derived only from L1 and it is in series with the intercostal and subcostal nerves. It emerges from the lateral border of the upper part of psoas major and crosses the front of quadratus lumborum. It then runs round the abdominal wall between the internal oblique and the transversus, just above the iliac crest. It gives branches to these muscles, a lateral cutaneous branch which supplies an area of skin over the side of the gluteal region, and ends by becoming cutaneous just above the pubis.

THE ILIO-INGUINAL NERVE (L I)

This is another branch of L1 which follows a similar course to that of the ilio-hypogastric nerve, but just below it. It does not have a lateral cutaneous branch and anteriorly it passes through the superficial inguinal ring, passes over the inguinal ligament and supplies an area of skin on the medial side of the thigh and the lateral side of the external genitalia, i.e. the scrotum in the male and the labia majora in the female. Since it runs just below the ilio-hypogastric nerve and has no lateral cutaneous branch, it may be regarded as the collateral branch of that nerve, corresponding to the collateral branches of the intercostal nerves.

THE GENITO-FEMORAL NERVE (L I, 2)

This nerve travels anteriorly through the substance of psoas major and emerges on its anterior surface. It descends on psoas and divides into two branches—genital and femoral—which supply respectively, the skin of the external genitalia and a small area of skin over the upper part of the femoral triangle on the medial side.

THE LATERAL CUTANEOUS NERVE OF THE THIGH (L2, 3)

This emerges from the lateral border of psoas, runs around the iliac fossa, lying on iliacus, passes beneath the lateral part of the inguinal ligament and enters the thigh. Here it supplies an extensive area of skin on the lateral side of the thigh down as far as the knee (Fig. 12.13).

THE OBTURATOR NERVE (L2, 3, 4)

Both femoral and obturator nerves arise from L2, 3 and 4, but their origins are not exactly comparable. The nerve roots divide into anterior and posterior (ventral and dorsal) divisions; the femoral arises from the dorsal divisions and the obturator from the ventral. This is because the front of the thigh is developmentally dorsal and the medial side ventral (Chapter 5). The obturator nerve is formed within psoas and emerges from its medial border. It enters the pelvis, running down on its lateral wall. It then passes through the upper part of the obturator foramen above obturator internus and divides into two branches as it pierces obturator externus (Fig. 12.9). The anterior division descends on the medial side of the thigh between adductor longus and adductor brevis. It supplies these two muscles and then continues medially to supply gracilis and, sometimes, gives a cutaneous branch which supplies a small area of skin on the medial side of the thigh. The anterior division also gives a branch to the hip joint and may also supply pectineus. The posterior division runs between adductor brevis and adductor magnus (the two divisions of the obturator nerve

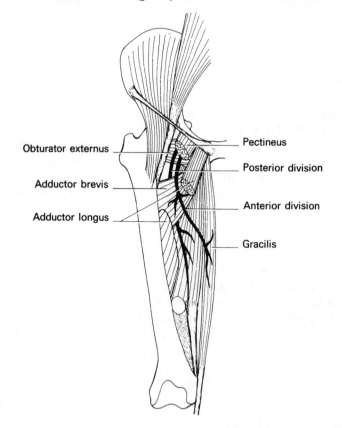

Fig. 12.9. The obturator nerve in the thigh, shown by removing adductor longus and pectineus.

thus straddle adductor brevis). It supplies both these muscles, although, in neither case completely, since adductor brevis is also supplied by the anterior division and adductor magnus (hamstring portion) is also supplied by the sciatic nerve. The posterior division also supplies obturator externus and gives a long descending branch to the knee joint. Since the obturator nerve thus supplies both hip and knee joint, pain in the hip joint may be referred to the knee.

THE FEMORAL NERVE (L2, 3 AND 4)

The femoral nerve arises from the dorsal divisions of the 2nd, 3rd and 4th lumbar nerves in the substance of psoas and leaves it at its lateral border. It is a large nerve which runs downwards and laterally on the surface of iliacus and then passes out of the abdomen under the inguinal ligament outside the femoral sheath and lateral to the femoral artery. After a course of only an inch or two in the femoral triangle, it breaks up into a large number of branches (Fig. 25.6).

Within the abdomen the femoral nerve itself, or its roots, supplies the psoas and iliacus. In the thigh it supplies pectineus and then divides into two main branches. The anterior division gives a branch to sartorius and then divides into the intermediate and medial cutaneous nerves of the thigh which supply the corresponding areas of skin down as far as the knee (Fig. 12.13). The posterior division supplies branches to rectus femoris and to the three vasti; the former branch supplies also the hip joint and the nerves to the vasti supply the knee joint—a good example of Hilton's Law (Chapter 10). The longest branch of the posterior division is the *saphenous nerve* which is sensory. It traverses the adductor canal (Chapter 25), gives an infrapatellar branch and then runs down the medial side of the leg, following the course of the great saphenous vein (see Chapter 14). It supplies sensory branches to the medial side of the leg and foot, ending up at the base of the big toe.

Lesions of the femoral nerve are not common; the main effects will be inability to flex the hip and to extend the knee. The knee jerk will, of course, be lost.

THE SACRAL PLEXUS

The sacral plexus is formed by the lower division of the 4th lumbar nerve, the 5th lumbar, and the 1st, 2nd and 3rd sacral nerves. The lower division of the 4th joins the 5th lumbar nerve to form a large trunk—the *lumbo-sacral trunk*—which crosses the sacro-iliac joint to join the 1st sacral nerve (Fig. 12.8). The roots of the sacral plexus give off branches which unite to form a number of named nerves most of which emerge from the greater sciatic foramen to enter the lower limb. The sacral plexus itself lies on the front of the sacrum and the pyriformis muscle, so that it is related to the pelvic viscera and may be involved in diseases of the bladder, uterus or rectum. The branches of the sacral plexus are described below—not all of them are important from the physiotherapist's point of view (Fig. 12.10).

THE NERVE TO PIRIFORMIS (s(1), 2)

This nerve leaves the roots of the plexus and enters the muscle inside the pelvis.

THE NERVE TO OBTURATOR INTERNUS (L5, S1, 2)

This leaves the pelvis via the greater sciatic foramen and enters again via the lesser sciatic foramen before supplying the obturator internus muscle and also the superior gemellus.

THE NERVE TO QUADRATUS FEMORIS (L4, 5, S1)

This is a small nerve which lies deep to the sciatic nerve and supplies both the quadratus femoris and the inferior gemellus.

THE POSTERIOR CUTANEOUS NERVE OF THE THIGH (S1, 2, 3)

This is a rather large sensory nerve which lies close to the sciatic nerve. It emerges from under cover of gluteus maximus, gives branches to the buttock and the perineum and supplies a large area of skin on the back of the thigh and the upper half of the calf (Fig. 12.13).

THE PUDENDAL NERVE [S2, 3, 4)

This leaves the pelvis below pyriformis, crosses the spine of the ischium and immediately enters the perineal region where it divides up to supply sensory nerves to the perineum, including the scrotum in the male, the motor nerves to the anal sphincter and the external genitalia.

THE SUPERIOR GLUTEAL NERVE [L4, 5, S1)

This, and the next two nerves to be described, are by far the most important branches of the sacral plexus. The superior gluteal nerve emerges from the greater sciatic foramen above pyriformis and comes to lie between gluteus medius and minimus. It supplies both of these muscles and then ends by supplying the tensor fasciae latae. It is therefore the nerve of the principal abductors of the hip although these muscles are more important as postural muscles, supporting the pelvis when the body weight is taken on one leg.

THE INFERIOR GLUTEAL NERVE (L5, S1, 2)

This nerve emerges from the greater sciatic foramen below pyriformis and immediately enters gluteus maximus, which it supplies. It is a surprisingly small nerve for such a huge muscle, but this is because gluteus maximus is concerned only with powerful movements at the hip joint and not with fine discrete movements, so that its motor units are very large.

THE SCIATIC NERVE (L4, 5, S1, 2, 3)

This is the largest nerve in the body and is almost as large as a rather flattened little finger (Fig. 12.10). It consists, in fact, of two nerves, which are usually fused together in the back of the thigh to form the sciatic but which, in the popliteal fossa, separate to become two completely separate nerves—the *tibial* (L4, 5, S1, 2, 3) and the *common peroneal* (L4, 5, S1, 2) nerves (Fig. 25.9). Often, however, the two nerves are separate right up to the gluteal region. In such cases, instead of a large sciatic nerve, the two nerves may be seen leaving the pelvis via the greater sciatic foramen, the sciatic nerve, as such, being absent. In any case, when the sciatic nerve is present, if the tibial and common peroneal

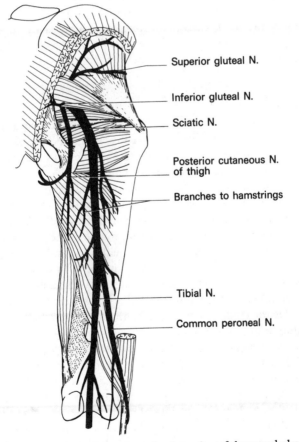

Superior gluteal N.

Inferior gluteal N.

Sciatic N.

Posterior cutaneous N. of thigh

Branches to hamstrings

Tibial N.

Common peroneal N.

Fig. 12.10. The sciatic nerve and other major branches of the sacral plexus.

nerves are pulled apart, the sciatic nerve may quite easily be split into its two component parts right up to the gluteal region.

The sciatic nerve leaves the pelvis via the greater sciatic foramen below pyriformis and deep to gluteus maximus. It descends the back of the thigh, more or less in its midline. At the upper angle of the popliteal fossa it (usually) divides into the tibial and common peroneal nerves. During its course it lies on the muscles of the gluteal region below and deep to gluteus maximus. In the back of the thigh it lies on adductor magnus and is crossed by the long head of biceps. During its course through the back of the thigh, the sciatic nerve gives articular branches to the hip joint and muscular branches to the hamstrings, including the 'hamstring portion' of adductor magnus. These muscular branches all rise from the tibial component of the nerve except for the branch to the short head of biceps which arises from the common peroneal component. The sciatic nerve may be represented on the surface by a broad line down the back of the thigh from a point slightly medial to the mid-point between the ischial tuberosity and the greater trochanter to the apex of the popliteal fossa.

The nerve is not commonly damaged except by severe trauma, in which case there will be complete paralysis of all muscles below the level of the lesion. The branches to the hamstrings, however, arise high in the thigh and may therefore escape. The roots of the nerve may be affected by pelvic tumours, especially carcinoma, causing severe pain.

THE TIBIAL NERVE (L4, 5, S1, 2, 3)

This is the larger of the two terminal branches of the sciatic nerve and it continues the midline course of its parent trunk. It runs downwards through the middle of the popliteal fossa lying on a plane superficial to that of the popliteal vessels which it crosses from the lateral to the medial side. In the calf it passes under the 'arch' of the origin of soleus and then lies deep to the large muscle mass of gastrocnemius and soleus and superficial to the three deeper muscles of the calf, lying particularly on tibialis posterior. It gradually inclines medially, emerging from under cover of the superficial muscles, and bends forwards into the foot, lying half-way between the medial malleolus and the tendo calcaneus and then between the medial malleolus and the heel. In this position, deep to the flexor retinaculum, it divides into its terminal branches: the *medial and lateral plantar nerves*. Behind the media malleolus it lies directly on the lower end of the tibia, lateral to the tendons of tibialis posterior and flexor digitorum longus and medial to that of flexor hallucis longus.

The tibial nerve supplies several articular branches to the knee joint and to the ankle joint. In the popliteal fossa it gives off its cutaneous branch: the *sural nerve*. This runs down the back of the calf and is joined by the *sural communicating nerve* which is a branch of the common peroneal. It then inclines laterally, enters the foot between the lateral malleolus and the heel and continues along the lateral side of the foot. It supplies the skin of the lower lateral and posterior parts of the leg and a strip along the lateral border of the foot. The tibial nerve also provides a number of large muscular branches in the popliteal fossa. These supply the gastrocnemius, soleus, plantaris and popliteus.

In the back of the calf the tibial nerve supplies the three deep muscles: tibialis posterior, flexor digitorum longus and flexor hallucis longus and also gives further branches to the deep surface of soleus. In the ankle region it supplies medial calcaneal branches which supply the skin on the medial side of the heel.

Of the two terminal divisions of the tibial nerve, the medial plantar is the larger (Fig. 12.11). It enters the sole of the foot by passing deep to abductor hallucis (this space is sometimes called the '*porta pedis*' (the gateway to the foot) because all the important structures which enter the sole pass through it). In the foot the medial plantar nerve runs forwards between abductor hallucis and flexor digitorum brevis and ends by dividing into a cutaneous *digital branch* to the medial side of the big toe and three *plantar digital nerves*, each of which divides into a pair of *digital branches* for the adjacent sides of the big and second, second and third and third and fourth toes. They not only supply the sides and

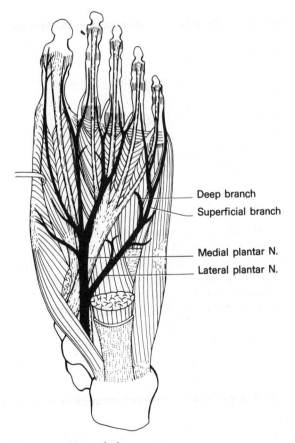

Deep branch
Superficial branch

Medial plantar N.
Lateral plantar N.

Fig. 12.11. The medial and lateral plantar nerves.

plantar surfaces of the toes but also the region of the nail bed on the dorsum. They thus closely resemble the digital branches of the median nerve in the hand. Other branches of the medial plantar nerve supply the skin of the medial part of the sole of the foot and also the abductor hallucis, flexor hallucis brevis, flexor digitorum brevis and the first lumbrical.

The lateral plantar nerve runs laterally from its origin, across the sole of the foot, between the first two layers of muscle, i.e. between the flexor digitorum brevis and the flexor digitorum accessorius. Near the base of the fifth metatarsal it divides into superficial and deep branches. The former supplies cutaneous digital branches to both sides of the little toe and the lateral side of the fourth toe and it also provides the motor nerve supply to the flexor digiti minimi brevis and the interossei of the fourth intermetatarsal space. The deep branch passes medially again, between the muscles of the third and four layers and supplies the 2nd, 3rd and 4th lumbricals, the remaining interossei and ends in adductor hallucis.

You will find it helpful to learn the nerve supply of the muscles of the hand and foot together, remembering that the median nerve corresponds to the medial plantar and the ulnar to the lateral plantar, and also that the flexor digitorum brevis corresponds to the flexor digitorum superficialis. You will find that, except for the 2nd lumbrical, the nerve supply of both skin and muscles in the hand and foot correspond, although there are no opponens muscles in the foot and no flexor digitorum accessorius in the hand. The tibial nerve may be marked on the surface by a line from the apex of the popliteal fossa to the midpoint between the medial malleolus and the tendo calcaneus. Throughout its course, the tibial nerve and its branches are accompanied by the popliteal and posterior tibial arteries and their branches and also, of course, the corresponding veins.

Lesions of the tibial nerve are uncommon and they produce severe disability, mainly owing to the loss of the plantar flexors.

THE COMMON PERONEAL NERVE (L4, 5, S1, 2)

This nerve is the smaller of the terminal branches of the sciatic nerve (Fig. 12.12). It leaves the tibial nerve and passes downwards and laterally along the medial edge of the tendon of biceps, where it can be palpated (Fig. 25.9). Following the biceps tendon to the fibula, it winds obliquely around the neck of that bone, piercing the origin of peroneus longus and then divides into the *superficial* and *deep peroneal nerves*. It gives articular branches to the knee joint, the *sural communicating nerve* (already described) and the *lateral cutaneous nerve of the calf* which supplies a small area of skin and is not important. The nerve may be marked on the surface by a line from the apex of the popliteal fossa, to the neck of the fibula, following the tendon of biceps. It may usually be palpated near the tendon of biceps and, rather more easily, where it winds round the neck of the fibula.

The superficial peroneal nerve descends on the lateral side of the leg between the peronei, which it supplies. It then emerges between the peronei and the extensor

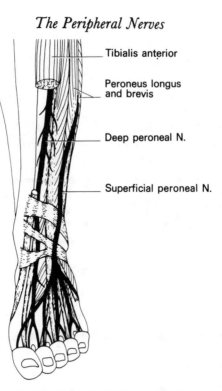

Tibialis anterior

Peroneus longus and brevis

Deep peroneal N.

Superficial peroneal N.

Fig. 12.12. The nerves on the front of the leg. They are, in fact, not as large as shown in the diagram.

digitorum longus in the lower third of the leg and breaks up into branches on the dorsum of the foot. These provide dorsal digital branches for the toes, except for the cleft between the big and second toes (deep peroneal), and the lateral side of the little toe (sural). Branches also supply the skin of the lower lateral part of the leg and the dorsum of the foot. The superficial peroneal nerve on the dorsum of the foot is occasionally visible as a sharp ridge on the surface.

The *deep peroneal nerve* continues the course of its parent nerve around the neck of the fibula to the front of the leg, piercing the origin of the extensor digitorum longus. It then descends the front of the leg, lying buried deeply between the muscles and accompanied by the anterior tibial artery. It supplies the four muscles of the front of the leg: tibialis anterior, extensor hallucis longus, extensor digitorum longus and peroneus tertius. It also supplies the ankle joint. On the dorsum of the foot the nerve breaks up into branches to supply the joints of the foot, extensor digitorum brevis, and a cutaneous branch to the cleft between the big and second toes. The nerve may be marked on the surface by a line from the neck of the fibula to a point approximately midway between the two malleoli, just lateral to the anterior tibial artery whose pulsations may be felt in this position.

The common peroneal nerve is exposed to injury as it winds round the neck of the fibula. This is about the height of a car bumper from the ground; it is also about the position of the upper anterior corner of a suitcase being carried

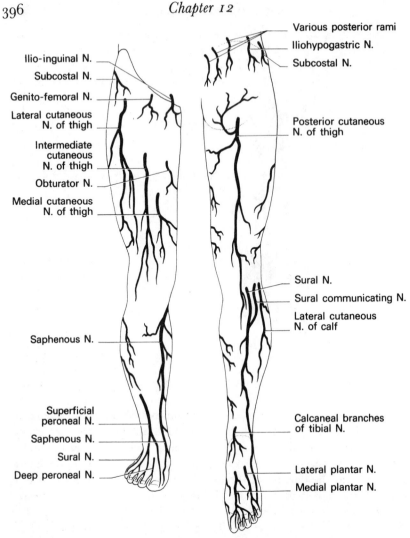

Fig. 12.13. The cutaneous nerves of the lower limb.

on the same side and this operation may provide a temporary, though agonising demonstration of the effects of damage to this nerve. Severe damage to the nerve will paralyse both the extensors and evertors of the foot giving rise to 'foot drop'. The effect of this is to prevent adequate dorsiflexion of the foot during the 'carry through' stage of walking so that as the affected foot is brought forwards the toes drag along the ground unless a high stepping gait is adopted.

DERMATOMES AND MYOTOMES

The detailed, and rather tedious, descriptions of the course and distribution of the various named nerves of the body, which have occupied the greater part of this chapter, will enable you to understand the effects of section of, or damage to, individual nerves such as the ulnar or common peroneal. Such nerves, as you have seen, are made up of branches from a number of spinal nerves. The effect of lesions of a single *spinal* nerve may involve more than one of the *named* nerves—for example, the 8th cervical nerve contributes to the ulnar, the median and the radial nerves. It is therefore important that you should know the areas of skin supplied by each individual spinal nerve (*dermatomes*) and the groups of muscles that are similarly supplied by individual spinal nerves (*myotomes*). The dermatomes are fairly well defined but the myotomes are rather more vague for the most part, so that only a few important examples will be given.

DERMATOMES

The dermatomes of the front and the back of the body are shown in Fig. 5.17. As was mentioned earlier, the thoracic nerves mostly have a very similar course and, except for the first, are not involved in the limb plexuses so that the dermatomes in the thoracic region are regularly arranged. It will help you to understand the arrangement of dermatomes in the limbs if you first revise your knowledge of the development of the limbs (Chapter 5).

There is a very high degree of overlap between the dermatomes supplied by consecutive spinal nerves. This is well seen in the case of the thoracic nerves, where a lesion of a single nerve may produce little or no sensory loss. In places, however, adjacent dermatomes are not supplied by consecutive spinal nerves— see, for instance, the boundary between C6 and T1 on the front of the upper limb in Fig. 5.17—and in these situations the area of sensory loss will be well defined.

MYOTOMES

From a functional point of view, it is best to consider the various *groups* of muscles which are supplied by a particular spinal nerve and we shall not, therefore, present here a list of all the individual muscles supplied by each nerve. Such lists are available in larger anatomical textbooks.

C5 (& 6)   — Abductors and lateral rotators of the shoulder. Flexors of the elbow. Supinators.

C6 (& 7)   — Adductors and medial rotators of the shoulder. Extensors of the elbow. Pronators.

C7 (& 8)   — Extensors of the elbow, wrist and fingers. Flexors of the wrist and fingers.

(C8) & T1 — The small muscles of the hand.

L2 & 3   — Flexors, adductors and medial rotators of hip.

L3 & 4   — Extensors of knee.

L4 & 5   — Dorsiflexors and invertors.

L5 & S1   — Evertors of foot, flexors of knee; extensors, abductors and lateral
              rotators of hip.
S1 & 2   — Plantar flexors.

It must be stressed again that the myotomes, as well as the dermatomes, are only approximate and are subject to considerable variations, so do not be worried if you see different accounts in different textbooks or if a patient with a nerve lesion does not present all the signs and symptoms which your anatomical studies have led you to expect.

Two lesions of the brachial plexus are particularly important. Lesions of the upper trunk (C5 and 6) may occur as a result of forcible depression of the shoulder or lateral flexion of the head to the opposite side. The upper trunk is also sometimes damaged in babies during birth. Since such lesions will affect particularly the abductors and lateral rotators of the shoulder, the flexors of the elbow and the supinators, the arm will be held in a position of adduction, pronation and extension of the elbow. This effect is sometimes known as *Erb's paralysis*. Lesions of the lower trunk of the brachial plexus may result from abnormalities in the region of the first rib or tumours in the apex of the lung. They produce wasting and weakness of the small muscles of the hand and of the flexors of the wrist and fingers. Occasionally the extensors may also be affected. The condition is called *Klumpke's paralysis*.

13 · Special Senses

The various types of ordinary sensation—touch, temperature, proprioception, etc.—have been described in Chapter 5, as have the pathways followed by the nerve impulses through the nervous system. The senses of sight, hearing, taste and smell are so different in the structure of their receptors and in the parts of the central nervous system concerned that they need a chapter to themselves. In addition to these four senses, one might add a fifth special sense— that of balance—which gives information about the position and the direction of any movement of the body or head relative to the surroundings. The anatomy of the organs concerned with this are so closely related to those of hearing that they will be described together.

THE EYE

The eyeball is a hollow structure filled with fluid under some pressure so that it feels rather firm. The eyeball is situated in the anterior part of the bony orbit and is embedded in loose fatty tissue so that its movements are quite free. The movements are carried out by six small muscles (Fig. 13.1). Four of the muscles

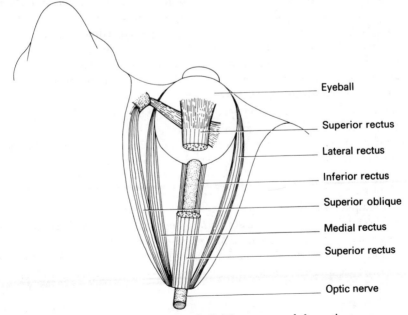

Eyeball

Superior rectus

Lateral rectus

Inferior rectus

Superior oblique

Medial rectus

Superior rectus

Optic nerve

Fig. 13.1. The muscles that move the eyeball. They surround the optic nerve.

399

are called recti (sing. rectus) and they are arranged above, below lateral to and medial to the eyeball. The other two muscles are called the superior and inferior oblique. The lateral rectus is obviously an 'abductor', i.e. it will turn the eye laterally, while the medial rectus has the opposite effect. The actions of the other muscles are a little more complicated and need not be described here. The muscles are supplied by three of the cranial nerves (3, 4 and 6). The muscles are capable of extremely fine movements with extraordinary co-ordination with the muscles of the opposite side. This is because it is essential that both eyes are directed towards the object that is being observed, otherwise double vision occurs. This means that when you are reading this line of print, *convergence* must be present, i.e. the two eyes are not directed straight ahead but each is turned slightly inwards. You can observe convergence if you ask a colleague to look at your finger while you bring it nearer and nearer to her nose. Convergence is a function of the left and right medial rectus muscles. Co-ordination between the two sides is necessary because if you are watching an object moving from right to left, the lateral rectus of the left eye and the medial rectus of the right eye have to contract together. If the object is also ascending, the other muscles also have to take part with different muscles contracting on the two sides, while if the object is also moving nearer to you, the two medial recti have to increase convergence at the same time. It has taken you quite a long time to read all this but when you are following a rapidly moving fly with your eyes, all these complicated muscles contractions have to take place simultaneously and have to be continuously changed and adjusted. All of this goes on when the head is not moving. Head movement, detected by the vestibular apparatus (pp. 405–408), is, through the vestibulo-ocular reflex mechanism, automatically and continuously accompanied by compensatory eye movements, so that the eyes, watching a stationary or a moving object, may keep their retinal images constantly on the fovea, the region of clearest vision. It is not surprising, therefore, that any excessive intake of alcohol which is a depressant of the nervous system, rapidly causes a loss of this remarkable muscle co-ordination so that double vision occurs.

The optic nerve emerges from the back of the eyeball and leaves the orbit though a circular foramen (the *optic foramen*). It is 3–4 mm in diameter (see Chapter 12). The orbit also contains numerous blood vessels, one of which gives off the minute, but very important, *central artery of the retina*. This enters the optic nerve and runs down its centre to reach the nervous layer of the eyeball, the *retina*. It is an end-artery (see Chapter 14) so that its blockage causes immediate blindness.

In the upper and outer angle of the orbit is the *lacrimal gland* which produces tears. The lacrimal secretion passes across the front of the eye from the lateral to the medial side and drains into two minute holes in the medial ends of the upper and lower eyelids. These can just be seen with the naked eye. They lead into the *naso-lacrimal duct* which, in turn, leads into the nasal cavity. An increased production of tears or a blockage of the naso-lacrimal duct will cause the tears to overflow and this can also happen if the facial nerve is damaged so that the

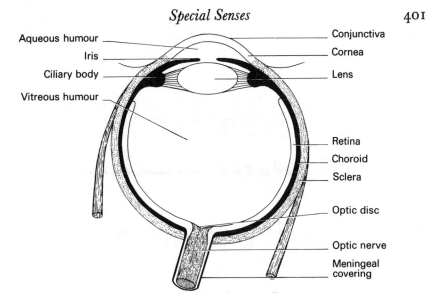

Aqueous humour — Conjunctiva
Iris — Cornea
Ciliary body — Lens
Vitreous humour —
Retina
Choroid
Sclera
Optic disc
Optic nerve
Meningeal covering

Fig. 13.2 Diagrammatic cross-section through the eyeball.

lower eyelid falls away from the eyeball owing to paralysis of the orbicularis oculi (Chapter 11).

The eyelids have a 'skeleton' of tough fibrous tissue, especially noticeable in the upper eyelid, and they contain skeletal muscle fibres. These form a part of orbicularis oculi and close the eyelids gently in sleep. On waking up, the upper eyelid is raised by a muscle which is supplied by one of the cranial nerves and also by the sympathetic system. Damage to either of these will cause drooping of the eyelid (*Ptosis*).

The eyeball itself consists of three coats (Fig. 13.2). The outermost coat, the *sclera*, is a very tough coat of fibrous tissue. At the front of the eye it is replaced by an equally thick but transparent layer called the *cornea*. The front of the cornea and of the eyeball is covered by a thin stratified squamous epithelium which is reflected off the eyeball on to the inner surface of the eyelids. This is the *conjunctiva*. It is very vascular and the vessels dilate rapidly in infection or in the presence of a foreign body, giving the appearance of a 'bloodshot eye'.

Inside the sclera is an extremely vascular layer called the *choroid*. Anteriorly, this coat leaves the sclera and forms the iris, the circular free edge of which forms the pupil. The iris is the coloured part of the eye and it contains smooth muscle which can constrict or dilate the pupil. This regulates the amount of light entering the eye by constricting the pupil in a bright light and dilating it in darker conditions. The pupil is also constricted for near vision, just as a camera lens has to be stopped down for a close shot. In both cases the optical properties of the system are improved.

The innermost layer is the *retina*. This consists of a layer of special receptor cells called *rods* and *cones* (Fig. 13.3). Surprisingly, other layers of nerve cells and fibres lie within the layer of rods and cones so that the light has to pass through all these layers before reaching the rods and cones themselves. From

Direction of light from lens

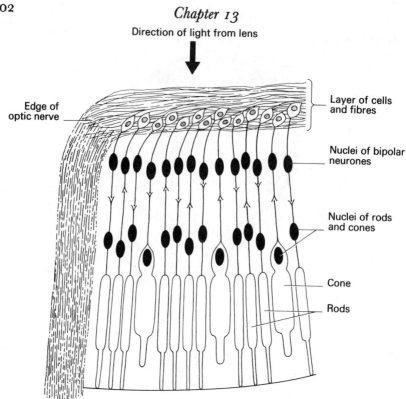

Edge of
optic nerve

Layer of cells
and fibres

Nuclei of bipolar
neurones

Nuclei of rods
and cones

Cone

Rods

Fig. 13.3. The layers of nerve cells in the retina. Note that the light has to pass through many layers to reach the rods and cones.

the most superficial layer of nerve cells, axons pass through the coats of the eyeball to form the *optic nerve*. The cones are mainly situated opposite the pupil whereas the more peripheral part of the retina contains a preponderance of rods. The latter are more sensitive to dim light than the cones, and at night it is therefore easier to see a faint light by looking at it indirectly. (Try looking for the very faint star right alongside the second star in the tail of the Great Bear.) The area of the retina where the nerve fibres leave the eye to become the optic nerve is called the *optic disc* and is devoid of rods and cones. It therefore forms a blind spot. It is important clinically because it can easily be seen with an ophthalmoscope and gives information not only about the state of the optic nerve but also about the pressure of the cerebro-spinal fluid, since an increased pressure produces a characteristic swelling of the optic disc.

The light, having passed through the cornea and the pupil, is focused by the transparent biconvex *lens*. This is enclosed in a very thin capsule which is attached to a thickening at the base of the iris called the *ciliary body*. Contraction of smooth muscle cells in the ciliary body slackens the lens capsule and causes the lens to increase in thickness, thus shortening its focal length and bringing the retinal image of a nearby object into sharp focus. When the ciliary muscle is

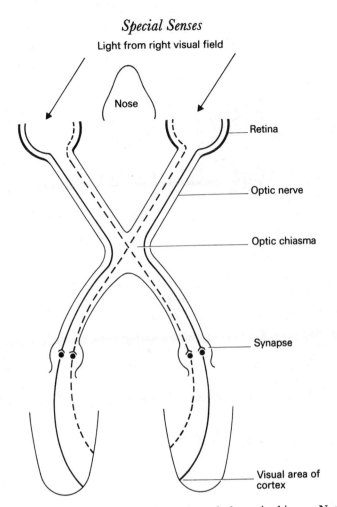

Fig. 13.4. The course of the optic nerve fibres through the optic chiasma. Note that the *right* visual field is registered on the *left* side of the brain.

relaxed, the eyes are focused on infinity, as occurs in day-dreaming or during a boring lecture. In front of the lens and its capsule, the eye is filled with fluid called *aqueous humour* which is secreted by the ciliary body and drains into the venous system. If the drainage fails, the intra-ocular pressure rises rapidly and can cause blindness. This condition is called *glaucoma*. The eyeball, behind the lens, is filled with a jelly-like mass, the vitreous humour.

THE VISUAL PATHWAY

The optic nerves are peculiar, not only by being enclosed in a layer of meninges but also because they join and separate again, giving the appearance of a crossover. This is called the *optic chiasma* and takes place inside the skull near the pituitary gland. In fact, only the medial fibres cross, the lateral fibres re-

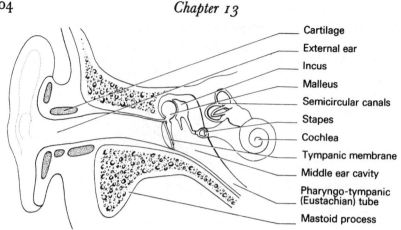

Fig. 13.5. Section through the temporal bone to show the ear.

maining on the same side (Fig. 13.4). Thus fibres from the *left* sides of the two retinae pass into the *left optic tract* while the two *right* sides pass into the *right optic tract*. After synapsing once, the axons of the second neurones pass to the occipital poles of the brain to end in the visual cortex (see Chapter 4). Since the *left* sides of the *retinae* receive light from the *right* field of vision, damage to the *left* occipital lobe will produce a *right-sided* blindness. Damage to the crossing fibres in the optic chiasma (by a pituitary tumour most commonly), will produce blindness on both sides of the visual field so that only objects directly ahead can be seen.

THE EAR

The ear, as a whole, may be divided into the *outer* (or external), *middle* and *inner* ear.

THE OUTER EAR

This is a canal, part cartilaginous and part bony, which leads from the *auricle* down to the ear drum or *tympanic membrane* (Fig. 13.5). It contains glands which secrete wax. The ear drum can be seen with an auriscope as long as excessive wax is not present. It is pearly grey in appearance and slightly concave.

THE MIDDLE EAR

This is a cavity within the petrous temporal bone, separated from the outside world by the tympanic membrane. It is connected by means of the *pharyngo-tympanic (Eustachian) tube* to the pharynx. This tube acts as a pressure regulator. If you climb a mountain, the atmospheric pressure drops and the ear drum bulges outwards causing slight deafness and feeling of fullness in the ear. Air then passes from the middle ear down the Eustachian tube in order to equalise the pressures on either side of the tympanic membrane, which resumes its normal shape with a sudden click. The middle ear also communicates posteriorly with the air cells in the mastoid process via a small cavity, the *mastoid antrum*.

The principal contents of the middle ear are three minute (less than 8 mm long) bones or ossicles called the *malleus, incus* and *stapes* (Fig. 13.5). The handle of the malleus (which means a hammer) is embedded in the tympanic membrane. The bones articulate with each other by means of tiny synovial joints and the footpiece of the stapes (the stirrup) is fixed in a small hole (the *oval window*) that leads into the inner ear. When sound waves enter the external ear, the tympanic membrane vibrates and the vibration is transmitted by means of the chain of ossicles to the oval window. The middle ear commonly becomes infected, particularly in children, and if the infection is not treated successfully, it may lead to the accumulation of pus in the middle ear. This often drains outwards by producing a perforation in the eardrum and it may also spread backwards into the mastoid antrum and the mastoid air cells or even further to surrounding structures.

THE INNER EAR

The inner ear is an extremely complex structure and only a simplified account need be given here. It consists of a series of minute spaces in the bone which form the *osseous labyrinth*. They comprise three *semicircular canals* which are arranged in three different planes, the *vestibule* and finally the *cochlea* which is a spiral canal rather like a snail shell. The osseous labyrinth contains a thin

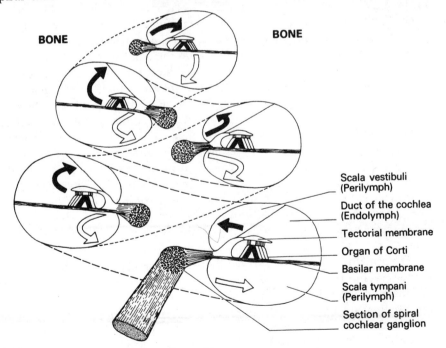

BONE

BONE

Scala vestibuli (Perilymph)

Duct of the cochlea (Endolymph)

Tectorial membrane

Organ of Corti

Basilar membrane

Scala tympani (Perilymph)

Section of spiral cochlear ganglion

Fig. 13.6. Cross-section through the cochlea.

fluid—the *perilymph*—within which is the *membranous labyrinth*. This consists of three *semicircular ducts* (within the semicircular canals), two thin-walled sacs in the vestibule, and the *duct of the cochlea* (within the cochlea). The membranous labyrinth contains another type of fluid—the *endolymph*.

The cochlea and its contents form the organ of hearing and it is supplied by the *cochlear branch* of the 8th cranial nerve—the *vestibulocochlear nerve*. The semicircular canals and the vestibule, together with their contents, are the organs of balance and they are supplied by the *vestibular branch* of the eighth cranial nerve.

The course of sound waves through the ear has already been described as far as the oval window. The oval window, closed by the footpiece of the stapes, leads into the cochlea and the vibrations of the oval window set up vibrations in the perilymph. These vibrations travel through the fluid up to the tip of the cochlea, through a small opening into another compartment of the cochlea and then down again. These two compartments are shown in Fig. 13.6, which depicts a transverse section through one of the turns of the spiral canal in the bone which form the cochlea. The upper and lower compartments are separated by the duct of the cochlea which is part of the membranous labyrinth. This is bounded by a thin membrane, the *basilar membrane*, and an even thinner membrane above. Upon the basilar membrane is the *organ of Corti*, which consists essentially of *hair cells*, leaning up against a pair of props. The sensitive hairs of the hair cells come into contact with a rather rigid *tectorial membrane*. Try to think in three dimensions and realise that the basilar membrane is actually a spiral structure like a spiral staircase with the steps smoothed out. In Fig. 13.6 you are only looking at a section of one portion of the staircase.

The vibrations in the perilymph will set up corresponding undulations in the basilar membrane and the point on the spiral membrane where the maximum undulations occur will depend on the pitch of the sound. At this point the hairs on the hair cells (which move up and down with the membrane) become pushed to one side by their contact with the tectorial membrane. This sets up action potentials which are transmitted along the cochlear nerve branches. The cell bodies of these nerve fibres are in the spiral *cochlear ganglion* (which corresponds to a posterior root ganglion) and the impulses then pass along the eighth nerve into the pons. After synapsing there, the cochlea fibres pass up the brainstem, some on their own side, others after crossing over to the opposite side. After a further synapse, higher up the brainstem, they are finally relayed to the auditory area of the cerebral cortex in the upper part of the temporal lobe (Fig. 5.6).

The sense of balance is subserved by the semicircular ducts and the membranous sacs in the vestibule. The three ducts lie at right angles to each other like three adjacent sides of a square box (Fig. 13.7). They contain endolymph because they are a part of the membranous labyrinth. At one end of each of the semicircular ducts there is a slight dilatation called the *ampulla* and this contains a number of hair cells whose hairs are embedded in a gelatinous mass of tissue. When the head is moved, currents are set up in the endolymph which moves

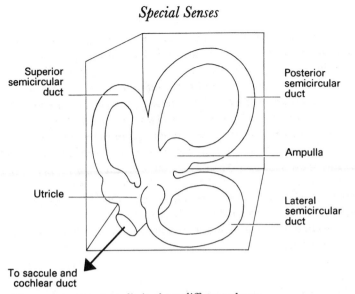

Superior
semicircular
duct

Posterior
semicircular
duct

Ampulla

Utricle

Lateral
semicircular
duct

To saccule and
cochlear duct

Fig. 13.7. The semicircular ducts lie in three different planes.

the gelatinous tissue and deflects the hairs. Since the ducts lie in three different planes, movement of the head in any direction is bound to set up currents in at least one of the canals but more probably in two or three of them. The movement of endolymph is relative to the ducts because what actually happens is that the ducts are moved while the endolymph, because of its inertia, remains still. Thus the hair cells move past the fluid rather than the fluid circulating in the ducts. This movement produces deflection of the hairs in one particular direction and this, in turn, sets up impulses in the vestibular part of the eighth nerve. (Note that it is the deflection of hairs that stimulate both cochlear and vestibular parts of the ear.) The bipolar cell bodies of these nerve fibres lie in the vestibular nerve itself within the temporal bone. Their axons synapse in the lateral part of the floor of the fourth ventricle (in the *vestibular nuclei*). From the nuclei, some fibres ascend to the forebrain and presumably, reach the cerebral cortex. Others pass to the cerebellum via the inferior cerebellar peduncle and to the nuclei of the muscles that move the eyeball. Finally, a large number of fibres from the main component of the vestibular nuclei complex—the *lateral vestibular nucleus*—pass down the spinal cord as the *vestibulo-spinal tract*.

The vestibule contains two sac-like components of the membranous labyrinth (the *utricle* and *saccule*) each of which contains a *macula*. These maculae each consists of an area several millimetres in diameter in which there are sensitive hair cells that project into a gelatinous mass of tissue similar to that of the semicircular canals. It differs from these, however, in that this tissue contains small calcareous crystals called *otoliths*. The maculae are arranged at right angles to each other so that when the head is tilted in any direction, the otoliths and the surrounding tissue pulls upon the hair cells and causes impulses to pass along the vestibular nerve. Thus they give information about the position of the head in space, unlike the semi-circular ducts which give information only about movements. In other words, the maculae tell you where you are and the ducts tell you where you are going. The course of these fibres of the vestibular nerve is similar to that of the fibres from the semicircular ducts.

The vestibular nerve thus carries information which is important for balance and posture and the movements of the eyes. Some of the impulses in the nerve pass to the cerebellum which, via the *rubro-spinal* and *reticulo-spinal* tracts, can influence muscle tone by varying the impulses arriving at muscles via their α and γ fibres. The vestibulo-spinal tract can similarly act upon the anterior horn cells directly. Connections to the neck and eye muscles can co-ordinate movements (particularly lateral movements) of the eyes and neck with rotatory movements of the head. One feature of certain diseases of the vestibular apparatus or of the cerebellum is *nystagmus* in which a rapid to and fro movement of the eyes occurs. These conditions are also associated with giddiness so that patients have to be carefully supported or they are liable to fall.

THE NOSE AND THE SENSE OF SMELL

The structure of the nose is shown in Figure 13.8. The two nasal cavities are separated by a midline *septum* although this is usually deviated to one or other side. In the side wall of the nose are 3 *conchae* on each side. These are curved thin bones covered with mucous membrane whose main function is to warm and moisten the inspired air. Various *paranasal sinuses* open into the nasal cavity the two most important of which are the *frontal sinus* which is in the frontal bone above the medial angle of the eye and the *maxillary sinus* (or *antrum*) which is a large cavity inside the maxilla. The whole nasal cavity and its offshoots except

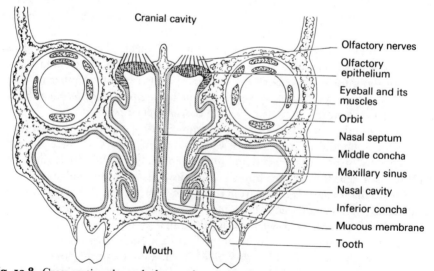

Cranial cavity

Olfactory nerves

Olfactory epithelium

Eyeball and its muscles

Orbit

Nasal septum

Middle concha

Maxillary sinus

Nasal cavity

Inferior concha

Mucous membrane

Tooth

Mouth

Fig. 13.8. Cross-section through the nasal cavity and surrounding structures.

for the olfactory region, but including the paranasal sinuses and the middle ear, is lined by ciliated columnar epithelium containing goblet cells. There are many mucous glands present. The *olfactory epithelium* which subserves the sense of smell, covers the uppermost part of the nasal cavity including the superior concha. The epithelium is thick and contains three types of cell of which the most important seem to be the *receptor cells*. These are, in fact, a form of bipolar neurone, the cell bodies being situated at the base of the epithelium. The dendrite extends to the surface of the epithelium while the axon joins other axons which run in the mucous membrane in bundles, finally to become one of the filaments of the *olfactory nerve* (the first cranial nerve). These filaments pass through the openings in the cribriform plate of the ethmoid, and, after a complicated series of connections, finally project to the cerebrum when the sense of smell is appreciated. The nerve endings are stimulated by airborne molecules but precisely how different types of smell produce their nerve impulse patterns is not understood. We discuss the non-olfactory nasal mucous membrane, its blood supply and its secretory activity in Chapter 19, on Respiration.

THE TONGUE AND THE SENSE OF TASTE

The tongue is, essentially, a bag of skeletal muscle. It is covered by mucous membrane which is rough on the upper aspect but smooth and thin underneath. The epithelium is stratified squamous in type. The muscle fibres are longitudinally, vertically and transversely arranged so that the shape of the tongue can

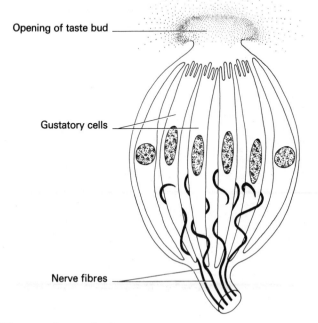

Opening of taste bud

Gustatory cells

Nerve fibres

Fig. 13.9. Diagram of a taste bud.

voluntarily be altered. Not everyone, however, can roll the tongue into a U-shape. This ability is genetically determined and you may be amused to find out how many of your colleagues are completely unable to do this. The upper surface of the tongue can be subdivided by a V-shaped furrow—the *sulcus terminalis*—into anterior and posterior parts. The anterior part (about two-thirds of the total surface area) is roughened by papillae which are of two types—some are narrow and pointed with a hard keratinised tip (*filiform papillae*) while others are flat-topped and deep red in colour (*fungiform papillae*). The latter are found particularly at the sides of the tongue and can easily be seen if you first dry the tongue with a handkerchief. Just in front of the sulcus terminalis is a V-shaped row of *vallate papillae*. These are much larger than the other papillae and they may reach 2 mm in diameter or more. Each is surrounded by a circular sulcus outside which is a wall (vallum). The posterior two-thirds of the tongue is smoother and contains much lymphatic tissue.

The organs of taste are the *taste buds* which are found in the sides of the vallate and fungiform papillae, in the posterior third of the tongue and in the mucous membrane of the palate. Each taste bud consists of a minute barrel-shaped structure opening on to the surface at one end. They contain *gustatory cells* whose apices reach the opening as well as other cell types (Fig. 13.9). Nerve fibres end in close relation to these cells and they are stimulated by different types of taste. The nerves which these fibres ultimately form, travel with certain cranial nerves as 'passengers' and end in a nucleus in the brainstem before being relayed to the cerebrum. Tastes have been classified into four main types, sweet and salt being appreciated at the tip of the tongue, acid at the sides and bitter at the back.

14 · The Structure of the Cardiovascular System

In very small animals, and in the human embryo at an early stage in its development, there is no need for a cardiovascular system since enough oxygen and nutrients can diffuse into the tissues from the outside world. As soon as the organism reaches a certain size, however, some sort of a circulatory system becomes essential. At first, quite a simple system suffices, in which a series of tubes reach all parts of the body while a portion of the system develops muscular walls and is able to pump blood all over the body. As the size and complexity of the organism increases, so more elaborate systems become necessary, and in the mammals (including man) the circulatory system is clearly divided into two parts, one of which is responsible for oxygenating the blood and removing carbon dioxide (the pulmonary circulation) and the other carrying oxygen and nutrients to the tissues and bringing back carbon dioxide and waste materials. In addition, the blood carries hormones ('chemical messengers'—see Chapter 26). Even the heart itself can be subdivided into two parts in this way and physicians often talk about the '*right heart*' and the '*left heart*', the former consisting of the right atrium and right ventricle (containing venous or deoxygenated blood) and the latter consisting of the left atrium and left ventricle (containing arterial or oxygenated blood).

A description of the circulation of the blood thus necessitates a description of the two circulations, which are quite separate, since the vessels do not communicate and the septa of the heart, which separate the right and left sides, are quite impermeable to blood so that no mixing can take place. To deal first with the *systemic circulation*, oxygenated blood is pumped by the heart out of the left ventricle into the aorta and thence into the arteries which supply all parts of the body (Fig. 14.1). Having traversed the smaller arterial vessels (*arterioles*), the blood passes through the microscopic *capillaries* from which arise small veins or *venules*. These unite with each other, forming larger veins, which themselves join with others until finally almost all the venous blood is delivered to the heart by the two largest veins—the *superior* and *inferior venae cavae* (sing. vena cava). These open into the right atrium. From this chamber the blood is passed into the right ventricle which, in turn, propels the blood into the *pulmonary trunk* which leads into the *pulmonary circulation*. The pulmonary trunk divides into the right and left *pulmonary arteries* which enter the lungs. Here, in the capillaries, the blood takes up oxygen from the inspired air and gives up carbon dioxide to it. The oxygenated blood is then returned to the left atrium of the heart via the four *pulmonary veins* (two on each side). From the left atrium the blood enters the left ventricle and the cycle begins again. The time taken by each stage of the cycle will be discussed in the next chapter. The action of the heart is assisted by the presence of a number of non-return valves

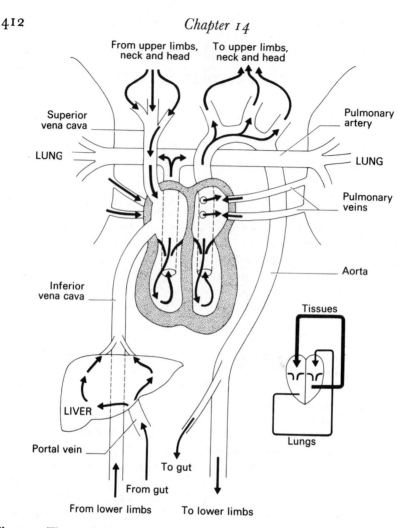

Fig. 14.1. The pulmonary and systemic circulations. The small diagram shows the basic pattern.

which lie between the atria and ventricles and at the origins of the aorta and the pulmonary trunk.

Finally, a brief mention must be made of a third system of blood vessels, namely, a *portal system*. Portal vessels are vessels which have capillaries or sinusoids (see below) at *both* ends. As has been described above, veins begin in a capillary network but at their other end they unite with larger veins until they eventually reach the venae cavae. In a portal system, however, veins begin in a capillary network, unite with other veins to form a larger vessel but then begin to branch again until they finally end once more in capillaries or sinusoids. The best example of a portal system is seen in the abdomen. Here, veins that drain blood from the intestine and from the spleen join together to form the large *portal vein* (sometimes called the hepatic portal vein). This runs up to the liver and then starts to divide again until its smallest branches form the minute, capillary-size *sinusoids* of the liver. Another portal system is found in the brain; this is called the *hypophyseal portal system* (see Chapter 26).

THE HEART

The heart lies between the two lungs in a compartment known as the *inferior mediastinum*. The space between the lungs as a whole is called the *mediastinum* and it is subdivided into two parts by a horizontal plane passing through the manubrio-sternal joint and the lower border of the fourth cervical vertebra. The *superior mediastinum* is the compartment above this plane and therefore lies behind the manubrium. It contains the arch of the aorta, the pulmonary arteries and various other structures to be described later. The inferior mediastinum is below the plane and contains the heart, the oesophagus and other smaller structures.

THE PERICARDIUM

The heart is enclosed in the *pericardium* which is a complex structure with three layers (Fig. 14.2). The outermost layer (the *fibrous pericardium*) is a simple bag of connective tissue which is shaped to contain the heart. Its floor is fused with the central tendon of the diaphragm while superiorly it tapers off and fuses with the adventitia of the aorta and pulmonary trunk. The *serous pericardium* consists of two layers. The outermost layer (the parietal layer) lines the fibrous pericardium from which it is reflected on to the outer surface of the heart itself where it forms the *visceral layer*, alternatively called the *epicardium*. The potential space between the two layers of serous pericardium contains only a very thin layer of fluid so that the layers are normally in contact and can slide easily over each other.

THE HEART

The heart wall consists of three layers. The outermost layer is a thin membrane, the *epicardium*, which is an alternative name for the visceral layer of serous pericardium. Beneath this is a certain amount of fat and some vessels and then comes the principal layer—the *myocardium*. As its name implies, this is composed of cardiac muscle. It is very thick in the ventricles (especially the left ventricle) but much thinner in the atria. The thickness of the myocardium is governed by the function of the various chambers. The thin atrial myocardium only has to propel blood into the ventricles so that no great power is needed. The right ventricle has to propel blood through the pulmonary circulation so that its myocardium is much thicker than that of the atria. The left ventricle is re-

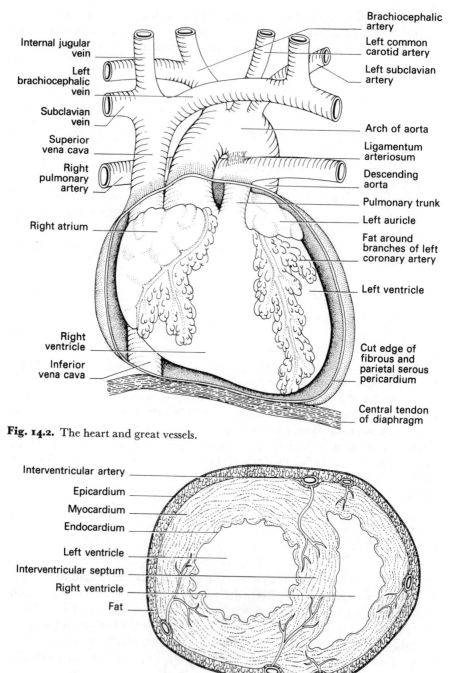

Internal jugular vein

Left brachiocephalic vein

Subclavian vein

Superior vena cava

Right pulmonary artery

Right atrium

Right ventricle

Inferior vena cava

Brachiocephalic artery

Left common carotid artery

Left subclavian artery

Arch of aorta

Ligamentum arteriosum

Descending aorta

Pulmonary trunk

Left auricle

Fat around branches of left coronary artery

Left ventricle

Cut edge of fibrous and parietal serous pericardium

Central tendon of diaphragm

Fig. 14.2. The heart and great vessels.

Interventricular artery

Epicardium

Myocardium

Endocardium

Left ventricle

Interventricular septum

Right ventricle

Fat

Fig. 14.3. Cross-section through the heart. Note the relatively thick wall of the left ventricle.

sponsible for pumping the blood throughout the whole of the rest of the body and it therefore needs a very thick myocardium. The relative thickness of the myocardium in the various parts of the heart is shown in Figure 14.3. Finally, the innermost layer of the heart wall is the *endocardium*. This is a very thin epithelium which lines the whole of the interior of the heart including the valves and other intracardiac structures. It is continuous with the *endothelium* which lines almost the whole of the vascular system and forms the innermost layer of the arteries, capillaries and veins. The endocardium (and the endothelium) is very smooth so as to allow the blood to flow freely over it and it also helps to prevent the blood clotting. Damage to the endocardium may lead to the formation of blood clots on the damaged area.

The shape of the heart as a whole is a little different from that shown in elementary textbooks of zoology or physiology where it is often depicted as having two atria above and two ventricles below. To convert this into a human heart, it should be laid on its side and then rotated as shown in Figure 14.4. As can be seen from this diagram, the right border of the heart consists entirely of the right atrium, which can also be seen from the front. The rest of the anterior surface of the heart (Fig. 14.2) consists of two-thirds right ventricle and one-third left ventricle with a small part of the left atrium (its auricle) just visible on the left border and the auricle of the right atrium projecting forwards and to the left. The inferior, or *diaphragmatic*, surface of the heart consists of one-third right ventricle and two-thirds left ventricle. The left surface of the heart, which is related to the left lung, is almost entirely left ventricle, with the auricle of the

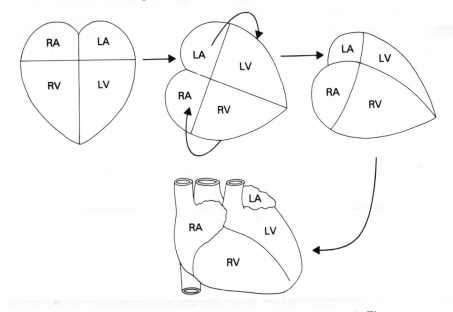

Fig. 14.4. The heart does not resemble diagrammatic hearts as seen in Figure 14.1. However, the true arrangement of the chambers may be remembered by lying a diagrammatic heart on its side and rotating it slightly.

left atrium above. The posterior surface or *base* of the heart therefore consists mostly of the left atrium, but also the posterior border of the right atrium. The *apex* of the heart is formed by the two ventricles and lies about $3\frac{1}{2}$ inches to the left of the midline.

THE ATRIA

The left and right atria are large, thin-walled cavities, each of which has a small *auricle* or *auricular appendage* attached. This is an offshoot of the main cavity with rather irregular walls and is supposed to look like a dog's ear, hence the name. (It is confusing to students that zoologists and also physicians often refer to the atrium as a whole as the auricle; but strictly speaking, the term auricle should be restricted to the offshoot from the main cavity.)

Opening into the right atrium are the two largest veins of the body: the superior and inferior *venae cavae* (Fig. 14.5). These do not have valves at their orifices although the inferior vena cava has a ridge of endocardium near it which is called a valve but it is non-functional. A much smaller venous channel which opens into the lower part of the right atrium is the *coronary sinus* into which drain most of the veins of the heart itself (see below). The four pulmonary veins open into the left atrium. The interatrial septum is thin and it has a particularly thin area near its centre which is called the *fossa ovalis*. This marks the site where the septum was deficient in the embryo so as to allow oxygenated blood returning from the placenta to pass directly from the right atrium to the left atrium and thence to the left ventricle and the systemic circulation. At birth, however, this opening normally is closed off so that the septum is impervious to blood. Occasionally, however, the closure mechanism is defective and the opening remains patent as an *atrial septal defect* (ASD). This is one form of the so-called 'hole in the heart', the other being a defect in the interventricular

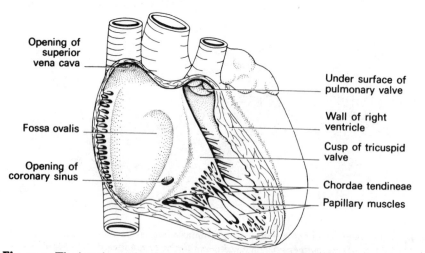

Fig. 14.5. The interior of the right atrium and ventricle.

septum. The popular term is misleading since it does *not* mean a hole in the wall of the heart opening into the pericardial cavity which would, of course be incompatible with life.

THE VENTRICLES

The ventricles are below and to the left of the atria (Figs. 14.2 and 14.5). The right atrium opens into the right ventricle, the orifice being guarded by the *tricuspid valve*, so called because it has three cusps. These are thin flaps of connective tissue covered by endocardium which are arranged around the atrioventricular opening in such a way that when the ventricle contracts, the cusps of the valve are pressed together and prevent any backflow from ventricle to atrium. There is a possibility, of course, that the high pressure developed during ventricular contraction might cause the cusps to be turned inside out and be blown back into the atrium. This is prevented by a number of cord-like strands of connective tissue covered with endocardium which pass from the edges of the cusps to the ventricular wall where they are attached to some conical muscular projections of the myocardium (Fig. 14.5). The cords are called *chordae tendineae* and the conical projections are the *papillary muscles*. Even away from the papillary muscles, the interior of the ventricle is not smooth but is raised into a number of muscular ridges called *trabeculae carneae*.

The left atrium opens into the left ventricle via the *bicuspid* or *mitral valve* although one needs a fairly vivid imagination to see the resemblance between this and a bishop's mitre. The two cusps are similar to the cusps of the tricuspid valve and they are restrained in the same way, by chordae tendineae and papillary muscles. The cusps are anterior and posterior in position. The anterior cusp is unique in that it is interposed between the atrioventricular opening and the aortic opening. It therefore has blood passing over both surfaces—from atrium to ventricle over its posterior surface and from ventricle to aorta over its anterior surface.

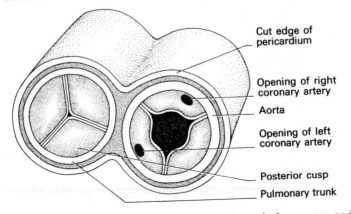

Cut edge of pericardium

Opening of right coronary artery

Aorta

Opening of left coronary artery

Posterior cusp

Pulmonary trunk

Fig. 14.6. The aortic and pulmonary valves. Although one is shown open and the other closed, in life both would be open simultaneously.

The pulmonary and aortic valves are placed, respectively, at the origin of the pulmonary trunk from the right ventricle and that of the aorta from the left ventricle. These valves also have three cusps but they differ in form from the tricuspid valve in that they are much smaller and of a semilunar shape. Each cusp consists of a very thin and pliable fold of endocardium containing a little connective tissue. They are concave upwards and when closed—i.e. when the ventricular muscles relaxes and the blood tends to regurgitate back into the ventricles—the edges of the cusps fit closely together to produce a completely watertight barrier (Fig. 14.6). Should you ever cook sheeps' hearts, it is worthwhile, when washing them, to experiment with a tap and the aortic and pulmonary valves to see how extremely efficient they are.

THE CONDUCTING SYSTEM

As you will read in the next chapter, the heart beat originates in the right atrium in a structure known as the *sinu-atrial (s-a) node*. This is a small nodule of specialised tissue situated in the right atrium near the opening of the superior vena cava. It is innervated by the vagus (parasympathetic) and by sympathetic nerves and it consists of specialised cardiac muscle which has an inherent rhythmicity of its own so that normally, under the combined influence of these nerves, about 72 impulses/min in an adult pass out into the surrounding atrial muscle causing atrial contraction. Having spread through the atrial muscle, each impulse reaches and stimulates another piece of specialised tissue close to the right atrioventricular opening, the *atrio-ventricular (a-v) node*. This node, through which impulses pass slowly, starts the impulse spread through modified cardiac muscle cells which convey it to the main mass of ventricular muscle. These cells are capable of a very rapid rate of impulse transmission. They are called *Purkinje cells* and the bundle of conducting tissue which they form is known as the *bundle of His* or (officially) as the *atrioventricular bundle*. After leaving the a-v node, the bundle of His runs into the interventricular septum, passing towards the apex of the heart. It soon divides into right and left branches, the former continuing down the right side of the septum and the latter piercing the septum to run down its left side. The bundles then break up into strands of Purkinje cells which spread out near the apex of the heart to run up again in the side walls of the ventricles. The distribution and timing of the wave of contraction through the *cardiac* muscle will be described in the next chapter.

THE BLOOD SUPPLY OF THE HEART

The first, and most important, branches of the aorta are the right and left *coronary arteries* which arise immediately above the aortic valve. Corona is the Latin for a crown (as in coronet) and the arteries have this name because they completely encircle the heart in the atrioventicular grooves. The right coronary artery (Fig. 14.7) runs round between right atrium and right ventricle, giving branches to each. It turns round the right border of the heart to reach the diaphragmatic surface where it anastomoses with the end of the left coronary

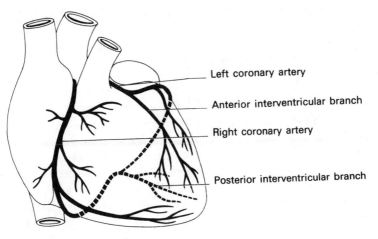

Left coronary artery

Anterior interventricular branch

Right coronary artery

Posterior interventricular branch

Fig. 14·7. The coronary arteries.

artery and sends an important *posterior interventricular branch* into the interventricular groove. The left coronary artery runs into the left interventricular groove and gives a very large *anterior interventricular branch* which descends towards the apex of the heart giving branches to both ventricles. The left coronary artery, now much diminished in size, continues around the left side of the heart and anastomoses on the diaphragmatic surface with the end of the right coronary artery. These arteries and their branches are accompanied by veins which have slightly different names but the names are not important. Almost all the veins open posteriorly into a large venous channel in the atrioventricular groove—the *coronary sinus*—which itself opens into the right atrium. In addition, a number of small veins also open directly into the right atrium and ventricle.

The anastomoses (intercommunications) between the various branches of the coronary arteries are extremely small so that if a major branch suddenly becomes blocked by, for example, a blood clot (coronary thrombosis), the anastomoses are not large enough to take over the blood supply of the affected region and a greater or lesser portion of the heart muscle will die. Like skeletal muscle, cardiac muscle has extremely limited powers of regeneration and if the patient does not die, healing can take place only by the formation of a fibrous scar. A very gradual blockage, however, will allow a collateral circulation to develop. The arteries therefore are *functional end-arteries* (see below) even though small anastomoses are present.

THE NERVE SUPPLY OF THE HEART

The heart is supplied by the autonomic nervous system. It receives branches from the cervical sympathetic ganglia and from the upper thoracic ganglia while its parasympathetic supply is derived from the cervical part of the vagus. These branches form the *cardiac plexus* of nerves which lies in relation to the arch of the aorta and the pulmonary arteries and is formed by a plexus of sympathetic and parasympathetic fibres and a number of parasympathetic ganglion cells. From the plexus, nerve fibres pass to supply the s-a and a-v nodes. Sympathetic effector fibres also supply the myocardium.

SURFACES MARKINGS OF THE HEART

The upper limit of the heart may be marked out by an oblique line passing from the second left to the third right costal cartilages at their sternal ends. The right border is just lateral to the right edge of the sternum and extends from the third to the sixth costal cartilages. The apex of the heart lies in the fifth intercostal space about 3½ inches from the midline. The inferior border is slightly concave and the left border convex (Fig. 14.2). The pulsations of the heart may clearly be felt at the apex (the *apex beat*) and may also often be seen (*cardiac impulse*).

THE BLOOD VESSELS

The histological structure of the arteries varies in different situations. The basic tissues of which the wall is composed are smooth muscle, elastic tissue and collagen fibres but the proportions vary. Thus, in the large arteries such as the aorta, carotid and subclavian arteries, there is a preponderance of elastic fibres, with only a relatively small proportion of smooth muscle cells and collagen. These vessels are therefore called *elastic arteries*. During contraction of the ventricles (*systole*), blood is forced out into the large arteries and distends their walls. During relaxation of the ventricles (*diastole*), the elastic recoil of

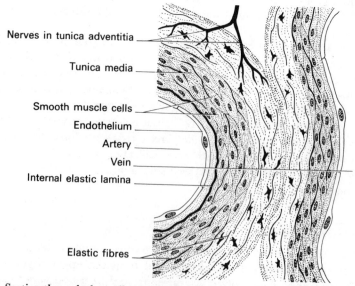

Nerves in tunica adventitia

Tunica media

Smooth muscle cells

Endothelium

Artery

Vein

Internal elastic lamina

Elastic fibres

Fig. 14.8. Section through the walls of an artery and a vein. Note the thinner wall of the latter and the absence of an internal elastic lamina.

the arterial walls continues to propel blood into the rest of the arterial tree, although under a reduced pressure (see Chapter 16).

The precise structure of the elastic arteries can be seen in Figure 14.8. The wall consists of three layers. The innermost layer—the *tunica intima*—consists of a lining endothelium, a little connective tissue outside it, and, finally, a dense layer of elastic tissue called the *internal elastic lamina*. The thickest layer, the *tunica media*, consists of a mixture of smooth muscle cells and collagen and is thickly interlaced with elastic fibres. Finally, the outermost layer—the *tunica adventitia*—consists mostly of connective tissue with some elastic fibres and small vessels which nourish the wall of the main artery. There is also a plexus of sympathetic nerves which supply the smooth muscle cells.

In smaller, but still fairly large arteries such as the radial or popliteal, the tunica intima is very similar but the media consists mostly of smooth muscle, with only a few elastic and collagen fibres. Such vessels are therefore known as *muscular arteries*. Their function is to conduct the blood to the finest vessels and their walls are distensible so that they increase in size during systole and decrease during diastole. If such an artery is injured, the smooth muscle in the wall can contract so as to narrow the lumen or even to close it off completely.

As the muscular arteries divide up into smaller and smaller branches, the amount of collagen and elastic fibres becomes less and less and the wall becomes thinner. Eventually, the vessels are very small (less than 0.3 mm in diameter) and are known as *arterioles*. These have a tunica intima composed only of endothelium, a media consisting only of a few smooth muscle fibres, and a thin adventitia. The small size of the individual arterioles, and their relatively small number, leads to these vessels being the principal site of resistance to flow in the circulation, and consequently to a very pronounced fall in pressure through this part of the vascular system. Contraction of the arteriolar smooth muscle throughout the body raises the resistance to flow—(the *peripheral resistance*) and therefore raises the general level of blood pressure, provided the output of blood from the heart does not change. This contraction is an effect of stimulation of the sympathetic nervous system. In addition, the arterioles in any tissue or organ are able to control the amount of blood passing into the capillary bed of that tissue. These effects will be discussed in more detail in Chapter 16.

The smallest vessels are the *capillaries* and it is in these vessels that the vital interchanges between the blood and the tissues takes place. The walls of the capillaries are therefore extremely thin and the diameter is only 5–15 μm, i.e. they are just about wide enough to allow a red blood cell to squeeze through. The wall of capillaries consists only of a single layer of endothelial cells with a few wisps of collagen outside and occasional external cells known as *pericytes*. It is possible that the pericytes or the endothelial cells are contractile but it is most likely that the flow of blood from arterioles into the capillary bed is controlled by the smooth muscle cells of the former (see below).

From the capillaries, blood passes into the *venules* and thence into small veins. These have a structure very similar to that of the arterioles except that the wall is much thinner. The small veins unite to form larger veins and here the basic pattern of the vessel wall remains similar to that of arteries of the same size but the walls of the veins are always thinner than the corresponding arteries because the pressure which they have to withstand is so much less. It is helpful to compare the main vessels which supply a limb, for example the femoral artery and vein. The wall of the former vessel has to withstand a pressure of about 120 mm Hg and it is therefore relatively thick. The latter contains blood at only a relatively low pressure so the wall, particularly the tunica media, is very much thinner. However, since the same amount of blood must leave the limb as enters it, and since the femoral vein is larger than the artery, the velocity of the blood must be lower. Furthermore, when the amount of blood flow through the limb has to be increased, for example as a result of muscular exercise, the vein enlarges in order to accommodate the extra blood flow. A space (the *femoral canal*—see Chapter 25) is therefore available for the necessary expansion of the vein.

Even the largest veins—the inferior and superior venae cavae— have relatively thin walls since the pressure of blood within them is still quite low. In fact, if the chest is severely compressed, as can happen when a large crowd of the aorta can be followed out into the neck and head, the upper limb, the trunk higher than the venous pressure so that venous return to the heart is reduced and fainting may occur.

THE BLOOD VESSELS IN GENERAL

As has already been described, blood is transported from the heart in arteries, from which it passes through a fine capillary network and is returned by the veins. The largest arteries are the aorta (for the systemic circulation) and the pulmonary trunk (for the pulmonary circulation). In the adult these vessels are relatively thick-walled (the aorta more so than the pulmonary trunk) and are easily large enough to allow the insertion of a finger into their lumen. The branches of the aorta can be followed out into the neck and head, the upper limb, the trunk and the lower limb, and these vessels branch progressively and gradually become smaller and smaller as they pass into the tissues and give off more and more branches. Most of these arteries have their own individual names but, with certain exceptions, you will only have to know the names of the larger and the more important arteries.

An important characteristic of the arterial system is that the arteries do not branch like the branches of a tree, but frequently rejoin each other to form a network. These intercommunications are known as *anastomoses* and they are important when a major artery becomes blocked or is tied by the surgeon as they offer an alternative route (*collateral circulation*) for the blood. Quite large vessels such as the axillary artery can therefore be occluded without endangering the life of the part that they supply (Fig. 14.9). Anastomoses between small arteries is especially common around synovial joints where the arteries form a *peri-articular plexus* from which the blood supply to the joint is derived. In a few

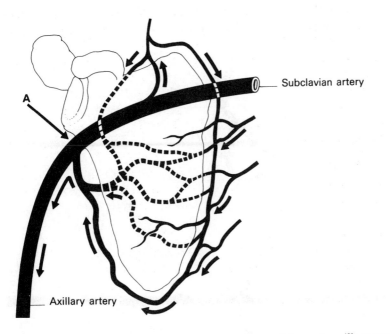

Fig. 14.9. Collateral circulation. In the event of a blockage of the main axillary artery at A, circulation could still be maintained in the upper limb by enlargement of the arterial anastomoses around the scapula.

places, however, there are no anastomoses so that if a vessel is occluded, the region that receives its blood supply from this artery will die. A good example of such an *end artery* is the central artery of the retina in the eye. This artery is just about large enough to be seen by the naked eye, but it is extremely important as it is (a) the sole arterial blood supply to the retina, and (b) an end artery. Occlusion of this vessel therefore leads to immediate and permanent blindness in the affected eye. There are a number of other important arteries which, although not true end arteries since they do anastomose with other arteries, are, nevertheless, vulnerable to blockage by, for example, a blood clot. This is because the anastomoses are relatively small so that these vessels may be described as *functional end arteries*. Such vessels are found in the heart (coronary arteries) and in the brain. The results of blockage of such vessels depends on the age of the patient and the rapidity with which the process occurs. In a young patient in whom the arterial wall is normal, and if the blockage is slow in onset, the anastomoses that are present will gradually enlarge and a full collateral circulation may be established. If, however, the arterial wall is diseased so that the lumen is narrowed, and if the onset is sudden, the collateral circulation will not develop and the result will be death of the region supplied by the affected artery. The region thus deprived of its blood supply is called an *infarct*.

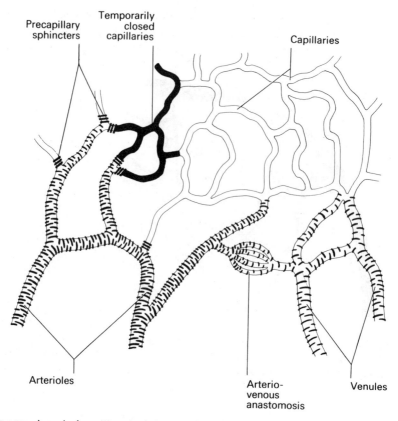

Precapillary sphincters

Temporarily closed capillaries

Capillaries

Arterioles

Arterio-venous anastomosis

Venules

Fig. 14.10. A typical capillary bed. See text for description.

The arterioles too, form a network and from this blood flows into the capillaries. There are very many anastomoses between arterioles and also between the arterioles and the venules. The latter are called *arteriovenous anastomoses* or *shunts* (Fig. 14.10) and they are important in certain regions such as the skin because they are capable of opening or closing. They can thus allow blood to pass through the capillary bed (when they are closed) or can short-circuit the capillary bed completely (when they are open). This occurs in the skin (Chapter 22) where the arteriovenous anastomoses play an important part in temperature control.

The form of the capillary network is usually adapted to the tissue in which they are found. Thus, in muscles, or in the medulla of the kidney, where the main structures run longitudinally, the capillary network has similarly elongated meshes but in glandular structures such as the exocrine pancreas the capillaries form a spherical basketwork around the acini. In the skin, the capillaries form a series of loops which project into the dermal papillae (Chapter 22).

The blood flow through a capillary bed is not constant, but is continually varying. Thus, blood may be flowing through some capillaries, while in others the blood is stagnant. A few minutes later flow begins in the stagnant vessels and stops in those in which it was previously flowing freely. At times a whole capillary bed may stagnate, blood only flowing through a very few channels from arterioles to venules (*preferential channels*). The way in which this alternation of flow and stasis in capillaries is produced is not absolutely certain, but is probably the result of contraction or relaxation of arteriolar smooth muscle, particularly those portions known as *pre-capillary sphincters*. Blockage of capillaries by single blood cells impacted in their arteriolar orifices has been seen. Relaxation of the sphincter allows the removal of the obstructing cell which passes down the capillary, thus allowing flow to be restored (Fig. 14.10).

From the capillaries, the blood passes into the venules and thence into veins. Some veins run in the superficial fascia while others, the deep veins, usually accompany the main arteries. This is well seen in the limbs, in each of which there are two major superficial veins and a number of deep veins, the two systems being interconnected by *communicating veins* which pierce the deep fascia. The superficial veins can be seen through the skin and can be studied on the back of your hand, particularly if the veins are distended by allowing your hand to hang down. You will notice that anastomoses between veins are very common— veins anastomose even more freely than arteries—and also that the pattern varies in different people. Thus variations in the course of smaller veins is very common, although the major veins are usually situated in the expected places. The blue colour of veins is due to refraction of light as it passes through the skin— venous blood is, of course, dark red in colour. The expression 'having blue blood in one's veins' is therefore misleading and is believed to originate from Spain where the peasants had dark sunburnt skin which made the veins almost invisible, whereas the pale skin of the nobility showed clearly the blue veins beneath the surface.

The larger deep veins accompany the corresponding arteries and have similar names (e.g. the femoral artery and femoral vein). The smaller arteries, however, are accompanied by two or more intercommunicating veins which are called *venae comitantes*. Thus, while there is an axillary artery and an axillary vein, the continuation of the axillary artery—the brachial artery—is not accompanied by a brachial vein but rather by a pair of venae comitantes.

An important feature of most veins is that they are provided with valves. These are semilunar in shape and are formed of a fold of endothelium. They are directed in such a way that they allow blood to flow only towards the heart. When the veins are compressed by muscle action, blood is thus forced only towards the heart, backflow being prevented by the valves. This *muscle pump* effect plays a vital part in venous return (see Chapter 16). Valves are not present in the large veins of the abdomen or in the veins inside the skull.

The two largest veins, into which blood passes from all parts of the body except the heart and lungs, are called the *superior* and *inferior venae cavae*. The former drains mainly the head, neck, upper limbs and thorax while the latter drains the abdomen and lower limbs.

THE ANATOMY OF THE MAIN BLOOD VESSELS

THE THORAX AND ABDOMEN

The largest arteries and veins are called collectively the *great vessels* and they are shown in Figure 14.11. Leaving the heart can be seen the *pulmonary trunk* and the *ascending aorta*. The former comes from the right ventricle and divides into the right and left *pulmonary arteries* whose subsequent course will be described in Chapter 19. The ascending aorta gives off the *coronary arteries* (already described) and then the *brachiocephalic, left common carotid* and *left subclavian arteries*. The region from which these three branches arise is called the *arch of the aorta* because of its shape. The brachiocephalic artery divides into the right common carotid and right subclavian arteries. The common carotid arteries supply the head and neck and the subclavian arteries the upper limb.

The arch of the aorta then becomes the descending (thoracic) aorta which runs in or slightly to the left of the midline on the bodies of the vertebrae. It gives off small branches to supply the chest wall and pierces the diaphragm at the level of the 12th thoracic vertebra to become the abdominal aorta (Fig. 14.11). This has three large ventral branches which supply the digestive

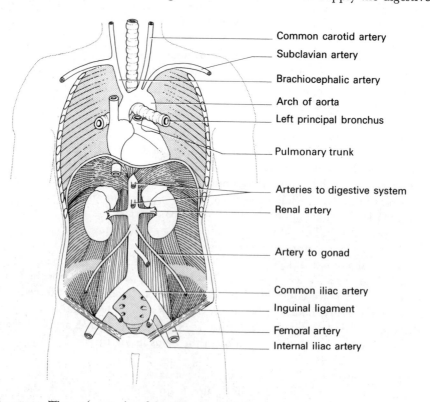

Common carotid artery

Subclavian artery

Brachiocephalic artery

Arch of aorta

Left principal bronchus

Pulmonary trunk

Arteries to digestive system

Renal artery

Artery to gonad

Common iliac artery

Inguinal ligament

Femoral artery

Internal iliac artery

Fig. 14.11. The main arteries of the thorax and abdomen.

tract, its associated glands, and also the spleen. It also has a number of lateral branches which supply the body wall and various viscera; note particularly the *renal arteries* which supply the kidneys and are so large that the two kidneys take about a quarter of the total output of the heart. The abdominal aorta ends by dividing into two *common iliac* arteries which, in turn, give rise to the internal and external iliac arteries. The former supplies mainly the pelvic viscera (bladder, uterus, rectum) while the latter passes under the inguinal ligament to enter the lower limb as the *femoral artery*.

The major veins shown in Figure 14.2 are the superior and inferior venae cavae and their tributaries. The *internal jugular vein* accompanies the common carotid artery and drains the head and neck. The *subclavian vein* accompanies the subclavian artery and drains the upper limb. On each side the internal jugular and subclavian veins unite to form the *brachiocephalic vein*. The left brachiocephalic vein crosses the midline behind the upper part of the manubrium sterni to join the right brachiocephalic vein, thus forming the superior vena cava. Certain other smaller veins drain the chest wall.

In the abdomen there are two distinct venous systems. The venous drainage from the intestine and its associated glands, and also the spleen, is into the *portal vein*. This enters the liver, its blood passes through the liver sinusoids and finally enters the inferior vena cava (see Chapter 24). The other (systemic) venous system drains the lower limb and abdominal wall. The veins have similar names to the arteries (iliac, renal, etc.), follow them closely and end up in the *inferior vena cava*. This pierces the central tendon of the diaphragm and, after a very short course in the thorax, enters the right atrium of the heart.

THE HEAD AND NECK

ARTERIES

The common carotid artery divides, at the level of the thyroid cartilage of the larynx (the 'Adam's apple') into *internal* and *external carotid arteries*, the former being the larger (Fig. 25.10). The course of the common and internal carotid arteries may be marked in the surface by a line from the sternoclavicular joint to the small gap between the sternomastoid and the angle of the mandible. Pulsation may be felt along this course, most easily at the anterior border of the lower part of sternomastoid.

The external carotid artery gives off a number of branches which go to supply the neck, face, scalp and the deeper structures of the facial region, in fact all the structures outside the cranial cavity. Most of these branches anastomose across the midline with the corresponding vessels on the opposite side. Some of these branches are worth mentioning by name since their pulsation can be felt on the surface. The *facial artery* arises in the upper part of the neck and enters the face by crossing the lower border of the mandible just in front of the insertion of the masseter. It runs across the face to the medial angle of the eye, giving branches to the upper and lower lips, the nose and other facial structures. The *superficial temporal artery* runs up just in front of the ear, where its pulsation can be felt, and it then divides into branches to supply the scalp. After severe exertion, the branches of this artery are often visible on the side of the forehead and in elderly people with a receding hairline the tortuous course of the branches can easily be seen.

The *internal carotid artery* enters the cranial cavity through a foramen in the base of the skull and after a complicated course it gives its first branch—the *ophthalmic artery*—to the orbit. The most important branch of this vessel is the *central artery of the retina* which enters the optic nerve to supply the retina of the eye. It is an end-artery. The internal carotid artery then divides into the *anterior* and *middle cerebral* arteries which supply the cerebrum except for the occipital lobe. Another artery—the *vertebral*—enters the skull through the foramen magnum and, for convenience, will be described here although it is actually a branch of the subclavian artery. The vertebral artery ascends the neck through

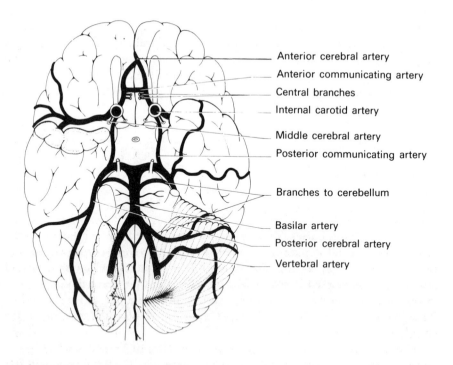

Anterior cerebral artery

Anterior communicating artery

Central branches

Internal carotid artery

Middle cerebral artery

Posterior communicating artery

Branches to cerebellum

Basilar artery

Posterior cerebral artery

Vertebral artery

Fig. 14.12. The arteries at the base of the brain. Two vessels on each side—the internal carotid and the vertebral—provide the blood supply. Note the arterial anastomosis (the circle of Willis) made up by the anterior cerebral and communicating arteries and the posterior cerebral and communicating arteries.

the foramina transversaria of the upper six cervical vertebrae and enters the skull through the foramen magnum. The two vertebral arteries join to form the *basilar artery* which travels up the ventral surface of the brainstem in the midline, giving branches to the medulla, pons, midbrain and cerebellum. It ends by giving off the right and left *posterior cerebral arteries* which supply the occipital lobe of the cerebrum. The three cerebral arteries on each side (anterior, middle and posterior) are united by communicating branches so as to form a complete arterial circle at the base of the brain. This is called the *circle of Willis* (Fig. 14.12) and it is important because the anastomoses are usually large enough to enable the circulation to both sides of the brain to continue even if one internal carotid artery is narrowed or occluded.

The branches of the cerebral arteries ramify on the surface of the brain (the *cortical branches*) and send branches which enter the brain more or less at right angles. There are also a number of deep central branches which arise in the region of the circle of Willis and enter the brain directly. One particular group of these vessels (the *striate arteries*) supplies the region of the internal capsule (p. 82) and haemorrage from, or occlusion of, one of these vessels will cause a 'stroke'.

The intracranial arteries have a very much thinner wall than other arteries of comparable size and they do, in fact, look very like veins. They are thus liable to local distension (*aneurysm*) which may rupture with serious results, causing a subarachnoid or intracerebral haemorrhage.

VEINS

In the neck the veins in general accompany the corresponding arteries but they anastomose more freely and, in places, form venous plexuses. Some of the veins that drain the scalp and face form the *external jugular vein* which ends in the subclavian vein. It is mentioned here because it is usually visible as it runs down on the surface of sternomastoid and it is a useful guide to the venous pressure. In heart failure, if blood is accumulating in the venous system, it will be distended when the patient is sitting up (it is *normally* distended when a person is lying flat). The *internal jugular vein* accompanies the common and internal carotid arteries. As you would expect, it has thinner walls and a wider lumen than these arteries. Its drainage area covers both internal and external carotid territory. It therefore receives tributaries from the neck, face, mouth, nose, etc., and also from the brain. Inside the skull there are two types of veins. Within, and on the surface of the brain, the veins are similar to those elsewhere although they do not accompany the arteries. These veins drain into a series of *dural venous sinuses*. These have relatively rigid walls, being formed from the dura mater and are lined by endothelium. The largest sinus is the *superior sagittal sinus* which runs in the midline over the top of the brain. It is into this sinus that most of the cerebro-spinal fluid drains by means of the arachnoid villi (p. 488). The venous sinuses intercommunicate and the largest of them combine to form the internal jugular vein which leaves the skull through a foramen in the base.

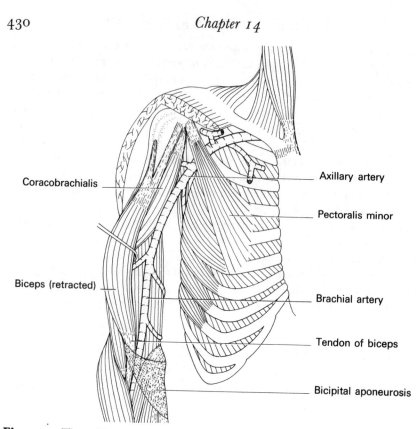

Coracobrachialis

Biceps (retracted)

Axillary artery

Pectoralis minor

Brachial artery

Tendon of biceps

Bicipital aponeurosis

Fig. 14.13. The axillary and brachial arteries.

THE UPPER LIMB

ARTERIES

The main artery of supply to the upper limb is the *subclavian artery*, which on the right side is a branch of the brachiocephalic, but on the left arises directly from the arch of the aorta. The artery leaves the thorax by arching over the first rib just in front of the lower trunk of the brachial plexus. Both these structures are liable to be affected by lesions at the apex of the lung or by such conditions as cervical rib. As the subclavian artery crosses the outer border of the first rib it becomes the *axillary artery* which extends down as far as the lower border of the teres major. While in the axilla it gives off a number of large branches, most of which accompany the local nerves (Fig. 14.13). They anastomose around the scapula with arteries of the chest wall and with branches of the subclavian artery. A collateral circulation is thus possible when the subclavian artery is occluded or divided distal to these branches (Fig. 14.9). The precise course of the axillary artery is described in Chapter 25.

At the lower border of teres major the axillary artery becomes the *brachial*

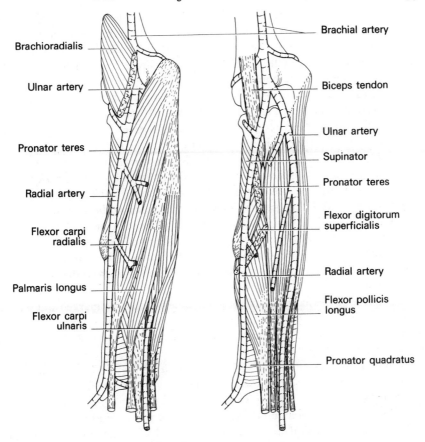

Fig. 14.14. The main arteries of the forearm.

artery. This runs down the medial side of the arm in the groove between biceps and brachialis, gradually getting closer to the midline of the arm which it reaches in the antecubital fossa—the slight depression in front of the elbow (Chapter 25). It gives branches to the surrounding muscles, to the humerus and to the peri-articular plexus of arteries around the elbow joint.

It enters the forearm in the midline and just below the elbow—at the level of the head of the radius—it divides into radial and ulnar arteries (Fig. 14.14). The ulnar artery passes down the forearm, moving towards the ulnar nerve which it joins under cover of flexor carpi ulnaris about one-third of the way down the arm. The two structures then travel together to the wrist. They pass into the hand superficial to the flexor retinaculum and just medial to the pisiform bone and the tendon of flexor carpi ulnaris. The ulnar artery then divides into *superficial* and *deep* branches. The superficial branch passes across the palm of the hand at the level of the outstretched thumb, lying immediately deep to the palmar aponeurosis. This branch joins one of the branches of the radial artery to form the *superficial palmar arch*. The deep branch passes through the muscles of the hypothenar eminence to form a *deep palmar arch*. This lies on the metacarpals and interossei and is completed by the termination of the radial artery. From this, *palmar metacarpal arteries* pass to the fingers, where they are joined by branches from the superficial palmar arch. From these vessels the digital arteries enter the fingers, lying close to the sides of the phalanges. Digital arteries to the thumb and to the radial side of the index finger come directly from the radial artery.

The radial artery is fairly superficial, being overlapped only by brachioradialis. It meets the superficial terminal branch of the radial nerve about one-third of the way down the forearm but, unlike the nerve, it continues down the front of the forearm to reach the lower end of the radius just lateral to the tendon of flexor carpi radialis. It then winds round to the back of the wrist, deep to the tendons of abductor pollicis longus and extensor pollicis brevis and thus finds itself in the anatomical snuffbox. Here it gives off a number of branches and then enters the palm by passing between the two heads of the first dorsal interosseous and ends by joining the deep branch of the ulnar artery to form the deep palmar arch, giving off branches to the thumb and the radial side of the index finger.

The course of the axillary and brachial arteries may be marked out on the surface by a line from the posterior wall of the axilla to the midpoint between the humeral epicondyles. The brachial artery divides at the level of the head of the radius and the radial and ulnar arteries may be represented by gently curved lines from the midline to points just lateral to the tendon of flexor carpi ulnaris and to the flexor carpi radialis. Pulsation of the arteries is easy to feel in the lateral wall of the axilla, down the faint surface groove between biceps and brachialis and in the antecubital fossa just medial to the tendon of biceps and just lateral to the median nerve. Both radial and ulnar arteries are palpable at the wrist in the regions already mentioned, the pulsation in the radial artery being commonly referred to as 'the pulse'. The radial artery may also be palpated in the anatomical snuffbox and the digital arteries may be palpated at the roots of the fingers. The main vessels of the upper limb are relatively superficial throughout the greater part of their course and they are thus rather exposed to injury, especially at the wrist, and to compression by tight bandages, etc. Particular care should be taken in applying bandages to the region of the elbow.

VEINS

The veins of the upper limb form superficial and deep systems. There are two main superficial veins, the *cephalic* and the *basilic*, both running in the superficial fascia (Fig. 20.3). The cephalic vein starts in the region of the anatomical snuffbox, draining blood from the veins on the back of the hand. It travels up the lateral side of the forearm and arm, often visible, and finally reaches the gap between the deltoid and pectoralis major where it passes deeply to joint the axillary artery. In the antecubital fossa it communicates, by one or two branches, with the basilic vein and with the deep veins. These superficial veins in the bend of the elbow are commonly used for obtaining blood samples.

The basilic vein travels up the medial side of the forearm and upper arm. It pierces the deep fascia half-way up the arm and it then joins the venae comitantes of the brachial artery to form the *axillary vein* (there is no brachial vein).

The deep veins of the upper limb accompany the corresponding arteries in the form of venae comitantes. The only named deep vein of the upper limb is the *axillary vein* which is formed at the lower border of the axilla by the union of the venae comitantes of the brachial artery and the basilic vein. The axillary vein accompanies the axillary artery and becomes the *subclavian vein* at the outer border of the first rib. This unites with the internal jugular vein to form the *brachiocephalic vein* on each side.

THE LOWER LIMB

ARTERIES

The major arterial supply to the lower limb is on the front of the thigh (Fig. 25.6) only relatively small branches of the internal iliac artery supplying the gluteal region by emerging through the greater sciatic notch. The *external iliac artery* passes beneath the inguinal ligament midway between the anterior superior iliac spine and the symphysis pubis. It then becomes the *femoral artery* which traverses the femoral triangle (Chapter 25) from the midpoint of its base to its apex. It is thus fairly superficial in this part of its course but at the apex of the triangle it passes deep to sartorius, thus lying in the adductor canal (Chapter 25). It finally pierces the adductor magnus muscle (Fig. 14.15) and enters the popliteal fossa to become the *popliteal artery*. In the femoral triangle the femoral artery gives a number of small branches to the lower part of the abdominal wall and the external genitalia and then gives off its largest branch (almost as large as its continuation) which is the *profunda femoris*. As its name implies, this passes deeply into the thigh, descending close to the femur and lying on adductor magnus at its insertion. It gives a series of branches which pass to the back of the thigh and which anastomose with each other, with the arteries of the gluteal region and with the popliteal artery. This important chain of anastomoses can therefore provide a collateral circulation (at least if the arteries are healthy) when the femoral artery is occluded.

The popliteal artery begins at the upper angle of the popliteal fossa (Chapter 25) and descends in the midline of the fossa as far as the lower border of the popliteus muscle. Since it enters the popliteal fossa through its floor (adductor magnus), it is the deepest important structure and lies close to the popliteal surface of the femur where it may be damaged in fractures of the lower end. The popliteal vein and the tibial nerve are closely related to the artery but lie superficial to it (Fig. 25.9). The popliteal artery gives off a number of large

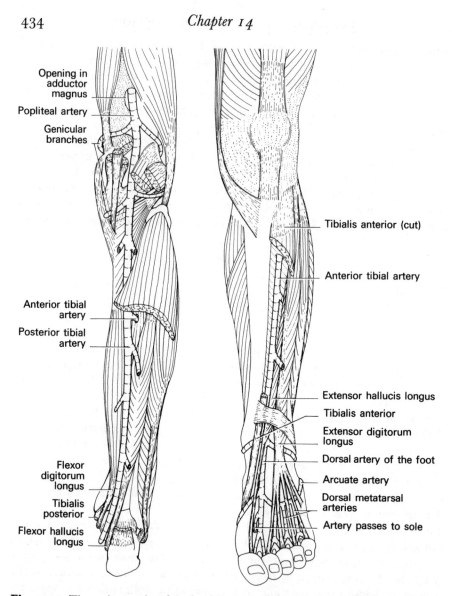

Opening in
adductor
magnus

Popliteal artery

Genicular
branches

Tibialis anterior (cut)

Anterior tibial artery

Anterior tibial
artery

Posterior tibial
artery

Extensor hallucis longus

Tibialis anterior

Extensor digitorum
longus

Dorsal artery of the foot

Arcuate artery

Flexor
digitorum
longus

Dorsal metatarsal
arteries

Tibialis
posterior

Artery passes to sole

Flexor hallucis
longus

Fig. 14.15. The main arteries of the leg (the femoral artery is shown in Figure 25.6).

muscular branches and also *genicular branches* which contribute to the peri-articular plexus.

At the lower border of popliteus, the popliteal artery divides into *anterior* and *posterior* tibial arteries. The anterior tibial artery reaches the front of the leg by passing between the tibia and the fibula above the upper border of the interosseous membrane. It then descends, lying on the interosseous membrane between tibialis anterior and the other muscles of the extensor compartment, giving off branches to the muscles as it does so. It also contributes branches to

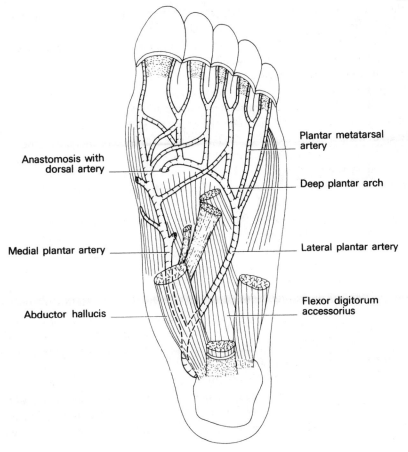

Fig. 14.16. The medial and lateral plantar arteries.

the peri-articular plexuses of the knee and ankle joints. At the ankle joint and midway between the two malleoli, the anterior tibial artery becomes the *dorsal artery of the foot* (dorsalis pedis). This travels forward to reach the space between the bases of the first two metatarsals where it passes between the two heads of the first dorsal interosseous muscle to reach the sole of the foot and completes the *deep plantar arch*. (Compare this to the course of the radial artery, remembering that the front of the leg corresponds to the back of the forearm.) It gives off a series of branches to the ankle and foot, including the *arcuate artery*. From this and from the dorsal artery itself, *dorsal metatarsal arteries* run forward to supply adjacent sides of the toes.

The posterior tibial artery continues the line of the popliteal artery, running down the back of the leg deep to gastrocnemius and soleus. It gives muscular branches and contributes to the plexuses around the knee and ankle joints. In the lower part of the leg it becomes superficial and lies about 1 inch in front of the medial border of the tendo calcaneus. It ends by entering the sole of the foot, passing deep to the abductor hallucis.

While lying in this position, it divides into *medial* and *lateral plantar arteries* (Fig. 14.16). The former runs along the medial side of the foot between the abductor hallucis and the flexor digitorum brevis. It supplies branches to the adjacent muscles and to the medial toes. The lateral plantar artery has a course very similar to that of the ulnar artery in the hand. It first runs towards the base of the fifth metatarsal between the first and second layers of muscles and then gives a superficial branch to the lateral toes. Its deep branch passes between the third and fourth layers of muscles to form the *deep plantar arch* which is completed by the termination of the dorsal artery of the foot. The deep plantar arch gives off *plantar metatarsal arteries* which divide to supply adjacent sides of the toes.

The surface marking of the femoral artery is the upper two-thirds of a line from a point midway between the anterior superior iliac spine and the symphysis pubis to the adductor tubercle. Pulsation of the artery may be felt at the former point. The popliteal artery may be marked out by a line in the popliteal fossa running from a point about a finger's breadth from the midline at the junction of the upper two-thirds and lower one-third of the thigh to the midline of the calf at the level of the tibial tuberosity. Its pulsation may be felt along this line, most easily after flexing the knee to relax the dense fascia which forms the roof of the fossa. The anterior tibial artery descends the front of the leg from the level of the tibial tuberosity to a point midway between the two malleoli. The dorsal artery of the foot passes from this point to the space between the bases of the first two metatarsals. Pulsation may be felt along the artery in the region of the ankle and foot. The posterior tibial artery descends to a point midway between the medial malleolus and the prominence of the heel, where its pulsation may be felt.

VEINS

The venous drainage of the lower limb is by means of superficial and deep systems. The principal superficial veins are the *great* and *small saphenous veins* (Fig. 20.4). The great saphenous vein begins by draining the venous network on the dorsum of the foot and it may usually be seen and palpated ¾ inch above and in front of the medial malleolus, where it lies in direct contact with the lower end of the tibia. It ascends the medial side of the leg, passes a hand's breadth behind the medial border of the patella and then moves more laterally. It pierces the deep fascia and enters the femoral vein at a point 1½ inches below and lateral to the pubic tubercle. The small saphenous vein begins on the lateral side of the ankle and ascends the calf, travelling towards the midline. It enters the popliteal vein by piercing the deep fascia of the popliteal fossa. Both veins receive numerous tributaries, and they are connected by one or more superficial veins. The great saphenous drains the lower part of the abdominal wall and the external genitalia as well as the lower limb.

The deep veins accompany the main arteries. There are thus *femoral*, *profunda femoris* and *popliteal veins* but the smaller arteries are accompanied by unnamed venae comitantes. The femoral vein lies immediately medial to the femoral artery as they pass under the inguinal ligament but lies behind it at the apex of the femoral triangle. The popliteal vein cross superficial to the popliteal artery from lateral to medial in the popliteal fossa.

There are important communications between the superficial and deep veins. These vessels pierce the deep fascia and they contain valves which are so directed as to allow blood to pass only from the superficial to the deep veins. In this way, blood from the saphenous veins, which is not propelled by the muscle pump is directed towards the deep veins in which the muscle pump is effective. In the condition of varicose veins, the valves in the superficial and communicating veins become ineffective and blood pools in the saphenous veins which become distended and tortuous.

15 · The Cardiac Cycle and Regulation of the Heart's Activity

The term cardiac cycle refers to the regular succession of events that occur in the heart as it beats. Each heart beat is a contraction of the whole muscle mass of the heart. It is also known as a *systole*, hence systolic (as applied to arterial blood/pressure). Between beats the heart muscle is at rest, or in *diastole*. The succession of events in each systole and each diastole, the cardiac cycle, can be studied in many different ways. First there is the electrical activity in the surface membrane of the heart muscle cells. In each cell a depolarisation of the membrane's resting electrical potential initiates contraction of that cell, and the depolarisation is conducted from cell to cell, spreading like a wave until the whole muscle mass of the organ has been activated and caused to contract. Next there is the contraction process itself, the development of tension and of shortening in the muscle fibre. This produces changes in pressure in the blood contained in the heart's chambers and vessels leading to and from the heart, and also bulk movement of blood between chambers or from the heart itself into major blood vessels. As pressure changes, valves open or close and in the latter event audible sounds are produced. The sounds, pressures and electrical events can all be recorded and studied and from these studies our knowledge of the action of the heart has been built up.

In this account we will look first at the electrical and mechanical events in the contraction of a typical heart muscle cell, for these differ in certain important respects from the description already given for skeletal muscle. We will then describe the initiation and spread of the wave of depolarisation of surface membranes that activates the contraction of the heart muscle. Then we will study the pressure changes, flow of blood and events at the heart valves that result from the heart's contraction. Finally, we must see how the rate and the force of the heart's contraction are regulated in response to changes in the living body. Before commencing these studies, we must first describe the special features of the structure and the electrical activity of cardiac muscle.

HISTOLOGY OF CARDIAC MUSCLE

Cardiac muscle is striated, as is skeletal muscle, having a similar arrangement of actin and myosin filaments. The cells themselves, however, are rather different, each being only 100 μm long and having a single central nucleus. At each end the cell branches and makes contact with two or three neighbouring cells. The regions of contact show as dark lines with the light microscope and are therefore called *intercalated discs*. With the electron microscope they can be seen to be a special type of intercellular junction which is peculiar in that the

438

depolarisation waves of the heart action potential can pass easily from cell to cell. Cardiac muscle cells resemble skeletal muscle cells in having a similar arrangement of transverse and longitudinal tubules and they also contain glycogen granules and mitochondria.

THE ELECTRICAL ACTIVITY OF HEART MUSCLE

RESTING AND ACTION POTENTIAL OF A TYPICAL HEART MUSCLE CELL

All heart muscle cells, like nerve and skeletal muscle cells, have an electrical potential charge on the surface membrane. It is about 85 mV, with the exterior positive to the interior. This charge is associated with differing concentrations of sodium and potassium ions on the two sides of the membrane with most of the sodium on the outside and most of the potassium on the inside. As in the other excitable tissues, these concentration differences are maintained by the resistance of the membrane to sodium ions, together with the sodium pump which actively extrudes the sodium ions that gain access to the cell's interior. Action potentials can develop and spread in heart muscle in the same way as in other tissues, but the propagation of the action potential over heart tissue varies from 0.2 to 4.0 metres per second, very much slower than that for nerve fibres, but similar to that in skeletal muscle.

The myocardial action potential differs in another fundamental way from nerve or skeletal muscle action potentials. After the phase of rapidly falling negative voltage within the cell (due to Na^+ entry as in nerve fibres) and the start of the fall back towards the resting potential level, this return of potential is slowed or practically ceases. Different bits of heart muscle tissue show differences in the exact nature of this delay, but some sort of delay or *plateau* of potential is seen throughout the heart. In nerve or skeletal muscle the potential fall is associated with an outward movement of potassium ions, so in cardiac muscle it is simple to assume that part of this movement is slowed or delayed. Experimental work confirms this assumption. Potassium movement falls drastically (or resistance of the membrane to potassium flow rises) as the intracellular potential becomes less negative in the initial phase of the action potential. The depolarisation plateau is also maintained by an inward movement of Ca^{2+} through special channels that open when Na^+ movement has brought the membrane potential up to -30 mV (inside) from the resting level of -85 mV. Potassium permeability then gradually recovers and potassium ion flow develops during the plateau, reaching a high level during the final recovery to normal resting potential (Fig. 15.1). As in skeletal muscle, contraction follows swiftly after depolarization, but cardiac muscle cells remain steadily contracted throughout the plateau phase of the action potential, only relaxing as the potential returns to resting level.

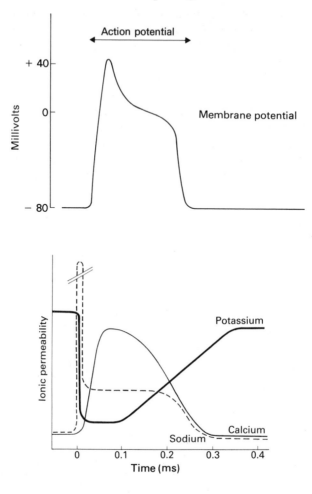

Fig. 15.1. Action potential and membrane permeability to sodium, calcium and potassium of a typical heart muscle cell.

RESTING POTENTIAL AT THE 'PACEMAKER' REGIONS

The cardiac muscle action potential is spontaneously generated in any piece of heart muscle separated from the heart and kept alive artificially. Nerve impulses are therefore not required to initiate heart muscle activity. There is, furthermore, a hierarchy in the rate of spontaneous activity, some regions generating action potentials sooner than others. The region that generates action potentials most frequently will inevitably transmit these to other regions since an action potential, once it starts somewhere in the heart, will be conducted over its whole mass. The region of the most rapidly generated action potentials thus becomes the leader of the whole heart. It is referred to as the heart's *pacemaker*. The reason for the spontaneous generation of action potentials is found in another peculiarity of heart muscle. The resting potential is not held constant, but falls steadily and slowly from its initial value, reached as potassium ions flow out rapidly at the end of an action potential. This slow fall in resting potential (or rise in intracellular potential) is due to a progressively developing imbalance between inward movement of sodium ions and outward movement of potassium ions in the resting membrane, the latter movement being gradually restricted. Eventually the fall in membrane potential (due to the net inward leak of positive charges on the sodium atoms) reaches the threshold value where fall in potential gradient causes a fall in sodium resistance of the membrane, causing a further fall in potential gradient, etc. The 'take off' of the new action potential is aided also by the progressive rise in potassium resistance as the resting membrane potential falls, thus further increasing the net charge change as sodium ions enter the cell.

Although all heart muscle *can* generate action potentials in this way, it has been noted above that the resting membrane potential change occurs more rapidly in some sites and that one of these becomes the pacemaker for the rest of the heart. This site is normally the s-a node (see Chapter 14). It is from here that the heart's action potentials start because it is here that the spontaneous change in resting potential occurs the fastest.

The s-a node is supplied with sympathetic, noradrenergic effector nerve fibres. Close to it are some parasympathetic ganglion cells, the short effector axons of which pass directly to the s-a node cells. Connector axons from the

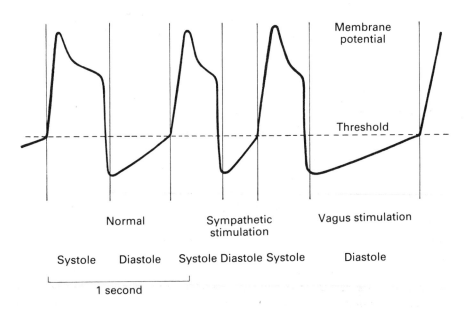

Fig. 15.2. Membrane potential and potassium permeability of a sinu-atrial cell.

vagus nerve supply these ganglion cells. Vagal stimulation slows the heart rate while sympathetic impulses increase its rate. The former potentiates potassium ion movement, while the latter, reduce potassium movement. Thus vagus nerve stimulation impedes while sympathetic stimulation accelerates the rate of fall of resting membrane potential (or rise in intracellular potential) produced by the continuing leaking of sodium ions (Fig. 15.2). Both vagus and sympathetic are continuously active, even in the resting person. If both nerve supplies are blocked, the s-a node 'fires' at about 100 beats/minute. This happens in some forms of heart transplant operation. Vagal impulses bring the rate of firing down to the resting 72 beats and the sympathetic can cause it to rise to 180 or 200 beats. Absence of the sympathetic tone at rest, leaving the vagus unopposed, results in a rate of about 50 beats/minute. Vagal and sympathetic stimulation effects are mimicked by application of acetylcholine and noradrenaline respectively.

THE TIME-COURSE OF THE ACTION POTENTIAL IN THE HUMAN HEART

The action potential in single s-a node cells lasts about 0.1 sec. In atrial fibres it lasts about 0.3 sec, travelling through the atrial wall at about 1.0 m/sec. About 0.1 sec after the s-a node has initiated the action potential, this reaches the a-v node (see Chapter 14), which also behaves differently from normal heart muscle. In the absence of the s-a node, the a-v node can act as a pacemaker, since its cells' resting membrane potential declines nearly as fast as does that of the s-a node. Action potentials in the a-v node cells are brief, but they travel slowly (0.2 m/sec.) through it. This node is situated on the fibrous tissue ring that separates atrial from ventricular muscle, close to the interventricular septum. From it runs the bundle of His, the strand of heart muscle tissue special-

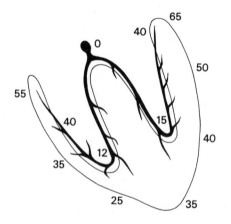

Fig. 15.3. The time-course of the spread of a myocardial action potential from the a-v node through the Purkinje tissue and the ventricular muscle of a dog's heart. Times are in msec.

lised for the conduction of action potentials. Its Purkinje fibre cells have a conduction velocity of 4 m/sec and in them the action potential lasts about 0.3 sec, Purkinje fibres spread in a branching network over the whole of the inner surface of the muscle of the two ventricles, including the interventricular septum and the papillary muscles. When action potentials leave the Purkinje fibres and enter the ventricular muscle mass their velocity is again slowed to 1 m/sec, but their duration remains at 0.3 sec. Figure 15.3 shows (in a dog's heart—similar results not being available in man!) the actual arrival of the impulse at different regions of the ventricular muscle.

THE ACTION POTENTIAL AND MUSCLE CONTRACTION

About 0.1 sec after the action potential passes over a muscle fibre's membrane, the contractile mechanism in the fibre goes into action. In the latent period the action potential spreads by the transverse tubules into the cell's interior, calcium ions enter the sarcoplasm both through the cell's external membrane and from the sarcoplasmic reticulum. The calcium frees the active sites on actin molecules so that the myosin heads can begin their sequence of attachment, deformation, and detachment reactions which is the basis of muscle contraction. In heart muscle the twitch contraction (the contraction that results from the passage of one action potential over a muscle fibre) lasts as long as the plateau phase of the action potential, about 0.3 sec for both atrial and ventricular muscle. The force exerted by a contracting piece of heart muscle is directly related to the initial length of the piece. In the intact heart, then, as the amount of blood that enters the heart in diastole varies, the force of its subsequent systole varies similarly, and the heart will therefore eject the same volume of blood as it received in the preceding diastole. This property, known as Starling's law of the heart, will be described later. In cardiac muscle, unlike skeletal muscle, the force exerted at any one resting length is

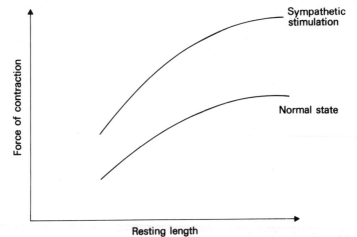

Fig. 15.4. The relationship between resting length and force of contraction (the Starling relationship) of normal and sympathetic nerve stimulated myocardium.

also increased by exposure to noradrenaline, which also shortens the duration of the contraction (Fig. 15.4).

Another special feature of heart muscle is that it remains completely re-fractory to further stimulation until after the completion of an action potential. The membrane potential must fall back to its resting level before a further action potential can be received and transmitted over a heart muscle cell. This means that a new contraction wave cannot start before the previous one is at its end. The heart cannot be made to contract continuously in tetanus.

THE ELECTRICAL ACTIVITY OF THE WHOLE HEART

It is about 100 years since Professor Waller stood his pet bulldog's feet in beakers of salt water and from these detected and recorded the voltages transmitted through the animal's body from its beating heart (Fig. 15.5). Since those early experiments, recording of the human heart's electrical activity (*electrocardio-graphy*) has developed into a major item of medical and scientific investigation. The relationship between the typical human electrocardiograph (E.C.G.) and the electrical activity of individual heart muscle cells is a complex one. For the purposes of this book it is best to relate the events recorded from electrodes placed on the body surface to the passage, as already described of the action potential from the s-a node through the atrial muscle, a-v node, bundle of His and Purkinje tissue, to the outer surface of the ventricular muscle. The typical human ECG takes the form shown in Figure 15.6 (the bottom trace). The rising and falling phases of the P wave are due to the spread of the action potential over the atrial muscle. It leaves the s-a node at the beginning of the P wave and reaches the a-v node at its peak. The bundle of His is small in

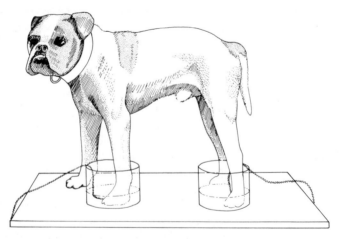

Fig. 15.5. Prof. Waller's dog, 'Jimmy', with the electrodes, vessels containing a weak salt solution, used in the very first recording of the electrical activity of the heart (reproduced with the permission of Prof. Richard Creese of the Physiology Department at St. Mary's Hospital Medical School, London W2, Great Britain).

bulk so the passage of the activity through it is unrecordable with electrodes at the body's surface (it can be clearly recorded with an electrode placed on a catheter tip inserted into the heart's chambers from a vein). The Q, R and S waves result as the action potential successively invades the septal, apical and basal muscle of the ventricles and the T wave results from the repolarisation of ventricular muscle.

EVENTS OF THE CARDIAC CYCLE

Now that we have described the heart's action potentials and the contraction of a single muscle fibre, it is possible to study the sequence of events in the intact beating heart in the body. Before this, however, the mode of action of the valves of the human heart must be described. You should refer repeatedly to Figure 15.6 while reading about the events in the cardiac cycle.

HEART VALVES

The position of the valve flaps allows a free flow of blood from atria to ventricles. Any rise in ventricular pressure over that in the atria (thus tending to cause a back-flow of blood) at once results in a closure of the valve cusps. These are prevented from 'turning inside out' by the presence of the chordae tendineae which through the contraction of the papillary muscle along with the remainder of the ventricular muscle, further support and protect the free margins of the valves as they come together. The cusps of the aortic and pulmonary arterial valves lie against the artery wall as blood leaves the ventricles in systole. When the ventricular muscle relaxes at the beginning of diastole, as soon as pressure within the ventricle falls below that in aorta or pulmonary artery, the valve cusps fall together and obstruct blood flow from the artery into the ventricle.

HEART SOUNDS

Typically, two sounds are usually heard from the normal heart, by the ear held close to the bared chest, or when the sounds have been transmitted through a stethoscope (originally a hollow tube of rolled paper, next a bored rod of wood, now a more or less elaborate device of rubber and curved metallic tubing with moulded plastic chest and ear pieces, designed so as best to transmit sounds from the chest wall to the listening ears). The first sound is synchronous with the closing of the atrioventricular valves as pressure rises in the ventricles at the beginning of their contraction. The second sound is caused by the closure of the aortic and pulmonary arterial valves at the end of ventricular contraction. It is described as being shorter in duration and higher in pitch than is the first sound. Modern analysis suggests that it is not the valve cusps meeting (like clapping hands) that produces the noise, but the sudden turbulence in the adjacent blood that actually produces the sounds.

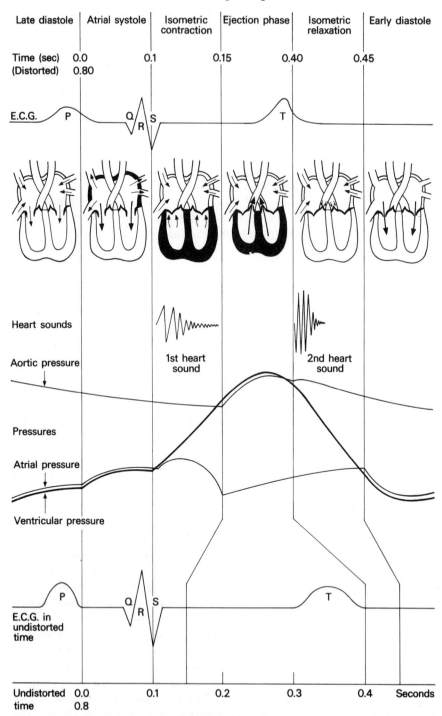

Fig. 15.6. The events of the cardiac cycle. Note the distortion of the (horizontal) time scale in all except the bottom record of the E.C.G.

PRESSURE AND VOLUME CHANGES IN THE HEART CHAMBERS

Figure 15.6 portrays many of the events of the heart beat in diagrammatic form. The sequence of these is followed from the beginning of one atrial contraction, time 0, to the beginning of the next, 0.8 second later.

The first thing to happen is the start, from the s-a node, of a spreading action potential through atrial muscle, shown on the ECG record as the P wave. This actually precedes the atrial contraction by about 0.05 sec. During the preceding diastole or resting phase (0.4 sec) the chambers, both atria and ventricles, have been passively filling with blood from the venae cavae and pulmonary veins. Tricuspid and mitral valves were open at this time. Pressures in all chambers are about equal to atmospheric pressure.

The atrial muscle contraction that follows the P wave by 0.1 sec 'tops up' the ventricles. Since there are no valves between atria and great veins, the rise in pressure consequent upon their contraction can be seen in the veins. The external jugular vein, under the skin on the side of the neck, is a favourite place for observing the standing pressure of blood in the right atrium and its rise and fall during atrial contraction or systole. It is possible, when this pressure is abnormally raised in heart disease, to measure the actual height of the column of blood above the heart's own level in the body and thus measure the atrial blood pressure directly. In a healthy person atrial pressure is usually about atmospheric pressure so this manoeuvre is not possible. While the atria are contracting the action potential is passing through the a-v node and bundle of His and its branches and invading the ventricular muscle. This is followed after 0.05 sec by the onset of ventricular contraction or systole. Immediately ventricular systole begins, the a-v valves (tricuspid and mitral) close and the first sound is heard. The pressure now rapidly rises in the ventricles until it exceeds aortic or pulmonary artery pressures and can then force open the valves guarding the openings to these vessels. During the phase of rising pressure, however, the ventricle is a closed chamber with all valves shut and blood unable either to enter or leave so that it does not alter in size. This stage is therefore called the *isometric contraction phase* and usually lasts 0.05 sec.

Eventually the intraventricular pressure exceeds arterial pressure and the aortic and pulmonary arterial valves open and blood begins to be ejected rapidly from the heart into the great arteries. These are elastic, therefore their distension causes a further rise of pressure in both ventricles and arteries until the peak pressure is reached about 0.15 sec after the beginning of the ejection of blood. By now the rate of ejection of blood from the ventricles is rapidly falling, and after a further 0.1 sec the contraction of ventricular muscle ceases. Those two stages, totalling 0.25 sec, have traditionally been called the *isotonic contraction phase* but it is better called the *ejection phase*. The total duration of ventricular muscle contraction is thus $0.05 + 0.15 + 0.1 = 0.3$ secs. Pressure in the ventricles falls rapidly and soon drops below that in the arteries. Immediately this occurs, the aortic and pulmonary arterial valves close, producing the second sound and, in the case of the aorta, a momentary rise of pressure can be recorded throughout the arterial system and felt by the examining finger on the radial artery at the wrist. This sharp change in pressure is known as the dicrotic notch of the falling phase in arterial pressure. It is seen in Figure 15.6 where ventricular and arterial pressures diverge at the end of systole. In peripheral arteries, since it is conveyed as a pressure wave through the blood and the elastic arterial wall, it arrives some time later. In the root of the aorta it is, of course, synchronous with the second heart sound. A measure of the time delay between this sound and the arrival of the dicrotic pressure wave at the wrist would indicate the elasticity of the artery wall, as does the rise of pressure during ejection of blood in ventricular systole. The smaller the rise in pressure and the slower the transmission of the dicrotic pulse wave, the more elastic is the arterial wall.

In the ventricles, now that the muscle has relaxed, the pressure falls rapidly. For 0.05 sec it is less than the arterial pressure, but still greater than the atrial pressure. The valves remain closed and the volume of blood within the ventricles does not change. This is known therefore as the *isometric relaxation phase*. During the whole of the time since ventricular systole began, blood has continued to enter the atria from the veins and the pressure there has been slowly rising. When atrial pressure exceeds ventricular pressure, the tricuspid and mitral valves open and all chambers of the heart fill again while the heart muscle remains quiescent in diastole for about 0.4 sec. The whole cycle of events has thus occupied 0.1 sec for atrial contraction + 0.3 sec for ventricular contraction + 0.05 sec for the isometric relaxation phase + 0.35 sec for diastole − 0.8 sec in all.

THE CARDIAC OUTPUT

In a man at rest, at the end of diastole each side of the heart (atrium and ventricle together) contains about 150 ml of blood (*end-diastolic volume*). During systole the ventricular contractions eject about 70 ml of this into the pulmonary artery and into the aorta, so that each ventricle contains about 80 ml at the end of systole (*end-systolic volume*), only about half of its original content of blood being ejected in systole. The volume ejected in each beat of the heart is usually called the *stroke volume*. The resting heart beats about 72 times a minute, so the volume of blood that enters the aorta (or the pulmonary artery) is 72 × 70 = about 5000 ml in each minute. Since the circulation is a closed-circuit system and the heart has not an unlimited reservoir of blood to draw upon, it follows that the output of the heart must be matched by the immediately preceding venous return. The overriding importance of the venous return will again be considered when the regulation of the heart's action is discussed.

MEASUREMENT OF THE CARDIAC OUTPUT

Of the various methods that have been devised to determine this, we will only describe two that are in regular general use in man.

In 1872 Fick outlined the principles of one method. If you know the total volume of oxygen removed from the air in the lungs by the blood in a given time, and also the oxygen concentrations of the blood before and after it has passed through the lungs, then the volume of blood can easily be calculated.

The resting person takes in 250 ml O_2 from the air each min.

> Blood arriving at the lungs contains 15 ml O_2/100 ml blood
> Blood leaving the lungs contains 20 ml O_2/100 ml blood
> 5 ml O_2 are picked up by 100 ml blood
> ∴ 1 ml O_2 is picked up by 20 ml blood
> ∴ 250 ml O_2 are picked up by $20 \times 250 = 5,000$ ml blood

Exactly the same calculation can be performed for carbon dioxide production from the body, though here typical figures might be 200 ml CO_2 produced per minute and the blood concentrations 52 and 48 ml CO_2/100 ml blood. The problem in the method was to devise a means of determining the amount of either gas in the blood just before it entered the lungs. This has been achieved by the use of a fine plastic catheter which can be threaded through veins from the elbow into the right atrium and thence into right ventricle and even into the pulmonary artery. This was first done in 1930 and 15 years later was being regularly performed. Alternatively, the CO_2 concentration of pulmonary arterial blood could be determined by breathing CO_2-rich air. After a short time a concentration was reached at which CO_2 neither entered nor left the blood as it passed through the lungs. This was then equated with the CO_2 of the blood in the pulmonary arteries, and the rest of the measurements made and calculations performed.

The second method depends on the rapid injection into the blood in the right atrium of a known quantity of a dye or some other 'indicator' substance, followed by the measurement of the concentration of this substance as it emerges fully mixed with the blood, in the systemic arterial blood. From the time taken for all the indicator to emerge and from its average concentration, the volume of blood mixed with dye and the time taken for it to pass through the heart can be calculated as follows:

> Amount of dye injected = 5 mg
> Time dye present in arterial blood = 40 secs
> Average concentration in arterial blood = 1.5 mg/litre of blood
> 1.5 mg dye was present in 1 litre of blood
> ∴ 1.0 mg dye was present in 1/1.5 litres of blood
> ∴ 5.0 mg dye was mixed with 5/1.5 litres of blood
> This volume ($3\frac{1}{3}$ litres) passed through heart in 40 seconds
> $3\frac{1}{3} \times \dfrac{60}{40} = 5$ litres passed through heart in 1 minute

REGULATION OF THE HEART'S ACTION

Regulation of the heart must be considered in two aspects, as must many other functions of the body. The first is the maintenance of a constant state in resting conditions, where random fluctuations of activity would disturb the state of other parts of the body. The second is the production and control of changes in the heart's activity that occur in response to changes in other bodily organs or systems, for example, in exercise. One must further remember that both sorts of regulation, the maintenance of the *status quo* of rest, and the controlled variation of activity required in exercise, are performed reflexly. Receptors are stimulated to varying degrees and the altered pattern of nerve impulses from these receptors brings about the appropriate alteration in the heart's activity.

In addition, within the heart, there is one further regulatory process, independent of reflex nerve activity, which is a function of one of the general rules of all muscular contraction. In the heart this is referred to as the *Starling mechanism* or *phenomenon*.

STARLING'S 'LAW OF THE HEART'

Professor Starling's original enunciation of this stated that the energy of contraction in heart muscle was proportional to the filling pressure to which the heart was exposed in diastole. In plain English, the more the heart is stretched as it fills, the more powerfully it contracts. This is an immediate, beat-by-beat response to variations in venous return and results in a constant end-systolic volume. The cause of this phenomenon is analogous to that of the free-weighted frog's gastrocnemius muscle (and to all muscles in the living human body). If they are stretched, or loaded, while at rest, the 'slack' in fibrous and elastic tissue in series with the contractile material is taken up before contraction occurs (see Chapter 7). More of the energy of contraction is then expended in bringing origin and insertion nearer together, less being expended in merely tightening up the connective tissue. Something comparable to this is the cause of the mechanism in the heart.

The stabilising action of the Starling mechanism was originally studied in an isolated heart, pumping blood through an artificial circulation with no nervous reflex control system operating. In the living body the situation, due to the reflexes, is very different, as will be described in subsequent sections of this chapter. However, since the Starling mechanism is active in real life and plays a part in the normal adaptations of the heart in health and in disease, we will now analyse its action further.

Variations may occur in peripheral resistance, causing a change in arterial blood pressure (as described in Chapter 16). If the heart were to contract with constant force against a changed blood pressure, then the stroke volume would vary inversely as the blood pressure, falling as the pressure rose and vice versa. This would result in the end-systolic volume rising and falling with pressure changes. Venous return might continue unaltered by the fluctuations in arterial pressure. In fact, in the isolated heart (and presumably in the heart *in vivo*), the Starling mechanism results in the force of contraction being higher when the end-diastolic volume rises and lower when the volume falls. The result is that stroke volume remains constant in spite of changes in blood pressure in the arteries, though there would, consequent upon changed end-diastolic volume, be a similarly changed end-diastolic pressure, transmitted back through the atria into the whole venous circulation. The mechanism also operates when the venous return itself changes. The consequent changes in end-diastolic volume cause corresponding changes in force of contraction, so stroke volume changes in parallel with the changing venous return.

EFFERENT NERVES AFFECTING THE HEART

The involvement of both the vagus (parasympathetic) nerves and the sympathetic in controlling the rate of the heart by their effects at the sinu-atrial node has already been described.

The connector axons of the vagal supply to the heart originate in the dorsal nuclei of the two vagi and run from the brain-stem, through the neck and into the thorax within the main vagus nerves. They pass from these nerves to the heart and terminate on effector neurones adjacent to the s-a and a-v nodes. The effector neurones have short axons that supply the adjacent nodal and, to a small extent, the conducting tissues.

The connector sympathetic axons originate in the lateral horns of the upper thoracic segments of the spinal cord, entering the sympathetic chain in the white rami communicantes. They pass cranially to terminate in the three cervical sympathetic ganglia, on effector neurone cell bodies. The effector axons pass back down into the thorax by the cardiac sympathetic nerves (three on each side), supplying again the nodal and conducting tissues of the heart.

As has already been seen, the effect of vagal stimulation is to reduce the heart rate, while sympathetic impulses increase the rate of heart beat. Vagal stimulation can also directly cause a failure of the a-v node to respond to impulses reaching it through the atrial muscle, or cause a delay in its response (*complete or partial heart block*). The sympathetic accelerates conduction through the a-v node and the conducting tissue (Purkinje fibres). Other actions that have been ascribed to sympathetic or vagal stimulation are either only seen in experimental conditions or are the result of 'spill over' of neurotransmitters or of the simultaneous release of adrenaline from the adrenal glands. Thus sympathetic stimulation and adrenaline cause an increase in the force of contraction of cardiac muscle by enhancing Ca^{2+} influx and also a shortening of the duration of contraction. Increased coronary blood flow also occurs (almost certainly the result of increased metabolic activity in the heart muscle).

The action of the sympathetic nerves and of circulating adrenaline in affecting the force of contraction of cardiac muscle is as important in regulating stroke volume as is the Starling mechanism. In the latter mechanism (already described), variations in *end-diastolic* volume cause corresponding changes in contraction force and thus in stroke volume. Sympathetic stimulation changes also cause variations in force of contractions and thus similar changes in stroke volume and in the residual *end-systolic* volume. In the latter case (sympathetic stimulation), just as in the former (Starling mechanism), there can be no sustained increase in cardiac output unless there has been a preceding change in the return of blood to the heart by the great veins.

Changes in heart rate and force of contraction are reflexly produced in response to receptor stimulation and it is best to classify them with reference both to receptors stimulated and function served.

THE BARORECEPTOR REFLEX

This is concerned with the maintenance of steady-state conditions in the cardiovascular system while steady-state conditions obtain generally in the living body. Its prime effect is to maintain a constant arterial blood pressure and thus a constant supply of blood to the vital organs, the brain and the heart itself. In general this constancy is achieved by the negative-feedback control loop arrangement. The detection of an alteration in blood pressure automatically produces responses that cancel out the alteration, returning blood pressure to its original level.

BARORECEPTORS

These are the detectors of blood pressure. The name mistakenly implies that these receptors are directly sensitive to pressure changes. In fact, they are stretch receptors and are placed in elastic artery walls so that they are stimulated as the artery wall becomes stretched when it responds to a pressure rise. In a static situation, where the artery is stretched by differing internal pressures, the response rate (impulses/second) from a single receptor has been shown to vary with the degree of stretch (Fig. 15.7). The rapidity with which the pressure rises also determines, for tenths of a second only, the rate of impulse production.

Arterial baroreceptors are found chiefly in the arch of the aorta and at the origins of the internal carotid arteries where the carotid sinus may be seen as a smooth spindle-shaped swelling of the artery. They are also found at other sites between the aorta and the carotid bifurcation along the courses of the brachiocephalic and common carotid arteries (Fig. 15.8). Nerve fibres from the aortic arch receptors enter the vagus nerve and those from the carotid sinus region run in the sinus branch of the glossopharyngeal (IXth cranial) nerve. These afferent fibres terminate on cells in the dorsal nucleus of the vagus, the axons of which are the connector fibres of the vagus supply to the heart. The synapses are excitatory so that a rise in arterial pressure, causing increased stimulation of the baroreceptors, causes increased firing of the vagus nerve efferent fibres to the heart. This slows the heart and reduces the cardiac output, which causes the blood pressure to fall again. At the same time as the impulses from the baroreceptors' afferent fibres stimulate the vagal cardiac efferent fibres, they inhibit, through inhibitory interneurones, a group of cells spread diffusely through the pons and medulla, which normally excite the cells of

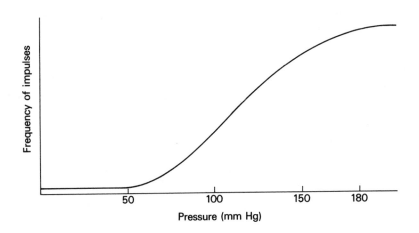

Fig. 15.7. The relationship between pressure in the arteries and the impulse frequency from an arterial baroreceptor.

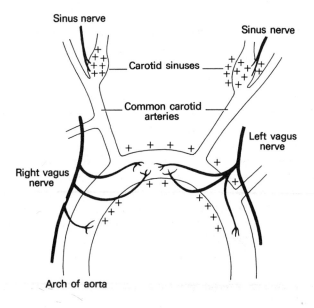

Fig. 15.8. The sites and innervation of the arterial baroreceptors.

origin of the cardiac sympathetic connector axons in the spinal cord. While the dorsal nucleus of the vagus along with other adjacent nuclei do truly contain the 'cardio-inhibitory' centre, it is not possible to be so precise anatomically about the 'cardiac accelerator' centre.

EXAMINING THE HEART

Cardiac catheterisation for measuring cardiac output has already been described. The catheter, once inserted, can be used for other investigations. The pressures in various chambers of the heart can be measured, as can that in the pulmonary arteries and veins, which provides information about the function of the heart valves. The composition of the blood can be determined. This, together with pressure movements, will indicate the presence and size of defects in the atrial or ventricular septa which normally separate completely the two halves of the heart. Either X-ray opaque contrast media and cine-radiography, or ultrasonic methods can be used to study the pumping action of the heart and the movements of its valves.

OTHER CARDIAC REFLEXES

While the baroreceptor reflex is the stabilising reflex in the cardiovascular system, it is not the only one that affects the heart. Others exist which are brought into action when various receptors are stimulated.

Low-pressure, or *volume receptors* exist in the terminations of the great veins and pulmonary veins in the right and left atria. Stimulation of these causes, among other effects, a stimulation of the sympathetic nerve supply to the heart and thus an increase in the heart rate. Thus, an increased venous return due, for example, to increased muscular activity (Chapter 23) will reflexly stimulate the heart to get rid of its increased load of blood by increasing its rate. This will also, of course, be assisted by the Starling mechanism. This effect was originally described by Bainbridge in 1915 but its reflex nature has been much disputed until the existence of both the receptors and the efferent sympathetic pathway were confirmed in the last decade (1965–75). Previous unsuccessful search for a vagal efferent pathway had led people to doubt the existence of Bainbridge's reflex and so for some time it was called the 'Bainbridge effect'.

Stimulation of the arterial chemoreceptors, pain receptors anywhere in the body, or the ears or eyes, or even neural activity resulting from emotional 'states of mind', can all affect the heart rate, mainly, as already described, through the alerting reaction centres in the hypothalamus and the midbrain (see Chapter 6). Usually these stimuli cause a rise in heart rate, but just occasionally the opposite occurs. Intense vagal stimulation may lower the heart rate to 30 or even 20 beats per minute and an accompanying vasodilatation is seen. The result is to lower the arterial blood pressure so much that blood supply to the brain is insufficient and the person 'faints'. This type of fainting attack is called a *vasovagal attack*.

16 · The Circulation of the Blood and its Regulation

A GENERAL DESCRIPTION OF THE CIRCULATION

In Man and other vertebrate animals the circulation of blood through the closed system of vessels pumped by the heart, the structure of which was described in Chapter 14, forms the main transport system for the living body. We must now study how the circulatory system works and how its functions can be varied to meet changing needs.

Since the system is a closed circuit, what happens in any one part intimately affects all other parts and it is therefore difficult to decide where to 'break in' to the circuit to begin a description of its function, for wherever one begins, one's description must continually refer both backwards and forwards to the other parts of the circuit. In this description we have decided to go straight to the heart of the matter and begin with the function of the heart as a pump. We will then proceed in a forward direction, following the blood as it flows successively through arteries, arterioles, capillaries, and veins and so back again to the pumping action of the heart.

THE CARDIAC OUTPUT

Each of the two ventricles of the heart in adult resting man, as already described, normally contain about 150 ml of blood just before they contract, and the ensuing contraction expels about 70 ml of this into the aorta and pulmonary artery. About 80 ml will remain in the chamber at the end of contraction. These three volumes, 150, 70 and 80 ml, are respectively called the *end diastolic volume*, *stroke volume* and *end systolic volume*. The *heart rate* is about 72 beats a minute at rest, so the volume of blood ejected each minute, the cardiac output, is about 5 litres.

ARTERIAL BLOOD PRESSURE

The systemic arteries convey blood to all parts of the body at a relatively high pressure, biologically speaking. As the left ventricle of the heart ejects blood into the aorta during systole, the pressure in this vessel rises to about 120 mm Hg. Since the aorta is elastic, it expands as the blood from the heart enters it. In elderly people, when the elasticity is reduced and the aorta more rigid, the systolic pressure may rise to 170 or 200 mm Hg.

During diastole of the heart, while blood flows away from the aorta and arteries into the smaller vessels, the pressure within them falls until, just before the beginning of the next systole, it reaches a level of about 70–80 mm Hg.

This is called the diastolic pressure and is due to the fact that during diastole the elastic tissue of aorta and arteries recoils. This helps to maintain the blood pressure since blood cannot re-enter the heart because the aortic valves are closed and it is therefore propelled onwards through the arterial system. In the less elastic vessels of the elderly, the diastolic pressure may be lower than in the young.

The rise of pressure in the arteries to the systolic value can be felt where large vessels run close to the skin. Well-recognised points for 'feeling the arterial pulse' are known, such as on the front of the wrist over the radius.

Systolic and diastolic pressures can be measured indirectly in arteries by an instrument called a *sphygmomanometer*. A hollow rubber cuff is wrapped around a limb, connected to a mercury or aneroid manometer and is then inflated by a hand-pump. When the pressure in the cuff exceeds the pressure in the arteries underneath it, these are collapsed and the pulse can no longer be felt in them. The pressure in the cuff at which the pulse disappears (or reappears if the cuff pressure is slowly released) is equal to the systolic pressure in the vessels.

At cuff pressures between systolic and diastolic blood pressures, the arteries under the cuff will be alternately opening and closing as the pressure in them rises and falls with the heart-beat. This causes turbulence in the blood flow and the turbulence produces noises that can be heard by a listener using a stethoscope over the artery. Blood pressure is normally measured in the brachial artery in the upper arm. The artery can be readily located on the front of the elbow (Fig. 25.5) and the bell of the stethoscope placed over it there. The sphygmomanometer cuff on the upper arm is then inflated to above systolic pressure and its pressure slowly reduced; a sharp tapping will be heard as the cuff pressure falls below systolic pressure. The sounds get louder and longer and become booming in quality as the pressure is progressively lowered. At the diastolic point, the sounds suddenly become quiet and muffled or cease altogether.

It will be appreciated that the systolic pressure will be affected partly by the stroke volume of the heart. Both systolic and diastolic pressures depend upon the elasticity of the arteries. The diastolic pressure is determined by the overall rate of 'run off' of blood from arteries through the arterioles into capillaries. Finally, it must be mentioned that the conventional method of recording the blood pressure is as a fraction, with the systolic pressure on top, e.g.

$$\text{B.P.} = \frac{120}{80} \text{ mm Hg.}$$

THE PERIPHERAL RESISTANCE

When we measure pressure along the arterial system, we find that it is much the same in all arteries (when corrected to heart level). In the arterioles it falls sharply so that in the capillaries it is only one-fifth of that in the arteries. This indicates that the arterioles are the major site of resistance to the flow of blood through the circulation. This resistance is due to anatomical considerations, the overall cross-sectional area of arterioles, their small calibre and their numbers being the important factors. The intrinsic viscosity of blood is also important in adding to its resistance to flow through the narrow arterioles. Physiologists, then, refer to the *peripheral resistance* (which is mainly due to the resistance to flow of the arterioles) and apply, to studies of blood pressure and flow, terms exactly similar to those of Ohm's Law in the study of electrical circuits. Arterial blood pressure is the driving force corresponding to volts in electricity. We measure flow rate, which is comparable to amperes and we talk of peripheral resistance units, pru's for short, analogous to ohms of resistance. Ohm's law is usually written: volts ÷ amperes = ohms. This is the same as amperes = volts ÷ ohms. In ordinary language, the current flow changes proportionately to the driving force, but in the opposite direction to changes in resistance. In the circulation, flow rate into the capillaries depends upon both arterial pressure and arteriolar resistance, increasing as pressure rises, but falling as resistance rises: flow rate = arterial pressure/arteriolar resistance. But one can just as well look back from the arteries to the heart and its output when arterial blood pressure becomes dependent upon cardiac output and total peripheral resistance. Now the same equation can be rewritten: arterial blood pressure = flow rate (cardiac output) × resistance. The student who needs a clear understanding of the circulation and how it is regulated must keep these two forms of the Ohm's Law-like equation clearly in her mind, and understand that blood pressure rises if either cardiac output or peripheral resistance rises and that flow into tissues rises with rising pressure, but falls as the (peripheral or arteriolar) resistance rises. Normally, of course arterial blood pressure remains constant (apart from the systolic-diastolic changes), flow into the arteries (cardiac output) being equal to flow through arterioles into the tissues of the body.

THE VENOUS RETURN

For understanding how the circulation works as a whole, the capillaries can be thought of as inert tubes joining arterioles to the veins. It is, though, very important to consider how the blood flows through the veins back to the heart, mostly against gravity. Various forces and mechanisms operate to overcome gravity and ensure that the venous return, as it must in a closed circuit with no reservoir or header tank, at all times balances the cardiac output. These are as follows:

Vis a tergo ('force from behind') is the Latin phrase given to the simple fact that, low though capillary pressures are when compared with arterial pressure, they still serve to drive blood continually into the peripheral veins. This entry of blood at their capillary end forces blood, then, along the veins towards the heart.

Venous tone, the state of partial contraction of the smooth muscle of the vein walls, helps to prevent a passive distension of veins which might then become engorged with blood. There is evidence that this contraction increases during exercise and it may increase whenever the venous pressure rises. This will reduce the capacity of veins and raise the rate at which blood returns to the heart after passing through the capillaries.

Venous valves permit only a one-way flow of blood in veins—towards the heart. Back-pressure in the veins cannot normally force blood backwards to the capillaries. When this happens due to defects in the valves, then varicose veins may be produced.

External compression of veins will also force blood along them towards the heart. Of particular importance is the contraction of skeletal muscles. Veins within and between contracting muscles will be compressed and the blood in them forced, because of the valves, towards the heart. So important is this action that it is called the *muscle pump* and the extensive venous plexus in soleus and gastrocnemius is particularly important in this respect. The raised intra-abdominal pressure that accompanies inspiration or vigorous expiratory efforts also acts in the same way.

Vis a fronte is the term given to two forward suction forces that aid venous return to the heart. The more important of these is the subatmospheric pressure within the chest, particularly in inspiration and most marked in rapid deep inspiration. This low pressure draws blood into the intrathoracic portions of the venae cavae from their more peripheral parts. Elevation of intrathoracic pressure, as occurs in a prolonged bout of coughing, will reduce or prevent the return of blood to the heart—this can be seen in the engorged vessels in the face during a paroxysm of coughing. As the ventricles of the heart contract in systole they pull down the fibrous ring between atria and ventricles and so enlarge the atria. This creates a suction force on the blood in these chambers and aids the entry of blood into them from the great veins of the chest. The more powerful the ventricular operation, the greater is this effect.

VENOUS RETURN AND CARDIAC OUTPUT

We have now come full circle and can see, at all times, the importance of the venous return in considering cardiac output. The heart has no reserve of blood (except the 80 ml in each ventricle at the end of systole) which it can expel into the arteries regardless of the venous return. It can only eject in one systole the blood that entered it in the previous diastole. This is why we have stressed in this account the closed nature of the circulatory system and the dependence of each part of the system upon that preceding (and in some cases, following) it. During the ensuing description of the regulatory mechanisms and factors that vary the performance of any part of the circulation, you should continually refer back to this description and bear in mind that the circuit is a closed one of various interlocking parts.

Having completed this description of the circulation, we can now discuss the various aspects of its regulation. This discussion is divided into four main areas:

(1) The first is not 'regulation' in the sense of active control of its function, but is an examination of the various factors that affect arterial blood pressure and so the flow of blood to the tissue of the body generally.

(2)   Then we will look at the ways blood pressure is regulated by factors acting on the whole circulation.

(3)   Next we will be concerned with effects that act on all regions of the circulation but are usually found or studied when acting locally or in specific regions.

(4)   Finally we will study specific different regional circulations where the functions served result in the development of special features in the local circulation and in its control.

FACTORS AFFECTING THE ARTERIAL BLOOD PRESSURE

The arterial blood pressure is the one factor of overriding general importance in determining the delivery of blood to each and every part of the body, so in this section we must study those factors that affect it.

THE CARDIAC OUTPUT

The volume of blood that enters the arteries from the heart (the cardiac output) is one of the two most important factors determining arterial blood pressure, the other being the peripheral resistance (see below). Cardiac output, as already described, is itself the product of heart rate and stroke volume and also depends on the volume of the immediate preceding venous return. Changes in cardiac output, however produced, will inevitably produce changes in arterial blood pressure *unless*, as almost invariably happens, they are but part of more general circulatory adjustments that themselves offset any possible change in blood pressure.

CIRCULATING BLOOD VOLUME

Blood volume changes would be reflected in changes in blood pressure in a non-regulated system, since the volume is contained in an essentially elastic system. Extreme changes in blood volume do, indeed, cause changes in arterial blood pressure, as is seen after haemorrhage or excessive sweating causing water loss from the body. Blood volume changes will, of course, also affect the cardiac output and hence the blood pressure.

ARTERY WALL ELASTICITY

The chief effect of changing elasticity is on the difference between diastolic and systolic blood pressure (the *pulse pressure*) as the heart ejects blood into the arterial system. The less elastic the arteries become (as we get older) the more does systolic pressure rise as blood is ejected from the heart. On the other hand, in diastole, when blood continues to leave the arteries for the tissues, if the arterial elasticity is reduced there will be less recoil effect in the arteries to maintain diastolic blood pressure. You must imagine a situation in which the tube system (the arteries) is receiving blood intermittently at one end (the heart), but losing it steadily at the other end (the capillaries). The more the system can stretch to accommodate the stroke volume from the heart, the less the pressure will rise. The more it can recoil upon the falling volume in the tubes in diastole, the less the pressure will then fall. Therefore, the average blood pressure will not be much affected by the change of elasticity (except for one very important effect, described later on p. 463).

PERIPHERAL RESISTANCE

As has already been described, the arterioles are the main region in the circulation where resistance is offered to blood flow. The chief determinants of the resistance are blood viscosity (which does not usually change much) and the average diameter of the arterioles. Narrowing of the arterioles raises their resistance to flow which, if the cardiac output continues unaltered, causes a rise in arterial blood pressure.

The arteriolar wall is largely composed of smooth muscle which is at all times in a state of partial contraction. This can be maintained without any nerve impulses reaching it, but in most arterioles of the systemic circulation, and in the larger arteries and all veins as well, the contraction of smooth muscle is increased by nerve impulses in the sympathetic effector, noradrenergic axons which supply them. This nerve supply is, then, a powerful way in which the peripheral resistance can be altered by the autonomic nervous system and forms the basis of the reflex regulation of arterial blood pressure which will now be described.

REGULATION OF ARTERIAL BLOOD PRESSURE

SYMPATHETIC INNERVATION OF BLOOD VESSELS

As has already been described, blood vessel smooth muscle is one of the major tissues innervated by the sympathetic division of the nervous system. This innervation is commonly known as the *vasomotor nerve supply* of blood vessels. Anatomical details of origins of connector axons, site of synapses in effector cell bodies and course of effector axons to the blood vessels are described in Chapters 6 and 14.

Arteries and veins are innervated, as well as arterioles, and contraction of smooth muscle in these larger vessels usually accompanies that of the arterioles. In arteries this adds a small contribution towards increasing total peripheral resistance. In veins, while not adding to flow resistance, contraction of the smooth muscle has the important result of reducing the capacity of the veins. In the resting person, two-thirds of the blood volume is in veins. Given a blood volume of 5 litres and a cardiac output of 5 l/min this means that 3⅓ litres are present in the veins and must take 40 sec to traverse them. Venous contraction can reduce their capacity to half the resting size so $1\frac{2}{3}$ litres can be made available for other purposes and the cardiac output of 5 litres can be achieved in 40 sec ($6\frac{2}{3}$ l/min). This alteration is not accompanied by any significant increase in resistance to flow which the naïve might think would accompany venous constriction.

In all vessels the sympathetic effector axons cannot be traced farther than the outermost part of the smooth muscle layer of the vessel wall. Here, simple neuromuscular junctions are seen, filled with vesicles containing noradrenaline. These junctions are predominantly α-type, though in some cases, in blood vessels of skeletal muscle there may be β-type noradrenergic junctions. Once action potentials have been increased by the nerve impulses to the innervated muscle fibres, these are transmitted to adjacent non-innervated fibres, much as the cardiac muscle action potentials are transmitted. Similar junctions to the intercalated discs or cardiac muscle have been seen in smooth-muscle tissues where they are called *nexuses*.

Under 'resting' conditions, the effector axons convey about 1 impulse/sec, but when intensely stimulated in a reflex response, they may carry up to 8 impulses/sec. Most of the noradrenaline released is subsequently taken up again by the nerve terminals and reused. Some may be locally destroyed and some leak into the local capillaries and thus enter the circulating blood.

The impulses in the vasomotor nerves can be traced back by various means to their origin. This is shown to be in a group of scattered neurones within the pons and medulla of the brain-stem. Experimental cutting of the brain-stem above this region does not affect the impulses in the vasomotor nerves; cutting in the lower medulla abolishes them. Stimulation in this region may enhance or diminish the vasomotor nerve activity. Physiologists speak of a *vasomotor centre* in the upper medulla, but this, though convenient, is not, from the anatomical point of view, a well defined mass of grey matter but a diffuse region sharing many functions of the ANS. However it is of use when considering reflex control of blood pressure which we will now proceed to do.

REFLEX REGULATION OF BLOOD PRESSURE

The baroreceptor reflex is of great importance here. The receptors and the afferent nerve pathway have already been described in the previous chapter. As well as passing to the dorsal nuclei of the vagus nerves, they pass to the vasomotor centre neurones, where the impulses coming from the receptors are inhibitory in function. The increase in impulse frequency caused by a rise in arterial pressure would thus, by inhibition, *decrease* the frequency of impulses emerging from vasomotor centre neurones, and thus reduce the activity of vasomotor nerve fibres. Arterial, arteriolar and venous relaxation follows. Arteriolar relaxation reduces the total peripheral resistance, allowing blood to flow more freely from arteries to capillaries and thus reducing arterial blood pressure directly. The venous relaxation reduces the rate at which this blood is returned to the heart, causing a consequent reduction in cardiac output (the heart rate and force of contraction will simultaneously have been reduced by the baroreceptor afferent impulses acting also on the vagal cardiac nerves). This also lowers blood pressure.

One everyday way in which this reflex is activated is seen in the changes that occur when we move from the lying to the standing positions. In the former, the leg veins are at about the same level as the heart so blood flows freely through the veins of the legs and abdomen on its way back to the heart. Similarly the carotid sinus baroreceptors are at heart level and the pressure in them will equal, during systole, the pressure in the heart. In the erect position, especially when the stance is such that muscular activity is minimal, gravity opposes the return of blood from most of the body to the heart and within a minute of standing up, a litre of blood may have accumulated in the veins. At the same time as this reduction in venous return and effective blood volume, the carotid sinus is raised some 25 cm above heart level and the pressure to which its baroreceptors are exposed consequently will be about 17 mm Hg lower than the actual pressure reached in the left ventricle. The reduced venous return results in a reduced stroke volume and so the pressure achieved in the arteries falls. The result of all these factors is a considerable fall in carotid sinus pressure and a smaller fall in that of the aortic arch. Both sets of baroreceptors reduce their rate of impulse production (since impulse production depends upon the level of pressure). The vasomotor centre neurones are released from the in-hibitory impulses from the baroreceptor afferent fibres. Consequently more impulses leave them and are transmitted by the sympathetic vasomotor nerves to arterioles (and arteries and veins) throughout the body. At the same time the stimulation of the dorsal nucleus of the vagus by afferents from the baroreceptors is diminished while the sympathetic cardiac neurones are also released from inhibition and so the heart rate and force are increased.

The arteriolar constriction raises peripheral resistance and so raises arterial blood pressure, the venous constriction reduces again the volume of blood pooled in the veins and the changes in the heart raise its effectiveness in dealing with the blood that it receives from the veins.

It takes a few seconds for the reflex to operate, as we all may experience when we get up suddenly. The slight dizziness and 'spots before the eyes' on these occasions are due to a transient reduction in brain blood flow consequent upon the fall in blood pressure. One finds usually that this lasts for about 5 seconds and then passes. A person with loss of autonomic nerve fibres, afferent or efferent (diabetes mellitus is a common cause), may lose his baroreceptor reflexes and the dizziness may be severe and incapacitating whenever he attempts to stand up.

OTHER CIRCULATORY REFLEXES

The medullary vasomotor centre is probably involved in the vasoconstriction that results from stimulation in the alerting reaction centres (described in Chapter 6), though the vasoconstriction in this case is not so widespread, the blood vessels of skeletal muscle not being involved in the response. Details about the nervous control of these blood vessels will be described later in this chapter. The coronary and cerebral circulations have little if any vasoconstrictor innervation, and that in the cutaneous vessels, although taking part in baroreceptor reflexes, is also involved in important reflexes of temperature regulation. All these will be discussed later in this Chapter.

HIGH BLOOD PRESSURE OR HYPERTENSION

At all ages a wide range of blood pressures may be found in healthy people, though in increasing age the average pressure rises (some individuals, however, have the same blood pressure at 60 as at 20 or 30). It is difficult, therefore, to define the normal blood pressure since many have a blood pressure above average with no ill effects. On the other hand, hypertension *can* produce definite pathological effects in patients, either as a disease in its own right the cause of which is not known (*essential hypertension*), or as a result of other diseases, such as chronic nephritis, toxaemia of pregnancy or tumour of the adrenal gland. Essential hypertension accounts for 90% of all people with a raised blood pressure.

Although many fit people, with nothing else wrong with them, live for many years with a blood pressure higher than usual, statistics do show that, on average, the higher the pressure, the shorter the life-span. Much research has therefore been done into the causes, consequences and treatment of high blood pressure and some facts have begun to emerge.

First, in a case of 'benign, essential hypertension'—high blood pressure of no known cause and producing no harmful effects—the baroreceptor reflex functions normally *but at the new raised pressure level*. Some degree of fibrous replacement of the elastic tissue of the large arteries is usual in such cases. It is thought possible that these two observations are related in that if elastic tissue is replaced partly by fibrous tissue in the aortic arch and carotid sinus, then a higher level of pressure would be required to produce the same number of impulses in the afferent nerves of the reflex arc. This would result in a resetting, by the baroreceptor reflex mechanism, of arterial blood pressure at a higher level.

The kidneys are concerned in elevation of blood pressure. Chronic kidney disease that leads to fibrous scar tissue formation is associated with a raised blood pressure. Normal kidneys contain and can release the enzyme *renin*, which acts on a protein in blood, splitting off a fragment called *angiotensin*. This material is the most powerful vasoconstrictor known, producing, molecule for molecule, about eighty times the rise of peripheral resistance as is produced by noradrenaline. But in smaller doses, angiotensin has other actions that might cause elevation of blood pressure. It causes, indirectly, the retention of salt and water by the kidneys and thus an increase in blood volume, itself capable of elevating blood pressure. Then it seems to have actions within the central nervous system, the end result of which may be a rise in blood pressure. The sodium retention also renders arteriolar smooth muscle more sensitive to the constrictor action of noradrenaline.

While all these have been observed in experimental conditions, the precise physiological mechanism of both benign essential hypertension and the hypertension of kidney disease remain undetermined. If the raised blood pressure, however caused, is thought to endanger the patient's health or life, then drugs affecting the sympathetic vasoconstrictor nerves may be used to lower it. Noradrenergic receptor blocking agents might be thought the drugs of choice, but for some unexplained reason, β-receptor blockers, acting perhaps primarily on the heart, are more effective than α-receptor blockers, acting on the arterioles. Drugs that encourage the kidneys to excrete more salt and water from the body are certainly effective in lowering blood pressure, presumably by lowering the circulatory blood volume or reducing the sensitivity of arterioles to noradrenaline. This observation has led to the suggestion that the primary cause of benign essential hypertension lies in the regulation of extracellular fluid volume and sodium content rather than in the altered baroreceptor sensitivity described above.

Thus one can see to what extent a knowledge of the physiological processes can be involved in understanding disease states and their treatment, although there are still considerable gaps in this field. This discussion of hypertension as a disease state leads on, inevitably, from a consideration of baroreceptor reflexes to the second main regulatory mechanism controlling blood pressure, regulation of the volume of circulating blood.

BLOOD VOLUME REGULATION

That blood volume is maintained, in the normal person, at around 5–6 litres is undoubted, but exactly how the regulatory process works is unclear. The kidneys' removal of salt and water from the blood is probably the final stage in the control process, but exactly how the agents (see chapters 21 and 26) that affect the kidneys are called into action is not known clearly. Renin production is one of these agents. The low-pressure or volume receptors of the great veins and atria of the heart may be another. Whatever the details of the mechanisms, they are bound to follow the same general pattern of negative feed-back control loops. Changes in blood pressure (or volume) will affect a detector that will produce a signal (? nerve impulses ? renin). This signal will activate some mechanism that produces change in blood pressure or volume opposite to that which originally affected the detector. All these steps always occur automatically, being 'built in' to the body's structure so that only the required reaction can be produced once the receptors have been stimulated.

LOCAL REGULATION OF BLOOD FLOW

AUTOREGULATION

In many of the different tissues and organs studied experimentally by physiologists, self regulatory phenomena in the supply of blood to the tissues have been discovered. What we find in experiments is that, when one deliberately alters blood pressure in the vessel(s) supplying an organ, the blood flow does not follow the blood pressure but remains, within limits, more or less constant. This happens equally in wholly denervated bits of the body as well as in innervated regions, so cannot be due to any way to vasomotor nerves. It can only be explained by the suggestion that arteriolar muscle tone varies with the arterial blood pressure, causing the local peripheral resistance to be high when pressure is high and low when it is low. Two possible mechanisms have been suggested to explain this alteration of arteriolar tone. One is that it is an inherent property of the smooth muscle to contract more vigorously when more severely stretched and vice versa. This is, in a sense, no explanation, but merely a description of what happens. The other is that altering the blood flow alters the supply of some essential nutrient, or the removal of some waste product, and that arteriolar smooth muscle is very sensitive to the concentration in the tissue of this nutrient or waste material, responding by contraction (causing reduction in blood flow) or relaxation (causing increase in blood flow) in such a way as to keep the concentration of this material at a constant level. There is little actual evidence for this hypothesis, though it can be seen that it fits the usual negative feedback control loop mechanism that we have seen operating elsewhere. In order to accept this as the mechanism that underlies autoregulation, we need to detect materials that are formed or used in metabolic processes, the concentrations of which vary in the tissue fluid adjacent to arterioles when blood flow alters, and to which the arteriolar muscle is sensitive.

REGULATION BY ALTERATIONS IN METABOLISM

In the case of the blood vessels of skeletal muscle and those of some glands there is clear, albeit not yet conclusive, evidence that metabolic changes cause changes in blood flow. *Bradykinin* is a substance that is formed by active salivary and sweat gland cells. It diffuses readily into the tissue fluid and is a potent relaxer of arteriolar smooth muscle. Through its action it is thought the blood flow through these, and perhaps other, glands is increased when they are metabolically active and therefore need an increased blood flow.

Bradykinin formation has not been detected in active skeletal muscle. In fact the cause of vascular dilatation in active muscle is a much researched area which still defies explanation. Some workers have produced evidence that a decrease in tissue-fluid oxygen combined with a rise in potassium are the cause, others champion a rise in inorganic phosphate, others the escape from active muscle of ATP or one of its breakdown products, while yet others claim that an overall rise in osmotic pressure in tissue fluid is the stimulus causing the arteriolar dilation. It is unlikely that CO_2 or heat, the primary waste products of metabolic activity, or lactic acid, produced when energy requirement exceeds the capacity of oxidative mechanisms of energy provision, are in any way responsible. Although they are the substances most likely to be concerned, experiments have failed to prove their involvement. As in the case of proving a chemical cause for autoregulation, concentration changes must be observed in tissue fluid of some substance that is shown to have the required effect on the arteriolar smooth muscle.

One thing that is certain about the changing blood flow in skeletal muscle is that while vasomotor and vasodilator nerves are both present, the major part of the change in blood flow is produced by some change in chemical composition of tissue fluid which is itself caused by changes in metabolic activity. More will be said about this in the section on the muscle circulation and in Chapter 23.

CHOLINERGIC AUTONOMIC VASODILATATION

Parasympathetic nerves supply the salivary glands and the effector neurone transmitter is acetylcholine. The sympathetic effector neurones supplying sweat glands are also cholinergic. Finally there is evidence (to be dealt with in more detail later) that the sympathetic system may supply cholinergic vasodilator effector neurones to the smooth muscle of blood vessels of skeletal muscles. In all of these cases it has in the past been suggested that acetylcholine-induced vasodilation accompanies the metabolic activity of the tissues, but in all cases such mechanisms are now seen to be irrelevant or non-existent. Experimentally administered acetylcholine is, though, a powerful vasodilator when applied to almost any arteriole in the body.

REGIONAL CIRCULATIONS

THE CEREBRAL CIRCULATION

The metabolic activity of the brain is constant and high. A large and relatively constant blood flow is necessary to sustain this metabolism. Cerebral vessels have little vasomotor innervation, so the cerebral circulation plays only a small part in the baroreceptor or other reflex responses. The cerebral vessels are responsive to changes in both the CO_2 and O_2 content of the blood. Increased CO_2 and lowered O_2 promote a vasodilatation and increased cerebral blood flow. The opposite changes in these blood gases produce a reduction in flow. It is in the cerebral circulation that autoregulation is most strikingly seen, the blood flow being independent over a wide range of the arterial blood pressure.

Further analysis of the action of CO_2 on the blood vessels reveals that it is not the blood CO_2 content, nor the blood acidity that causes the arteriolar muscle to dilate (remember that CO_2 in blood, or any watery medium, reacts with water, producing carbonic acid, H_2CO_3). It is when the CO_2 diffuses into the brain tissue fluid and cerebrospinal fluid and forms H_2CO_3 in these fluids that it affects the blood vessels, causing dilatation. Since these fluids have little buffering power, a small change in CO_2 content will produce much larger changes in acidity than in blood, where the proteins (haemoglobin in particular) react with the H_2CO_3 to neutralise it (for further discussion of this subject see Chapter 18).

Voluntary or emotional overbreathing while at rest results in reduction of blood CO_2 content. This causes constriction of cerebral blood vessels and a reduction of supply of O_2 and nutrients to the brain. The result is dizziness and a clouding of consciousness. This can be made much worse by various manoeuvres to reduce the cerebral circulation, for example squatting, holding the breath and then standing up suddenly. Unconsciousness and even convulsions due to cerebral anoxia can result from these dramatic but largely harmless procedures, which are not, however, recommended. Simple breath-holding attacks also may produce unconsciousness and convulsions, though here the accumulation of CO_2 will result in dilated blood vessels and a high cerebral blood flow. Breathing invariably starts again due to the CO_2 build-up (Chapter 19) before any damage has occurred. Finally, the true fainting attack causes unconsciousness and occasionally convulsions due to reduced cerebral blood flow. Intense emotional or painful stimulation may suddenly cause a reversal of the usual response of cardiac acceleration and peripheral vasoconstriction. The heart rate falls to 20–30 beats/min, and all sytematic arterioles are dilated. Blood pressure falls to very low levels; so low that the autoregulation process in the cerebral circulation cannot maintain an adequate blood supply to the brain, even though the blood composition is normal and the tissue and cerebrospinal fluids are accumulating CO_2. People suffering from this sort of faint, also known as a vasovagal attack, are seriously at risk, unlike those affected by the two conditions previously described (voluntary overbreathing and breath holding) and action must be taken *at once* to establish an adequate blood supply to the brain. This can only be achieved by placing the victim flat on the floor, head to one side and, if possible, raising the legs on a chair or bench. Monitoring the heart rate will indicate when it is safe to restore the person to a normal posture, which may be several minutes after he has regained consciousness. He should not be moved until a normal heart-rate is recorded, except for such emergency procedures as stopping bleeding (which may be the cause or a consequence of the fainting) or ensuring a clear air-way from lips to trachea. A failure of cerebral circulation lasting more than 2–3 minutes will result in permanent brain damage. After 4–5 minutes deprivation, so much cerebral damage has occurred that death usually follows.

CORONARY (CARDIAC) CIRCULATION

Of equal importance to the brain in sustaining life is the heart. Coronary vessels, like cerebral vessels, do not have a functional vasomotor nerve supply. Coronary blood flow is closely linked to heart rate and contraction force, presumably through some product of metabolic activity. A further complication is that the flow of blood through the smaller arteries and in capillaries embedded in the heart muscles is severely limited when the cardiac muscle contracts. Most of the coronary artery blood flow and therefore nutrition of the heart muscle can occur only in diastole. A heart beating 70 times/min spends 40 sec of that minute in diastole, but one beating 150 times/min spends only 22.5 sec in diastole. Doubling the metabolic activity is thus accompanied by halving the time available for coronary blood flow. Any mechanical interference with blood flow in the coronary arteries will therefore matter particularly when the heart rate is fast.

SKELETAL MUSCLE CIRCULATION

Some of the material in this section is repeated and amplified in Chapter 23, on the effects of exercise. No tissue changes in its metabolic rate, and therefore its requirements for O_2 and nutrients, to an extent that is more than a small fraction of the changes seen in skeletal muscle. In extreme activity the metabolic rate may be 100–150 times its resting value. It is very difficult to obtain measures of blood flow in active muscle but it is thought to rise to 10–20 times the resting value. Certainly the number of open capillaries can increase by that factor. Furthermore, the extraction of oxygen (and presumably of other materials) from each unit volume of blood can be doubled during exercise.

In man there are two sets of sympathetic effector neurones supplying the blood vessels of skeletal muscle. The first is the noradrenergic vasoconstrictor pathway which is involved in the baroreceptor reflexes, along with most other vascular beds in the body. The second is a cholinergic pathway, which is also anatomically within the sympathetic system. Impulses in this pathway originate in the alerting reaction centres and result in a modest (up to 4 × resting) increase in blood flow. This pathway is activated by the anticipation of exercise and also by emotional stresses such as simulated oral examinations in which the candidate is always wrong! Naturally, not many studies have been performed to confirm the existence of this pathway; its cholinergic nature has been defined by giving atropine which abolished the stress-induced muscle vessel dilatation. Some people however, claim that this vasodilator innervation is not cholinergic, but beta-receptor mediated noradrenergic, since beta-receptor blocking drugs also abolish its activity.

As has already been described in the section on circulatory regulation by alterations in metabolism, by far the most important factor in causing muscle blood flow to increase is the change in tissue fluid concentration of some metabolism-related chemical material. This substance has not been identified, nor can it with methods at present available. We know it is not the two most likely candidates, CO_2 and lactic acid. Blood CO_2 changes similar to those caused by exercise can easily be produced experimentally by breathing CO_2 rich air. Carbon dioxide being freely diffusible, is at all times present in tissue fluids in concentrations equal to those of blood. Such changes do not evoke any significant increase in muscle blood flow. The vasodilatation is brisk in mild and moderate levels of exercise in which no lactic acid is produced. Therefore something other than lactic acid must be causing this rise.

Failure of the circulation to maintain metabolism at an adequate level has two main effects. Muscular weakness develops simply because energy in a form suitable for the contractile machinery (ATP) cannot be made as fast as it is used. The second effect is a sensation of pain. This is initially a dull ache, but crescendos to a severe cramp-like pain which rapidly stops further attempts at exercise. This state is seldom reached in normal people in whom breathlessness is the usual limit to exercise, but may be seen when arterial disease restricts the muscles' blood supply.

THE CUTANEOUS CIRCULATION

The metabolism of skin is small and constant, and usually the blood supply is more than sufficient for its needs. Close under the stratum basale of the skin there is a rich network of small venous channels. Blood enters these not only by capillaries but also by arteriovenous anastomoses (Fig. 14.10). Through the latter, when dilated, large amounts of blood can flow rapidly into the venous plexus, by-passing the capillaries. Both the normal arterioles and the A-V anastomoses are supplied with sympathetic effector, noradrenergic vasomotor nerve fibres. The connector cells of these, in the lateral horns of spinal cord grey matter, receive impulses, not only from the medullary vasomotor centre but also from the temperature regulatory centre in the posterior hypothalamus. This centre is excited by cold-sensitive receptors, found mainly in the skin. It is, however, inhibited by heat-sensitive receptors present in the anterior hypothalamus. Heating these receptors (which occurs normally whenever body temperature rises) will inhibit the posterior hypothalamus cells and thus reduce the excitation of the connector sympathetic neurones of the skin blood vessels. (It may be that there is an inhibitory pathway direct from the heat-sensitive receptors to the connector cells in the spinal cord, but it has not yet been defined.) The cold skin receptors excite the posterior hypothalamic neurones and thus the connector neurones. These two sets of receptors thus control skin circulation which can vary (in the hand) from < 1.0 to about 100 ml blood per 100 ml hand tissue per minute. These variations will cause similar variations in delivery of heat to the skin and so heat loss by radiation, conduction and convection from the skin to the environment. The deep (hypothalamic) heat receptors can detect changes of < 0.1 °C in blood temperature and with this change cause marked changes in hand skin blood flow. Changes in mean skin temperature of the trunk on the borderline of perceptibility can also cause marked changes in hand blood flow. The latent period for this reflex has been found to be about 12 seconds.

In all regions of the body inflammation causes the same changes in blood flow, but these are most easily studied in human skin. Damage provokes the '*triple response*', the three components of which are capillary dilatation, arteriolar dilatation and wheal formation. This will be fully described in Chapter 22.

Later aspects of the inflammatory reaction include the movement of leucocytes into the inflamed region, sometimes death of tissues and leucocytes leading to pus formation and frequently the final formation of a knot of fibrous scar tissue. All these stages of the inflammation reaction can be seen in the skin after a minor injury, in an acne pustule or in a boil, where they are but examples of inflammation which may occur in any tissue of the body with very similar results.

BLOOD FLOW IN THE KIDNEYS

Although the renal blood vessels have rich a vasomotor nerve supply, they also show autoregulation in their blood flow. Two main functions are served by the kidney's blood circulation. The first is that of providing an ultra-filtrate of plasma, the second is the provision of materials for the metabolically active tissues.

Filtration proceeds best under pressure and each renal arteriole is interrupted by a network of capillary vessels, the glomerulus, in which the pressure may be as much as 45 mm Hg. Filtration of plasma occurs in the glomerulus, a more or less protein-free filtrate being formed. The arterioles leaving the glomeruli open into capillary networks surrounding the kidney tubules, in which the filtrate from the glomeruli is altered and concentrated to form urine.

PORTAL CIRCULATIONS

A *portal vein* is one which, having been formed by the joining of capillaries to form venules, smaller and larger veins, conveys blood directly to another organ, in which it breaks up into a second set of capillaries. Thus the blood that has flowed through the intestinal epithelium, and absorbed the products of food digestion is collected into the *hepatic portal vein*. This vein then enters the liver, where it breaks up once again, and the blood passes along with the hepatic arterial blood, into the hepatic *sinusoids*—wide, capillary-like channels (not lined by endothelium) between columns of liver cells. In this second capillary circulation the blood gives up materials it has absorbed from the intestine to the liver cells and poisonous substances are removed from it.

Blood vessels in the gut and spleen have a rich sympathetic vasomotor supply, but arterial flow in the liver is controlled by autoregulation. There is no control of the portal venous supply to the liver, that organ receiving all the blood that has flowed through the gut and spleen. About one quarter of the total cardiac output passes through the liver and one quarter of the liver blood flow comes via the hepatic artery and three-quarters via the portal vein.

Another portal circulation is found between the hypothalamus and the pituitary gland at the base of the brain. The pituitary is an endocrine gland (Chapter 26), many of whose functions are controlled by nerve cells in the hypothalamus. These cells liberate from their axon terminals polypeptide materials which enter the local capillaries. These join to form veins which convey the blood from the hypothalamus into the pituitary gland. In this they break up into capillaries again and the materials liberated from the hypothalamic nerve terminals exert their controlling effect upon its activity.

ARTERIAL DISEASES

Blood flow to various regions may be disturbed by diseases of the arterial wall. These have already been mentioned in some specific instances. We wish here to make some general comments on the most important of these conditions.

Atheroma can affect any artery, from the aorta down to vessels about 1 mm in diameter. It is a disease of the intima, characterised by thickening with the appearance of fatty deposits, fibroblasts and collagen deposition, endothelial damage and platelet adhesion leading to thrombus (clot) formation on the lesion. Many factors contribute to its development. These include:

1   Turbulent blood flow, inducing vibration of the vessel wall.
2   Eating of foods, probably those containing saturated fatty acid fats and cholesterol that both increase platelet adhesiveness and damage the intima.
3   Cigarette smoking.
4   Raised arterial blood pressure, intermittent or continuous.
5   'Hereditary Factors', including the sex. Males are more often affected than females, in whom the disease develops after the menopause.
6   Lack of exercise.

The effects of the condition are:

1   The increase of blood flow to exercising skeletal and cardiac muscle is restricted. Maximal exercise performance is therefore also restricted.
2   Basal blood flow to tissues is limited, causing damage. In the skin (usually of the feet) gangrene may develop. In the brain, the loss of neuronal function may cause a variety of changes, incorporation of new material into memory and control of movement being most affected. The state is known as atherosclerotic or senile dementia.
3   Thrombosis (clot formation) may occur (see p. 482), producing sudden deprivation of blood supply. This is most frequent in coronary vessels (heart attacks) or cerebral vessels (strokes and 'transient ischaemic attacks'), though any other vessel may be affected with equally disastrous results.

INVESTIGATING ARTERIES

Injection of X-ray opaque materials and cine-radiography is still the most widely used method for locating the site of arterial disease. It has the disadvantage that the diseased vessel must be perforated and foreign material injected into it. Alternative, non-invasive techniques are being developed. The Doppler ultrasound method is the best of these, since it can be used to determine the diameter of a vessel and the average linear velocity of red blood corpuscles into it, and hence the volume flow of blood in the vessel. The possible use of electromagnetic flow meters has also been explored since the late 1950s but these are still not available 'off the shelf'.

THE PULMONARY CIRCULATION

Pressures throughout the pulmonary circulation are lower than in the systemic, particularly on the arterial side. Systolic pressure in the pulmonary artery trunk is only 20–25 mm Hg and diastolic pressure 6–12 mm Hg. Pulmonary capillary pressure is 5–10 mm Hg at all times lower than capillary osmotic pressure. There can thus be no tissue fluid formation by filtration through capillary walls. Left atrial pressure is normally about 5 mm Hg. All these pressures are, of course, relative to atmospheric pressure, the zero pressure reference level from which they are measured.

This has two consequences in lung blood flow. The first is that the bases of the lungs, in the upright position, receive substantially more blood than do the apices. Pressures relating to the height of the column of blood above or below the roots of the lungs will have to be added or subtracted from the pulmonary arterial trunk pressures, in order to determine the true perfusion pressure, and therefore blood flow, of the different portions of the lungs.

The second consequence is that if artificial respiration is carried out with a positive pressure, i.e. blowing air from a pump into the lungs, then this must not be so high (10 mm Hg or over) as to embarrass pulmonary blood flow.

The pulmonary circulation is supplied with sympathetic vasomotor nerves. These take part in the baroreceptor reflexes constricting the vessels when pressure falls and allowing them to dilate when it rises, just as in most systemic vascular beds. The response to chemoreceptor stimulation (by oxygen lack) is a brisk vasoconstriction, which is backed up by a direct excitatory effect of O_2 lack on pulmonary arteriolar muscle in the absence of nervous impulses. Details concerning the pulmonary circulation and gas exchange in the lungs are left to Chapter 19.

17 · Blood, Lymph and other Body Fluids

The living body contains about 60% water in men and 50% in women. As well as being the medium within cells in which chemical substances exist and react, it bathes all living cells and is the medium through which nutrients and waste products must move to and from cells; finally water is the medium in which the circulating fluids of the body (blood, lymph, cerebrospinal fluid and aqueous humour) transport substances to and from different organs and systems of the living body.

The water of the body can thus be divided into separate *compartments*. A compartment in the physiological sense is the term used to denote the spaces in which a particular type of fluid exists. The *intracellular compartment*, for example, means the sum total of all the water within all the cells of the body. About 40% of the whole body mass is intracellular water. *Tissue fluids* surround living cells. For the most part, tissue fluid does not exist as a free watery pool in which cells float, but is enmeshed in a network of connective tissue fibres and may in fact be a thin gel rather than a free flowing liquid. It makes up about 15% of body mass. *Blood and lymph* are similar, but not identical, to tissue fluid in composition; they are contained, in the mammal, in a closed circulatory system. Together their water accounts for 5% of body mass. These are the forms in which water is the medium for the bulk transport of materials between the different organs and systems of the body. Finally, there are the *transcellular fluids*, which have been secreted by living cells. They may serve a transport function (cerebrospinal fluid and aqueous humour) or be lubricating in function (synovial fluid of joints and the fluid of serous spaces such as the peritoneal, pleural and pericardial cavities). These account for less than 1% of body mass.

MEASUREMENT OF VOLUMES

For the major three compartments, the technique of indicator dilution is used. A known mass of a substance is added, time allowed for complete mixing and then a sample taken from the fluid. From the concentration of the substance and the known amount added, the volume of the fluid can be determined. Of course the substance must be harmless, be detectable by a chemical or physical means, be unaltered and remain solely within the compartment being measured. In fact, the last criterion is never achieved completely because all substances, when introduced into the body, no matter how harmless or how similar to normal materials, will be eliminated gradually from the body. We will describe now one example in detail and indicate the way in which all the compartments' volumes are determined.

Cane sugar or sucrose will not enter or pass through cells except those of blood capillaries so if it is injected into the blood it will eventually become distributed solely in blood plasma (the watery medium of blood), lymph, and tissue fluids. It is, however, slowly lost to the body in the urine. One hundred and fifty mg of sucrose is injected into the circulation of a 70 kg man. When mixing with tissue fluids is complete, the concentration of sucrose will now be found to be 1.0 mg/100 ml plasma water and 10 mg will have been lost in the urine. 140 mg remain in the body. They must therefore be distributed in 140 × 100 ml water. This volume, 14 litres, is the total of blood and tissue fluid water, which form together 20% of the body mass.

The dyestuff, Evans' blue, when injected into blood, becomes tightly bound to the protein of blood plasma, so it can be used similarly to measure plasma volume. If we know the proportion (usually 45%) of the cells in blood it is then possible to measure the total blood volume. This, for the standard healthy 70 kg. man, living near sea level, is 5 litres.

Total body water can be measured by similar means, but chemicals that diffuse freely into cells and thus become distributed throughout the whole water volume of the body are not common, except water itself—and therefore 'labelled' water may be used to measure total body water. The labelling consists of replacing some of the common hydrogen atoms with those of the 'heavy' isotope of hydrogen (*deuterium*) or of the still heavier radioactive atoms, called *tritium*. Chemically, these are just like ordinary water but by physical means their presence can be determined. In the small doses needed, both are perfectly harmless.

Once the whole body water, extracellular water and blood plasma volumes have been determined, then both intracellular and tissue fluid volumes can be rapidly calculated by simple arithmetic.

COMPOSITION OF BODY FLUIDS

While the details about the composition of each fluid will be described in the separate sections of this chapter, some general statements can be made here. It is customary to give the concentrations of the dissolved substances in *molar* terms. A molar solution of a substance is that solution that contains one *gram molecular weight* of that substance dissolved in one litre of solution. The gram molecular weight is the sum of the atomic weights of the materials' component atoms, with the atomic weight of hydrogen taken as 1 dalton. Thus the molecular weight of common salt is 58, since the atomic weight of sodium (Na) is 23 and that of chlorine (Cl) is 35. The sodium chloride (NaCl) concentration of blood is about 0.110 molar or 110 mM. (This is in fact close to 6.4 g NaCl/l for those who have not yet converted to S.I. units.) Table 17.1 gives, in millimoles, the average cation and anion contents of intracellular and extracellular fluids, though such a table should be cautiously used. Thus it makes no mention of such non-ionised, but dissolved, substances as glucose and urea. It treats 'protein' as if all proteins were the same and behaved entirely as acidic ions. It fails to take note of the very different protein contents of blood plasma and tissue fluid. However, it will draw your attention to some important points. First there is a balance between positive and negative ions in each fluid. Both fluids are thus approximately neutral chemically. Potassium is the chief positive ion within cells and sodium in extracellular fluids. Protein and phosphate are the chief intracellular anions, while chloride and bicarbonate (from dissolved CO_2) are the main extracellular anions. While it is believed that the composition of extracellular fluid might represent sea water in the remote geological past, no one has attempted to decide why the intracellular water is just so. As has already been seen, the different concentrations of sodium and potassium outside and within cells is concerned in producing cell membrane electrical potential differences and the action potential of nerve impulses.

The various transcellular fluids, being secretory products of cells, are different from both extra- and intracellular fluids. Cerebrospinal fluid and the aqueous humour of the eyes are almost protein-free, have less glucose than extracellular fluid and significantly different ionic composition. The synovial fluids of joints, tendon sheaths and bursae contain complex carbohydrate (glycosaminoglycan) molecules which act as lubricating material.

Table 17.1 Electrolyte concentrations in millimoles/litre of body fluids

Extracellular				Intracellular			
Cations		Anions		Cations		Anions	
Na^+	145	Cl^-	110	Na^+	12	Cl^-	0–2
Ka^+	5	HCO_3^-	27	K^+	155	HCO_3^-	8
Ca^{2+}	2	$Prot^-$	15	Ca^{2+}	2*	$Prot^-$	60
Mg^{2+}	2	HPO_4^{2-}	2	Mg^{2+}	15	HPO_4^{2-}	90
						Others	24
Total + 154		Total − 154		Total + 184		Total − 184	

*Most of the intracellular calcium is sequestered in various intracellular organelles such as mitochondria.

REGULATION OF FLUID VOLUMES

The processes, apart from individual cells' metabolic activity, that regulate intracellular water volume are not known. Facts are now accumulating on the regulation of blood and tissue fluid volumes. They are largely concerned with the kidneys' excretion of water and sodium chloride, and each of these processes is controlled by a hormone. Thus *aldosterone*, which is produced by the adrenal gland, increases the retention of sodium chloride, and the *antidiuretic hormone* from the posterior pituitary gland increases water retention. Aldosterone production is stimulated by a fall in venous blood volume (which would follow a fall in tissue fluid volume, as will be seen later), whereas antidiuretic hormone production is stimulated by a rise in solute concentration of the circulating volumes remain remarkably constant by the action of this system. This is yet another example of negative feedback control.

The exchange of fluid between blood and tissue fluid, while not an actively regulated process, will be seen later to follow automatically upon any change in the volumes of either, and to act always so as to tend to restore normal volumes in both compartments.

Since the transcellular fluids are secretions, their volumes are normally controlled by the secretory processes. However, in the cases of both the cerebrospinal fluid and the aqueous humour of the eyes, the pressure of the fluid is *not* part of the regulatory process. If reabsorption of either is diminished, secretion carries on, albeit at a lower rate, and pressure builds up, producing headache or eye pains and ultimately mechanical and anoxic damage to the structures as normal blood flow is impeded by the raised fluid pressure.

BLOOD

Blood has been described as a fluid connective tissue (see Chapter 3). It contains two principal types of blood cell: red cells and white cells, also known as *erythrocytes* and *leucocytes* respectively (Fig. 17.1). It is the medium in which many chemicals are transported within the body. Foodstuffs and waste products, oxygen and carbon dioxide, hormones to regulate cells' functions and the excess heat produced by metabolism are all carried to and from tissues in the blood. Many substances are in simple solution but some are bound to proteins (oxygen to haemoglobin and fatty acid molecules to plasma albumin, for example).

RED BLOOD CELLS (Erythrocytes)

These form about 46% of the volume of the average healthy male person's blood, and about 43% in the female's blood. There are about 5 million in each microlitre (μl) of blood (5.4 in the male and 4.8 in the female), making a total of 25 million million in the 5 litres of blood. Each one lives about 100 days, so some 250 thousand million new ones are formed each day. We leave it to you to continue the calculations and learn how many red blood cells are produced and die in each minute of your own life. Each cell is a flattened disc with a thickened rim and thinned out central region (Fig. 17.1). It is 7 μm in diameter and 1–2 μm thick. Its volume is about 90 μm^3. It has the normal type of cell membrane, but is far simpler than most cells in its interior. The most prominent features are the solution of the protein *haemoglobin* in the cytoplasm, and the absence of a nucleus. There are no mitochondria and none of the protein-forming endoplasmic reticulum. Glycolytic enzymes are present in the cytoplasm, also enzymes that maintain the iron atoms of the haemoglobin molecule in the bivalent state (trivalent iron will not form the same reversible compound with oxygen and so is useless for oxygen transport). Finally, an enzyme—*carbonic anhydrase*—is present in red cells, which is responsible for carbon dioxide transport in the blood. Thus the red blood cell, as a cell, is stripped down to the barest essentials, keeping, in its mature state, only those aspects of normal cellular function that are needed for its major function, that of oxygen transport.

RED BLOOD CELL FORMATION

In embryonic life various tissues undertake the task of blood cell production, but by the time of birth this is confined to red bone marrow. In adults the marrow of vertebrae, ribs, the sternum, pectoral and pelvic girdle bones, the skull and the ends of the long bones are the sites of blood cell formation. If cell formation becomes increased, or these sites are invaded by some disease process, then other regions, e.g. the shafts of the long bones, may again become the site of blood cell production and change in colour from yellow (fatty) to red (haemopoietic) marrow. Red blood cells start life as undifferentiated *haemocytoblasts* which also give rise to white blood cells (see below). These cells live in the red marrow, growing and dividing from time to time. They are 2–3 times the size of red blood cells. Of the daughter cells, some continue as haemocytoblasts, while others become *pro-erythroblasts*. These cells do not divide or form haemoglobin but the nuclear chromatin begins to condense. The next stage is known as the *normoblast* because, by repeated cell division, it approaches the mature red blood cell in size. Normoblasts also start forming haemoglobin. The production of haemoglobin changes the staining reaction of the cytoplasm. In the earlier stages (as in the haemocytoblast) it is pale blue or mauve when stained by routine methods. Haemoglobin is acidophil and stains red with eosin dye. Therefore in cells stained with this material, addition of haemoglobin to the cytoplasm results in progressive change in colour from blue, through purple, to a red colour. Eventually, at the end of the normoblast stage, the nucleus disappears, leaving only the endoplasmic reticulum. While this is still present, it can be stained in the living red blood cell which is called a *reticulocyte*. Eventually this, too, is lost and the cell becomes a mature erythrocyte. About 1 % of circulating red blood cells are reticulocytes, and the numbers of these cells can increase to 35 % during episodes of rapid extra blood formation, as occurs after a large haemorrhage.

CONTROL OF RED BLOOD CELL FORMATION

It has been known for over 50 years that removal to a place where there is little oxygen, high in a mountainous country, will result, in a few weeks, in the development of extra RBCs, the count rising to 6–7 million /μl, and the total cell volume to about 60 % of blood. It is not so generally appreciated that the same response occurs in the sufferer from chronic bronchitis, in whom passage of oxygen from air to blood in the lungs is impaired. In both, the state is called *reactive polycythaemia*.

How shortage of oxygen acts on the bone marrow is still being studied. It appears that oxygen lack stimulates the kidneys to liberate an enzyme (*erythrogenin*) which acts on a component of the plasma proteins, causing the liberation of a glycoprotein fragment, called *erythropoietin*. This substance acts on the haemocytoblast—pro-erythroblast transformation. Without it, or if the bone marrow does not respond to it, there would be failure of RBC formation. This may be one cause of the rare condition of aplastic anaemia in which RBC production does fail completely. However, people who have had their diseased kidneys removed and are maintained on dialysis machines do not develop aplastic anaemia. The story is thus, as yet, incomplete. Erythropoietin is also responsible for the increased reticulocyte count 24 h after a haemorrhage.

Occasionally bone marrow cells emerge which go on to produce pro-erythroblasts, normoblasts and RBCs without the stimulus of erythropoietin. RBC formation is uncontrolled and a polycythaemia develops without the hypoxic stimulus described above.

REQUIREMENTS FOR RBC FORMATION

Haemoglobin is an iron-containing protein, so it is not surprising that its formation is ultimately dependent upon dietary iron and protein intake. While protein intake in Western countries is seldom deficient, an imbalance between iron intake and iron loss can develop in any person with excessive iron loss (heavy menstrual periods, for example), or in whom iron absorption is defective. Iron is stored in the body and deficient haemoglobin production will not occur until the stores are exhausted. In iron deficiency, the RBCs contain less than the normal quantity of haemoglobin. They are small, and fewer are found in the blood.

In the earlier normoblast stages, the cells divide repeatedly. This requires the synthesis of fresh nucleic acid material. Two vitamins, folic acid and cobalamin (vitamin B_{12}) are required in nucleic acid synthesis and both may be absent in the diet (if this is almost or entirely vegetarian in origin). The absorption of B_{12} depends also on the production of a special substance from the stomach, known as *intrinsic factor*. Absorption therefore fails in some stomach diseases or after total gastrectomy. Deficiency of either vitamin results in failure of normal normoblast division. Few RBCs are produced, they are larger than normal (*macrocytes*) and the abnormal precursor cells (*megaloblasts*) can be seen in the bone marrow. The macrocytes break up earlier in the circulating blood which adds to the sufferer's anaemia, there being both deficient production and excessive destruction of his RBCs.

Other specific factors that are needed for normal RBC production include ascorbic acid (vitamin C) and thyroxine. Production is reduced in any chronic infective or other disease state, e.g. in theumatoid arthritis. Chronic renal failure causes an anaemia, the chief characteristic of which is small RBCs, few in number, but containing the normal concentration of haemoglobin. It is hard to equate this with a partial breakdown of the erythropoietin mechanism. The functions of the red blood cells will be dealt with in the next chapter.

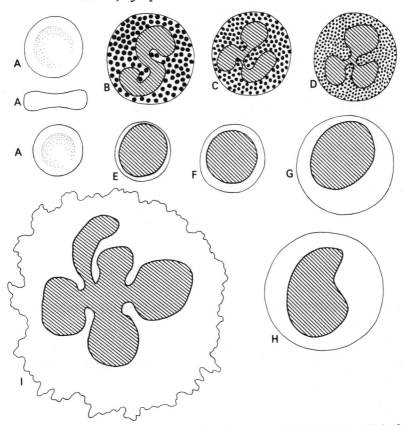

Fig. 17.1. Blood cells. A. Red cells; B. Basophil (blue granules); C. Eosinophil (red granules); D. Neutrophil; E, F and G. Lymphocytes; H. Monocyte; I. Megakaryocyte.

WHITE BLOOD CELLS (leucocytes)

The leucocytes are, in general, larger than the red blood cells and there are fewer of them—about 7,000–8,000/μl. Unlike the red cells they are nucleated and they are of a number of different types. They may be subdivided into two main groups, depending on the presence or absence of stainable granules in the cytoplasm. Those with granules—the *granulocytes*—have a multi-lobed nucleus (Fig. 17.1) and are therefore also called *polymorphonuclear leucocytes* (more commonly called 'polymorphs' in conversation). They have a diameter of about 10–15 μm. They are of three types, depending on the staining reactions of the granules, namely, *basophil leucocytes*, *eosinophil leucocytes* and *neutrophil leucocytes* The leucocytes without granules are called *agranulocytes* and they have a large spherical or slightly indented nucleus. They are subdivided into two main types—*monocytes* and *lymphocytes*—the latter being small, medium or large. These cell types will now be described individually.

BASOPHIL LEUCOCYTES

These form only 1–2 % of the total number of leucocytes. As their name implies, the cytoplasm contains numerous, rather coarse granules which stain heavily with basic dyes, i.e. they appear blue with the usual stains for blood. Their function is uncertain but they are believed to be related to mast cells (Chapter 3).

EOSINOPHIL LEUCOCYTES

These cells, comprising 2–3 % of the total leucocyte population, have granules stainable with acid dyes in the cytoplasm. The granules stain pink or red with the usual blood stains and are smaller than the granules of basophils. Eosinophils are capable of phagocytosis and can migrate from the capillaries into the tissues. They are related in some way to the immune reaction (Chapter 20). They are increased in allergic conditions such as asthma and hay fever and are also increased when the body is infested with parasites such as tapeworms.

NEUTROPHIL LEUCOCYTES

These are by far the most numerous of the leucocytes, making up 60–70 % of the total. The granules are not noticeably red or blue but appear to be some intermediate colour. The cells are actively phagocytic and can engulf microorganisms which are then digested by the cell's lysosomes. They can leave the blood vessels and are found in the tissues in very large numbers when infection occurs. The number of circulating neutrophils in the blood rises steeply in acute infections (*leucocytosis*).

MONOCYTES

Monocytes are agranular and comprise 2–8 % of the total number of leucocytes. They are large, measuring 15–20 μm in diameter, and much of the cell's interior is occupied by a large kidney-shaped nucleus. Monocytes are highly mobile and are phagocytic. They are very probably identical to the macrophages which are found in the tissues.

LYMPHOCYTES

These, as already stated, are very variable in size, ranging from 5 to 15 μm. They form about 25 % of the total leucocyte population. There is relatively little cytoplasm as most of the cell is occupied by the large spherical nucleus. Lymphocytes play a very important part in the immune reaction so their functions will be discussed in Chapter 20.

FORMATION OF LEUCOCYTES

Leucocytes are formed in the red bone marrow, at least in the adult, and they probably all differentiate from a primitive stem cell, the *haemocytoblast*. These are large cells, up to 20 μm in diameter, and they divide to give rise to more haemocytoblasts, to red blood cell precursors (see above) and to the precursors of the various types of white cells. In the case of the granulocytes, the first precursors are known as *myeloblasts* whose granular endoplasmic reticulum produces the basophil, eosinophil or neutrophil granules, forming the three types of *myelocytes*. Subsequently, these cells become smaller and their nucleus acquires its characteristic shape. Monocytes and lymphocytes also pass through a 'blast' stage, when they are called respectively, *monocytoblasts* and *lymphoblasts*.

BLOOD PLATELETS (Thrombocytes)

These bodies, of which there are $\frac{1}{4}-\frac{1}{2}$ million/μl of blood, are not classified as red or white cells but they are extremely important cellular components of the circulating blood. They are very small oval bodies, 2–4 μm in length and they contain various cytoplasmic organelles but no nucleus. This is because they are formed in the bone marrow by the pinching off of small pieces of the cytoplasm of large cells known as *megakaryocytes*. They play an important role in blood clotting (see below), and bone marrow damage can therefore lead to problems with the clotting mechanism.

PLASMA

This forms the remaining 55 % of blood. The composition of plasma has already been described, so it is sufficient to say here that the two chief components are the 6.4 g/l of NaCl and the 70 g/l of proteins. Plasma is the medium in which the soluble requirements and waste products of body tissues are transported. These include glucose and lactic acid, amino acids, urea and creatinine, the water soluble vitamins and hormones, and other organic and inorganic requirements of the tissues or waste products of tissue cells. Some components (e.g. calcium ions) have to be maintained within narrow limits if cellular function is to proceed normally. If there is too little calcium in the plasma, nerve cells become over-excitable: if too much, they become unresponsive.

PLASMA PROTEINS

These are divided into two groups of compounds, the *albumins* and the *globulins*. The albumins (45 g/l) are smaller, having molecular weights of 70–100 thousand. They are chiefly concerned with transport of the water-insoluble materials that cells need. These include fat-soluble vitamins and hormones and fats themselves. While fatty acids are somewhat soluble in plasma water, much of the fat used by tissues is transported combined with albumin and other substances as *lipoprotein* molecules, some of which can be very large, with molecular weights of over 1 million and in size not far below the resolving power of the light microscope.

The globulins (25 g/l), which have molecular weights between 90,000 and 160,000, include antibody molecules and fibrinogen, which plays an important role in blood clotting (see below).

Since the plasma proteins are dissolved in the plasma and cannot, to any great extent, pass through capillary walls into tissue fluid, they will exert an osmotic effect, sucking water from the tissue fluid space into capillaries with a force equivalent to 25 mm Hg. This plays an important role in the circulation of tissue fluid (see below). The proteins in plasma, together with the RBCs, give to blood its characteristic viscosity, and also form an important buffering system which helps to keep the pH of the blood constant.

COAGULATION (CLOTTING) OF BLOOD

When blood is shed, in a few minutes it will clot, being transformed from a liquid into a firm jelly-like solid known as a blood clot or *thrombus*. This is the result of the conversion of a soluble protein, *fibrinogen*, into sticky threads of *fibrin*, which form a network entrapping the RBCs. The formation of fibrin from fibrinogen is the last stage of a series of reactions involving some thirteen different substances (Fig. 17.2). These are commonly referred to as *factors* and numbered I to XIII, fibrinogen being factor I.

Clotting begins when blood is exposed to damaged tissue or to a water-wettable surface. Normal blood vessel endothelium is not water-wettable, so blood does not clot in healthy blood vessels. When blood is exposed to a water-wettable surface, to damaged endothelium or to other tissues of the body, various chemicals are released from inert precursors in plasma. Platelets adhere to the abnormal surface and liberate a *phospholipid* from their interior. This platelet adherence is inhibited by *prostacyclin*, formed in endothelial cells. It is enhanced by *thromboxane*, formed in the platelets themselves. Both these compounds are *prostaglandins*, metabolic products of unsaturated fatty acid metabolism. Damaged tissues also liberate phospholipids. All these (see Fig. 17.2), together with calcium ions, convert *prothrombin* to thrombin, an enzyme that changes the soluble *fibrinogen* to insoluble, sticky fibrin threads. Prothrombin is formed in the liver and its molecule incorporates that of *vitamin K*. This is a fat-soluble material, normally formed within the small intestine by the resident bacteria. Absorption of vitamin K depends upon the presence of bile salts. Liver disease may interfere with blood clotting by a reduction in both vitamin K absorption and pro-thrombin formation.

Blood that has clotted subsequently *retracts* (shrinks) to form a firm mass of fibrin threads and RBCs. The fluid squeezed out in this process is called *serum*. It is essentially plasma without its fibrinogen, but containing most other

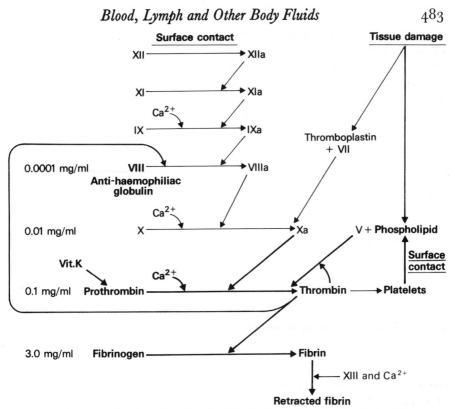

Fig. 17.2. The cascade of reactions involved in the formation of a blood clot. The most important factors and agencies are shown in heavy type.

constituents of the original plasma. Once a blood clot has formed *in vivo*, further substances are produced, as a side reaction of one of the blood clotting factors, that will dissolve fibrin and so eventually remove the clot.

The clotting process can be slowed or stopped by *heparin*, by interfering with prothrombin formation (by giving a person synthetic vitamin K-like substances), or by removing ionised calcium with citric or oxalic acids (either of which may be used to prevent blood from clotting that has been withdrawn from the body).

From Figure 17.2 you will see that many more factors than have been described in the text are involved in the clotting process. Formation of most of these is genetically determined, and absence of any may produce defects in the clotting process. Haemophilia is the commonest of these defects. A substance called *anti-haemophiliac globulin* (AHG, also known as Factor VIII) is absent from a haemophiliac's blood. Its formation depends on a gene carried on the X chromosome. Heterozygote women, containing one normal X chromosome and one with the absent or abnormal gene make sufficient AHG to get by, but a man with one X chromosome only can make no AHG when his X chromosome does not contain the normal gene. Thus half of a heterozygote woman's sons will be haemophiliac since she is healthy but passes the trait on to her sons. She is commonly known as a *carrier* and half of her daughters (if her husband is normal) will, like her, be carriers. All of a haemophiliac man's daughters will likewise be carriers but his sons will be entirely normal. In the rare event of a carrier woman marrying a haemophiliac man, then not only will all sons be haemophiliac, but all the daughters will be carriers. However, it seems that any female embryo inheriting the defective X chromosome from both parents does not usually survive until normal birth, so genuine women sufferers from haemophilia have rarely been detected.

The importance of this disease to physiotherapy lies in the fact that spontaneous bleeding into joints frequently occurs. Forceful movements to the limits of a joint's normal range may cause this bleeding. Free blood within a joint provokes an acute, painful reaction therein, with accompanying muscle spasm, as in any other acute arthritic condition. Repeated haemorrhages into joint will produce permanent structural alterations in it so that all movements become both limited and painful. Concentrated AHG preparations are now available and can be injected into a haemophiliac's blood stream whenever bleeding starts. It is hoped that this treatment will reduce the amount and severity of joint trouble in this disease.

HAEMOSTASIS

The control of bleeding *in vivo* depends on both the initial cascade reaction leading to fibrin formation, on the subsequent retraction of the fibrin threads, which might help to draw cut ends of blood vessels together, and also on the contractile power of vessels themselves. This is particularly the case when very small arteries or arterioles are involved. The smooth muscle in their walls can contract so as to completely close the vessels for a few minutes. This enables clot formation to proceed and the typical story of bleeding in a haemophiliac is that it starts some minutes *after* the injury. Minor damage to capillaries does not lead to bleeding from them if the intercellular materials that bind the endothelial cells together are normal, but excessive capillary fragility may develop in some disease and vitamin deficiency states, e.g. scurvy.

BLEEDING TIME AND CLOTTING TIME

Haemophilia is one known defect of blood clotting, and prothrombin deficiency is another. Further defects are known. But excessive bleeding from wounds or other minor forms of trauma can occur when the blood itself is normal, but a defect exists in the endothelial cell walls of capillaries. The *bleeding time* test is performed by finding for how long a deliberate puncture of the skin will continue to bleed. This time is largely determined by the ability of capillary beds to close down (Chapter 14) and block off punctured capillaries. In practice this is 2–6 minutes. The *clotting time* test is performed by withdrawing a small sample of blood and putting it in a glass tube and observing how long it takes to clot, usually 6–12 minutes. This test, naturally, is a measure of the efficiency of the blood's intrinsic clotting mechanism when exposed to a foreign, water-wettable surface.

BLOOD GROUPS

RBCs, like other cells, carry a variety of antigenic molecules on their surface membranes. For practical purposes, the most important of these are substances A and B. People are divided into the 4 groups: A, B, AB or O, according to whether they possess the enzymes that manufacture either A or B substances, both of them, or neither. Possession of these enzymes is determined genetically by inheritance from one's parents. Individuals with no A or B substance automatically produce the corresponding antibody, which is always thus present in their plasma. Thus a person with A substance, but no B, has anti-B in his plasma, while someone with B on the cells has anti-A. Someone with neither A nor B (group O) has both anti-A and anti-B, and a person with A and B has neither antigen. If A-containing cells are mixed with anti-A containing plasma, as occurs when giving group A blood to a person of group B or O, an antigen–antibody reaction occurs. The cells clump together (*agglutination*) and then their membranes rupture (*haemolysis*) liberating the contents, including the haemoglobin, into the plasma. This peculiar arrangement of having pre-formed antibody, without the need for previous 'sensitisation' of the T-lymphocytes (see Chapter 20) is the basis for the practical importance of the ABO series of blood groups.

Blood from group O people, with cells containing neither A nor B substances, can be transfused safely into those of the other three groups for the cells cannot be agglutinated by the recipients' anti-A or anti-B factors (the group O blood plasma content of both anti-A and anti-B can be ignored when a relatively small amount is transfused into another person). Group O blood is therefore known as *universal donor* blood. A person with cells containing A and B substances does not have either anti-A or anti-B plasma factors. This person can be therefore transfused with blood of any group and is called a *universal recipient* accordingly.

There are many other antigenic materials which *may* be present on RBC membranes. If any one of these are given to a person not having it already in his cells, he may start making the corresponding antibody and if he is then given a second dose of cells from the same person, the antibody that has now been formed will react with, agglutinate, and destroy the given cells. One example of the way in which this happens naturally is provided by the case of the *rhesus factor* (Rh). Fifteen per cent of Caucasian races are deficient in rhesus factor and are said to be rhesus negative (Rh—). If a Rh— woman marries a Rh+ man their children may be Rh+. Small numbers of fetal blood cells can cross the placenta into the mother's circulation during pregnancy and particularly during labour. If these cells are compatible so far as substances A and B are concerned, they will remain in her circulation for long enough to stimulate the production of anti-Rh antibody. This antibody will persist in her plasma, perhaps for the rest of her life, just like any other antibody produced by foreign living cells. It is small enough to cross the placenta and break up the cells of any future Rh + infant the Rh − mother may bear. The infants will be born jaundiced or perhaps dead, depending on the degree of the attack. One effect of this intravascular haemolysis is that the excess bilirubin formed from the porphyrin portion of the haemoglobin molecule may itself damage the new born infant's brain.

Prevention of this trouble is simple. If the Rh + cells from the infant can be rapidly destroyed in the Rh − mother's circulation as soon as she has been delivered, then they will not cause the formation of anti-Rh antibody. This destruction is readily achieved by giving the newly delivered mother a dose of anti-Rh antibody. This will harmlessly destroy the Rh + cells and so prevent anti-Rh formation.

TISSUE FLUID

The composition of tissue fluids is very similar to that of blood plasma, the only significant difference being the much lower concentration of protein—less than 1 % compared with the 8 % of plasma—this being due to the capillaries being relatively impermeable to protein. The total volume of tissue fluid is about 10 litres in the 'normal' 70 kg man and it is usually regarded as exerting a zero hydrostatic pressure.

The molecules of water and of solutes in tissue fluid are freely exchangeable with those of blood due to the physical process of diffusion. If a concentration difference exists then diffusion 'down the gradient' from higher to lower concentration will exceed diffusion uphill, the net diffusion rate being proportional to the concentration difference. If metabolising cells produce such concentration gradients between blood and tissue fluid then the substances will move through the tissue fluid and through capillary walls between these cells and the blood. Large amounts of water diffuse freely in both directions, but in addition some will filter through capillary wall membranes when the pressures (either hydrostatic or osmotic) are different on the two sides.

TISSUE FLUID CIRCULATION

The concentration of protein in the plasma maintains at all times an osmotic pressure effect of 25 mm Hg, which acts so as to suck water into the blood vessels. At the arterial end of a capillary the hydrostatic pressure might be about 35 mm Hg and at the venous end 12 mm Hg. These pressures will tend to drive water out from capillaries to the tissue fluid. The final net filtration force will be the difference between the capillary hydrostatic blood pressure and the plasma protein osmotic pressure. Thus at the arterial end of capillaries there is a 10 mm Hg (35–25 mm Hg) filtration force driving water out and at the venous end a 13 mm Hg (25–12) force sucking water back from the tissue fluid into the capillaries. It should be emphasised that the volume of water filtered in this way to and from the tissue fluids is only a tiny fraction of the amount that diffuses in both ways at the same time.

Arteriolar dilatation, as well as allowing an increase in capillary blood flow, will also result in a rise of pressure throughout the length of capillaries. Therefore filtration outwards is increased and the return of fluid from tissues to blood reduced. If, at the same time as arteriolar dilatation, capillary walls become permeable to proteins and their concentration in tissue fluid rises, then the effective protein osmotic pressure falls. This reduces the return of tissue fluid to blood. Histamine and bradykinin, both released from damaged tissue cells in the inflammatory reaction, can both dilate arterioles and increase permeability of capillaries to protein. It is to be expected, therefore, that one important feature of inflammation is an increase in tissue fluid formation and that the inflamed area becomes swollen, red hot and painful (classically tumor, rubor, calor and dolor).

LYMPH FORMATION AND FLOW

Blind-ended lymphatic capillaries penetrate into most tissues. The lymph they contain has a similar composition to the tissue fluid from which it is derived. It moves slowly along the lymphatic vessels ultimately being discharged into systemic veins at the root of the neck. The lymphatics thus serve, in all tissues, to drain off surplus tissue fluid, including its protein content, that has not re-entered capillaries.

In addition to this there are two other important functions served by the lymphatic system. One is the absorption of fat from the digestive canal. Much of the fat is passed from the cells lining the gut into lymph channels rather than into the blood. Because the finely divided fat droplets give these lymph vessels a milky appearance they are called *lacteals*. After a meal containing fat, this milky-looking lymph can be traced as it passes from the intestinal lymphatics to the *cisterna chyli* on the posterior abdominal wall and thence up the main lymph vessel of the body, the thoracic duct, which finally joins the systemic veins in the angle between the left internal jugular and the subclavian veins.

The other main function of the lymphatic system is related to defence against bacterial infection. This topic will be dealt with in detail in Chapter 20 so nothing more will be said here.

INTRACELLULAR WATER

In the standard 70 kg man there are about 29 l of water within cells. This is the medium in which the cell's chemical reactions occur, which bathes the active sites of all the many enzymes within a cell, and through which molecules needed or produced by these enzymes move between their sites of entry or production and the sites of their use or ejection from the cell. In living tissues as we know them the whole chemistry of life goes on in the water of the cell's cytoplasm. In fact, water and carbon compounds *are* life as we know it. This does not deny the science fiction writer's dream of another form of life, based on a different chemistry and using a solvent other than water, but the only life form we know is carbon- and water-based and it is in the intracellular water that life goes on. Ultimately, then it is upon intracellular water that life depends. Remove water and the cell dies (or at best its life is suspended), even though all of its solid structural elements may remain.

Beyond that, what is there to be said about intracellular water? It is present as a gel, not as free-flowing liquid water. It contains, as has already been seen, dissolved salts and other substances. It is neutral in reaction (pH is 6.9 at $37°C$). All cells have the means to keep constant the nature and amounts of dissolved substances and indeed their total water content, so they neither swell or shrink. The processes that maintain the constancy of the intracellular environment require energy. In fact, a large part of the body's total energy production is needed for this maintenance.

TRANSCELLULAR WATER

This component of body water is water that has been secreted by cells into a hollow organ or space. It is conventional to consider only cerebrospinal fluid, aqueous humour and synovial fluid in this category, but the definition also, in fact, includes the water of the secretions of the gastro-intestinal tract, for this has passed through secretory cells and, almost entirely, is destined to be reabsorbed into the body. Inflammation in the region producing any of these fluids is likely to cause over-production of fluid, which has a composition approaching that of plasma.

CEREBROSPINAL FLUID (CSF)

This is secreted by the choroid plexuses of the cerebral ventricles, and in adult man about 0.2 ml/min are formed from these. The choroid plexuses are within the brain ventricles (see Chapter 4). CSF leaves the lateral ventricles via the interventricular foramina, entering the mid-line 3rd ventricle. It passes thence through the aqueduct of the midbrain, entering the 4th ventricle, situated between the pons and medulla ventrally and the cerebellum dorsally. Three foramina in the roof of the 4th ventricle, connect it with the subarachnoid space which completely surrounds the brain and spinal cord within the bony cranium and vertebral canal.

Most of the CSF is absorbed back into the cerebral veins through the *arachnoid villi and granulations*. The mechanism of this reabsorption has long been in doubt, but it is now thought to be valvular, fluid passing freely into the veins through one-way valve channels when the CSF pressure exceeds the venous pressure.

RAISED CSF PRESSURE AND HYDROCEPHALUS

Congenital abnormalities, tumours or infections of the brain or meninges may block the circulation of CSF and hinder its reabsorption. As secretion continues, the pressure above the block will rise. The normal pressure varies from 20 to 30 mm Hg (it is usually measured directly in a vertical U-tube manometer as mm CSF) depending on posture and respiratory effort. Elevation of this pressure above 35 mm Hg causes a reduction of cerebral blood flow by progressive collapse of the cerebral veins and capillaries. The resultant fall in O_2 tension and rise in CO_2 tension within cerebral tissue lead to an arteriolar dilatation and a consequent rise of capillary blood pressure and thus a maintenance of adequate blood flow. There are limits, however, to this process and a progressive rise in CSF pressure leads to persistent cerebral ischaemia with resultant changes in neurological and mental function. The ventricles become dilated but since the brain is enclosed in the rigid bony cranium there will be no increase in size of the brain or the head, unlike conditions in the infant which are described below.

Since the optic nerves are surrounded by prolongations of the subarachnoid space, and the retinal veins share in the compression and obstruction experienced by cerebral veins when CSF pressure is raised, consequent changes can be seen in the retina. These are given the name *papilloedema*. The retinal veins are enlarged, the point of entry of the optic nerve into the retina, the *optic disc*, appears swollen and there may be visible excess of tissue fluid (exudate) on the retina's surface.

In infants, before the cranial bones have fused, the rise in intracranial pressure may cause an overall expansion of the head (*hydrocephalus*) and a reduction in the actual quantity of cerebral tissue formed. It is not surprising, therefore, that hydrocephalus is associated with reduced mental powers and abnormal neural function.

FUNCTIONS OF CSF

The main function that can be ascribed to this fluid is the provision of a support for the delicate brain. This organ, weighing about 1400 g in air, weighs no more than 50 g when suspended in CSF. Furthermore, in a sudden movement of the head, the brain is cushioned from impact or distortion by the fluid surrounding it. Severe blows to the cranium might damage the brain in three ways:

1   Laceration of the brain following a fracture.
2   Rupture of vessels in the outermost layer of the brain's covering, the *dura mater* leading to a rise in intracranial pressure.

3 A *contre-coup* bruising of the brain at a site opposite to the original blow. It is this last that is largely prevented by the CSF.

AQUEOUS HUMOUR

This is secreted by the ciliary body epithelium, behind the iris of the eye. Aqueous humour is chemically similar in its content of dissolved salts, etc., to blood. It contains, however, large amounts of non-polymerised glycosamino-glycans. It is the osmotic pressure exerted by this substance (the source of which is not clear) that maintains the continuous formation of the fluid. It flows over the anterior surface of the lens, through the pupillary aperture and is reabsorbed in the angle between the front surface of the iris and the corneo-scleral junction. As in reabsorption of CSF, this process is thought to be a free flow through one-way valve channels between the anterior chamber of the eye and the veins.

SYNOVIAL FLUID

Wherever there are synovial membranes, in joints, bursae and tendon sheaths, small amounts of synovial fluid are formed. This is a lubricating fluid, by virtue of the glycosaminoglycan molecules it contains. These are of different lengths and when small quantities of synovial fluid are trapped between opposing load-bearing articular cartilage surfaces, the short-chain molecules and water are preferentially squeezed out, leaving the more viscous longer molecules. Thus the greater the pressure the 'heavier' in engineering terms becomes the lubri-cating fluid between the bearing surfaces.

As with CSF and aqueous humour, inflammation of synovial membrane leads to excessive production of fluid, and this fluid may contain sufficient plasma protein to form a clot. Injury, with or without haemorrhage into the joint space, infections and *auto-immune* processes such as rheumatoid arthritis may all cause joint, bursa or tendon sheath swellings with pain and limitation of movement, as a result of excessive synovial fluid production. Sometimes the original injury may be trivial and escape observation, but the inflammatory reaction persist for many days or weeks. An extreme example of this process is the result of joint haemorrhage in haemophilia, already described in this chapter.

18 · Gas Transport in the Blood

While small amounts of nitrogen and traces of other gases are present at all times dissolved in the water of blood, this chapter is only concerned with the two gases of major physiological importance, oxygen and carbon dioxide.

This account must begin with a look at the quantities of these two gases present in blood and the major changes in blood gas contents that occur in the body. Arterial blood usually contains about 20 ml of oxygen and 48 ml of carbon dioxide in each 100 ml blood. When this blood has been around the body and has been collected by the great veins and ejected from the right ventricle of the heart towards the lungs, in a resting person, each 100 ml of blood will be found to contain 15 ml of oxygen and 52 ml of carbon dioxide. In its passage through the organs and tissues of the body each 100 ml of blood will by then, on average, have lost 5 ml of oxygen and gained 4 ml of carbon dioxide. But different organs have differing blood supply and differing metabolic rates, so their venous blood can vary in composition. Thus blood leaving the skin (low metabolism) or the kidneys (high blood flow) still resembles arterial blood in its gas composition, while blood in the coronary sinus, the internal jugular vein, or skeletal muscle veins in exercise may contain only 10 ml O_2 and 56 ml CO_2 per 100 ml. We must now see how these amounts of gases are carried in the blood and how the exchanges between alveolar air, blood and tissues are effected.

SOME PHYSIOLOGICAL CONSIDERATIONS

In the next chapter, on respiration, the partial pressure of the component gases of the alveolar air will be considered in detail. If the dry gases in the lungs contain 5.6% CO_2 and the effective pressure of the dry gases is 760 mm Hg (atmospheric pressure), less 47 (water vapour pressure at body temperature) (i.e. 713 mm Hg) then the *partial pressure* of the CO_2 is $\frac{5.6}{100} \times 713$, = 40 mm Hg. The S.I. unit for pressure which is replacing the mm Hg is the Pascal (Pa), and normal atmospheric pressure is about 100 kilo Pascals (kPa). This means that gas containing 5.6% CO_2 has a partial pressure for CO_2 of 5.6 kPa. S.I. units will be used in the rest of this chapter. If now CO_2 dissolves in water that is in contact with the alveolar air until this water solution of CO_2 reaches equilibrium with the gas mixture, then the *tension* of CO_2 in the solution is also 5.6 kPa. Every gas has its own inherent solubility in water, expressed as the number of volumes of gas that will dissolve in a unit volume of water per unit of partial pressure. For CO_2 at a tension of 5.6 kPa, 2.5 ml are dissolved in 100 ml of a water solution. The oxygen partial pressure in the lungs is 14.0 kPa, and 0.22 ml O_2 are contained in simple solution in 100 ml of a watery medium that is in equilibrium with alveolar air and at 37°C.

In this chapter we will refer many times to gas tensions. Not only does the tension of a gas determine the quantity dissolved in water, but it directly determines the diffusional movement of the gas through water or between the water of blood and the alveolar air, and also the amount of the gas that enters into combination with other chemical substances. Each gas has its own inherent diffusibility through water, just as it has its specific solubility (the more soluble gas being also the more diffusible one). Rates of diffusion will depend on distance, the specific diffusibility and the difference in tensions, or the *tension gradient* along which the substance is moving.

Having made these introductory points, one can now consider the means by which oxygen and carbon dioxide enter and leave the blood and the forms in which they are carried by the blood between lungs and tissues.

OXYGEN

As has already been seen, the amount of oxygen that can dissolve in the water of blood plasma is only 0.2 ml/100 ml blood. If this were the sole means of carrying oxygen to the tissues, then 125 litres of blood would have to flow through the lungs each minute to pick up enough oxygen to maintain the resting body's supply of oxygen. The cardiac output is, in fact, only 5 litres a minute. Blood contains the very remarkable protein, haemoglobin, each gramme of which can combine with 1.34 ml O_2. Each 100 ml blood contains 15 grammes of haemoglobin; therefore, this volume can carry 20 ml O_2 and the O_2 needs of the resting body are met, four times over, by the cardiac output of 5 litres a minute. We must now study how haemoglobin is made and how it achieves this remarkable feat.

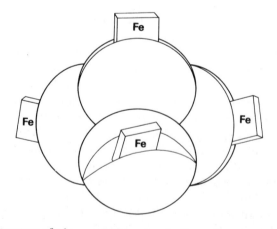

Fig. 18.1. The structure of a haemoglobin molecule. It consists of 4 roughly spherical ball-shaped protein units; in a slot in each of them is the square, plate-shaped porphyrin molecule containing an iron (Fe) atom.

THE STRUCTURE OF HAEMOGLOBIN

The haemoglobin molecule is made up of four sub-units. Each sub-unit contains a protein chain folded into a rough ball and a simpler organic molecule, called a *porphyrin*, which itself contains a single iron atom (Fig. 18.1). This atom is in the bivalent or ferrous state. Each iron atom can combine (loosely) with one molecule of oxygen, without altering its bivalent state. For this reason we speak of 'oxygenation' of haemoglobin and not of 'oxidation' or of 'oxidised' haemoglobin. When the iron atom is oxidised to the trivalent state, it will no longer behave as does the bivalent iron of normal haemoglobin and it is useless to the living body. Oxygenation of one of the four iron atoms of a haemoglobin molecule subtly changes the shape of the protein sub-units, which makes oxygenation of a second iron atom occur more readily. This further alters the protein chains, until the affinity of the whole molecule for its fourth oxygen molecule is about twenty times that for the first. This difference in affinity is the cause of the very peculiar relation between the oxygen tension of a haemoglobin-containing fluid and the degree of oxygenation of the haemoglobin. This is shown in Fig. 18.2. It can be seen that at the oxygen tension of alveolar air, the haemoglobin is 97% fully saturated. At an O_2 tension of 5.6 kPa, it is still 75% saturated, but as the tension falls further, oxygen is more rapidly given off, and at a tension of 2.4 kPa only 25% of the haemoglobin still has oxygen bound to it.

This curved line, shown in Fig. 18.2, is known as the *oxygen haemoglobin dissociation curve*. Its shape is due to the decreasing affinity of haemoglobin for oxygen as the gas leaves the haemoglobin. The curve is also affected by other

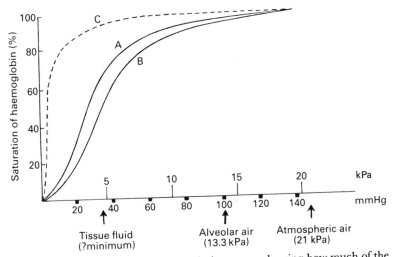

Fig. 18.2. The haemoglobin–oxygen dissociation curve, showing how much of the haemoglobin (% saturation) is combined with oxygen at different levels of oxygen partial pressure. A and B indicate the Bohr shift (see text). C is the corresponding curve for the muscular oxygen storage compound, myoglobin, which gives red meat its colour (p. 000).

factors. Both a rise in temperature and a rise in H^+ concentration reduce the affinity of haemoglobin for oxygen at all tensions, though these effects are most marked in the middle range of the dissociation curve, when the haemoglobin is between 25 % and 75 % saturated. The concentration of haemoglobin is also an important factor, increasing concentration reducing the affinity for oxygen. Here we should digress for a moment. The haemoglobin concentration of the blood of a healthy adult human, living at less than 3,000 metres above sea level is about 15 gm/100 ml blood. This is all contained within the red blood cells, themselves occupying 45 % of the blood volume. Thus the concentration of haemoglobin within the cells is 33 %. The changes in dissociation curve produced by all three factors (temperature, H^+ concentration, haemoglobin concentration) have little effect on the combination of oxygen with haemoglobin in the lungs, but raise the tension at which a given proportion of oxygen is released as the blood passes through the tissues. The effect of elevating H^+ concentration is of particular importance in exercise, Several of the products of exercise are acid and their entry into the blood raises the dissociation of oxygen from haemoglobin, making more available for the exercising muscle. This change is known as the *Bohr shift* of the dissociation curve, named after its first discoverer. Thus curve A in Fig. 18.2 is that of blood with lower CO_2 content or more alkaline, and curve B is that of blood with raised CO_2 or more acidic.

FACTORS AFFECTING UPTAKE OF OXYGEN IN THE LUNGS

Blood takes about 0.8 sec to traverse the pulmonary capillaries in a resting person. It takes only 0.1–0.2 sec for enough oxygen to diffuse from the alveolar air, through the alveolar and capillary walls, through the plasma, the red cell membrane, and to combine with haemoglobin, to bring blood and alveolar air virtually into equilibrium, so that the O_2 tension of the blood, well before it leaves the pulmonary capillaries is almost 100 mm Hg. A small amount of blood, in a healthy person, by-passes the alveolar capillaries, but enters the pulmonary veins (much of this is blood that has supplied the bronchi). When this relatively O_2-poor blood mixes with the O_2-saturated blood from the capillaries, it steals oxygen, mostly dissolved oxygen, from the plasma, so the final state of mixed arterial blood is that the haemoglobin is 95–97 % saturated, and the tension may be as low as 90 mm Hg. A more exaggerated version of the same change, known as *venous admixture*, occurs in conditions in which alveoli are underventilated (or not ventilated at all) but in whose walls blood flow still proceeds. Owing to the shape of the dissociation curve (Fig. 18.2) a marked fall in O_2 tension occurs when the (relatively) small amount of non-oxygenated blood mixes and shares the available oxygen with the larger amount of oxygenated blood, but there may be only a small change in the saturation of the haemoglobin of the mixed blood when it is pumped from the heart to supply the body.

The following example illustrates what happens:

1 Suppose that so many alveoli are under/non-ventilated so that 20% of the mixed venous blood is effectively not reoxygenated.

2 This blood therefore reaches the left ventricle with the haemoglobin only 75% saturated at a tension of 5.6 kPa.

3 The remaining 80% of the blood is 95% saturated with oxygen at 100 mm Hg tension.

4 If the blood contains 15 gm/100 ml haemoglobin, then:

    (a) the oxygenated blood contains 19 ml O_2/100 ml, and

    (b) the non-oxygenated blood contains 15 ml O_2/100 ml

5 Each 100 ml of mixed blood contains:

$$\frac{80}{100} \times 19 + \frac{20}{100} \times 15 = 15.2 + 3 = 18.2 \text{ ml}$$

It is therefore *91% saturated with oxygen.*

6 From the oxygen–haemoglobin dissociation curve, one finds that at this saturation the *tension is* 9.1 kPa.

Although, in the body at rest, blood is fully oxygenated in the first 0.1–0.2 sec of its 0.8 sec passage through the pulmonary capillaries, much of this temporal reserve is required in really vigorous exercise. The maximal rate at which oxygenated blood can be supplied to muscles and the muscles can use this oxygen is about sixteen times the resting value. A healthy pair of lungs, by increased ventilation with fresh air, and by an increase in effectively perfused alveoli, can readily achieve as much oxygenation of blood as is required of them in vigorous exercise, for ventilation can increase by thirty times the resting value whereas O_2 delivery to tissues increases by 16–20 times the resting level.

THE DELIVERY OF OXYGEN TO THE TISSUES

The oxygen-using enzymes of mitochondria have a much greater affinity for oxygen than has haemoglobin. The oxygen tension around them is about 0.1 kPa Hg. There is thus a gradient of tension (from the approximately 12.5 kPa in the blood at the arterial end of capillaries) of about 12.4 kPa driving oxygen from the red blood cells, through plasma, capillary endothelium, tissue fluid, cell wall and cytoplasm to the mitochondria. As oxygen leaves the blood this gradient falls until a minimum value is reached, beyond which it is too small to drive the oxygen over the distance that separates haemoglobin in the blood cells from the mitochondria. This minimal value will depend on this distance, and it varies from one tissue to another, and within one tissue as the number of patent capillaries varies from time to time. An average minimal value, that fits resting muscle fairly closely for the oxygen tension gradient is 4.9 kPa. At this tension then, found towards the venous end of the capillaries, the haemoglobin might be 66% saturated, if one looks at a convential dissociation curve. It is in fact only 60% saturated, for, as the oxygen has diffused into the tissue, carbon dioxide has diffused out and has shifted the dissociation curve (the Bohr shift) so that the haemoglobin gives off some additional oxygen. In vigorous metabolism when more CO_2 is being produced, and perhaps lactic acid also, the Bohr shift is even more marked and more oxygen is able to leave the haemoglobin and diffuse down the same oxygen tension gradient to the mitochondria. If the arterial blood arrives with an O_2 tension of only 9.1 kPa, as in the example above when 20% of blood flow through the lungs passed through unventilated alveoli, the gradient for diffusion is reduced, and so the rate of oxygen delivery to mitochondria will be correspondingly lessened, even though the oxygen content of the blood is only slightly affected.

FORMATION AND FATE OF HAEMOGLOBIN

Red blood cells are formed in the red bone marrow (Chapter 9) of ribs, sternum, vertebral column, limb girdles and cranial bones and as they grow there, they produce their own haemoglobin. The protein chains are formed from their constituent amino acids, the porphyrin molecules from simpler substances including the iron atoms. The metabolic machinery for producing the compound depends, ultimately, on the cells' inherited nucleic acids. Abnormalities in these, the genes, can lead to the production of haemoglobins, with chemical and physical properties differing from those of the normal compound (see below). Amino acids and the substances needed for making porphyrin are drawn from the body's metabolic pool. The body carries some stores of iron, but an iron-poor diet, or excessive loss of iron in chronic bleeding states (even in normal menstruation) can drain the stores, so there is no longer sufficient iron to make the normal amount of haemoglobin and anaemia develops (it should be noted that in the 1980s people are increasingly being forced by economic necessity on to an iron-poor diet). Various accessory factors are needed which were described in Chapter 17.

A red blood cell lives for about three months in the blood. It is then removed by 'scavenger' cells lining parts of the circulation in the liver, spleen and other tissues of the reticulo-endothelial system. The haemoglobin molecule is split up. The protein is broken down to amino acids which can be re-used. Some of the iron returns to store but some is lost *en route* in the urine. The porphyrin is not reused, but is converted to *bilirubin* and *biliverdin* which are excreted by the liver cells into the bile. These substances are coloured and their excessive formation or impaired removal lead to the characteristic colour of jaundice, which may be due to excessive red blood cell destruction, liver disease or biliary tract obstruction.

ABNORMAL HAEMOGLOBINS

In the normal haemoglobin molecule the four protein chains are of two types, called for convenience α and β. There are two of each chain in each haemoglobin molecule. A special haemoglobin with raised oxygen affinity is present in the fetus (to enable it to function in the lowered oxygen tensions on the fetal side of the placenta). In this the β chains are replaced with 'gamma' chains. In gamma chains, 37 of the 146 amino acids of the β chain are different.

An altered haemoglobin molecule has been found to be the cause of the inherited disease, *sickle cell anaemia*. In this condition, when the cells become de-oxygenated, they become distorted in shape, sticky and fragile. In the haemoglobin, one single amino acid of the β chain is altered. The alteration is produced by a small change in the nucleic acid gene molecule (two only of its 430 links are changed), but the change in the haemoglobin molecule is such that, when it has lost its oxygen molecules, the β chains of neighbouring molecules can fit into each other and massive haemoglobin polymers are formed which distort and split the cells containing them. Among the consequences of this are damage to synovial membranes of joints which become affected in a manner reminiscent of the joint damage of haemophilia.

MYOGLOBIN

This compound gives the red colour to the dark meat of the turkey leg (and other muscles!). Chemically it is similar to one of the four sub-units of haemoglobin, but it has a higher affinity for oxygen. At a tension of 20 mm Hg it still retains 80 % of the oxygen that is needed to saturate the myoglobin molecule. Its role in muscles is to act as a reserve oxygen supply when activity out-runs the ability of the circulation to supply oxygen.

CARBON DIOXIDE (CO_2)

Carbon dioxide is present in blood at all times as dissolved CO_2 gas, as carbonic acid, as sodium and potassium bicarbonate and as *carbamino* compounds in which it is directly combined with protein (see below). Addition of acid (lactic acid, citric acid or hydrochloric acid) will split the combined forms, producing more CO_2, and will reduce the solubility of the dissolved CO_2 and thus cause CO_2 gas to be liberated. The process is exactly like the addition of lactic acid (in soured milk) to baking soda in various cooking processes. We find that from 100 ml of arterial blood, 48 ml CO_2 gas can be driven off with acid, whereas from mixed venous blood 52 ml CO_2 are formed. The CO_2 tensions are 5.6 and 5.9 kPa respectively for arterial and mixed venous blood. During the passage of the blood through the pulmonary capillaries, the 1.1 kPa gradient is more than enough to remove the extra 4 ml CO_2 carried in each 100 ml of blood entering the capillaries.

DISSOLVED CO_2

As has already been mentioned, at a tension of 5.6 kPa, about 2.5 ml CO_2 dissolves in each 100 ml of plasma water. This will react very slowly with the water to form carbonic acid, a very weak acid $(H_2O + CO_2 \rightarrow H_2CO_3)$. At 5.9 kPa tension the dissolved CO_2 rises to 3.0 ml/100 ml blood.

CARBAMINO COMPOUNDS

CO_2 can react with any spare amino ($-NH_2$) groups that may be present in the amino acids of a protein (two of the twenty amino acids have two amino groups in their molecules, one of which does not join in the formation of the protein chain molecule). Most plasma proteins have a few free amino groups, but haemoglobin has an abundance of them, and their affinity for CO_2 is raised when the haemoglobin becomes de-oxygenated. In arterial blood about 2.5 ml CO_2 are carried as carbamino compound in each 100 ml. blood, and in mixed venous blood the figure rises to 3.5 ml/100 ml blood.

BICARBONATE ION

When CO_2 dissolves in water a small amount changes to carbonic acid. Inside red blood cells carbonic anhydrase enormously speeds the process. The carbonic acid dissociates into hydrogen ions and bicarbonate ions, $H_2CO_3 \rightarrow H^+ + HCO_3^-$. The former get absorbed, in a non-ionic form by the haemoglobin which acts here as a hydrogen ion buffer, so a lot of bicarbonate ion (HCO_3^-) may be present in the blood with no disturbance in its acidity. When red blood cells are present and active in this respect, 43 ml CO_2 are present in 100 ml arterial blood as HCO_3^-, and 45.5 ml in mixed venous blood. Seventy per cent of this HCO_3^- is, in fact, present in the plasma, though it was formed from CO_2 in the red blood cells, and can normally only be converted back into CO_2 in these cells. Table 18.1 summarises these facts concerning CO_2 in blood.

Table 18.1

	Arterial blood	Mixed venous blood
Simple solution (mainly in plasma)	2.5	3.0
Carbamino compound (mainly in red cells)	2.5	3.5
Bicarbonate in plasma	30.0	31.5
Bicarbonate in cells	13.0	14.0
Total	48	52

THE ADDITION OF CO_2 TO BLOOD

We must now follow the sequence of reactions that occur as, in metabolising tissues, CO_2 enters the blood from the tissue fluid.

The first point to note is one not involving CO_2 at all, but the oxygenation of haemoglobin. Oxygen first acquired its name in the beginning of scientific chemistry because in its combination with other elements (sulphur, nitrogen, phosphorus and carbon) it produced acidic products. The *oxygenation* of the bivalent iron atoms of the haemoglobin molecule, though not a chemical reaction in the same sense as the *oxidation* of the other free elements listed above, is accompanied by a similar, though small, change in the haemoglobin molecule. Oxygenated haemoglobin behaves in cell fluid as a weak acid and requires alkaline ions to neutralise it. Potassium ions are available, and it would be correct, though cumbersome, to refer to oxyhaemoglobin as *potassium oxyhaemoglobinate*. When haemoglobin releases oxygen to metabolising tissue, it also releases some of its bound potassium, which can be replaced in the protein molecule by hydrogen.

When CO_2 diffuses into the red blood cells, carbonic anhydrase acts upon it and water to produce carbonic acid, or H_2CO_3. Some of this ionises, producing ionic hydrogen (H^+) and bicarbonate ions, HCO_3^-. An exchange of hydrogen with potassium can occur, since the haemoglobin is simultaneously losing its oxygen, with the result that potassium bicarbonate accumulates within the cells. The equilibrium state between cells and plasma is now upset and some of the HCO_3^- moves out into the plasma, being replaced inside the cells with Cl^-, moving in the opposite direction. This movement of Cl^- into the cells as CO_2 enters blood has been christened the *chloride shift*. You should look upon it as the last of a series of chemical reactions which accommodate the

Table 18.2

Tissue cells	Tissue fluid	Plasma	Red blood cells	Process no.
CO_2 diffusion down tension gradient $\longrightarrow$				1
	$CO_2 \rightarrow H_2CO_3$	$CO_2 \rightarrow H_2CO_3$	$CO_2 \rightarrow H_2CO_3$	2
	spontaneously	slowly	rapidly, due to carbonic anhydrase	
			$H_2CO_3 \rightarrow H^+ + HCO_3^-$ spontaneous ionisation	3
			$K\,HbO_2 \rightarrow K^+ + Hb^- + O_2$	4
			$Hb^- + H^+ \rightarrow H\,Hb$ neutralisation of H^+ by haemoglobin	5
	$Cl^- \xrightarrow{}$	$\xleftarrow{} HCO_3^-$ re-establishing equilibrium between plasma and red blood cells		6

extra CO_2 gained by blood as it flows through the capillaries of metabolising tissues. The chain of reactions is summarised in (more or less) conventional chemical terms in Table 18.2.

Naturally, when the blood reaches the lungs and gives off CO_2, the whole process operates in reverse.

THE BUFFERING POWER OF BLOOD

In the account just given, one can see an example of what chemists call *buffering reactions*. These are reactions that automatically offset the changes in hydrogen ion concentration of a reagent mixture that might occur on the addition of strong acidic or alkaline substances. We will take as an example a mixture of a weak acid, acetic acid (from vinegar), and its sodium salt, sodium acetate. Such a mixture will itself be slightly acidic. If a small amount of a strong acid is added, like hydrochloric acid, it will react with the sodium acetate, producing neutral sodium chloride and more of the weakly acidic acetic acid. Thus the addition of a strong acid produces only little effect, an equivalent quantity of the weak acid being formed. No buffer system can be perfect. Their reaction is such as to minimise the effects of adding the strong acid substance to the mixture, not to totally abolish it. You will realize that the word 'buffer' has nothing to do with railway buffers. The word is of German origin, and means a tampon or sponge—it 'mops up' excess H^+ ions.

In the body fluids there are various pairs of buffering substances. One of the most important of these, present in all parts of the body, is the bicarbonate–carbonic acid system. In red blood cells there is the 'potassium haemoglobinate'–haemoglobin system, and in all tissue fluids there is a system formed from small amounts of the two phosphate salts. When an acid, like lactic acid (see Chapter 23), enters body fluids it is a far stronger acid than carbonic acid, and the hydrogen ions it could liberate would have serious consequences for many aspects of bodily function. In tissue fluid and plasma, lactic acid reacts with sodium bicarbonate, producing sodium lactate and carbonic acid, which, as an acid is far weaker than lactic acid and scarcely produces any hydrogen ions. It has a further advantage in that carbonic acid splits up to form CO_2 and water. The CO_2 stimulates respiration and is thus lost into the atmosphere.

The one potentially acidic product of metabolism that this bicarbonate buffer system cannot neutralise is, of course, CO_2 itself. Here it is the reaction between carbonic acid and potassium haemoglobinate, that buffers the carbonic acid formed in the red blood cells by carbonic anhydrase when extra CO_2 enters the cells. *Hydrogen haemoglobinate* is a far weaker acid again than is carbonic acid and its formation thus minimises the change in hydrogen ion concentration that would otherwise occur when CO_2 enters the blood.

You should not read and attempt to understand this chapter in isolation, but should refer constantly to the chapter on respiration that follows it, and also to the general information on cell metabolism, in Chapters 2 and 23. The blood is but an intermediary between the alveolar gases and the metabolising tissues; it is not possible therefore, to consider the transport of gases in the blood without, at the same time, considering their exchanges at lung and tissue capillaries, the use of oxygen and production of CO_2 and other substances within tissues, and the exchange of these gases between alveolar and atmospheric air. In writing this book we are forced, inevitably, to treat the different stages of the process in different chapters, but you, the reader, must seek out the way the different sections link together into a complete picture.

19 · The Respiratory System

Respiration is a term used to cover two very different processes. *External respiration* is the exchange of oxygen (O_2) and carbon dioxide (CO_2) between a living body and its environment; it is the concern of this chapter. *Internal respiration* includes exchange of gases between single cells and the extracellular fluids, and also the intracellular reactions involving O_2 utilisation and CO_2 production. Internal respiration is included in the chapter on gas transport (Chapter 18) and in the chapters concerned with general metabolism (Chapters 2 and 24).

The subject of external respiration falls into five sections. First we will describe the anatomical means whereby atmospheric air is brought into close proximity to the blood—the structure of the lungs and the bronchi which connect them to the exterior, and also the structures of the chest wall, the movements of which alternately draw air into and expel it from the lungs. Then we will describe the pressure and volume changes within and surrounding the lungs during their inflation and deflation. After that we will consider the alterations that take place as air is drawn into the lungs and the composition of the considerable volume of air that is always contained in the lungs in life. Next comes the final stage, diffusion of O_2 and CO_2 between air in the lungs and blood in the capillaries of the pulmonary circulation. Finally, in describing the normal process of external respiration, we must consider how these processes are regulated, so that ventilation is continually adjusted to the body's rates of O_2 usage and CO_2 production. Throughout this chapter, where it is relevant, we will show how different disease states interfere with the normal functions, and how treatment measures may offset the effects of the diseases.

THE ANATOMY OF THE RESPIRATORY SYSTEM

The respiratory system comprises not only the lungs, the conducting airways and their associated blood vessels but also their covering—the pleura—and the structures associated with respiratory movements, i.e. breathing. These include the skeleton of the thorax, the muscles of the thoracic wall, the diaphragm and certain other muscles which are called into play when particularly powerful respiratory movements are required. Respiration in general is due to elevation of the ribs which increases the transverse and antero-posterior diameters of the thorax (thoracic respiration) and to movements of the diaphragm which compresses the abdominal contents and increases the vertical dimension of the thorax (abdominal respiration).

THE CHEST WALL

If you look at a complete mounted skeleton you will notice that the ribs slope downwards from their attachment to the vertebral column. With a few exceptions, the costal cartilages ascend from the end of the rib to reach the sternum. Thus, in general, the lowest point along the whole length of each rib and costal cartilage is laterally placed, while the anterior end of each rib–costal cartilage unit is lower than the posterior end. These observations are important in considering how movements of the ribs increase the diameters of the thorax.

The movements of the ribs during ordinary quiet respiration are a little complicated. In the case of the first, and to some extent the second, rib, little or no movement takes place so that the first rib and manubrium sterni form a

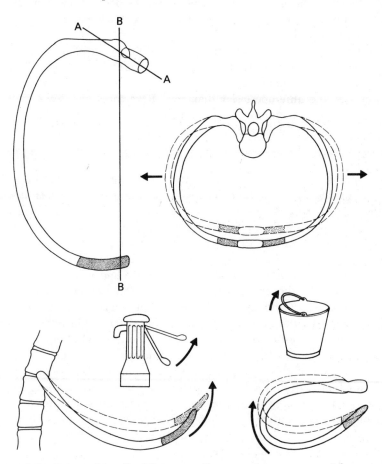

Fig. 19.1. Movements of the ribs. The upper ribs move about two axes. Movement about the axis A will elevate the anterior end, producing the 'pump handle' type of movement. Movement about the axis B will elevate the middle of the rib, producing a 'bucket handle' type of movement. The lower ribs open out like callipers, as shown in the diagram at top, right.

more or less stationary unit. The remaining true ribs (3–7) move around two axes. The first type of movement is sometimes called the '*bucket-handle*' movement in which the rib may be regarded as pivoting around its anterior and posterior ends (Fig. 19.1). The rib (and costal cartilage) then moves like the handle of a bucket, the most lateral part being elevated or depressed. Since the central part of the rib–costal cartilage unit is lower than either of the two ends, elevation of the 'bucket-handle' type will increase the side-to-side diameter of the thorax. The second type of movement is the so-called '*pump-handle*' movement. Here, the axis runs along the head of the rib so that movement of this type will elevate the anterior end of the rib and costal cartilage and thus elevate the body of the sternum, since the anterior end of the costal cartilage is lower than the posterior end of the rib. This movement increases the anteroposterior diameter of the thorax. The manubrium does not move, so that the body of the sternum hinges on the manubrium, the movement taking place at the secondary cartilaginous joint between manubrium and body.

The lower ribs (7–10) move in a rather different fashion because of the shape of their articular facets. They open out laterally like a pair of callipers (Fig. 19.1). This movement increases the side-to-side diameter of the thorax but decreases the anteroposterior diameter. Remember, however, that these ribs enclose mainly abdominal viscera rather than the lungs, so that this type of movement is helpful in accommodating the displaced abdominal contents during descent of the diaphragm. The two floating ribs have little effect on respiration.

The spaces between the ribs—the *intercostal* spaces—are filled in by three strata of muscles. The most superficial stratum comprises the *external intercostals* and the *external intercostal membrane*. These muscles are necessarily short and the fibres run downwards and forwards from the lower border of one rib to the upper border of the next. In the anterior part of each space however, the muscle fibres are replaced by the external intercostal membrane. The whole complex is thus very similar to the external oblique muscle and its aponeurosis in the abdomen. Deep to the external intercostals lie the *internal intercostals* whose fibres run downwards and backwards, i.e. at right angles to the external inter-costals. Posteriorly, the muscle fibres are replaced by a thin *internal intercostal membrane*. This layer of muscles corresponds to the internal oblique muscles of the abdominal wall. Finally, a third layer of muscle fibres is incomplete, only the anterior part of this layer being easily recognised. This is the *transversus thoracis* which arises from the back of the sternum and is inserted into the costal cartilages of the lower 3 or 4 true ribs. This incomplete stratum of muscles corresponds to the transversus abdominis, the nerves and vessels being found superficial to this layer in both cases. The intercostal muscles are supplied by branches from the intercostal nerves.

The actions of the intercostal muscles are extremely difficult to determine. There is some theoretical evidence that the external intercostals elevate the ribs and electromyography shows this to be the case in at least some of the intercostal spaces. The reason for this action is shown in Figure 19.2. When the

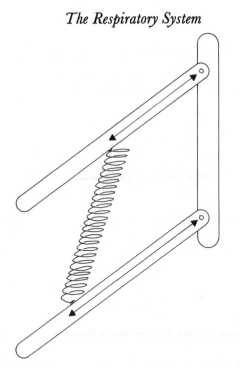
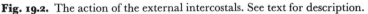

Fig. 19.2. The action of the external intercostals. See text for description.

muscle fibres contract, the leverage on the lower rib is greater than that on the upper rib so that the lower rib is elevated rather than the upper rib depressed. According to this theory, the internal intercostals should have the opposite action but there is little evidence that this is the case. On the whole, therefore, the most that one can say is that the intercostal muscles certainly play an important, but not indispensible, role in respiration but that their precise action is still uncertain.

THE DIAPHRAGM

The diaphragm is a rather unusual muscle in that it is composed of skeletal muscle fibres which are under the control of the will so that the rate of respiration can be voluntarily changed or, if, for example, your hobby is diving, the respiration can be stopped for some time. On the other hand, when you are asleep or deeply anaesthetised, respiration (fortunately) continues by means of rhythmic discharges from the respiratory centre (see below).

The diaphragm arises (Fig. 19.3) from the inner surfaces of the lower six ribs (interdigitating with the origin of transversus abdominis) from the bodies of the upper two lumbar vertebrae (and also from the third on the right) by means of the right and left *crura* (sing. *crus*) and from fibrous arches over psoas and quadratus lumborum (the *medial* and *lateral lumbocostal arches*). Between the

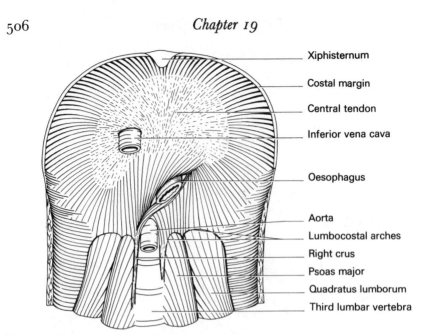

Xiphisternum

Costal margin

Central tendon

Inferior vena cava

Oesophagus

Aorta

Lumbocostal arches

Right crus

Psoas major

Quadratus lumborum

Third lumbar vertebra

Fig. 19.3. The under surface of the diaphragm.

two crura, a fibrous arch bridges over the aorta, and diaphragmatic fibres also arise from this. Anteriorly, there is also a small origin from the back of the xiphoid process. From this more or less circular origin, the fibres ascend and converge to be inserted into a substantial *central tendon* which is trefoil in shape. The upper surface of the central tendon is fused with the fibrous pericardium, while the lower surface is in contact with the liver. There are three main openings in the diaphragm, for the *inferior vena cava* (in the central tendon), for the *oesophagus* (just to the left of the midline) and for the *aorta* (in the midline between the two crura), and these openings lie at the levels of the 8th, 10th and 12th thoracic vertebrae respectively. The motor nerve supply to the diaphragm is from the *phrenic nerve* (C3, 4 and 5, mainly C4). The nerve thus arises in the neck, comes to lie on the anterior surface of scalenus anterior and then has to travel the whole length of the thorax to reach the diaphragm. The two phrenic nerves lie on either side of the mediastinum. The dome of the diaphragm rises on each side to the level of the 5th intercostal space during quiet respiration so that the lower part of the rib cage is separated by the diaphragm from abdominal rather than thoracic viscera. The liver, spleen, upper parts of the kidneys and the fundus of the stomach are thus direct relations of the diaphragm. For this reason, pain in the lower oesophagus or upper part of the stomach is often referred to by laymen as 'heartburn' since the heart and fundus of the stomach are separated only by the thickness of the diaphragm.

For the understanding of the action of the diaphragm it is important to remember that it is not a transverse partition but is dome-shaped, the muscle fibres having to ascend to reach the central tendon. The peripheral part of the diaphragm in fact is almost vertical (Fig. 19.4). When the muscle fibres contract,

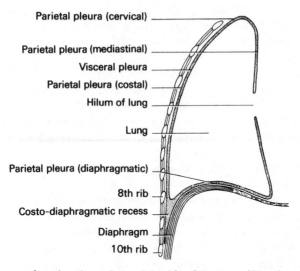

Parietal pleura (cervical)

Parietal pleura (mediastinal)

Visceral pleura

Parietal pleura (costal)

Hilum of lung

Lung

Parietal pleura (diaphragmatic)

8th rib

Costo-diaphragmatic recess

Diaphragm

10th rib

Fig. 19.4. A coronal section through the right side of the chest. The pleural cavity is stippled, although in life it contains only a thin film of fluid.

the attachments to the lower six ribs may be regarded as the origin of the muscle, so that the initial effect is to lower the whole central tendon by about 1.5 cm without greatly altering its shape. This increases the vertical dimensions of the thoracic cavity but also compresses the abdominal contents. This latter factor hinders the final descent of the diaphragm so that the central tendon, curved over the upper surface of the now stationary liver, becomes a fixed structure. The central tendon thus becomes the origin of the muscle fibres and their continued contraction will elevate their insertion, i.e. the lower ribs, the dome of the diaphragm becoming flattened.

The diaphragm has other actions besides that of respiration. It contracts during any expulsive action of the abdominal wall such as occurs in micturition, defaecation, childbirth, etc. The descent of the diaphragm together with the contraction of the abdominal muscles raises the intra-abdominal pressure. This is particularly noticeable in vomiting, when the rapid inspiration that precedes this activity is usually audible. The diaphragm probably also helps to prevent reflux of stomach contents up the oesophagus.

THE MOVEMENTS OF RESPIRATION

It is now possible to understand the mechanics of breathing. In quiet breathing, the first rib and manubrium (and to some extent the second rib) form a fixed unit from which the intercostals can act to elevate the ribs in inspiration. The upper ribs move both upwards and outwards by means of their 'pump handle' and 'bucket handle' actions while the lower ribs expand the thoracic cavity laterally by their 'calliper' action. At the same time, the diaphragm descends to increase the vertical diameter, until the final stages of its contraction when it helps to elevate the lower ribs. At the end of inspiration, the muscles relax and expiration is brought about by the elastic recoil of the large amount of elastic tissue in the lungs and the surface tension of the fluid lining the alveoli (see also p. 514). Expiration is thus normally a purely *passive* process. From the physiotherapist's point of view it is important to realise that it is not only possible voluntarily to breathe using the diaphragm alone or the intercostals alone, but also that a good physiotherapist can train a patient to emphasise movements of particular parts of the rib cage so that the lung can be expanded selectively in these regions.

In forced or energetic respiration, a number of other muscles take part and the first rib and the whole sternum are elevated by the neck muscles—the scalenes and the sternomastoid. The intercostal muscles are also active in expiration as well as inspiration, and other muscles which are attached to the ribs come into play. In order to do this, the upper limbs must be fixed (by holding on the side of the bed, for example) and then such muscles as latissimus dorsi and the pectoral muscles are able to work from insertion to origin. Serratus anterior seems an obvious accessory muscle of respiration but it is not altogether certain whether this is the case. In forced expiration, (including coughing), there is vigorous contraction of the abdominal muscles which, by increasing the intra-abdominal pressure, push the diaphragm upwards. These muscles also assist expiration by depressing the lower ribs.

Respiration is possible without diaphragmatic movement, *or* without the intercostal muscles (but not without both). Interruption of one phrenic nerve causes the diaphragm to rise and become immobile on the affected side and there is an initial fall of about 20 % in the vital capacity. Even if both phrenic nerves are interrupted, the functional disability is slight—in fact, the diaphragm is largely immobile in late pregnancy for obvious reasons. On the other hand, in young babies and in patients with emphysema, the ribs are almost horizontal—notice the barrel shaped chest—so that thoracic respiration is hardly possible and breathing is carried out mainly by the diaphragm. If both the diaphragm and the muscles of the chest wall are immobilised, as may happen in a lesion of the spinal cord above C3 or 4, all breathing stops and artificial ventilation is essential for life.

THE PLEURA

The inside of the chest and the outside of the lung are covered by a thin membrane called the *pleura*. During early development the lung pushes its way into the pleural cavity and thus becomes covered by the membrane which, where it is attached to the lung, is called the *visceral pleura*. The outer layer, which lines the chest cavity is the *parietal* pleura. The two layers are continuous at the hilus of the lung. The parietal pleura is further subdivided into the *mediastinal pleura* which covers the sides of the heart and great vessels, etc., the *diaphragmatic*

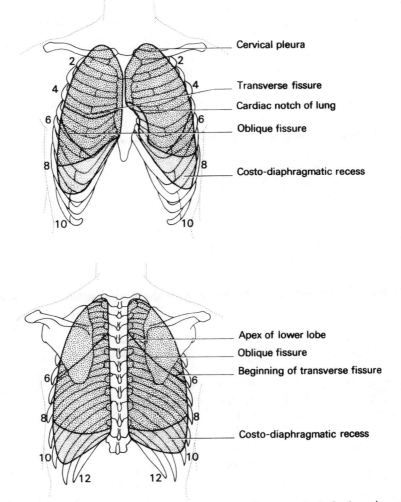

Cervical pleura

Transverse fissure

Cardiac notch of lung

Oblique fissure

Costo-diaphragmatic recess

Apex of lower lobe

Oblique fissure

Beginning of transverse fissure

Costo-diaphragmatic recess

Fig. 19.5. The surface markings of the lungs and pleura. Note particularly the apices of the upper and lower lobes.

pleura which covers the upper surface of the diaphragm, the *costal pleura* which lines the chest wall and the *cervical pleura* which covers the apex of the lung and rises above the medial third of the clavicle. Figure 19.4 is very diagrammatic for, in life, the partietal and visceral layers are in contact, with only a thin film of fluid between them. During respiration the two layers move gently over each other. When the pleura becomes infected however, deep breathing may cause a sharp pain as the layers tend to stick together, and in chronic lung diseases there are often extensive adhesions between visceral and parietal layers. The surface markings of the pleural cavities are easy to remember—the even numbers from 2 to 12 are a guide (Fig. 19.5). Starting behind the sternoclavicular joints,

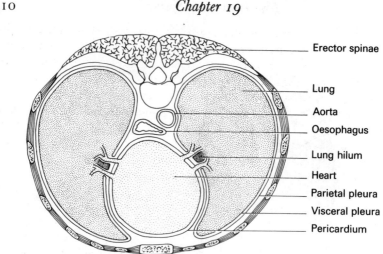

Fig. 19.6. Cross-section through the chest.

draw lines downwards and medially to reach the midline at the level of the *2nd* costal cartilage. From here, continue both lines downwards together to the level of the *4th* costal cartilage. On the right side, this line continues to the *6th* costal cartilage but on the left it diverges to reach the lateral border of the sternum at the *6th*. On both sides, the lines now cross the *8th* rib in the mid-clavicular line, the *10th* rib in the midaxillary line and the *12th* rib just before turning upwards at the level of the 12th thoracic vertebra about a finger's breadth from the midline. The cervical pleura rises to a level of the neck of the 1st rib, i.e. about two finger's breadth above the medial third of the clavicle.

The lungs, with their covering of visceral pleura, do not completely fill the pleural cavity. The indentation in the left lung between the 4th and 6th costal cartilages (the *cardiac notch*) is rather larger than that of the pleura. Also the lower border of the lung is about two intercostal spaces above that of the pleura. Thus the lower borders of the lungs cross the 6th rib in the mid-clavicular line and the 8th rib in the midaxillary line. There is thus a part of the pleural cavity below the lower border of the lung—the *costodiaphragmatic recess*—which is 'empty'. That is to say, the diaphragmatic and costal pleura are in contact with no intervening lung. However, during a deep inspiration, the lower edge of the lung expands into this potential space and separates the layers.

THE LUNGS

The lungs have surfaces which correspond to those of the parietal pleura. There is a rather concave medial surface which is separated from the bodies of the vertebrae and the mediastinal structures by the mediastinal pleura. The concavity is much more pronounced in the left lung than the right since it has to accommodate the heart which extends more to the left of the midline than to the right. The left lung also, for this reason, has a cardiac notch (Fig. 19.5). The diaphragmatic surface is also concave and here the left lung has a deeper concavity than the right lung because the right dome of the diaphragm is slightly higher than the left due to the presence of the liver on that side. The costal surface is extensive and leads smoothly into the rounded posterior border which fits into the hollow produced by the posterior ends of the ribs and the vertebral bodies. The anterior border, on the other hand, is sharp since it has to fit in between the heart and the anterior chest wall (Fig. 19.6). The apex of the lung extends above the clavicle to the same extent as the cervical pleura.

The lungs are divided into lobes by one or two fissures. The right lung has an *oblique* and a *horizontal fissure* which subdivide it into *upper*, *middle* and *lower lobes* (Fig. 19.8). The oblique fissure runs downwards and forwards and may be marked on the surface by a line from the fourth thoracic spine round the side of the chest to reach the lower border of the lung near the junction of the 6th rib with its costal cartilage, 7–8 cm from the midline. As a rough approximation, the oblique fissure may be marked out by a line along the medial border of the scapula and its downward prolongation, when the arm is abducted by placing the hand behind the head. The portion of lung below the oblique fissure is the *lower lobe*. The shorter horizontal fissure may be marked on the surface by a line from the oblique fissure in the midaxillary line to reach the anterior border of the lung at the level of the 4th costal cartilage. This marks out a rather triangular *middle lobe* and the remaining part of the lung above the horizontal fissure is the upper lobe, whose *apex* projects above the medial third of the clavicle. If you look at the lung from behind, you will realise that the lower lobe, as well as the upper lobe, has an *apex* which reaches its highest point at the level of the 4th thoracic spine.

In the left lung there is only an oblique fissure whose position corresponds approximately to that of the corresponding fissure of the right lung although it may be slightly more vertical (Fig. 19.9). The left lung therefore only has two lobes. There is often a small forward projection of the superior lobe at the lower border of the cardiac notch called the *lingula*. This region of the lung corresponds to the middle lobe of the right lung and the arrangement of the bronchi which pass to this region is similar on both sides (see below).

The *hilus* of the lung is the region where various structures enter or leave the lung. It is enclosed in a sleeve of pleura which is intermediate between parietal and visceral pleura. In the hilus of each lung are found the principal bronchi (see below), the pulmonary artery, two pulmonary veins, a plexus of autonomic nerves (the pulmonary plexus) derived from the sympathetic system and the vagus nerve (parasympathetic), lymphatics and a number of lymph nodes. The latter are usually black in the adult lung owing to their content of carbon particles which have been inhaled and carried to the lymphatic system by phagocytic cells (*dust cells*).

THE AIR PASSAGES

The inhaled air passes through the nose (or mouth if the nose is obstructed) into the pharynx. It then passes through the larynx (the 'voice box'), past the vocal cords, and then into the trachea. The trachea bifurcates into right and left principal bronchi which enter the corresponding lungs before dividing further.

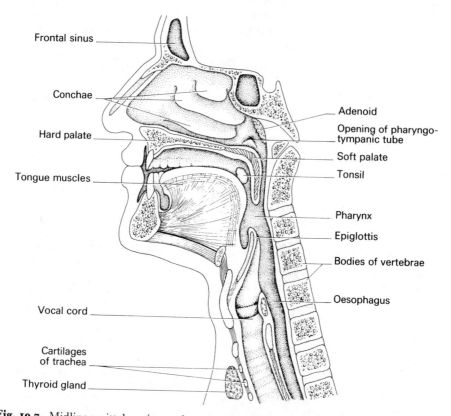

Fig. 19.7. Midline sagittal section to show the upper respiratory tract and the mouth.

THE NOSE AND PHARYNX

These are shown in section in Fig. 19.7. The nose contains a number of thin bones (*Conchae*, sing. concha) covered with mucous membrane. The epithelium is of the columnar ciliated type and the sub-epithelial tissue contains numerous mucus-secreting glands. It is also so vascular as to deserve the name, *erectile tissue*. The blood vessels are innervated by both divisions of the ANS. The physiology of the nasal mucous membrane is described below on p. 518. A number of cavities, the *paranasal sinuses*, open into the nose and were described in Chapter 13.

THE LARYNX

At the upper end of the trachea is situated the *larynx* which contains the vocal cords. The opening into the larynx is guarded to some extent by an almost vertical leaf-shaped piece of elastic cartilage, covered by mucous membrane, called the *epiglottis*. This lies in front of the opening into the larynx. When fluids are swallowed, it acts as a form of breakwater, so that the fluid is diverted into two streams which pass on either side of the opening. At the same time, respiration is temporarily halted so that inhalation of the fluid does not occur. During swallowing too, the larynx is elevated (this is easily visible), and muscles around the inlet of the larynx perform a sphincter-like action in narrowing the inlet and pulling the epiglottis backwards, again protecting the inlet to the larynx during swallowing. The functions of the larynx are therefore not only concerned with voice production, they also include preventing the entry of food or liquid into the respiratory tract. If, in spite of the mechanisms to prevent it, this should occur, the sensory nerve supply of the larynx and trachea is stimulated by the inhaled material and sets up a vigorous *cough reflex* which usually rapidly expels the offending foreign body.

THE TRACHEA

The trachea is about 12 cm long and extends from the lower end of the larynx in the neck to just below the level of the manubrio-sternal joint (5th and 6th thoracic vertebrae). It is a little less than 2 cm in diameter and is held open by a series of incomplete cartilaginous rings which give it a characteristic appearance (Fig. 26.5). The rings are deficient posteriorly where the trachea is in contact with the oesophagus so that the free passage of food down the latter is not obstructed. The deficiency is filled in by connective tissue and smooth muscle. The trachea is lined by pseudo-stratified ciliated columnar epithelium with scattered goblet cells and the mucous membrane also contains glands which secrete mucus. The secretions of the glands emerge on the surface and are moved by the cilia upwards and out of the respiratory system, along with any small small foreign particles such as inhaled dust. Larger quantities of secretion may be removed by coughing.

THE PRINCIPAL BRONCHI

At the level of the 5th or 6th thoracic vertebra (just below the manubrio-sternal joint) the trachea divides into right and left *principal bronchi*. The former is slightly wider than the latter and is more in line with the trachea, so that inhaled foreign bodies are more likely to be found in the right lung than the left! The bronchi pass to the hila of the lungs where they divide further into *lobar* and *segmental bronchi*. The bronchi, like the trachea, are surrounded by incomplete cartilaginous rings.

FURTHER SUBDIVISIONS OF THE AIR PASSAGES

The *right bronchus* divides into three divisions, one for each lobe. A little over 2 cm from its origin, the right bronchus gives off the *upper lobe bronchus* from its lateral aspect. About a similar distance farther along its course, the *middle lobe bronchus* is given off and passes downwards, forwards and laterally and the main bronchus then continues downwards and rather backwards to become the lower lobe bronchus. The *left bronchus* bifurcates about 5 cm from its origin into *upper* and *lower lobe bronchi*. The former, after a short course gives off a *lingular bronchus* which corresponds to the right middle lobe bronchus although, of course, there is no separate middle lobe in the left lung. It passes downwards, forwards and laterally.

Each of the subdivisions mentioned above divides further into *segmental bronchi* (see below). Each of these, in turn, divides further into subdivisions which themselves branch repeatedly until the smallest bronchi are reached. These are called *bronchioles* (just as the smallest arteries are called arterioles). The bronchioles give rise to *respiratory bronchioles*, so called because a number of the actual respiratory sacs of the lung—the *alveoli*—are given off from their side walls. The respiratory bronchioles end in a cluster of very thin-walled *alveolar ducts* which in turn open into *alveolar sacs* from each of which a number of *alveoli* arise.

The microscopic structure of the air passages becomes less and less complex as they are traced peripherally. The intrapulmonary bronchi do not have cartilage rings, but merely irregular pieces of cartilage in their walls with intervening smooth muscle and even this cartilage disappears by the time the bronchioles are reached. The pseudostratified ciliated columnar epithelium of the trachea and bronchi gives way to a lower columnar epithelium, then cubical non-ciliated epithelium and finally, in the alveoli, to an extremely thin squamous epithelium. Some of the non-ciliated cells are secretory in character and these, like occasional cells in the alveoli, are believed to secrete *surfactant*. This is a substance which reduces the surface tension of the alveoli and helps to prevent their complete collapse during a deep expiration. It is also important in the newborn baby in helping the alveoli to expand. Smooth muscle fibres continue down to the respiratory bronchioles; the whole of the interstitial tissue of the lung contains large numbers of elastic fibres.

The alveoli themselves are small saccules with extremely thin walls. They are lined by very thin epithelial cells outside which lies a small amount of connective tissue. Beyond this again is the capillary network which surrounds the alveoli. The blood in the capillaries is thus very close indeed to the air in the alveoli, the total thickness of the alveolar wall (i.e. alveolar epithelial cells, connective tissue and capillary endothelial cell) being as little as 0.2 μm in places (about one-fortieth the diameter of a red cell).

THE PULMONARY ARTERY

The main pulmonary trunk leaves the right ventricle alongside the aorta and divides under the arch of the aorta into right and left pulmonary arteries (Fig. 14.2). Each pulmonary artery enters the hilus of its lung with the main bronchus and thereafter divides repeatedly within the lung, the divisions corresponding to those of the bronchi. Each segmental bronchus is thus accompanied by a branch of the pulmonary artery as are the bronchioles, right down to their smallest divisions. When these small arteries and arterioles finally reach the alveoli, they break up to form a rich capillary network around each alveolus. From this, venules arise which unite to form veins. The veins do not, however, accompany the corresponding arteries and bronchi but rather lie in between the segments (see below). Finally the veins of each lung unite to form two pulmonary veins on each side which leave the lung hilus below and in front of the artery and enter the left atrium of the heart.

BRONCHOPULMONARY SEGMENTS

As was mentioned above, each principal bronchus divides within the lung into branches for the lobes and within each lobe, further subdivision takes place into *segmental bronchi*. The segmental bronchi are fairly constant in direction and in distribution although, of course, variations do occur quite commonly. They are however constant enough to be named. Each segmental bronchus subdivides into smaller and smaller air passages and eventually alveoli so that the whole lung can be subdivided into segments, each of which is supplied by a segmental bronchus. There are no communications between the air passages of one segment and those of the next so that should one segmental bronchus become blocked, the corresponding segment of the lung will lose its air supply and eventually collapse. Each bronchus and its subdivisions are accompanied by branches of the pulmonary artery so that the lung tissue of each segment is supplied by a segmental artery. The whole complex is called a *bronchopulmonary segment*. The segmental bronchi and the position of the segments are shown in Figs. 19.8 and 19.9. The veins, however, are intersegmental so that each segment is drained by more than one vein and each vein drains more than one segment.

A knowledge of the bronchopulmonary segments is important both to the physiotherapist and to the clinician. It is possible for a surgeon to ligature a segmental bronchus and its accompanying artery and excise the whole of one segment for a localised lesion. If a portion of the lung is found to be collapsed, either by clinical examination or by x-ray, it is possible to deduce which segmental or lobar bronchus is effected. For the physiotherapist, however, the main

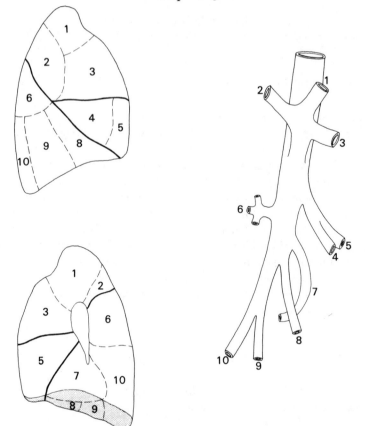

Fig. 19.8. The bronchopulmonary segments of the right lung, lateral and medial surfaces. The segmental bronchi, seen on the right of the diagram, are viewed from the *lateral* side.

interest lies in the possibility of *postural drainage* of secretions from an affected segment or lobe. When a bronchus is occluded by mucus or muco-pus it is extremely important that it should be cleared as soon as possible, either by coughing or by putting the patient in such a position that gravity will assist the activities of the physiotherapist in draining the secretions. Also, in conditions such as bronchiectasis the ciliated epithelium may be destroyed and the bronchi become dilated and rigid so that fluid pools in the air passages.

A study of Figs. 19.8 and 19.9 will suggest appropriate positions for postural drainage. Details of this technique will be found in clinical textbooks but in general it may be said that the upper lobes drain well in the upright position although in the case of the lingula and of the right middle lobe the patient should lie on his back with the foot of the bed raised, and then turn to the right or the left. For the lower lobes the patient must be positioned with the head down. The young and

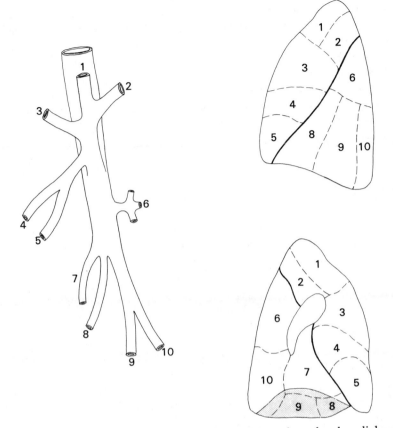

Fig. 19.9. The bronchopulmonary segments of the left lung, lateral and medial surfaces. The segmental bronchi, seen on the left of the diagram, are viewed from the *lateral* side.

otherwise fit patient may lean over the side of the bed with some support for the arms on a low table but in the elderly such an extreme position needs to be modified.

RIGHT LUNG

The superior *lobe bronchus* is rather short and runs upwards and laterally. It divides into three segmental bronchi. The *apical segmental bronchus* continues upwards and laterally, the *posterior segmental bronchus* runs backwards and laterally and the *anterior* segmental bronchus runs anteriorly and downwards.

The *middle lobe bronchus* divides into two after a short course downwards and forwards. These are the *lateral* and *medial segmental bronchi*.

The *inferior lobe bronchus* continues downwards posteriorly and laterally. Its first branch is the *superior* or *apical* bronchus which arises from its posterior surface and runs backwards to supply the uppermost part of the lower lobe. It then gives a *medial basal segmental bronchus* which runs downwards and medially to supply a rather small segment below the hilus. After a further short course, the principal bronchus finally breaks up into three more branches: the *anterior basal, lateral basal* and *posterior basal segmental bronchi* which pass downwards and anteriorly, downwards and laterally, and downwards and posteriorly respectively. There is commonly an additional segment in the lower lobe. This is the *subapical segment* and is supplied by a subapical bronchus which arises from the posterior surface of the main bronchus a little below the apical bronchus.

THE LEFT LUNG

The *superior lobe bronchus* soon divides into two branches which correspond to the superior and middle lobe bronchi of the right lung. The upper division is distributed in a manner very similar to the superior lobe bronchus of the right lung except that the apical and posterior segmental bronchi usually arise by a common stem which is then called the *apico-posterior segmental bronchus*. The lower division is the *lingular bronchus* which divides into *superior* and *inferior lingular* segmental bronchi.

The *lower lobe bronchus* runs downwards, backwards and laterally before giving off a *superior* or *apical segmental bronchus* similar to that of the right lower lobe. It then usually divides into two. Two segmental bronchi arise from each of these: the *medial* and *anterior basal segmental bronchi* from the anteromedial division and the *posterior* and *lateral basal segmental bronchi* from the posterolateral division.

THE UPPER RESPIRATORY TRACT

The nasal passages, trachea and bronchi not only conduct atmospheric air into the depths of the lungs, where gaseous exchange with the blood can occur, but they also ensure that the air that reaches alveoli is not able to cause any damage. It must be warmed, humidified and cleared of any dust particles or micro-organisms.

In the nose the very vascular, mucus-secreting mucous membrane covering the conchae ensures that the incoming air reaches body temperature and is fully saturated with water vapour. The large surface area of the conchae aids this process. The columnar epithelial cells are ciliated, the cilia ensuring that any dust particles, etc. that are trapped in the mucus are removed by swallowing or expectoration. Sympathetic nerve stimulation causes constriction of the erectile vascular component and so shrinking of the mucus membrane, while parasympathetic stimulation and inflammatory processes, colds and hay fever have the opposite effect. The ANS effects are controlled from a special hypothalamic centre which ensures a regular alternation in vasoconstriction and thus in resistance to air flow of the two sides of the nose. The ANS also reflexly ensures that the upper nostril remains open when you sleep on one side by maintaining its erectile tissue in the collapsed constricted state. The receptors for this reflex are pressure receptors in the axilla, and their effect can be simulated by using a crutch. The whole of the nasal mucous membrane becomes more constricted in vigorous exercise, thus reducing total resistance to air flow. It should be noted that much of the heat and water added to air as it enters the respiratory tract, causing cooling and drying of the nasal mucosa, is returned to the mucous membrane as the warm and moist air is again exhaled through the same passages.

In the trachea and bronchi the ciliated columnar epithelium continually sweeps a layer of mucus, secreted by its goblet cells and simple glands, upwards towards the glottis. This helps to clear the lungs of any dust particles, etc. that get beyond the nose (larger amounts of secretion are removed by coughing). These processes are not entirely successful as the black lungs of a person living in an urban environment show, in whom the sooty particles have been cleared from the alveoli by macrophages and deposited in the tissue spaces of lung parenchyma and the pulmonary lymph nodes. Some irritant particles, silica and asbestos, may also, if small enough, enter the alveoli and lung parenchyma, eventually causing serious damage to the lungs. Inflammation, produced by viral or bacterial infection, by irritant gases (including tobacco smoke) or by allergic reactions, causes an increase in thickness of the mucous membrane, due both to vascular dilatation and raised capillary permeability. The membrane secretes more mucus and its ciliary movement may be impaired. After abdominal surgery pain limits a patient's willingness to cough. If allergy plays a part in causing the inflammation it will also cause the bronchial muscle to contract. All these results of inflammation in the airways lead to narrowing or to complete blockage of the airway and are included under the general heading of *obstructive lung disease*. When this is of limited segmental distribution (for instance, in apical segments of the lower lobes in a patient lying on his back after surgery), it is obvious why the physiotherapist needs to know how best to assist, by gravity, the drainage of mucus from an affected lung segment by correct positioning of the patient, as well as by promotion of proper respiration and vigorous coughing.

THE PHYSIOLOGY OF THE RESPIRATORY SYSTEM

PRESSURE AND VOLUME CHANGES

INTRA-PLEURAL PRESSURE

Both the lungs and the chest wall contain elastic elements in their structure The elastic tissue, in life, is stretched and the lungs thus always tend towards shrinking down or collapsing. The chest wall is always tending in the opposite direction, since its inherent elasticity provides a force opposite to that of the lungs' elastic force. The effect of this is clearly seen when an injury creates an opening between the pleural cavity and the exterior. The contained lung collapses and the surrounding chest assumes the barrel shape normally seen in maximal inspiration.

The equilibrium point for the chest, also known as the resting or relaxation point, or the *end-expiratory position*, is the position assumed when no muscles are affecting the chest volume by their contraction, so that no air is entering or leaving the lungs. It follows that at this point the elastic forces of chest wall and lungs exactly balance each other. Since they are pulling in opposite directions on the pleura, the pressure in the fluid in the pleural cavity is lower than atmospheric pressure by about 0.3 kPa. If the muscles that expand the chest contract so that the ribs and sternum move outwards and the diaphragm descends, the lungs follow this movement, drawing air into their substance. The increased volume of the lungs increases the stretch of their elastic tissue and thus its tendency to collapse the lung. As a consequence the sub-atmospheric pressure, already seen in the pleural space at the equilibrium point, becomes even more sub-atmospheric, reaching -0.8 kPa at the peak of a quiet inspiration and as low as -4 kPa in maximal inspiratory effort. Relaxation of the inspiratory muscles is followed immediately by a return of the chest wall to its original resting state when the elastic recoil forces once again balance. This allows the intrapleural pressure to rise towards -0.3 kPa. Contraction of muscles that further reduce intrathoracic volume will reduce still further the elastic collapsing force exerted by the lungs. This raises the intra-pleural pressure towards or even above atmospheric pressure.

INTRA-PULMONARY PRESSURE

As the lungs are pulled out in inspiration, there will be a slight subatmospheric pressure within their substance, so that air will enter through trachea and bronchi to fill them. In quiet breathing this pressure will be of the order of -0.1 kPa only. A corresponding $+0.1$ kPa pressure above atmospheric is seen during quiet expiration. It must be realised that in respiration in the healthy person, air is *sucked* into the lungs and not *pushed* into them. The situation is thus different from that of blowing up a balloon or football bladder with a pump, in which process the air pressure within is always *greater* than atmospheric and the air is *pushed* into it by a yet greater pressure. In the lungs it is a sub-atmospheric pressure that sucks air in, and the result of the increased stretch of the lungs is to render the pressure in the surrounding intra-pleural space also increasingly subatmospheric (again quite unlike that of the space between football bladder and surrounding leather skin, which is always above atmospheric pressure). In vigorous breathing the intra-pulmonary pressure shows greater change. Indeed, the raised intra-pulmonary pressure in forced expiration when there is any obstruction to air flow in the bronchi or trachea, may exceed capillary blood pressure and thus obstruct blood flow in the lungs. Coughing can do this, as can asthma or bronchitis.

LUNG VOLUMES

The volume of the lungs obviously varies in breathing and a number of technical terms are used to describe the corresponding volumes. The description below seems complicated but is easily understood if you make constant reference to Fig. 19.10.

Lung volumes are measured from two different starting or zero points. The first of these is when the lungs are completely collapsed and emptied of all their air and only the rigid trachea and bronchi contain any air. (This volume is termed the *anatomical dead-space volume*, it is about 150 ml.) This first zero point is never reached again from the moment the newborn baby takes its first breath, unless a lung is completely collapsed by free movement of gas or liquid into a pleural cavity. The second zero point is the *end-expiratory volume*, the equilibrium point already described, where the opposing elastic forces of lungs and chest wall are equal. This is the point to which the chest returns when the respiratory muscles are all relaxed and not causing any change in chest volume. At this point normally there are about 2.2 litres of air present in the lungs.

TIDAL VOLUME

In a resting person, contraction of inspiratory muscles usually draws about 500 ml of air into the lungs, to mix with the 2,200 ml already present. This is called the *tidal volume*. When the contraction of the muscles ceases, then the out-of-balance elastic forces in lungs and chest wall immediately cause the return to the resting point, the 500 ml of air being again expelled. During the inspiration of this 500 ml of air, the first 150 ml that move on from the dead space into the functional parts of the lung (respiratory bronchioles, alveolar ducts and alveoli) are similar in composition to the 2.2 litres of air already present in the lungs (alveolar air) since they were the last 150 ml to be expelled from the lungs at the end of the last breath. Three hundred and fifty ml of fresh atmospheric air then enter and mix with the 2.35 litres of alveolar air, leaving, at the end of inspiration, 150 ml of atmospheric air in the dead space. Thus, during normal respiration in a resting person, 2.35 litres of alveolar air are mixed at each breath with 0.35 litres of atmospheric air. The implications of the various magnitudes of these volumes will be discussed later.

INSPIRATORY RESERVE VOLUME

More vigorous and prolonged contraction of the inspiratory muscles can further increase the lung volume by 2–3.3 litres, the amount varying with sex, size and athletic training. This volume is appropriately termed the *inspiratory reserve volume*, and when its limit is reached by a maximal voluntary effort, the lungs contain their *total lung capacity*. The *inspiratory capacity* is the sum of resting tidal and inspiratory reserve volumes, or the difference between the volume

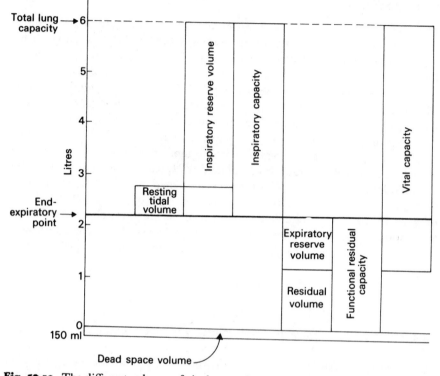

Fig. 19.10. The different volumes of air that can be voluntarily drawn into or expelled from the lungs. Note that at the end expiratory point, when all muscles are relaxed and elastic forces balanced, the lungs contain 2.2 litres of air and a further 0.15L is accommodated in the rigid trachea and bronchi.

in the lungs at the end-expiratory and the total lung capacity points. It should be noted that when using the whole of the inspiratory reserve volume, 2.35 litres of alveolar air are being mixed in the functional part of the lungs with 2.35–3.65 litres of fresh atmospheric air which now penetrates right down into individual alveoli (in quiet breathing it hardly gets into them).

EXPIRATORY RESERVE VOLUME

The muscles of expiration are not used at all in quiet breathing of the resting person (unless he is anxious). They cannot cause, when they do contract, as great a change in lung volume as the inspiratory muscles, the *expiratory reserve* volume being only 0.7–1.0 litre. It is clear then that, when all the expiratory reserve volume of air has been expelled from the lungs, about 1.2 litres will remain. This volume, always present in the healthy lungs, is called the *residual volume*. It cannot be removed from the lungs by voluntary effort during life, but only by the entry of fluid (air, blood or inflammatory exudate) into the

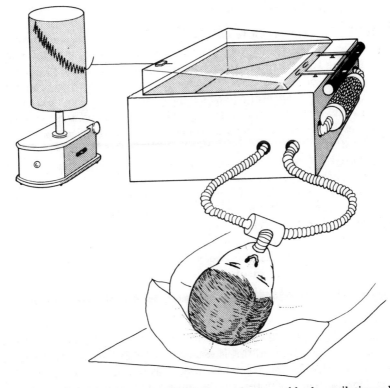

Fig. 19.11. A simple spirometer which can be used to record both ventilation volumes, per breath or per minute, and rates of oxygen consumption.

pleural space around one or both lungs which allows the elasticity of the lung tissue to drive out the remaining air.

The combined expiratory reserve volume + residual volume is known as the *functional residual capacity*. It is the volume of air present in the lungs when they are at the resting or relaxed lung volume point, 2.2 litres in men and 1.8 litres in women.

From the foregoing description of lung volumes and capacities it should be clear that for any person the three capacities, the *total lung capacity*, made up of the *functional residual capacity* and *inspiratory capacity* are of fixed size, being determined by the anatomical size of the person and the point at which the lung and chest wall elastic forces balance each other (the end-expiratory volume point). The volumes, however (except for the residual volume), can vary according to the degree of stimulation of the respiratory muscles by their controlling centres in the brain (see below). When these are stimulated, the tidal volume increases at the expense of both the inspiratory and expiratory reserve volumes, so that fresh atmospheric air is drawn into the depths of the functional part of the lungs.

MEASUREMENT OF LUNG VOLUMES

With suitable arrangements for replenishing O_2, removing CO_2 and recording the movements, the respiratory volumes can be determined by asking a person to breathe into and from a recording *spirometer* (Fig. 19.11). Essentially this is an enclosed air-space of several litres capacity, the volume of which can freely alter with no pressure change, and to which the person's mouth can be connected through a suitable mouthpiece and tubing. Tidal volume and the two reserve volumes are easily measured using a spirometer, as is also the *vital capacity*. This is the sum of the tidal volume + inspiratory + expiratory reserve volumes. This is measured by asking the person to breathe in fully and then to make as large an expiration as possible.

These measurements, useful though they are, do not tell us all we need to know about the maximal capacity to ventilate the lungs that is needed in vigorous exercise. Further measurements must be performed, such as voluntarily breathing as rapidly and deeply as possible (the *maximal voluntary ventilation test*) into and from a recording spirometer, or recording the *forced vital capacity*. In this the person inspires maximally in his own time and reaches the total lung capacity. He then exhales as rapidly and deeply as possible into the spirometer, while this volume change is recorded. The total vital capacity volume is determined and also the fraction exhaled in the first second of the exhalation. In a healthy person the total vital capacity is between 3.1–6 litres (dependent upon sex and size, etc.), and 85% of this is exhaled in the first second of the forced vital capacity (see Tables for normal values). This test supplies the same information as the maximal voluntary ventilation test, but is far simpler to perform and, for the person with respiratory disease, much less uncomfortable.

COMPOSITION OF ATMOSPHERIC, ALVEOLAR AND EXPIRED AIR

Atmospheric air contains about 21% O_2, 79% nitrogen (N_2), which includes 1% of rare gases, and 0.05% CO_2. These values assume that the air is dry (which it never is!). However, the water vapour content is so variable (anywhere between 1 and 3% of the total) that dry gas percentages are usually given. The gas composition is also often expressed as *partial pressures*. If the total barometric pressure of the atmosphere is 760 mm Hg or 100 kiloPascals (kPa), then the partial pressure of the O_2 in it is 21% of 760 = 159 mm Hg. (This is, of course, a numerical statement of Dalton's law of partial pressures.) Twenty one per cent of 100 kPa is of course 21 kPa (one distinct advantage in adopting the S.I. system of units). The partial pressures of N_2, CO_2, etc., can be similarly determined.

Water forms a special case. Its partial pressure is largely temperature dependent and at any temperature there is a limit, rising with rising temperature, to which the air can 'carry' water vapour. This limit is expressed as the partial

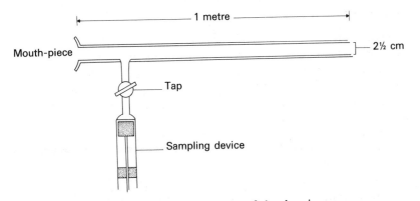

Mouth-piece

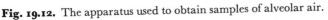

Fig. 19.12. The apparatus used to obtain samples of alveolar air.

pressure at this limit, which is referred to as the *saturated water-vapour pressure* (S.V.P.) at that temperature. Since the chemical gas analysers always operate with the gas sample saturated with water vapour, as is the air within the lungs or breathed out from them, the S.V.P. has always to be subtracted from observed barometric pressure before partial pressure calculations are preformed. This S.V.P. value must, of course, be that appropriate to the condition being studied, so for alveolar air at body temperature it is 47 mm Hg or 6.2. kPa.

ALVEOLAR AIR

This is the air present in the lungs at all times, i.e. at the end of expiration. It therefore corresponds to the functional residual capacity. It is now possible to consider the composition and partial pressures of the gases of alveolar air. A sample of this air is usually obtained by voluntarily exhaling all of the expiratory reserve volume down a long tube (Fig. 19.12) and collecting the last 30–50 ml of this. In a trained subject (and it is not always easy to get good values for this determination), alveolar air is remarkably constant in composition at 5.6% CO_2 and 14% O_2, the rest being N_2, etc. Once again these are dry gas values. If the total barometric pressure is 760 mm Hg (100 kPa) then gases account for 760−47=713 mm Hg (100−6.2=93.8 kPa), and the partial pressures are 5.6% and 14% of 713=40 and 100 mm Hg (5.6% and 14% of 93.8=5.3 and 13.1 kPa) respectively for CO_2 and O_2.

Alveolar air composition changes only slightly during the respiratory cycles of alternating inspiration and expiration at rest (Fig. 19.13). As some atmospheric air enters the lungs the O_2 content rises slight and the CO_2 content falls a little. The changes are small since only 350 ml of atmospheric air enter a volume of over 2 litres. Most of the exchange of O_2, moving inwards towards the blood vessels and of CO_2 moving in the opposite direction, is due to diffusion movements of individual gas molecules through the air in the alveoli from regions of higher concentration to lower concentration, and not to bulk movement of gas. The situation is very different in the respiration of vigorous exercise. Now the tidal

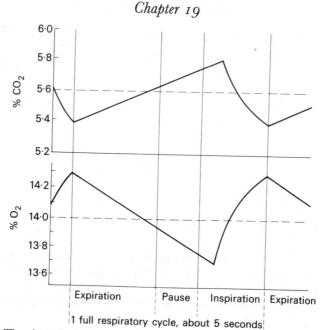

Fig. 19.13. The alterations that occur in alveolar air composition during lung ventilation at rest. In exercise similar, but much larger, alterations probably occur.

volume may exceed 3 litres and the volume in the lungs at the start of inspiration may be only 1.5 litres, since much of the expiratory reserve volume is included in tidal volume. The addition of 3 litres of atmospheric air to 1.5 litres of alveolar air will result in considerable changes in the compostion of the latter. The size of these changes has not yet been determined, but their significance lies in the possible consequent changes in blood composition and the effect this may have on the control of respiration in exercise (p. 581).

EXPIRED AIR

Expired air, being a mixture of alveolar and dead-space air, will have a composition reflecting the relative proportions of the two in the final mixture. Its composition is also affected slightly by the volumes of air being breathed and the rates at which O_2 is used and CO_2 produced by the body. In normal life, however, a very close balance is maintained between these, so that the *respiratory minute volume* (the tidal volume multiplied by the number of breaths per minute) always changes in parallel with changes in the O_2 consumed and CO_2 produced by the body. In really vigorous exercise (p. 581) the balance is broken, since lactic acid produced by the muscles stimulates ventilation in addition to oxidative metabolism. Under all other circumstances, the balance results in expired air containing 3–4% CO_2 and about 17% oxygen.

EXCHANGES OF GASES BETWEEN ALVEOLAR AIR AND BLOOD

THE PULMONARY CIRCULATION

All the blood from the right ventricle of the heart is delivered through the pulmonary arteries to the lungs. The pressure in these vessels is low. In the main pulmonary artery it is about 25 mm Hg in systole and 10 mm Hg in diastole, while in the capillaries it is about 10 mm Hg. Since this is less than the plasma protein osmotic pressure of 25 mm Hg, there is no filtration of fluid out from pulmonary capillaries in the healthy person. Filtration will only occur if the blood pressure in the pulmonary capillaries rises above 25 mm Hg, if the plasma protein content falls or if inflammatory processes in the lungs, e.g. those produced in severe bronchitis or in pneumonia cause a leak of plasma protein through the capillary walls.

Pulmonary capillaries are closely applied to about half of the surface area of the alveolar walls, so that, while the total surface area of the lungs is given as 144 sq metres, the area available for gas exchange with blood is about 72 sq metres. At rest, a blood corpuscle takes about one second to pass through a pulmonary capillary. In vigorous exercise, when the output of the heart may rise from 5 to 25 litres/min, the transit time of blood through the capillaries would be expected to be correspondingly reduced, to $\frac{1}{5}$ second. In fact it falls to $\frac{1}{3}$ second, because some increase in capacity of the pulmonary capillaries occurs.

BLOOD COMPOSITION

At rest, the blood entering pulmonary capillaries contains 52 vol CO_2/100 vol blood, which exerts a tension of 6.4 kPa. It contains 15 vol O_2/100 vol blood, the haemoglobin being about 75% saturated. The dissolved O_2, in equilibrium with that combined with haemoglobin, has a partial pressure of 5.6 kPa. When the blood leaves the pulmonary capillaries it contains only 48 vol CO_2/100 vol blood at a tension of 5.6 kPa, but 19.8 vol O_2/100 vol blood. The haemoglobin is 97% saturated and the partial pressure is 13.6 kPa. How do these changes in blood gas content occur?

ALVEOLAR-CAPILLARY TRANSFER OF GASES

It was noted earlier that, in a gas, molecules will diffuse from regions where they are more concentrated to regions where they are fewer of them. In other words, they move down their concentration gradients. In yet further words they move down their partial pressure gradients. In fact, as we normally think of pressure as a force which drives things, this is a very useful way of looking at gas diffusion. Gases diffuse from regions of high partial pressure to regions of lower partial pressure, and the rate of diffusion will depend on the difference of pressures and the distances over which this difference occurs—the partial pressure gradient.

Both CO_2 and O_2 can be dissolved in water (CO_2 much more readily than O_2, hence soda-water and all other fizzy drinks from champagne to lager). The amounts that will dissolve depend also on the partial pressure of the gas to which the water is exposed, and on the water temperature. At body temperature and a tension of 5.6 kPa, 100 ml water would dissolve 2.5 ml CO_2 and only 0.12 ml O_2. At a tension of 14 kPa water would contain 0.2 ml O_2. The reader should note how very different, especially for O_2, these values are compared with O_2 and CO_2 in a gas mixture, such as alveolar air, at the same partial pressures.

However, if pressure gradients exist the gases will diffuse through water, watery liquids, cytoplasm of cells and cell membranes just as they do in a mixture of gases, and their ability to diffuse is comparable to their ability to dissolve in the medium. Thus, from the figures just given above, CO_2 is 20 times as soluble in water as is O_2. Its diffusion is correspondingly greater than that for oxygen.

When blood enters a pulmonary capillary there is a pressure gradient for CO_2 of 6.4–5.6 kPa driving it from blood into the alveolar air and one for O_2 of 14–5.6 kPa driving this gas in the opposite direction. The distances over which these gradients apply are so short that it only takes 0.2 sec for diffusion alone to bring about complete equilibrium of partial pressures between blood and alveolar air for both gases. No other process, secretion or active transport, is required for this to occur. With a total blood flow of 5 litres/min through the lungs' capillaries (the resting cardiac output), $5,000 \div 60 = 83$ ml traverse the capillaries every second. In 0.2 sec this blood loses $\frac{83}{100}$ of 4 ml CO_2 and gains $\frac{83}{100}$ of 4.8 ml O_2 from the alveolar air. These work out as 3.3 and 4 ml respectively. During the remaining 0.8 sec no further exchange takes place between blood and alveolar air. In vigorous exercise the pulmonary blood flow rises to 25 litres/min. Therefore, $25,000 \div (60 \times \frac{10}{3}) = 125$ ml traverse the capillaries in 0.3 sec. During this time 20 ml O_2 and about the same amount of CO_2 are exchanged between blood and alveolar air.

The reason for the wasted 0.8 sec while blood is in the lung capillaries at rest, should now be apparent. It allows a reserve of diffusion capacity for the much greater rate of gas transfer that must occur in vigorous exercise, just as the inspiratory and expiratory reserve volumes allow a great reserve of ventilatory volume in the air-filled spaces of the lungs, so that the alveolar air can at all times supply the O_2 to, and remove the CO_2 from, the blood. The size of the reserve of diffusion for O_2, the *diffusion capacity* of the lungs, is of course, a vital factor in a person's capacity to perform exercise. Disease states limit O_2 diffusion much more readily than CO_2 diffusion since O_2 diffuses so much less readily than CO_2. Determination of O_2 diffusion capacity is thus part of the essential study of lung function in chronic or progressive lung diseases.

THE CONTROL OF RESPIRATION

The muscles of respiration are all striated or skeletal muscle, and contraction is caused by nerve impulses leaving the spinal cord in motor nerve fibres originating in anterior horn cells. Like all other skeletal muscles, the muscles of respiration *can* be contracted, or inhibited from contracting, at will. Normally, though, their alternate contraction and relaxation proceeds automatically and regularly throughout the whole of life, and is of such a degree as to balance the metabolic use of O_2 and the production of CO_2. In studying respiratory control, we have two aspects to consider. First, there is an *intrinsic brain-stem control mechanism*, which at all times maintains the rhythmic alteration of inspiration and expiration. Secondly, there are the *chemical controls*, which modify the activity of the intrinsic system, so that it balances the metabolic O_2 use and CO_2 production.

THE INTRINSIC BRAIN-STEM MECHANISM

In the reticular formation of the pons and medulla of the brain-stem are groups of neurones which, by a complex interaction of excitation and inhibition maintain an alternation between contraction and relaxation of the respiratory muscles by a periodic stimulation of their anterior horn cells. Neurones determining inspiration, that are spontaneously active in inspiration or which cause inspiration when artificially stimulated, are found bilaterally in two sites in the upper medulla. These neurones can collectively be called the *inspiratory centre*, but you must always remember that using the word 'centre', does not imply an anatomically defined group of nerve cells. Expiratory neurones are found on both sides of the medulla caudal to the inspiratory centre. The inspiratory centre is now thought to consist of two types of neurones, R-alpha ($R\alpha$) and R-beta ($R\beta$) cells. $R\alpha$ cells drive the spinal motoneurones of the inspiratory muscles. They also excite the $R\beta$ cells which in turn inhibit $R\alpha$ cells. Thus the isolated medulla maintains an irregular and gasping rhythmic respiration. The $R\alpha$ cells are further excited by reticular formation neurones of the lower pons and inhibited by cells of the upper pons and by afferent fibres in the vagi (see below, under Hering-Breuer reflex). The two pontine regions are known respectively as the *apneustic* and *pneumotaxic* centres, though these words are currently falling from common usage among the experts. In *resting* man, the respiratory cycle begins with a burst of activity, lasting about 2 sec in the $R\alpha$ cells of the inspiratory centre. This stimulates the anterior horn cells of the inspiratory muscles and also the $R\beta$ cells and pneumotaxic centre neurones. Together these then inhibit the inspiratory centre, thus stopping the contraction of the inspiratory muscles. Passive expiration then follows over 2 sec, there is a 1 sec pause and the inspiratory centre, no longer inhibited by the $R\beta$ cells and the pneumotaxic centre, starts a new respiratory cycle. In this way, cycles repeat at 5 sec intervals or 12/min. If respiration is stimulated by factors outside this mechanism, then the more intense excitation of the inspiratory centre causes more rapid and deeper inspiration and through the $R\beta$ cells and the pneumotaxic centre, an earlier inhibition of this activity, and a simultaneous stimulation of the expiratory centre and contraction of the muscles of expiration. Not only are inspirations deeper, but they follow more quickly on each other so the cycle may last 2 or even 1 sec only, instead of 5 sec of quiet breathing. Figure 19.14 shows diagrammatically these sequences in quiet and vigorous breathing.

NEURAL CONTROL OF THIS MECHANISM

Impingeing on these brain-stem centres are afferent impulses from stretch receptors in the lungs and from the muscle spindles of the inspiratory muscles. The former afferents convey impulses via the pulmonary plexuses and the vagus nerves, and the latter via posterior roots and the spinal cord to the medullary respiratory centre. The increased frequency of impulses from both sets of receptors (lungs and skeletal muscles) during inspiration inhibit the inspiratory centre neurones and may stimulate those of the expiratory centre. This mechanism is called the Hering-Breuer reflex. Four different types of mechanoreceptors with vagal afferent fibres have been found in the lungs. Those involved in the Hering-Breuer reflex control of breathing are static stretch receptors, their impulse production increasing as the degree of stretch or depth of inspiration increases, and they do not adapt to a maintained stimulus. Their afferent fibres are of large diameter and conduct impulses rapidly.

If either the pneumotaxic centre or the stretch receptor afferent fibres are cut off from the medullary centres, rhythmic respiration continues, but is slower and deeper. If the medulla is severed from both controls, then the rhythm becomes more irregular and breathing may be gasping in nature.

CHEMICAL CONTROL

The function of this brain-stem mechanism is to maintain ventilation at a level that results in O_2 and CO_2 exchange between blood and air, just balancing the exchanges between blood and the metabolising tissues of the body. For it to perform this function, the brain-stem mechanism must be informed of the levels of O_2 and CO_2 in the blood. If metabolic exchange in the tissues exceeds exchange in the lungs, then the O_2 will fall and CO_2 will increase in the blood leaving the lungs, i.e. in the arterial blood. If ventilatory exceeds metabolic exchange, the CO_2 content of arterial blood will fall. O_2 content, in this case, will not rise significantly, for in the balanced state the blood is already carrying 95 % of the theoretical maximal amount of O_2.

Receptors, sensitive to both O_2 and CO_2 levels are supplied with arterial blood. These thus monitor blood gas composition *after* the blood has passed through the lungs. The theoretician might argue that it would be better to have the receptors before the lungs, on the pulmonary arteries, so that ventilation might adjust to the metabolic load of CO_2 and O_2 deficit before the blood reaches the lungs. The actual site of some of the receptors may, however, be determined by our fishy ancestry, for the analogous position of the receptors in these organisms would be between the heart and the gills, where external respiration occurs.

THE AORTIC AND CAROTID BODY CHEMORECEPTORS

Close to the arch of the aorta and to the carotid bifurcation are situated a number of solid, 5 mm diameter bodies, richly supplied with blood and sensory nerve fibres. The nerve fibres from the *aortic bodies* pass to the brain via the vagus

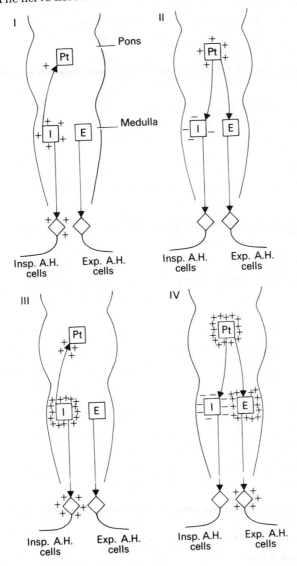

I and II Quiet breathing III and IV Vigorous breathing

Fig. 19.14. A simplified diagram showing how the pneumotaxic centre in the pons might affect the activity of the medullary respiratory centres. The modern view is that the Rβ cells of the inspiratory centre itself perform the function here ascribed solely to the pneumotaxic centre.

nerves, while those from the pair of *carotid bodies* pass via the glossopharyngeal nerves. The latter are the more important in man.

These chemoreceptor organs are primarily sensitive to O_2 lack. They produce impulses at a low rate when the arterial blood is in equilibrium with normal alveolar air with an O_2 partial pressure of 13–14 kPa. These impulses are reduced to zero if pure O_2 is breathed, which increases the blood partial pressure to 67–80 kPa. They are increased markedly by reducing the O_2 content of the air breathed so far that the arterial blood has a partial pressure of O_2 of 4 kPa (remember that at this level haemoglobin is still 55% saturated with O_2—Fig. 18.2). Further stimulation occurs if blood CO_2 is raised, but CO_2 alone has little effect on these chemoreceptors (see Fig. 19.15). The role of the impulses from these receptors in the control of ventilation is difficult to assess.

THE BRAIN-STEM CHEMORECEPTORS

Areas on the anterior surface of the medulla, exposed to cerebrospinal fluid, have been found in animal experiments to be very sensitive to changes in the CO_2 content of this fluid. The receptors and their afferent nerve fibres have not actually been identified, but the response in ventilation to CO_2 stimulation has been studied directly.

CO_2 diffuses freely from blood to CSF, where it is in equilibrium with H_2CO_3 (carbonic acid). Some of the H_2CO_3 dissociates electro-chemically into H^+ (hydrogen ion) and HCO_3^- (bicarbonate ion) as happens with all acid substances, and with H_2CO_3 in the blood. In the CSF, however, much more H^+ is formed per unit of CO_2 than in blood, because CSF lacks the buffer materials that in blood react with the H^+. It is this CSF H^+ that actually stimulates the receptors when blood CO_2 rises and the CO_2 passes into the CSF. Increase in blood H^+ alone is not a very good stimulator of these receptors, for it does not penetrate to the CSF so easily as does CO_2. The response to CO_2 is so sensitive that a 10% rise in CO_2 content of arterial blood results in a doubling of ventilation.

THE RESPIRATORY RESPONSES TO CHANGES IN BLOOD GAS COMPOSITION

The role of the impulses reaching the inspiratory centre from both the arterial and the brain-stem receptors in the control of respiration is difficult to assess. The former group, being directly exposed to blood in the carotid and aortic bodies, and containing phasic as well as static receptors, respond to fluctuations as well as to sustained changes in the blood. Then, any stimulation of breathing that might occur due to O_2 lack will result in the ventilatory loss of CO_2 exceeding its metabolic production and thus to a fall in blood CO_2. As will be described below, this has a powerful depressant effect on ventilation, which overcomes the stimulation of O_2 lack. It used, therefore, to be taught that arterial blood O_2 partial pressure had to fall to 8 kPa before respiration was stimulated. If the experiments are performed in such a way as to maintain a constant, normal blood CO_2 content, then the ventilatory response to blood O_2 levels follows a curve similar to that for the impulse production of the arterial receptors, stimulation of ventilation appearing at any change of O_2 from the normal value (Fig. 19.15). Change in arterial blood content *in either direction* will result in a doubling or halving of ventilation as CO_2 rises or falls. These changes are achieved by equal increases in respiratory rate and tidal volume. If, however, the CO_2 level in arterial blood rises towards twice the normal level (and this can be seen in severe chronic lung disease), then the stimulating effect is masked by a general anaesthetic action known as *CO_2 narcosis*, in which the patient becomes drowsy and confused and his respiratory rate and depth fall. Chronic lung disease, with raised blood and CSF CO_2 levels, also results in a slow loss of acid from the CSF, so that CO_2 liberates less H^+ and the stimulation of the CO_2 receptors is weakened. In such a situation, the patient relies upon stimulation of the carotid and aortic bodies by O_2 lack to maintain ventilation. Giving him O_2-rich air may be disastrous, for this ventilatory drive is cut off, the blood CO_2 rises to the narcotic level and the patient may well die as a direct result of the therapeutic measure!

PERIPHERAL CHEMORECEPTORS

For many years evidence has been slowly accumulating that suggests that chemoreceptors may exist within skeletal muscle, and that afferent impulses from these pass in small-diameter sensory nerve fibres, when the muscles are

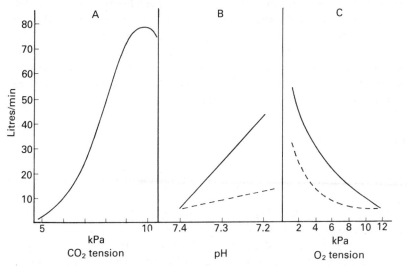

Fig. 19.15. The ventilatory responses to changes in blood gases. A, the response to CO_2. B, the response to changes in pH and C, the responses to a fall in O_2. In B and C, the dashed lines are the responses when CO_2 is allowed to fall and the continuous lines are the responses when CO_2 tension is held at 5.3 kPa.

stimulated to contract, to the spinal cord and the brain. Some people suggest that these impulses are partly responsible for the increase in respiration seen in muscular exercise (p. 581). The anatomical existence of these receptors has not been shown, nor whether they are O_2- or CO_2-sensitive.

RESPIRATION AND METABOLISM

We have already discussed some aspects of this topic, including the balance between metabolic gas exchange (internal respiration), gas exchange between blood and alveolar air, and ventilation of the lungs (external respiration). We must now consider what can actually be learnt about internal respiration from a study of volume and gas-exchanges of external respiration. So long as such studies are performed in the *steady state*, so that internal respiration, blood and body fluid O_2 and CO_2 levels, and external respiration are all constant, reliable information can be obtained.

The overall level of metabolic activity of a living body in steady-state conditions, whether at rest or in mild to moderate activity, is closely reflected by the O_2 uptake of that body from its respired air. It is simple to measure this in either one of two ways.

THE SPIROMETER

The same apparatus is used as for the measurement of lung volumes. Now it is filled with O_2, a CO_2 absorber put in the breathing-tube circuit, and the person breathes from the spirometer for several minutes. During this time, as he uses O_2, the volume in the spirometer steadily falls, and the rate of fall per minute can be calculated from the record (Fig. 19.11), corrected to its corresponding dry gas at standard temperature and pressure, and then, from the known chemical relationships between O_2, oxidised foodstuffs and energy liberated, the energy turnover rate can be calculated.

COLLECTION AND ANALYSIS OF EXPIRED AIR

This is more complex, but more information can be obtained from this method. Expired air can be collected into a large bag or spirometer over a known period of time. The volume is measured and corrected to standard conditions. A sample is analysed for its O_2 and CO_2 contents. From these and the known composition of atmospheric air, the volume of O_2 used and CO_2 produced can be determined. In oxidation of carbohydrate, exactly 1 vol CO_2 is produced for every 1 vol O_2 consumed, and 21 kJ of energy are produced/ml O_2 consumed. In oxidising fat, only 0.7 vol CO_2 is produced for every 1 vol O_2 consumed and 17 kJ of energy are produced. Again, the chemists have given us these figures, but they apply equally to the living body and to the apparatus of a chemist's laboratory. So, from the composition of expired air and its volume we can obtain precise knowledge of how much of which of the two main food-stuffs is being used by the body, and how much energy is being produced. It was the use of such methods, that showed us that, at rest and in mild to moderate exercise, our muscles obtain their energy mainly by oxidising fat. That is another story which will be taken up in detail in a later chapter of this book (Chapter 23).

The subject of tissue metabolism (internal respiration) will also be studied more fully in Chapter 24. It is sufficient to say here that the methods just described are the ones that have been used in most of the investigations of energy turnover in the living body, whether at rest, during normal daily activity or in experiments on controlled exercise.

20 · The Lymphatic System

The lymphatic system consists of a system of fine channels similar to, but much smaller than, those of the blood vascular system and containing *lymph* rather than blood, along with a series of *lymph nodes* (or glands) and islets of lymphatic tissue elsewhere such as the *Peyer's patches* in the small intestine. In addition, it includes the lymphocytes that circulate in the blood and the phagocytic cells that occur in most of the tissues of the body. The latter cells are often classified as a separate system—the *reticulo-endothelial system*—but they are so closely bound up with the lymphatic system that they will be included therein for the purposes of this Chapter. This is, of course, an anatomical description of the lymphatic system and you may find it rather unsatisfactory because the system seems so diffuse. Physiologically, too, the system is difficult to define. Briefly, it may be said that the lymphatic system has two main functions, namely the removal of some tissue fluid and of large molecules from the tissues and the defence of the body against invasion by foreign materials such as bacteria and their products and other foreign proteins. These functions will be considered in more detail later in this chapter, but a few commonplace examples will serve to illustrate these functions for the moment.

You will remember the mode of formation and removal of tissue fluid by the capillaries (Chapter 17) in which the differences in the hydrostatic and osmotic pressures inside and outside the capillaries caused fluid to be extruded from the arterial end of the capillaries and reabsorbed into the venous end. The endothelium of the blood capillaries, however, is not completely impermeable to proteins and some molecules leak through. These, and some of the water of the tissue fluid are picked up by the lymphatic capillaries. Swelling of the tissues can therefore be caused by lymphatic obstruction (*lymphatic oedema*). In infections, the endothelial cells become more permeable and a good deal of protein (and even some blood cells) escape, and swelling of the tissues occurs. The protein and much of the fluid that accompanies it are also taken up and removed from the tissues by the lymphatic capillaries. In local infections, too, bacteria may invade the lymphatic system from the tissues but they are usually prevented from invading the body extensively by the protective action of the lymph nodes. These contain phagocytic cells and they are placed in such a way that the lymph coming from the tissues has to pass through one or more lymph nodes before entering the bloodstream. In the lymph nodes, the bacteria are taken up by the cells of the reticuloendothelial system and, if all goes well, are destroyed. Most people, at some time or other, have experienced a sore throat or tonsillitis along with enlarged and painful lymph nodes in the neck, while occasionally a septic condition in the skin such as a boil may lead to inflammation of the superficial lymphatic channels (*lymphangitis*) so that a red line may be seen extending from the lesion to the nearest lymph nodes which are enlarged and tender.

So much for the removal of tissue fluid and mechanical protection against invasion by micro-organisms. The other protective function concerns a chemical or *humoral* protection against invasions by micro-organisms, foreign proteins and other large molecules. When micro-organisms enter the body, for instance, the cells of the lymphatic system are stimulated to produce *antibodies* which are able to kill or inactivate the organisms themselves or to neutralize their toxins. This phenomenon is known as the *immune reaction* and is the reason why in many diseases, one attack confers immunity to a second. Advantage is taken of this to immunise susceptible subjects to diseases such as smallpox, typhoid, diphtheria, etc., by injecting either dead or damaged bacteria, closely related but harmless organisms, or bacterial poisons (toxins). Unfortunately, the immune reaction occurs in response to most other foreign proteins in addition to bacteria and their products and this produces the most important difficulty in the grafting of organs such as the kidney and the heart or even skin. Tissues from one person act as antigens (see below) when they are grafted into another, so that unless the two subjects are closely related (e.g. identical twins) or their tissue types are closely similar, antibodies will be produced by the host and the graft will be rejected.

Before studying the function of the lymphatic system in detail, or discussing its role in disease, it will first be necessary to describe the anatomy of the system in some detail.

THE ANATOMY OF THE LYMPHATIC SYSTEM

The microscopic anatomy of the lymphatic vessels is very similar to that of the small veins except that the walls of the vessels are thinner and contain even less smooth muscle. The lymphatic capillaries are found in most tissues of the body (cartilage and the cornea of the eye are notable exceptions and corneal grafting can therefore be carried out very successfully between unrelated people). The capillaries consists of simple endothelial tubes which begin as blind ended channels and form a network in the tissues. Rather special lymphatic capillaries are found in the villi of the small intestine where they are known as *lacteals;* these are important in the absorption of fat from the intestine (see Chapter 24). The capillaries then join up to form larger lymph vessels (or *lymphatics*) which have a little smooth muscle in the walls. These combine still further to form main lymphatic channels which have a thin coat of smooth muscle outside the endothelium and a little connective tissue outside that again. They possess valves very similar to, but much more numerous than, the valves in veins and they have a similar function.

The main lymphatics pass centrally from the periphery and eventually enter a lymph node, in company with a number of other lymphatics. The lymph filters through the complex interior of the node and then emerges in a

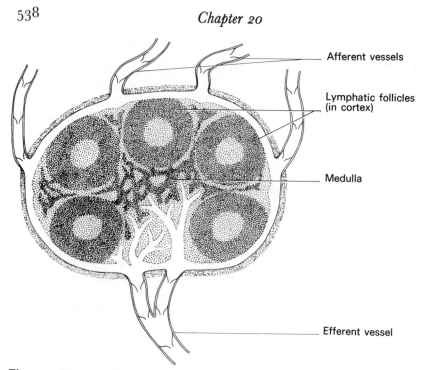

Fig. 20.1. Diagram of a lymph node.

single outgoing or *efferent* vessel which continues the course centrally of the original lymphatics towards the root of the neck.

Each lymph node is surrounded by a capsule of connective tissue and has a rich blood supply. They are usually oval or kidney-shaped and up to 1½ inch in length (Fig. 20.1). The interior of the node has a basic 'skeleton' of special fibres and cells which form a network or *reticulum* (in the nineteenth century, ladies carried a primitive type of handbag called a *reticule* because it was made of netting). As will be seen, a reticulum of cells and fibres is found also in other organs and, together with phagocytic cells lying free in the tissues, forms the *reticulo-endothelial system*. The reticulum pervades the whole node but is not everywhere recognisable because of the large number of *lymphocytes* embedded in it. These are similar in appearance to the lymphocytes of the blood and, as will be seen later, are interchangeable with blood lymphocytes. In a histological section, a lymph node, like the kidney or the suprarenal, has an outer *cortex* and an inner *medulla*. The lymphocytes in the cortex are aggregated to form large spherical masses or *lymphatic follicles* while in the medulla the lymphocytes are less closely packed so that the medulla stains more lightly. The lymphocytes in the centre of the lymphatic follicles are in a state of active mitosis and are thus producing new lymphocytes. This process is particularly active during infections when the node may become enlarged. Lymph which enters the lymph node from the incoming (*afferent*) lymphatics has to percolate through the reticulum and other cells of the node before leaving via the efferent

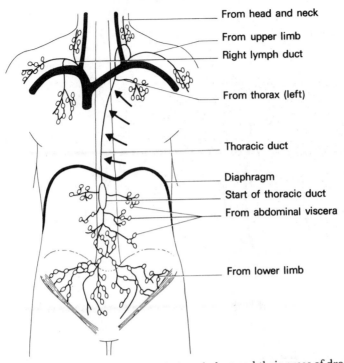

From head and neck
From upper limb
Right lymph duct
From thorax (left)
Thoracic duct
Diaphragm
Start of thoracic duct
From abdominal viscera
From lower limb

Fig. 20.2. The thoracic duct and the right lymph duct and their areas of drainage.

lymphatics. The cells of the reticulum are phagocytic and are able to remove foreign bodies such as bacteria from the lymph. The lymphocytes of the node produce antibodies as a result of the trapping of the antigens by the phagocytic cells.

The larger lymphatic channels finally drain into the blood stream, either via the *thoracic duct* or by the *right lymph duct* (Fig. 20.2). The thoracic duct begins in the upper abdomen near the midline. It passes up through the thorax, lying on the front of the bodies of the vertebrae and ends by entering the venous system in the root of the neck. It actually joins the junction between the left internal jugular vein and the left subclavian vein. Into the thoracic duct drain lymphatic vessels from the lower limbs and abdomen, the left upper limb, the left side of the head and neck and the left side of the thorax. The lymphatics from the right side of the thorax, the right upper limb and the right side of the head and neck all enter the junction between the right internal jugular and right subclavian veins, either separately or after joining to form the right lymph duct.

You will need to know the position of the main groups of lymph nodes and the direction of the main lymphatics because, as a physiotherapist, you can do a great deal to promote the free drainage of lymph from the tissues. Also, because as well as providing a pathway for infections to travel to the local lymph nodes, the lymphatics also provide a channel whereby the cells of malignant disease can travel to, and colonise, the regional lymph nodes, which thus become the

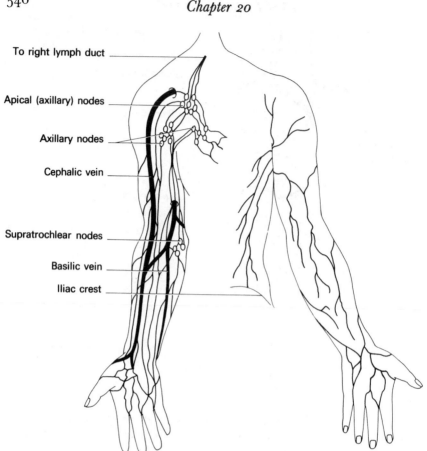

To right lymph duct

Apical (axillary) nodes

Axillary nodes

Cephalic vein

Supratrochlear nodes

Basilic vein

Iliac crest

Fig. 20.3. The superficial veins and the lymphatics of the upper limb.

site of secondary tumour formation (*metastases*). In general terms, the lymph nodes are found on the flexor aspects of the limbs, while the main lymphatic channels follow the veins, both superficial and deep.

LYMPHATICS OF THE UPPER LIMB

There are a few small lymph nodes just above the medial epicondyle (*supra-trochlear*) but the main lymph nodes of the upper limb are situated in the axilla where they lie in relation to the vessels. The highest group of nodes is at the apex of the axilla (*apical*). The deep lymph vessels run alongside the main arteries and their corresponding veins or venae comitantes (Fig. 20.3). They finally enter the axillary lymph nodes whose efferent vessels end in the thoracic duct or right lymph duct at the root of the neck. The superficial lymphatics start in a plexus of capillaries in the skin and superficial tissues and the larger lymphatics pass towards the medial or lateral borders of the forearm after which they accompany the cephalic or basilic veins. A few pass up the midline

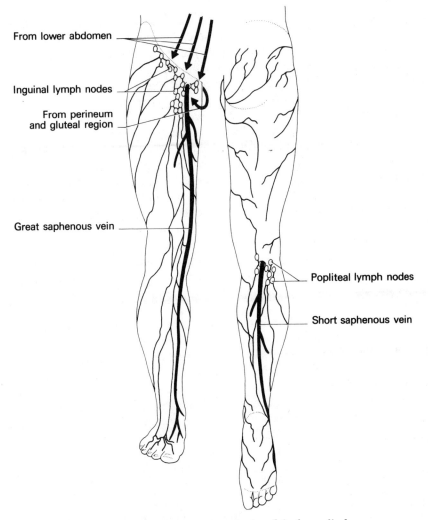

From lower abdomen

Inguinal lymph nodes

From perineum and gluteal region

Great saphenous vein

Popliteal lymph nodes

Short saphenous vein

Fig. 20.4. The superficial veins and the lymphatics of the lower limb.

before inclining medially towards the basilic vein. The posterior lymphatics turn round the medial and lateral borders of the limb to join the lymphatics on the flexor aspect. The lymphatics accompanying the basilic vein are interrupted by the supratrochlear nodes but otherwise all the superficial lymphatics follow the veins to the axilla where they enter the axillary lymph nodes (Fig. 20.3). Note that the axillary nodes do not only drain the arm. They drain an area corresponding roughly to the area covered by the muscles of the whole upper limb, i.e. the superficial tissues of the back down as far as the iliac crest and, most important, the breast. The breast also drains to lymph nodes inside the thorax behind the sternum.

LYMPHATICS OF THE LOWER LIMB

The groups of lymph nodes in the lower limb are situated in the popliteal fossa and in the inguinal region. The *popliteal lymph nodes* are only a few in number and are related to the vessels in the fossa. The *inguinal lymph* nodes lie in a chain below and parallel to the inguinal ligament and also along the upper part of the femoral vessels and the great saphenous vein (Fig. 20.4).

Like those of the upper limb, the lower limb lymphatics are both superficial and deep. The superficial lymphatics form a plexus of vessels in the superficial fascia. Most of them tend to run medially to join the major lymphatic vessels which accompany the great saphenous vein to the inguinal group of nodes. Some run laterally and then follow the small saphenous vein to the popliteal nodes. The deep lymphatics follow the major vessels, those running up the back of the leg being interrupted by the popliteal nodes. All the deep lymphatics end in the inguinal nodes, the efferent vessels from which pass under the inguinal ligament to enter the abdomen.

The lymphatics from the lower limb join those from within the abdomen and after passing through a number of intra-abdominal nodes, empty into the thoracic duct. Note that the inguinal lymph nodes drain the lower abdominal wall (below the umbilicus), the buttocks and the perineum in addition to the lower limb.

There are a large number of lymph nodes inside the abdomen, the main groups being related to the aorta and the subsidiary groups to its branches. The nodes alongside the aorta (*para-aortic*) drain lymph from the iliac lymph nodes (alongside the iliac arteries) which themselves receive lymph from the lower limb and from the pelvic viscera. The para-aortic nodes also drain the laterally placed organs such as the kidneys, testes, etc. The nodes lying in front of the aorta (pre-aortic) drain the digestive system. All abdominal lymph finally ends up in the thoracic duct. The lymph from the gut is milky in appearance on account of the absorbed fat droplets that it contains.

In the thorax there are lymph nodes associated with vessels in the thoracic wall and also large groups around the trachea and bronchi. All lymphatic vessels eventually end in the thoracic duct or right lymph duct.

In the head and neck there are a number of outlying groups of small nodes such as the submandibular, occipital, etc., but they all finally drain into two main groups: the *upper* and *lower deep cervical nodes*. These are situated respectively in the angle between the sternomastoid and the lower border of the mandible and in the angle between the sternomastoid and the clavicle. All lymph from the neck and head eventually drains into the thoracic duct or right lymph duct.

THE SPLEEN AND THE THYMUS

Both these organs contain mainly lymphatic tissue and the thymus, particularly, is closely involved with the immune reaction and they will therefore be described in this chapter.

The *spleen* is about the size of a bun and lies on the left side in relation to the 9th, 10th and 11th ribs and the diaphragm. The spleen is rather similar in structure to a lymph node since it contains a reticulum of fibres and cells in the interstices of which are dense lymph follicles which produce lymphocytes. The spleen has a rich blood supply but the complicated vascular arrangements in the spleen need not concern us here.

The *thymus* is situated behind the upper part of the sternum in relation to the great vessels. In the fetus it is relatively huge, occupying much of the upper part of the chest. It reaches its greatest size at puberty but is then relatively smaller in relation to the surrounding structures which have grown more. After puberty it slowly diminishes in size and is gradually replaced by fat.

The microscopic structure of the thymus, is at first sight, very similar to that of the spleen or a lymph node in that it is composed of a reticulum of cells and fibres impregnated with small lymphocytes. The cells, however, are of a different type to those found in the other organs.

THE RETICULO-ENDOTHELIAL SYSTEM

Now that the anatomy of the lymphatic tissues has been described it is possible to give an overall description of the functions of these tissues both with regard to the reticulo-endothelial cells and the immunological system which involves the lymphatic and related cells.

The reticulo-endothelial system is a diffuse collection of cells which occur throughout the body (in much the same way as one might describe a 'physiotherapy system' consisting of a widespread system of physiotherapists found in many different environments throughout the world but all belonging to the same 'type' and all having a similar function). The cells of this system are essentially *phagocytic*, i.e. they can take up foreign particles by *phagocytosis*. They are found as *reticular cells* forming (along with fibres) the reticulum of lymph nodes, spleen, bone marrow, tonsils and certain other structures; as *macrophages* in connective tissue; as *large mononuclear cells* (monocytes) in the blood; and as isolated cells in the liver (*von Kupffer cells*). These cells can be recognised by their often bizarre shape due to their numerous processes (*pseudopodia*) and by their content of ingested particles and micro-organisms. In the living body their function is to provide a first line of defence against invading agents by engulfing foreign particles and bacteria. In the lung, for instance, cells of the reticuloendothelial system known as *dust cells* remove inhaled particulate matter from the alveoli and transport it to the nearest lymphatics. Cells of the system also take up haemoglobin from dead red cells and break it down to haem and globin (Chapter 17).

IMMUNOLOGY

The function of the lymph vessels in removing large molecules and fluid from the tissue spaces and fat from the intestine has already been discussed. The function of the cells of the lymphatic system and other related cells is to provide a second line of defence in addition to the phagocytic action of the reticulo-endothelial system. This mechanism forms the basis of the subject of *immunology* which is concerned essentially with the recognition of 'self' and 'non-self'. You may wonder, for example, why it is that in a severe burn, the damaged skin may be replaced by a graft from another part of the body, whereas if an attempt is made to graft skin from another person to the same area, the graft will almost certainly die and be rejected. How does the body recognise that the skin in the first case is 'self' and in the second is 'non-self'? The next few pages will give an extremely brief and elementary account of the immune response and its mechanism but this can only be the merest outline partly because this is an enormous and extremely complex subject and partly because much of it is still unknown and is the subject of intensive research.

Foreign substances which are capable of setting up an immune response when they enter the body are known as *antigens* and the result of their entry is to set up an *immune response*. There are two types of such reaction, *humoral* and *cell-mediated*. In the first place, the result of their entry may be to stimulate the body to produce an antagonistic substance called an *antibody* which circulates in the blood and combines with, and neutralises, the stimulating antigen. It is important to remember that the reaction is *specific*, and that such *humoral antibodies* will only neutralise antigens of the same type as that which originally stimulated their production. Thus an attack of measles will result in the production of an antibody which will make it unlikely that a child will ever suffer from measles again but the possibility of an attack of whooping cough is un-diminished. The other type of immune reaction is *cell-mediated immunity* in which lymphocytes which have been stimulated or sensitized by the antigen develop 'cell-bound' antibodies on their surface.

The antibodies themselves are proteins (*globulins*) and are therefore called *immunoglobulins* (Ig). There are five main types and these are distinguished by letters: IgG, IgM, etc. Immunoglobulin G is present in the largest quantities.

Two types of small lymphocytes are involved in the immune response. These are the *thymic-processed* (T) lymphocytes and *bursa equivalent* (B) lympho-cytes.

T-lymphocytes are so called because although formed in the bone marrow, they migrate via the bloodstream to the thymus where they divide repeatedly to produce large numbers of daughter cells. These lymphocytes pass back into the bloodstream whence they may return to the bone marrow or may colonise lymph nodes, the spleen or the lymph follicles in various viscera, such as the Peyer's patches. T-lymphocytes are involved in cell-mediated immunity.

The 'bursa' referred to in bursa-equivalent has nothing to do with the bursae with which you are so familiar. The term refers to a peculiar lymphoid organ found in birds which was the subject of much early (and current) work on immunity. In birds, the bursa is the organ which controls the maturation of lymphocytes for the production of *humoral antibody*. In man, a large group of lymphocytes carries out this task and are therefore called B-lymphocytes even though the organ(s) corresponding to the bursa in birds has not been identified for certain. It is probably represented by lymph nodes, tonsils or other lymphoid organs. B-lymphocytes are produced in the bone marrow and pass in the blood-stream to the lymphoid tissues. When stimulated by the appropriate antigen, B-lymphocytes proliferate and can transform themselves into *plasma cells* which manufacture the humoral antibodies in their extensive granular endoplasmic reticulum.

To sum up this rather complicated story, the bone marrow produces large numbers of lymphocytes which are not, however, capable of reacting to antigen at this stage. After processing by the thymus (T-lymphocytes) or by the other lymphoid tissues (B-lymphocytes), each becomes competent to react to a specific antigen, the T-lymphocytes carrying out cell-mediated immune responses and the B-lymphocytes transforming to plasma cells and producing humoral antibody.

21 · The Urinary System

The urinary system consists essentially of a pair of kidneys (which secrete the urine), a pair of ureters (which conduct the urine to the urinary bladder), the bladder itself (which stores the urine) and the urethra (which is the communicating channel between the bladder and the outside world. Almost the whole of the urinary system is lined by transitional epithelium which is waterproof and able to stretch during distension of the organs which it lines. In this chapter we shall first describe the general anatomy of the urinary system, which is relatively uncomplicated, and then give a fairly detailed description of the structure and function of the kidney, which is rather more difficult to understand.

THE GENERAL ANATOMY OF THE URINARY SYSTEM

Each kidney is roughly the size of a clenched fist and lies in the loin. The indentation, or *hilum* of the kidney faces medially and it is in this region that the large renal artery enters the kidney and the ureter and the renal vein leave (Fig. 21.1). The kidneys have an enormous blood supply, receiving between them about one-quarter of the cardiac output (about 1300 ml) each minute. The renal arteries are therefore short, wide, and travel directly from the aorta to the kidneys. Each kidney is surrounded by a fibrous capsule, outside which is a large amount of fat that helps to hold it in position. The hila of the kidneys lie on a level corresponding to the first lumbar vertebra but the right kidney is a little lower than the left (because of the presence of the liver on the right). The upper half or so of each kidney is therefore protected by the last one or two ribs and is related to the diaphragm.

The ureter emerges from the hilum of the kidney, the proximal part being dilated to form the pelvis. This is situated partly inside and partly outside the kidney (Fig. 21.2). The ureter is about 10 inches long and is about the diameter of a small drinking straw. It is lined by transitional epithelium and has a wall of smooth muscle. It descends on the surface of psoas major, enters the pelvis by crossing the common iliac vessels and then enters the bladder. It runs very obliquely through the thick muscular wall of the bladder and this fact helps to prevent reflux of urine up the ureter when the bladder contracts.

The bladder has a comfortable capacity of about 220 ml in the male (but rather more in the female). When empty, it lies entirely within the pelvis, behind the body of the pubis. As it fills, however, it rises up above the pubis and when pathologically over-distended may even reach as high as the umbilicus. In shape, it is rather like a boat (Fig. 21.3) when empty but becomes rounded or oval as it fills. The triangular *base* of the bladder is related posteriorly to the rectum in the male and the uterus and vagina in the female (Fig. 27.5). Internally, the triangle is represented by a smooth area called the *trigone* which has the ureters opening into its lateral corners and the urethra leaving at the lower angle. The bladder is lined

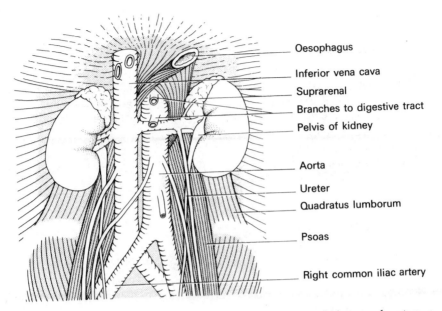

Fig. 21.1. The posterior wall of the upper abdomen to show the kidneys and ureters.

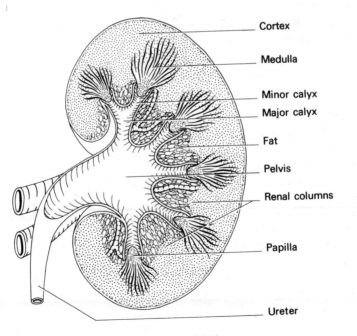

Fig. 21.2. Diagrammatic section through the left kidney.

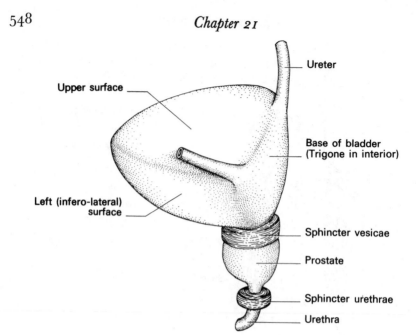

Fig. 21.3. The male bladder.

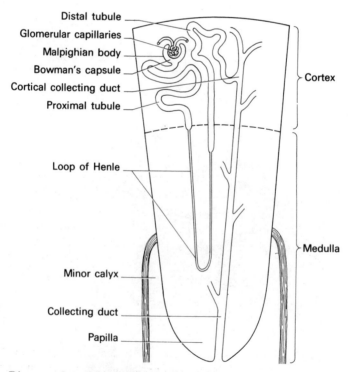

Fig. 21.4. Diagram of a nephron and its collecting duct.

by transitional epithelium and has a powerful coat of smooth muscle whose fibres are arranged both circularly and longitudinally. At the neck of the bladder there is a thick ring of smooth muscle which surrounds the orifice of the urethra and is called the *sphincter vesicae*. The bladder and the sphincter vesicae are supplied by both sympathetic and parasympathetic nerve fibres, both systems containing sensory and motor fibres. If this nerve supply is lost, normal micturition becomes impossible.

The urethra is only about $1\frac{1}{2}$ inch in length in the female but is long and rather more complicated in the male. In the female it leaves the bladder and runs downwards to enter the vestibule. As well as the upper sphincter of smooth muscle (sphincter vesicae) already mentioned, it also has a lower sphincter of skeletal muscle called the *sphincter urethrae* which is supplied by an ordinary somatic nerve. In the male the urethra is surrounded by the sphincter vesicae at the bladder neck and it then passes through the middle of the *prostate* (*not* prostrate) gland. With increasing age, this gland often becomes hypertrophied so that micturition may become progressively more difficult. The next part of the male urethra is surrounded by a sphincter of skeletal muscle (the sphincter urethrae) and it then enters the central buried portion of the penis (the *corpus spongiosum*). After turning through a right angle it traverses the penis itself to open at the tip.

Urine is produced by the kidneys at an average rate of 1.5 litre per day, although this, of course, varies with the fluid intake and with the amount of fluid lost in the sweat, etc. In very hot weather, only a little urine may be passed. The urine passes down the ureters, assisted by peristaltic contractions of the smooth muscle, and enters the bladder where it is stored. At a convenient time and place the sphincters relax and the urine is expelled, helped by contraction of the smooth muscle of the bladder wall and by a rise in intra-abdominal pressure produced by contraction of the diaphragm and the muscles of the abdominal wall (see Chapter 11).

THE SECRETION OF URINE BY THE KIDNEY

In previous chapters you have seen over and over again how important it is that the internal environment (the '*milieu interne*') be kept constant, and a number of mechanisms have been described which help to maintain the *status quo*. The kidney, in addition to its very obvious functions of excreting waste substances such as urea, also helps to regulate the composition of the tissue fluid (and indirectly that of the cells) with particular regard to the salt and water balance and the pH of the blood. Before discussing the way in which these functions are carried out, it will be necessary to describe the detailed anatomy of the interior of the kidney.

THE STRUCTURE OF THE KIDNEY

The human kidney consists of a series of lobes and each lobe is made up of a *cortex* and a *medulla* (Fig. 21.2), the former being wrapped around the latter so

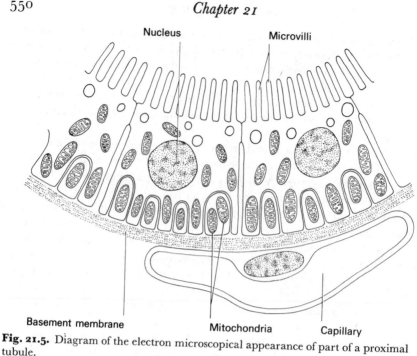

Fig. 21.5. Diagram of the electron microscopical appearance of part of a proximal tubule.

that the cortical tissues of two adjacent lobes fuse at the sides of the lobe to form the *renal columns*. The innermost part of the medulla of the lobes is conical in shape and is called the *papilla*. Each papilla projects into the end of a narrow cavity called a *minor calyx*. Several minor calyces unite to form a major calyx and the major calyces open into the *pelvis* of the kidney. Urine leaves the renal substance via openings in the tip of the papillae, passes into the calyces and pelvis and thence down the ureter to the bladder.

The unit of renal tissue is called the *nephron*, which is shown in Figure 21.4. It begins with a *glomerulus*, or *Malpighian body* of which there are about one million in each kidney. The glomerulus consists essentially of an anastomosing network of capillaries that are surrounded by special, very thin, epithelial cells. This capillary network projects into a spherical capsule—*Bowman's capsule*—also formed of thin epithelial cells. The space of Bowman's capsule is continuous with the lumen of the *proximal convoluted tubule*. This has an extremely tortuous course before finally heading towards the medulla. Its cells are modified for the transport of large quantities of salt, water and other substances from the lumen to the peritubular capillaries (see below). The cells are approximately cubical and have an enormous number of tiny processes (*microvilli*) which project into the lumen (Fig. 21.5), and increase the surface area of the cell. The base of the cells contain a large number of mitochondria whose enzymes provide the energy necessary for the active transport of salt and other substances. The proximal tubule leads into the *loop of Henle* which is a thin-walled tubule that passes down into the medulla before performing a hairpin bend and returning towards the cortex. Its terminal part becomes thick-walled and is continuous with the *distal convoluting tubule*. This also contains many mitochondria and, after a shorter course than the proximal tubule, it finally opens into a *cortical collecting duct* which itself joins other similar ducts from other nephrons to form a *collecting duct* (which is common to many nephrons). The collecting ducts run down through the medulla, joining each other as they go to form a smaller number of larger ducts. The large collecting ducts finally open at the tip of the papilla.

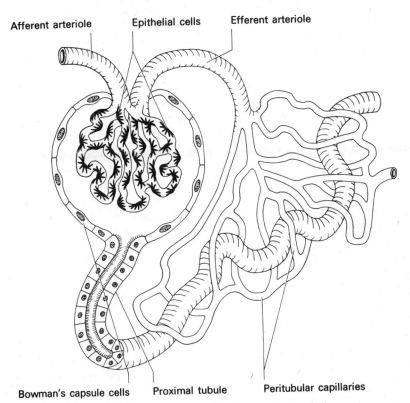

Fig. 21.6. A glomerulus and the blood supply of its associated tubules.

The renal arteries branch repeatedly after entering the kidney and the smallest branches finally end up as minute *afferent arterioles* each of which supplies one glomerulus (Fig. 21.6). The arteriole breaks up to form the glomerular capillary network and from this and the blood drains into the *efferent arteriole* which leaves the glomerulus near the afferent arteriole but soon breaks up again into a network of *peritubular capillaries*. These form a dense network around the proximal and distal tubules. In the case of the glomeruli nearest to the medulla, the efferent arterioles break up to form the capillary plexus around the loops of Henle and collecting ducts in the medulla.

THE FUNCTIONS OF THE KIDNEY

The formation of the *glomerular filtrate* is rather similar to the formation of tissue fluid. Part of the blood plasma (except for the proteins, whose molecules are too large), is filtered through the capillary walls and their covering of thin epithelial cells, into Bowman's capsule. The composition of the filtrate is therefore exactly similar to that of blood plasma minus its proteins. The driving force for this filtrate is the blood pressure in the capillaries and filtration is opposed by the osmotic pressure of the plasma proteins. The total quantity of the glomerular filtrate is enormous—approximately 180 litres per day—so that to prevent the body becoming a mere puff of dust within 24 hours, it is essential that almost all this water and salt is reabsorbed, so that only about $1\frac{1}{2}$ litres is finally lost in the urine. However, you have no doubt noticed that if you are very thirsty, or if you have lost a lot of fluid by sweating in hot weather, the urine is small in amount, while within a very short time of drinking a few pints of fluid, a correspondingly large volume of urine is passed.

The reabsorption process is carried out in two stages. As the glomerular filtrate passes along the proximal tubule, about seven-eighths of its water and sodium chloride is reabsorbed by the cells of the tubules and is passed via the peritubular capillaries back into the bloodstream. This happens whether you are extremely thirsty or have drunk as much as you are able, and is therefore known as *obligatory reabsorption*. In the distal tubule and collecting ducts, further reabsorption of salt and water takes place but the reabsorption of water is variable and under the control of the antidiuretic hormone (ADH) which, as will be described in Chapter 26, is produced by the pituitary gland. The reabsorption of salt is also under the control of hormones. This variable reabsorption is therefore known as *facultative reabsorption* and will be described below. As well as the reabsorption of salt and water, both obligatory and facultative, a number of other substances are also reabsorbed from the filtrate as it passes along the tubules and certain substances, potassium for example, even pass in the opposite direction and are excreted into the tubular fluid. Various control mechanisms exist for this reabsorption and excretion but their main objective is to maintain the constancy of composition of the body fluids. For example, glucose is normally totally reabsorbed from the glomerular filtrate by the proximal tubule cells so that none is found in the urine, but if the level of glucose in the blood is excessively high for any reason (see Chapter 26), the excess will not be reabsorbed but will be excreted (or 'spill over') into the urine. Other substances, particularly waste and toxic substances, are not reabsorbed at all but are excreted in the urine and expelled from the body. The chief waste products are the acidic and the nitrogen-containing products of protein metabolism. Although these are all water-soluble, a minimal daily volume of 500 ml of urine is required to remove the quantity that is formed each day in a healthy person.

The proximal and distal tubules also help to maintain the pH of the blood. You have seen in Chapter 18 how the pH is kept constant by buffer systems, particularly by the carbonic acid—bicarbonate system. In the tubules, H ions

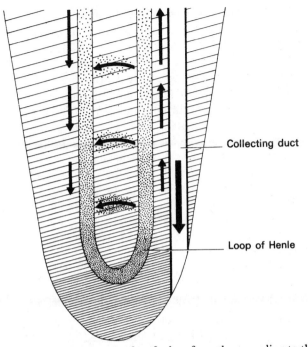

Fig. 21.7. To show how the active transfer of solute from the ascending to the descending limb of a loop of Henle can produce an osmotic gradient in the medulla.

are excreted from the blood into the lumen of the tubules where the excess of H ions may be got rid of in various ways by combination with certain other ions such as phosphates and ammonium ions.

No mention has yet been made of the loops of Henle or of the control of the concentration of urine (i.e. the amount of water reabsorbed) by ADH. You might therefore deduce that the loops of Henle are involved in the concentration process, especially if you know that only birds and mammals are able to produce a concentrated urine and only birds and mammals have loops of Henle. The function of the loops of Henle is to produce a highly concentrated environment in the medulla which they do by a process known as a *countercurrent multiplication system*. This is a complicated process but the essential step is the *active* transport of sodium chloride, without water, out of the ascending limb and into the descending limb of the loop (Fig. 21.7). Thus salt which might have left the medulla is transported across to the descending limb and recirculated and as this happens at all levels, the concentration of salt will gradually increase towards the tip of the papilla, not only within the loops of Henle but also in the tissue between the loops. The process is so efficient that fluid leaving the loops of Henle has a slightly lower salt content than that entering it, and further salt is removed from the distal tubule (see above). The collecting ducts, on their way from the cortex to the tip of the papilla, have to pass through this zone of greatly increased osmotic pressure and as the fluid in them is dilute owing to the removal of salt, there is a tendency for water to be withdrawn from the collecting duct lumens. For this to happen, however, the collecting ducts would have to be permeable to water and this is where ADH exerts its control. In the presence of ADH the collecting ducts are permeable so the water (without salt) is reabsorbed from the collecting duct urine and the urine becomes concentrated. In the absence of ADH, the collecting ducts are impermeable and they conduct the urine safely through the zone of increased osmotic pressure so that it remains dilute. The system is called a *countercurrent system* because there is a countercurrent flow through the loops—i.e. fluid in the ascending limb and that in the descending limb are flowing in opposite directions. It is called a *multiplier system* because the effect of active transport of sodium near the top of the loops is multiplied over and over again by the transport of more and more sodium as the loop approaches the tip of the papilla. The longer the loop, the greater the concentration gradient produced so that desert animals such as hamsters, have a longer papilla than animals which have plenty of water to drink such as beavers.

It may be helpful at this stage to combine parts of this chapter and Chapter 26 in order to describe the effects of water deprivation. If the intake of water is insufficient, there will be a tendency for the blood to become more concentrated and to have a higher osmotic pressure. As the blood circulates through the hypothalamus, this condition will be detected by special receptors known as *osmoreceptors*. As a result, nerve impulses will pass down the pituitary stalk to the posterior lobe of the pituitary and more ADH will be produced. This is carried by the bloodstream to the kidneys where the increased ADH level will increase the permeability of the collecting ducts. Water will be reabsorbed by osmotic pressure from their lumens because of the concentration gradient in the tissues around the collecting ducts in the medulla. This water will be taken back into the bloodstream but salt and other substances will continue to be excreted in the now concentrated urine.

To sum up the functions of the kidneys, they maintain at a constant level the salt and water content of the body and also that of other substances such as glucose, etc. They maintain constant the pH of the blood and other tissues of the body and they excrete waste substances (especially urea) from the body.

The kidney also functions as an endocrine organ and is involved in calcium metabolism by means of 1.25-vitamin D3 (p. 634); in regulation of the blood pressure and the level of salt in the body by means of the renin-angiotensin system (p. 634) and on red cell production by means of erythrogenin and erythropoietin (p. 635). It also produces a variety of *prostaglandins* which have important effects on smooth muscle, on the blood pressure, and on salt and water balance. The precise functions of this group of compounds have yet to be elucidated.

MICTURITION

The urine, having passed from the kidneys to the bladder via the ureters, is now stored there until it is convenient for it to be voided. As the bladder fills, the smooth muscle in its wall is stretched without any increase in its tone, so that there is very little increase in its tension or in the intravesical pressure. It is not until the bladder contains 150–200 ml of urine that the desire to micturate is felt, although it does not become urgent until more than about 400 ml have accumulated.

The micturition reflex is mainly, if not wholly, under the control of the parasympathetic system, the fibres travelling in the 2nd, 3rd and 4th sacral nerves. The sensory impulses from the distended bladder travel in these nerves to the sacral segments of the spinal cord, while the efferent effector pathway is along these same nerves to the *detrusor muscle*, which contracts, and the *sphincter of the bladder* (sphincter vesicae) which relaxes. Impulses from the cerebrum can inhibit this reflex arc so that the detrusor muscle remains relaxed and the sphincter contracted. Additional control is provided by the skeletal muscle of the *sphincter urethrae* and the muscles of the pelvic floor. When micturition commences, the pelvic floor muscles relax, the detrusor muscle contracts to open up the bladder neck and the sphincters relax. Once urine has begun to flow down the urethra, the detrusor contractions become more powerful as a result of reflex stimulation from sensory nerve endings in the urethra, but micturition can still be interrupted if necessary by a voluntary contraction of the sphincter urethrae and the pelvic musculature. The pressure rise in the bladder is reinforced by voluntary contraction of the abdominal muscles. At the end of micturition the urethra is emptied by gravity in the female and by the contraction of certain muscles in the penis in the male.

A study of the anatomy of the urethra in the male and in the female (Figs. 27.2 and 27.5) will make it obvious why the commonest difficulty in micturition in the male is obstruction while in the female, it is incontinence. Disturbances of micturition occur in diseases of the nervous system affecting the reflex arc and its cerebral control; by obstruction by an enlarged prostate in the male; by alterations of anatomical relations in the pelvis, e.g. by prolapse in the female; by nervousness (especially in males); and by pain produced by an abdominal operation.

22 · The Skin and Body Temperature Control

The skin is the part of the body—indeed, the organ of the body—of which one is most aware since every square inch of it is visible (although some regions need a mirror for detailed inspection) and it is the bodily component most exposed to injury. Every day brings a number of minor traumata to the skin and relatively major incidents involving an actual breach of the skin happen to everyone. Familiarity is apt to breed contempt and the skin may be dismissed simply as the paper which covers the parcel and 'keeps everything in'. In fact it is a very complex organ whose correct functioning is essential to life. It has the extremely useful properties of being more or less waterproof (but see later in this chapter), self-replacing and self-healing, capable of becoming thicker if exposed to wear and tear, able to become darker in colour to protect against excessive sunlight, and capable of producing an important vitamin—vitamin D. In addition, it plays a vital part in the regulation of the body temperature, a function so important that this subject will be included in this chapter since the skin provides one of the most important factors which keep the body temperature constant.

Immediately deep to the skin lies the superficial fascia, deep to which, in turn, lies the deep fascia. These tissues are intimately related to the skin and play an important role in its mobility over the underlying structures. When you have read this chapter, therefore, you will find it helpful to read again the accounts of the superficial and deep fascia which were given in Chapter 3.

THE STRUCTURE OF THE SKIN

The skin is composed essentially of two principal layers, the *epidermis*, and the *dermis*, the former being the most superficial layer. The epidermis is a rather complex stratified squamous epithelium of which the most superficial layer consists of a relatively thick zone of dead, keratinised cells (i.e. they are composed almost entirely of the mechanically tough protein *keratin*). The dermis, on the other hand, is composed largely of dense connective tissue containing various cells, lymphatics, blood vessels and nerves.

THE EPIDERMIS

Five different layers can be distinguished in the epidermis of thick skin but the relative thickness of the layers varies from place to place. The deepest layer of cells, which is in contact with the dermis, is the *stratum basale* or basal cell layer. These cells are capable of rapid mitotic division, some of the daughter cells being pushed towards the surface so that they provide a constant replenish-

556

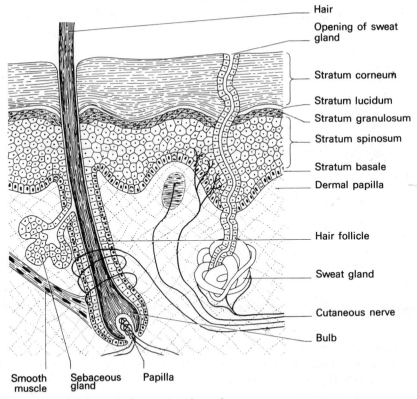

Hair

Opening of sweat gland

Stratum corneum

Stratum lucidum

Stratum granulosum

Stratum spinosum

Stratum basale

Dermal papilla

Hair follicle

Sweat gland

Cutaneous nerve

Bulb

Smooth muscle   Sebaceous gland   Papilla

Fig. 22.1. Section through skin to show its major components.

ment for the cells which have been lost from the surface (Fig. 22.1). The next layer is the *stratum spinosum*, or prickle cell layer. These are cells derived from the basal cell layer by mitosis, but they are a little different in form since they are not so tightly packed, being separated from each other by narrow spaces. They are shaped rather like holly-leaves (hence the name) and they are attached to each other by the fine projecting processes. The next two layers are thin and the cells show the accumulation of keratin in their cytoplasm. These form the *stratum granulosum* whose cells contain granules of a precursor of keratin, and the *stratum lucidum* whose cells are clear, flattened and without nuclei. Finally, the *stratum corneum* consists of many layers of very flattened, scale-like, dead cells composed largely of keratin. These cells are shed from the surface as flakes. The thickness of the epidermis is variable as you can easily tell if you compare the skin of your eyelid with the skin of the sole of your foot. In the thin regions, the stratum granulosum and stratum lucidum cannot be distinguished, and the stratum corneum is thin. In the sole, or palm of the hand, the stratum corneum is extremely thick and if the skin is particularly liable to wear and tear (for example, in the hands of manual workers) it can become even thicker in response to the demands made on it.

Scattered among the deeper cells of the epidermis are a number of branching cells called *melanoblasts*. These are capable of producing the pigment *melanin* which darkens the colour of the skin on exposure to ultra-violet light. Pigmentation of the skin which occurs as a result of such exposure over a fairly long period should not be confused with the bright red faces, backs and shoulders which are to be seen on a beach on Bank Holidays. This is an acute vascular reaction caused by the harmful effect of ultra-violet light on relatively unpigmented skin.

The junction between the dermis and epidermis is not flat. The dermis projects into the deeper layers of the epidermis in a series of ridges like the surface of a ploughed field, or in a number of conical projections called *dermal papillae*.

THE DERMIS

The dermis consists largely of dense connective tissue. It contains numerous elastic fibres which give the skin its resilience, but with increasing age the elastic fibres diminish in number and the skin becomes wrinkled and slack. The dermis is a vascular tissue and it contains a network of blood vessels derived from the plexus of vessels in the superficial fascia. Capillaries extend into the dermal ridges or papillae but do not enter the epidermis. Lymphatics are also found in the dermis, as are nerve fibres. The latter form a plexus of nerve bundles from which individual efferent fibres supply blood vessels, sweat glands and smooth muscle (see below). All the efferent nerve fibres belong to the autonomic nervous system. Most of the nerves in the skin, however, are afferent somatic fibres, carrying the different forms of sensation. They innervate the nerve endings which have been described in Chapter 4 and some also end in the deeper layers of the epidermis. The latter are pain endings so that although an injury that is confined to the epidermis will not bleed, it may nevertheless be painful.

HAIRS

Most parts of the skin are covered with hair—either the coarse type of hair found on the scalp and pubic regions (and on other parts of the body in the male) or fine downy hair. No hairs are found on the palms of the hands and soles of the feet, the terminal and usually the middle phalanges of the fingers or on parts of the external genitalia. The *roots* of the hairs are found in the dermis. The deepest part of the root is expanded to form a *bulb* within which is a small vascular *papilla*. The root consists of layers of cells rather like the layers of the epidermis and, by division, the cells lay down keratin which is pushed towards the surface through a channel lined by cells that is called the *hair follicle*. Opening into the follicle are one or more glands. These are the *sebaceous glands* and they secrete an oily material called *sebum*. Attached to the side of each follicle is a small bundle of smooth muscle, innervated by the sympathetic nervous system, whose function is to make the hair 'stand on end'. It is unimportant in human skin.

The colour of the hair depends on the amount and type of pigment present.

SWEAT GLANDS

The main part of these glands is situated in the dermis but the duct spirals towards the surface through the epidermis. The body of the gland consists of a tightly convoluted tube lined by cubical epithelium and surrounded by a network of capillaries. In the axillae, and around the external genitalia, are found specialized sweat glands called *apocrine glands*. Sweat glands are innervated by the sympathetic nervous system. Sweat contains certain waste products such as urea and also electrolytes, principally sodium chloride. In very hot climates, therefore, the salt loss caused by sweating may have to be made good by taking salt tablets. This also explains the great importance of salt in trade, etc., from Roman times onwards. The word *salary* comes from the Roman habit of paying soldiers in salt.

THE NAILS

The nails are developed from the stratum lucidum of the nail bed and, like the stratum corneum, they are composed of densely packed keratinised cells.

THE BLOOD SUPPLY OF THE SKIN

As was mentioned above, the skin has a rich blood supply in the form of a complex network of vessels in the dermis from which the capillaries arise. There are also a large number of arteriovenous anastomoses (see Chapter 14) in the skin. Some comment is needed on the venular plexus. The capillaries and a-v anastomoses drain into a sub-papillary plexus of venules which contains far more blood than the capillary loops themselves. It is the blood in this venular plexus that, together with the skin pigments, gives the characteristic colour of the skin. The venules are more readily filled by blood flowing through the dilated anastomoses than through the capillaries. If the arterioles and the anastomoses are dilated, the venules are filled, and the skin will be pink or red. If the supply of blood is reduced the skin will be pale if venular constriction is also marked, or, if the venules contains much stagnant blood, the skin will appear blue, purple or grey. You can observe this yourself if you take a hot bath on a cold day. Before entering the bath your skin will probably be pale, or even blue if you are very cold. The arterioles are only allowing a trickle of blood through the capillaries and arterio-venous anastomoses into the venular plexus. What little blood does enter the veins, and stagnates there, contains mostly reduced haemoglobin. After a soak in the hot bath, however, you will emerge with a pink flushed skin, owing to the large amounts of blood flowing rapidly into and through the venules. These blood flow and consequent colour changes are due both to local tissue temperature changes and to changes in whole-body temperature (see below), when the sympathetic nervous system is involved. However, stimulation of the sympathetic nervous system due to fear can also cause constriction of the arterioles and subsequent emptying of the capillaries and venules, so that a person may be 'pale with fright'. The arterioles in the face can dilate as a result of self-consciousness in the form of a blush, and Darwin is said to have asked an artist friend to observe, the next time he drew from the nude, how far down the model's blush extended when she removed her clothes. A blotchy dilatation of cutaneous blood vessels is commonly seen in such stressful situations as a viva voce examination.

THE TRIPLE RESPONSE

The vascular system of the skin plays an important part in its reaction to non-penetrating injury. Three stages are recognised, according to the severity of the trauma. These are known collectively as *the triple response* which consists of the *red reaction, flare* and *wheal*. The red reaction consists of a reddening of the skin over an area corresponding to the exact site of the injury. It is caused by release of *histamine* or some similar substance from the damaged tissues which causes capillary dilatation. This dilatation is due, as was described in Chapter 14, to relaxation of precapillary sphincters. With more severe injury, the red mark becomes surrounded by an irregular *flare* of red skin which is due to arteriolar dilatation. The flare is caused by a local *axon reflex*, illustrated diagrammatically in Figure 22.2. The injury sets up nerve impulses in the fibres associated with the perception of pain. These impulses also travel directly through another axon branch to the arteriolar smooth muscle, without proceeding to the central nervous system. Substance P is the transmitter released at the nerve ending on the vessels which allows the smooth muscle to relax and the vessel to dilate. The skin temperature is raised in the area of the flare. In cases of even more severe injury to the skin, damage to the capillaries increases their permeability and plasma passes out into the tissues. This causes a swelling or *wheal* to appear.

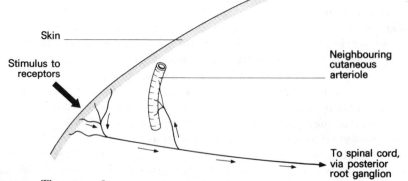

Fig. 22.2. The axon reflex.

The triple response can occur as a result of many different types of trauma such as burns, chemical damage or mechanical damage and it is also produced by the injection into the skin of histamine or other related substances.

THE FUNCTIONS OF THE SKIN

Many of the functions of the skin have already been mentioned and its role in temperature control will be considered in detail below, but it may be helpful to present here a list of the skin's most important functions:

MECHANICAL PROTECTION

This is largely the function of the stratum corneum, which is thickest where it is most exposed to trauma. When small areas of skin are lost by injury, rapid division of the cells of the stratum basale around the wound can replace the lost skin quite quickly, the new skin encroaching on the defect from the periphery so that the wound diminishes in size. The process is helped by contraction of the fibrous scar tissue that is formed. Large defects in the skin cannot be closed in this way, so that skin grafting is necessary.

BARRIER FUNCTION

The superficial layers of the skin are fairly waterproof so that it is possible to bathe in fresh or very salt water without upsetting the osmotic balance of the interior of the body. The skin is not, however, completely impervious and about 600 ml of water each day may be lost from the body by diffusion of water from the tissues followed by evaporation. This fluid loss is sometimes known as '*insensible perspiration*' (a misleading term that we prefer to avoid) but it is quite different from sweat since it is only water which is lost, not urea or electrolytes. The skin also forms a barrier to microbial invasion. Many types of bacteria are found on the skin, including harmful species, and a breach in the skin is always in danger of becoming infected.

SENSATION

The skin is an extensive sensory organ and although one is not consciously aware of it, the constant stream of afferent impulses from the skin plays a most important part in everyday activity. This aspect has been covered in more detail in Chapter 5.

VITAMIN D FORMATION

7-dehydro-cholesterol is made in the body along with the common cholesterol, of which we hear so much nowadays. When the 7-dehydro compound is exposed to ultra-violet radiation, it is converted into *cholecalciferol* which can replace the dietary sources of this substance. The pale skin of the European in his temperate climate, with only a moderate amount of ultra-violet light, makes as much cholecalciferol as does the dark-skinned Indian or African, exposed to the stronger tropical sunlight. When people of the latter races migrate to temperate climates, or in areas of high industrial activity where atmospheric smoke-haze forms a barrier to solar ultra-violet light, the skin can no longer make sufficient cholecalciferol (see also Chapters 24 and 26).

TEMPERATURE REGULATION

This is such an important function that it needs to be considered in detail and in conjunction with other factors affecting body temperature.

BODY TEMPERATURE CONTROL

MECHANISMS OF LOSS OF HEAT

If the skin is warmer than the environment, and it usually is, then the physical processes of *convection, conduction* and *radiation* will transfer heat to cooler objects or media in contact with the warm skin. Heat is conducted from my skin to my clothes, the chair I am sitting on and to the air. These processes are relatively slow for none of these materials is a good heat conductor. If I were lying directly upon a metal operating table, that would be a different matter, for the metal would conduct heat well, and it would feel cold to the touch. When air is warmed it becomes lighter, and moves upwards through the surrounding cooler air. This is called a *convection current* and such currents, carrying heat away from the skin by convection, are present at all times. *Radiation* is the direct transfer of heat from warmer to cooler objects by *infra-red* waves of the electromagnetic wave spectrum. Heat radiation does not depend directly on the temperature between a hot body and its cold environment, but upon the difference multiplied by itself four times (i.e. to the power of 4). Small changes in the temperature difference can thus cause large changes in radiant heat loss. In the case of the human body, 75 % of all heat loss from the body occurs in these ways (the remainder is via water-evaporation); 80 % of this loss is by radiation and only 20 % by conduction and convection. Control of the skin surface temperature is therefore very important in the control of heat loss from the body.

Since the skin's own metabolic rate is very low, its temperature is only raised by the flow of warmer blood through the cutaneous blood vessels. The arterioles and the arteriovenous anastomoses are supplied with noradrenergic sympathetic effector neurones, impulses in which cause contraction of the smooth muscle walls and restriction of blood flow. The temperature of the skin is thus affected by this sympathetic vasoconstrictor control, an increase in nerve impulses lowering skin temperature and thus heat loss to the environment and vice versa. Skin temperature is directly related to the amount of blood present in and flowing through the sub-papillary venous plexus, the a-v anastomoses being particularly effective in controlling the amount of blood in these veins.

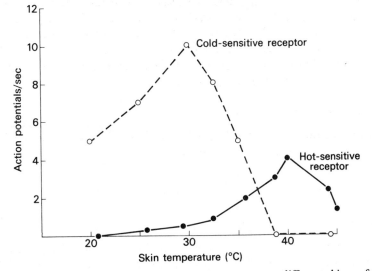

Fig. 22.3. The responses of two skin temperature receptors to different skin surface temperatures.

Some heat is also lost as a result of the water that diffuses through the skin. As this water evaporates it needs extra energy. This energy is known as the *latent heat of vaporisation*. This is 2.5 MJ per litre so that the evaporation of 600 ml of water requires about 1.5 MJ of energy, or about 20% of all body heat losses. You will see that we have accounted for 95% of the heat lost from the body. The remainder is due to evaporation of some of the watery film lining the respiratory tract into the inspired air.

The evaporation of secreted sweat is the second controllable method by which heat is lost from the body through the skin. The sweat glands produce a watery secretion that contains about 30% of the NaCl that is found in blood and tissue fluids and also small amounts of the other components of these fluids. The glands are inactive unless stimulated by cholinergic effector neurones of the sympathetic nerve supply to skin, and secretion is accompanied by an increase in blood flow in the sweat glands. Sweat secretion rates of up to 1.7 litres/hr. have been recorded, but a prolonged rate of 12 litres/day is seldom exceeded. In a dry hot environment, when the skin is heated by a high rate of cutaneous blood flow, sweat secretion and evaporation is an effective way of removing excess heat from the body. In conditions of high humidity, however, sweat does not evaporate so easily and hot humid weather is therefore much more uncomfortable than even hotter but dry weather.

REFLEX CONTROL OF BODY TEMPERATURE

Receptors are the natural starting-point in describing any reflex regulatory system, and body temperature control is no exception, even though we have here started with a description of two of the effector mechanisms, skin blood flow and sweat secretion.

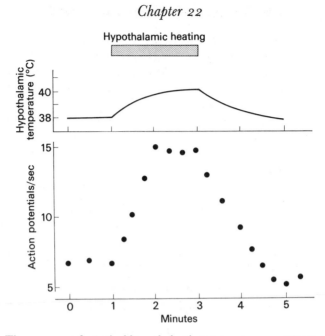

Fig. 22.4. The response of a typical hypothalamic temperature receptor to warming of the local tissue.

Two sets of receptors are involved in these reflexes. One set is found in the anterior part of the hypothalamus where neurones have been found that respond to local changes in temperature. Eighty per cent of all these sensitive neurones respond to an elevation of temperature, and 20% to a fall. These receptors are, naturally, exposed to the temperature of the deeper parts of the body, its *core*, and to the circulating blood. Their activity is thus determined by the *core body temperature*. The other receptors are in the skin. Detailed studies have shown that both warm-sensitive and cold-sensitive receptors are present in the skin of the whole body. However, from the point of view of the temperature-regulating reflexes, only the cold-sensitive cutaneous receptors are important. The average skin temperature of a 'comfortably' dressed person, or a person who is naked but in a comfortable environment, is about 35°C. A receptor, such as the cold receptor in Figure 22.3, with its peak discharge of impulses at 30°C, would vigorously stimulate the appropriate reflex mechanism at that skin temperature, as would a hypothalamic receptor when its temperature is raised from 38°C to 40°C (Fig. 22.4). Naturally the skin receptor would cause reflex vasoconstriction in the skin, lower the temperature and so reduce radiant heat loss. It would also inhibit the production of sweat, stimulate shivering in skeletal muscles and other metabolic, heat-producing mechanisms, and cause an elevation of hairs (ineffectual in man, but producing the characteristic goose flesh appearance). The impulses from the hypothalamic heat receptors inhibit the skin vasoconstrictor nerves, but stimulate those supplying the sweat glands. The blood flow increases, warming the skin and so raising radiation heat-loss. Sweat is secreted and readily evaporates, drawing heat from the skin already warmed by the increased blood flow. Stimulation of the hypothalamic receptors inhibits shivering and metabolic heat production.

The centres that control the effector mechanisms just described, are all found in the posterior hypothalamus and the body temperature will depend upon the balance between heat-producing mechanisms (metabolism, muscle activity and shivering), heat conserving mechanisms (vasoconstriction in the skin and insulation by subcutaneous fat) and heat-losing mechanisms (vasodilatation in the skin and sweating).

OTHER ASPECTS OF BODY TEMPERATURE

Now that we have proceeded from a description of skin and its functions to the very important part played by skin in body temperature regulation, it is time to consider generally the temperature of the body, why it is what it is, where we measure it and factors that cause variations in it.

Man has a 'normal' body temperature of 37°C, as do most other mammals. Many birds maintain a temperature of about 40°C. Both groups of animals maintain this constant temperature regardless of the environment and so are termed *homeothermic*. Reptiles and amphibia cannot control their temperatures and are termed *poikilothermic*. Many reptiles, however, function best if the sun is able to warm their bodies up to about 37°C when they are as warm blooded as we are! That is why you find lizards and snakes basking in the sun, and why they are able to move so quickly when disturbed. There are many theoretical explanations for choosing a temperature in the 37–40°C area. One of the simplest is that the hotter the faster and the faster the better, until the limit is reached that is set by the chemical nature of proteins, many of which undergo irreversible changes in the 42–43°C range.

The whole of the body is not at a temperature of 37°C. The human body can be divided into a *core* in which the constant temperature is maintained, and a *shell* in which the temperature may vary. The core consists of the pelvic, abdominal and thoracic viscera, the buccal and cranial cavities, the deeper tissues of the proximal parts of the limbs and those of the musculo-skeletal wall of chest and abdomen. The shell consists of the more superficial and distal parts of the limbs and the skin of the whole body. The testes in the scrotum are within the shell of the body. For some utterly obscure reason, spermatozoa will not develop at the 'normal' core body temperature in any mammal. The core-shell boundary is not constant, retreating inwards in a cold environment (thus the proximal limb and girdle muscles, so important in swimming, suffer a reduction in temperature and therefore in contractile speed and power when the swimmer is in cold water), and extending out to the skin, even of hands and feet when heat is being rapidly produced in exercise, or loss to the environment is reduced.

Since the core is the important temperature, this is the one that should be measured. The best simple approach to the body's core is the mouth, though recent hot or cold drinks, talking or mouth-breathing can all disturb mouth temperature measurements. The temperature in a stream of urine *as it is being voided from the bladder* is a very good index of core temperature, as is that of the external auditory meatus protected by an insulating pad.

When we measure the core temperature, we do not find it is always at 37°C, even though we are careful, using one site only and always observing the same precautions. The temperature varies with the time of day. It is at its highest between 2 and 6 in the afternoon and at its lowest before we begin to waken in the morning. This, the *diurnal* or *circadian* rhythm, goes on all the time, even through feverish illnesses like typhoid. We suspect that it may be related somehow to activity, for on changing from day to night shift working, a new rhythm is established after a few days, to coincide with the new sleeping and waking pattern. If you lie still on a couch from, say, ten in the morning, the normal rise in temperature may be stopped. The size of the diurnal rhythm varies, but is usually 0.5°C above and below the average level.

Ovulation, coming in the middle of the menstrual cycle, is followed within a day by a rise of about 0.5°C in the average temperature, but there is no change in the diurnal rhythm which just continues about the new higher level.

Exercise causes the liberation of vast amounts of heat—up to 20 times the basal level—so it is not surprising to find a rise in core temperature and a stimulation of the heat-losing mechanisms during exercise. However, these are not stimulated as much as they are when the same temperature rise is caused by other means. It appears that exercise partially inhibits the actions of the heat-sensitive receptors, rendering their stimulation of the heat-losing responses less effective.

Finally, one must add that bacterial and virus infections, surgery or other forms of major tissue damage can all cause the temperature to rise. Chemicals released from body cells after contact with micro-organisms or after damage cause an inhibition of the effects of the heat sensitive receptors, a consequent release of the heat producing and conserving mechanisms and so a rise in body temperature.

23 · Effects of Exercise

We are all affected in ourselves by, and concerned in others with the need for, and consequences of, muscular exercise. Most people are worried by anything that limits their ability to exercise, whether in work or in play. Muscular exercise produces far more profound and far-reaching changes in the functions of the living body than does any other of its many activities. You will be continually meeting, in your professional career as a physiotherapist, the effects diseases have on muscular exercise, and you will be producing, in your patients when you ask them to exercise, the changes in body function that accompany the exercise.

In this chapter we will consider first the exercising muscle itself, the stimulation of its contractile process (again) and the provision of energy to drive that process. Since the raw materials for the energy-providing system are brought in the blood, which also removes waste products, there must be both local and general circulatory adjustments in exercise. Energy production and respiration are linked closely, and any changes in tissue respiration will inevitably be linked with corresponding changes in external respiration. Substrates for oxidation in the energy-producing system are withdrawn from their stores and finally waste products, other than CO_2, must be disposed of.

MOTOR UNITS IN EXERCISE

As you have already learned, muscles consist of individual units, each consisting of an anterior horn cell in the spinal cord or the equivalent in the brain-stem, its axon and the muscle fibres innervated by the axon. In general, there are two types of muscle fibre known as fast and slow (these terms are used because the twitch duration is brief or prolonged) which are white or red respectively, in colour. The former are used in short-lasting contractions and the latter in more sustained contractions. In man, most muscles are mixed, slow and fast, but some method exists of selecting units for the type of exercise undertaken. Another factor of motor unit function concerned in exercise is the manner in which the stimulation shifts from motor unit to motor unit. This allows each unit to alternate between periods of activity and periods of rest, which is probably an important factor in preventing *fatigue*.

Muscular fatigue is obviously an important subject, but is one in which much imprecise thinking goes on. A physiological definition might be 'the awareness that an increasing degree of effort is required to maintain the same degree of performance'. Alternatively, 'contracting muscle(s) become progressively weaker, so that their maximal performance is reduced'. The former definition implies an *in vivo* situation, but the second one applies equally to a muscle excised from the body and studied as an isolated preparation. In both cases, something develops *in time*, the muscular weakness becoming more pronounced as work is continued. It is most tempting to relate the development of fatigue to metabolic changes and this will be pursued later. First we must look at the implications of this hypothesis (that fatigue is metabolic in origin) in relation to the number of motor units available.

My maximal hand-grip strength is about 50 kg, which I can maintain without loss for a few seconds. A contraction of the same muscles, exerting a force of 5 kg, can just be held for 6 minutes. Presumably, only 10% of the available motor units are active at one time and each unit is only working for 10% of the total time. In this situation fatigue develops in 6 minutes. What would happen if, due to an attack of poliomyelitis, or the development of a motor neuropathy or a muscular dystrophy disease, I lost half of the motor units? The maximal grip strength would be reduced to about 25 kg, and to maintain a grip of 5 kg, 20% of the units would be active or each available unit would be active for 20% of the time. Fatigue would now set in much earlier at this grip strength. This, in fact, is what happens. A person with motor unit deficiency will tell you that at any one work level his muscles tire earlier and the time he can continue working at this level is reduced. When you test the maximal output from the affected muscles you find this is also reduced.

Training can alter the behaviour of motor units. Quite apart from the confused and anecdotal statements that abound, frequently based on poorly controlled studies of human athletic performance, there is evidence that training can double the maximal force exerted by a muscle group in voluntary contraction with little change in overall muscle mass. In training regimes (weight-lifting, for example) where there is an increase in muscle mass, this is largely due to an enlargement of existing fibres, for muscle fibres have only an extremely limited ability to reproduce themselves. The effect of voluntary training in man has been mimicked in animal experiments. In particular it has been extensively studied by implanting electrodes around motor nerves and then stimulating the muscles to contract at determined rates over several days. Not only does the metabolic activity and enzyme content of the muscles increase, changing fibres from white to red varieties, but the capillary density and blood flow also rise in these chronically stimulated muscle fibres.

METABOLISM IN SKELETAL MUSCLE

The use and provision of living tissues' energy exchange material, adenosine triphosphate (ATP), is central to the study of exercising muscle's metabolism.

ATP is used in the re-extrusion of Na^+ that enters the sarcoplasm during the passage of the muscle action potential. It is also used in returning calcium from the sarcoplasm to the tubule system at the end of the activation of contraction. But even more is used in the contraction process itself.

To recapitulate this, it is the splitting of an ATP molecule that transfers potential energy to a myosin head in the extended position. When calcium has exposed the myosin-binding sites on the actin molecules, myosin heads bind and release their potential energy, moving into their flexed position and thus pulling the actin (thin) filaments along the myosin (thick) filaments. At this stage, replacement of the ADP attached to the myosin head with fresh ATP causes both its release from actin and its return to the extended position. Thus each ATP molecule performs three functions in the myosin–actin interactions: splitting the myosin–actin link, straightening the myosin head, and, as the ATP molecule is itself split into ADP and free phosphate ion, imparting to the myosin head the energy which will be used when it subsequently binds to another actin molecule and again assumes the bent form. It is the last action that causes the shortening of the muscle. At best about 40% of the energy of the ATP molecules is converted through muscle contraction into useful external work.

THE IMMEDIATE SOURCE OF ATP

There is sufficient ATP present in resting muscle to sustain only a few twitch contractions, yet in exercising muscle its concentration hardly falls. This is because the ATP is reformed from ADP as fast as it is used, by the transfer of phosphate groups from creatine phosphate (CP). Sufficient CP is stored in muscle to maintain ATP level unaltered through many twitch contractions, so during exercise free creatine and inorganic phosphate accumulate, both within the muscle fibres and in the tissue fluid surrounding the fibres. The use of CP to supply energy for reforming ATP is seen principally in the first 1–2 minutes of exercise, before the alternative mechanisms have gone into action. After the end of a bout of exercise, or during a motor unit's rest periods, ATP is used to resynthesise CP from the free creatine and phosphate present (see Fig. 23.1 for diagrams of these processes). There is always, even in resting muscle, a small loss of creatine, which passes into the blood and, after being changed into creatinine, is lost from the body in the urine. Creatinine excretion rate is increased when muscle tissue atrophies following damage to motor nerve fibres.

OXIDATIVE METABOLISM AND ENERGY PROVISION

The general processes of oxidation of foodstuffs and the trapping of the energy thus liberated are described in Chapter 24. The process is 'switched on' by the accumulation of ADP, and builds up, over 1–2 min of sustained exercise until it is producing extra ATP as fast as this is being used.

In muscle, as in all other tissues, the principal site of oxidation is in the mitochondria, which use a substance called acetyl Co-A. This substance is the end product of the non-oxidative breakdown of carbohydrates and of the splitting up of fatty acid chains. The mitochondria themselves cannot discriminate between the two origins of acetyl Co-A. In fact, in resting muscle

a) In muscle contraction

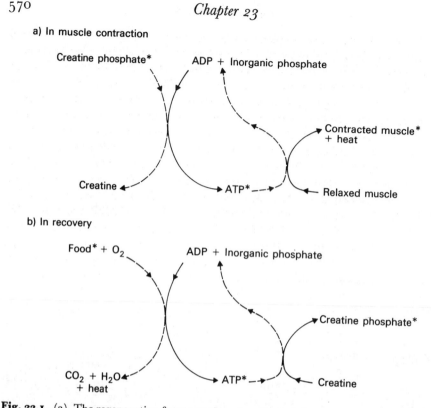

b) In recovery

Fig. 23.1. (a) The regeneration from creatine phosphate of ATP in muscle contraction. (b) The regeneration, by oxidation of foodstuffs, of creatine phosphate in recovery after muscle contraction.

this comes almost entirely from fatty acids, which are broken down in the mitochondria themselves. The situation is unchanged in mild and in moderate exercise. It appears, then, that the preferred fuel for oxidative metabolism in muscle is the fatty acids, which are transported in the blood, attached to the plasma albumin, from the fat stores of the body.

Oxidative metabolism can supply energy at rates up to 25 times the resting rate for human muscle, though as the rate of oxidation increases, carbohydrate forms an increasing proportion of the foodstuff oxidised. In either case, only about 50 % of the energy of the oxidised material is retained in ATP molecules, the rest appearing as heat. Since only 40 % of the ATP energy is converted into external work, the body's musculo-skeletal system is only 20 % efficient in terms of energy transformation from potential energy of foods into mechanical work, a value that is comparable with steam or internal combustion engines.

GLYCOLYSIS

When contraction is using more ATP than can be supplied by oxidation, a further source of ATP is the non-oxidative breakdown of *glycogen* to form *lactic acid*. Skeletal muscle cytoplasm contains up to 1% of glycogen, stored in granules. Glycogen is formed by the union of 100–400 separate glucose molecules in a branched chain *polymer* structure. It is formed after meals when the circulating blood glucose concentration is high. Even at rest glycogen breakdown, or *glycolysis*, continually occurs at a slow rate. In fact the normal end product of glycolysis, *pyruvic acid*, is needed at all times to 'prime' the oxidative system in the mitochondria. In vigorous exercise the rate of glycolysis is greatly increased and it can, in extreme situations, provide energy at 150 times the resting level, though this can only be continued for a short time, being limited by the amount of glycogen present. About 10% of the energy of glycogen is liberated in glycolysis, and of this half is retained in the ATP molecules formed from ADP. Thus, in terms of energy of ATP molecules, glycolysis is only 5% efficient. While glycolysis can occur in all muscle fibres, the white fibres depend primarily on this process for re-forming ATP. They have a high glycogen content while at rest and an abundance of cytoplasmic glycolytic enzymes. They are larger than the red fibres and contract more rapidly and with greater force. These are the fibres used in powerful short-term, 'explosive', efforts, in contrast to the red fibres which rely on oxidative energy production for less powerful but sustained efforts. You sprint with your white fibres, but keep going all day or run marathon races on your red fibres.

The lactic acid formed in glycolysis reacts with sodium bicarbonate ($NaHCO_3$), present in the tissue fluid, forming sodium lactate and CO_2. When exercise ceases, most of the lactic acid is reconverted back into glycogen. This process requires some energy, obtained by oxidative metabolism. In fact it has been found that 90% of the lactic acid formed in exercise can be restored to glycogen, using the energy obtained by oxidising the remaining 10%. Since much of the lactic acid (as sodium lactate) escapes from the exercising muscle into the blood, it is taken up and metabolised (oxidised or converted into glycogen) by many tissues in the body, principally the liver, but also the heart and all skeletal muscles. The sodium released combines with CO_2, forming $NaHCO_3$ again. This process takes 30–60 minutes, depending upon how much lactic acid was produced.

TOTAL RATES OF ENERGY CONVERSION

At rest, human muscles use 0.25–0.30 ml O_2/100 ml muscle/min, which is equivalent to an energy production of 5–7 Joules. This, for the whole muscle mass of man, is about 20% of the resting metabolic rate. In sustained mild or moderate exercise this rate can be increased by 5–25 times. Carrying a 10 lb load of shopping increases energy use in the forearm hand-grip muscles by 5 times, and walking probably does the same for the muscles involved. Vigorous shivering may raise muscle metabolism by 30 times resting level. In vigorous exercise, muscle metabolism may be raised to 50–150 times resting level. These levels are reached in efforts such as the 100 yard dash or explosive throwing or jumping efforts and cannot be maintained for more than a few seconds.

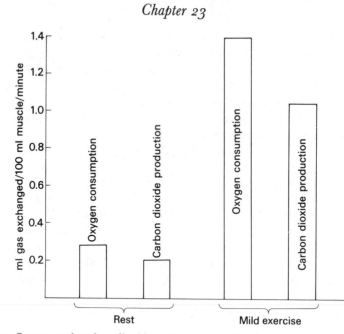

Fig. 23.2. Oxygen and carbon dioxide exchanges of skeletal muscle at rest and during mild exercise.

METABOLISM AND FATIGUE

As was said earlier, there is much unfounded speculation on the nature of fatigue. Its causes may vary in different experimental situations and in different work rates. Studies performed on human subjects in the early 1970s in which small samples of muscle were removed during exhausting exercise showed that the weakness and pain of severe fatigue were associated with a decline in ATP concentration. Since this substance is the source of energy of contraction and it is used to maintain so many other vital functions in all tissues, it is to be expected that its decline is related to fatigue. There was in these studies, no direct link between fatigue and lactic acid production. On the other hand fatigue develops more readily in moderate exercise in a person who cannot form lactic acid from glycogen (an inherited disorder called McArdle's disease) than it does in normal people. Fatigue may also develop after several hours of moderate level exercise in association with low blood glucose levels and a high level of intermediate fatty-acid breakdown products. It is not known whether this sort of fatigue is due to reduced ATP formation or to some other metabolic defect or accumulation, or even to effects on conscious will to continue the exercise.

INTERNAL RESPIRATION AND EXERCISE

At rest, with fatty acid oxidation predominating, the *respiratory quotient* (CO_2 production $\div$ O_2 consumption = RQ) of muscle is about 0.75. Thus for every

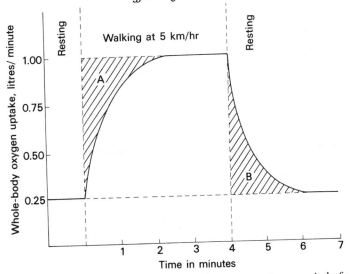

Fig. 23.3. Oxygen uptake by the lungs during and after a 4-minute period of walking at 5 km/hr. A=oxygen deficit; B=oxygen debt.

0.28 ml O_2 used by 100 ml muscle in a minute, 0.21 ml CO_2 are produced. In mild, sustained exercise (a hand-grip of 5 kg force), both O_2 uptake and CO_2 production increase by the same proportion, and the RQ is unaltered. This observation, suggesting that fatty acid remains the chief substrate of muscle in mild exercise, has been supported by the far more difficult direct estimation of the fatty acid uptake of the exercising muscle (Fig. 23.2).

Comparable studies of external respiration in man also support these direct measurements of internal respiration. Thus the respiratory exchange ratio of CO_2 loss in expired air to O_2 consumption of the whole body remains at the resting level of about 0.8, until the exercise has caused O_2 uptake to rise from 250 ml/min to 2 litres/min. This is the state in running at 9 km or 6 miles/hour. This is the fastest running rate that can be maintained for long periods, without any progressive lactic acid accumulation in the body. Once lactic acid is being formed and reacting with $NaHCO_3$, then extra CO_2 is formed. This raises the respiratory exchange ratio from 0.8 to 1.0 or even 1.2, as this CO_2 of non-metabolic origin adds to that still being formed by oxidative metabolism. Once glycolysis starts, as it does when whole body exercise causes an O_2 uptake greater than 2 l/min, it is thus no longer possible to equate tissue RQ with the respiratory exchange ratio. Oxidative metabolism can rise, in more vigorous exercise, to a peak rate of 4–5 l/min, but glycolysis occurs at an increasing rate in this form of exercise. Direct studies of metabolism in animal muscle show that at these rates of work much of the substrate is no longer fatty acid, but glucose.

In both these types of exercise, without and with glycolysis, there is an initial period of exercise during which O_2 uptake is rising towards its final value; that is, there is an O_2 deficit during this period (Fig. 23.3). In exercise without

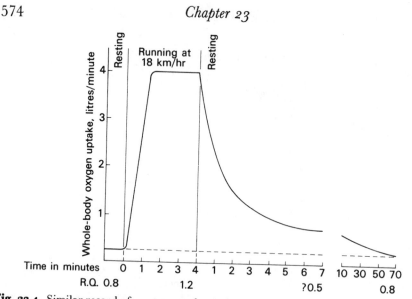

Fig. 23.4. Similar record of oxygen uptake during and after a 4-minute run at 18 km/hr. (= 1,500 metres or 1 mile in 5 minutes).

glycolysis this lasts about 2 minutes. After the completion of exercise the O_2 uptake takes a comparable time to return to its resting level as the O_2 *debt* is repaid. This O_2 debt is due to the use and re-forming of CP which, we as saw earlier, was the immediate source of phosphate and energy for maintenance of ATP concentration.

When lactic acid is formed and accumulates during exercise this, like CP, provides energy for making ATP which is not accompanied by oxidative metabolism. In the rest period which follows exercise with lactic acid production, oxidation continues at an enhanced rate until the excess lactic acid has been removed. In fact the energy obtained by oxidising 10% of the lactic acid is roughly equal to the energy liberated from glycogen, without oxidation, by glycolysis during the exercise. Thus the oxygen debt, realised by the accumulation of lactic acid during exercise, is repaid, over 30–60 minutes after the end of exercise as the lactic acid is itself metabolised. During this period, as has already been mentioned, the sodium released as the lactic acid is removed combines with CO_2, forming $NaHCO_3$. This CO_2 is of metabolic origin. Once it has become $NaHCO_3$, it cannot be lost from the body in the expired air, so during lactic acid oxygen debt repayment, the respiratory exchange ratio will fall below tissue RQ. It may be as low as 0.5 for a while after severe exercise (Fig. 23.4).

MYOGLOBIN

This is a haemoglobin-like molecule that is found in skeletal muscles. It forms the same sort of loose association with O_2 as does haemoglobin, but it remains fully saturated with O_2 until the partial pressure of dissolved O_2 falls very low.

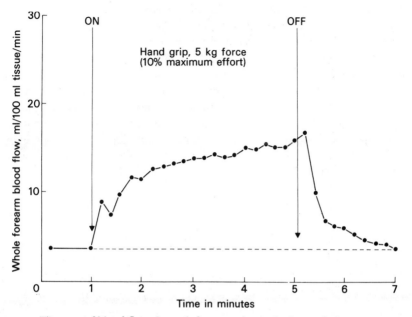

Fig. 23.5. The rate of blood flow through forearm tissues during and after a 4-minute hand-grip at 10% of the subject's maximal voluntary strength.

It can thus provide a store of O_2 which is used at the commencement of exercise and repaid with the other components of the oxygen debt. The maximal concentration of myoglobin in human muscle is about 0.4 g/100 g tissue. Therefore its oxygen capacity is just over 0.5 ml/100 g tissue. This is about two minutes' supply of oxygen for the muscle *at rest*. Even during mild exercise the myoglobin can only supply a small fraction of the oxygen required for only a few seconds.

MUSCLE BLOOD FLOW

In any tissue the volume flow rate of blood through the capillaries, and not the velocity of its movement down any particular capillary, is the parameter that determines exchange of materials across capillary walls. The volume flow rate is determined by the average arterial blood pressure and the resistance to flow offered by the muscular arterioles. The blood pressure in exercise will be considered in the next section.

THE SYMPATHETIC NERVOUS SYSTEM

Arteriolar smooth muscle is caused to contract, and thereby to raise resistance to flow, by stimulation of its α-type of catecholamine receptors by noradrenaline released from the sympathetic vasoconstrictor nerves. This is part of the generalised vasoconstrictor mechanism that affects most systemic arterioles in the body and is mainly concerned in arterial blood pressure stabilisation. A different set of sympathetic nerves supplies arterioles in skeletal muscle. Some people say these liberate adrenaline which acts on β-receptors, while others say that they liberate acetylcholine. Whichever it is, stimulation of these nerves causes smooth muscle of arterioles in skeletal muscle to relax, and thus lowers resistance to flow through these arterioles. In exercise it seems that both systems may be involved, the vasoconstrictors being inhibited and the vasodilators activated. However, blood flow through exercising muscle rises in the complete absence of the sympathetic nerves, so they can safely be ignored, for all practical purposes, when considering blood flow changes in exercising muscle. Either system may be active in the anticipatory period immediately before exercise begins.

CHEMO-REGULATION OF MUSCLE BLOOD FLOW

In mild sustained and isometric hand-grip contractions, the forearm blood-flow is seen to rise during 1–3 minutes and then reach a steady-state level, the change from resting flow being proportional to the grip strength and thus to the extra metabolic activity of the contracting muscles. A 10% maximal contraction may double the whole forearm blood flow, which represents a 4-fold rise in flow through the active muscles (Fig. 23.5). Since the time-course of the blood flow change is similar to that of oxidative metabolism (compare Figs. 23.3 and 23.5) it has been fashionable to look for 'the' vasodilator material among the chemicals involved in metabolism. Many candidates have been sought in the venous blood leaving muscles, or mixed and perfused with the arterial blood supplying muscles. But the arteriolar muscle can be affected only by substances reaching it from the surrounding tissue fluid, far removed from arterial or venous blood. Unfortunately it is not yet practicable to study the chemical composition of muscle tissue fluid directly, at rest or in exercise, though guesses can be hazarded from changes seen in venous blood. Furthermore, it is unlikely that a single substance will be concerned but it is probable that several of the materials that are affected by exercise interact in their effects on arteriolar smooth muscle.

The possible candidates include O_2 lack, and increases in CO_2, lactic acid or acidity *per se*, increases in free phosphate, potassium, creatine or ATP breakdown products. All have their supporters and detractors and it is not possible to do more at present than to add further plausible candidates to the list and remove materials like lactic acid (for long a favourite candidate) since muscle blood flow increases markedly in mild or moderate exercise when no lactic acid is formed, and also in McArdle's disease. At present the best candidates offered by experimental evidence are free phosphate ions and the breakdown products of ATP. Since all will be formed at the same time, it seems that argument about which is the active material has little point! One can only say that the fall in arteriolar resistance is *not* dependent upon the sympathetic nerves, and that it accompanies the metabolic changes of exercise.

Table 23.1

	Untrained		Trained	
	Rest	Exercise	Rest	Exercise
Rate, beats/min	72	180	40	180
Stroke Volume, ml	70	100	140	190
Cardiac output, l/min	5	18	5.6	35
End diastolic volume, ml	145	160	200	220
End systolic volume, ml	75	60	60	30
Diastolic time, sec	0.53	0.13	1.1	0.08
Systolic time, sec	0.3	0.2	0.4	0.2

Some of the values given for the trained person in exercise are somewhat speculative!

THE CAPILLARIES IN MUSCLE

Several true capillaries arise from each arteriole, and there may be a ring of smooth muscle around each capillary's orifice from the arteriole. There is physiological evidence also for the existence of vessels conducting blood from arteries to veins without having passed through capillaries exposed to the metabolising muscle, though they are not anatomically distinguishable (see chapter 14). It is through the capillary walls that metabolic exchanges between blood and tissue fluid and therefore the sarcoplasm occur. At rest there may be 30 open capillaries/mm^2 muscle cut in transverse section (they tend to run longitudinally, parallel to the long axis of muscle fibres). In exercise this figure rises to as much as 400 or 500/mm^2. It was once thought that the capillaries that did not carry flowing blood at rest were actually closed and collapsed. This is now known to be untrue, at least for muscle capillaries (but see Chapter 14). The force that would be required even to open the precapillary sphincter region, once this has closed, is far too great. Inspection *in vivo* has shown that the closed capillary orifices are in fact plugged by blood cells, which are then washed down the newly opened vessels as the precapillary sphincter region dilates.

THE GENERAL CIRCULATORY CHANGES

VENOUS RETURN TO THE HEART

The greatly increased volume flow rate through the capillaries of exercising muscle results in much more blood entering the venous system from the periphery, a rise in peripheral venous pressure and a consequent increased flow rate towards the heart. This is further aided by various factors. Venous smooth muscle contracts more in exercise, reducing the volume of blood that can be contained in the veins (but not the passage of blood through them). The veins in the limbs are compressed by contracting muscles. This compression,

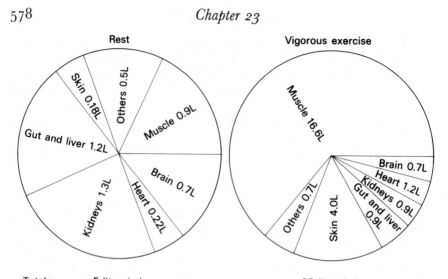

Fig. 23.6. The distribution of the cardiac output at rest and during vigorous muscular exercise.

because of venous valves that prevent backflow towards the periphery, also forces blood towards the heart (the muscle pump effect, already described). Respiration, also increased by exercise, has a pump-like action on the thoracic and abdominal great veins. Inspiration lowers intrathoracic pressure and raises intra-abdominal pressure which together drive blood towards the heart, with valves preventing backflow into the veins of the legs. The greater vigour of the cardiac ventricular contraction increases the suction force exerted within the atria on blood entering them from the veins. All these factors, working together, can raise the venous return to the right atrium of the heart from its resting value of 5 litres/min to as much as 25 or 35 litres/min in vigorous exercise.

THE HEART IN EXERCISE

The changes in cardiac performance are shown in the accompanying table (Table 23.1).

The increase in the heart rate is caused by a reduction in vagal and an increase in sympathetic cardiac stimulation. The origin of these impulses is from the sensory receptors responsible for the onset of exercise, acting via regions in the reticular formation in the posterior hypothalamus and the midbrain which are collectively called the *alerting reaction centre*. This inhibits the vagal cardiac centre while stimulating the sympathetic centre. It is also possible that the alerting reaction centre is stimulated by the postulated chemo-receptors of skeletal muscle. It should be noted that this mechanism for controlling vagal and sympathetic nerves to the heart is entirely distinct from that baro-receptor control mechanism for maintaining the resting heart's performance constant in stable conditions.

The increases in stroke volume can only be caused by an increase in force of contraction. This is caused in part by the Starling mechanism since the end-diastolic volume is raised. It must also be partly due to the sympathetic nerve stimulation, since the end-systolic volume is decreased by exercise. Thus, as is shown in Table 23.1, the cardiac output, in litres/min is increased 2-fold by the rise in stroke volume and $2\frac{1}{2}$–3 times by the rise in heart-rate. The Starling mechanism can act so as to exactly balance output (stroke-volume) to venous return, while the changes in vagus nerve and sympathetic impulses are in response to the stimulation of various receptors.

ARTERIAL BLOOD PRESSURE AND PERIPHERAL RESISTANCE

During rhythmic exercise the systolic blood pressure commonly rises, but diastolic pressure remains unchanged or falls. Average pressure is unchanged or rises slightly. The rise in systolic pressure is due to the increased stroke volume. The small or non-existent change in diastolic pressure indicates that, though there has been a very large decline in vascular resistance in the muscle vascular bed, the total peripheral resistance has fallen to a smaller extent. In other regions, therefore, the resistance must have risen. In fact, the same sympathetic stimulation that affects the heart, also affects the vasoconstrictor nerves throughout the body, and blood flow is reduced to all tissues, apart from the exercising muscle itself, the heart and the brain (Fig. 23.6). In the muscle, the increased metabolic activity, causing a fall in arteriolar resistance and so an increased blood flow, far outweighs the vaso-constrictor nerve activity, while the vessels of heart and CNS have few, if any, vasoconstrictor nerves. The action of vasoconstrictor nerves is augmented by stimulation of the adrenal gland medulla, which secretes adrenaline, itself vaso-constrictor to vessels that have mainly catecholamine α-receptors. The origins of the impulses in the sympathetic system have already been considered in the description of the changes in the heart beat.

ISOMETRIC CONTRACTIONS OF SMALL MUSCLE MASSES

Voluntarily controlled isometric contractions of, say, the hand-grip muscles in the forearm have been found to produce changes in heart rate and blood pressure that are large in proportion to the change in muscle metabolism that involves them. There is an elevation, during these contractions, of heart rate and of both systolic and diastolic blood pressures, but no change in cardiac output. The result is that the heart is doing more work by ejecting the same volume of blood, but raising it to a greater pressure. At the same time, due to the raised heart rate, there is less time for coronary artery perfusion which only occurs in diastole. This is of no significance if the heart and its arteries are healthy, but in coronary artery disease such efforts may cause further damage. A common situation involving isometric muscle contractions is in driving a car, where the situation is further complicated by the degree of overall alertness continually required.

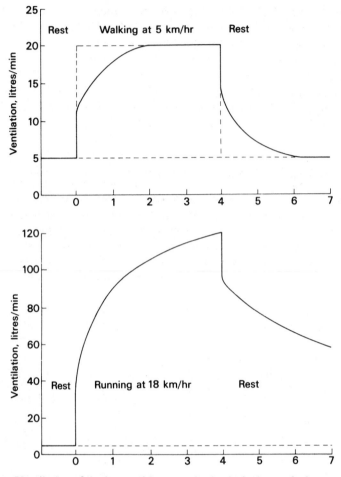

Fig. 23.7. Ventilation of the lungs with atmospheric air during and after a 4-minute period of walking at 5 km/hr (above), and a similar period of running at 18 km/hr (below).

EXERCISE AND ARTERIAL DISEASE

The arteries supplying the legs and the coronary arteries of the heart may be affected by *atheroma*. This causes an irregular hardening and narrowing of arteries. Since the altered state of the artery wall facilitates clotting of blood in the affected area the blood clots formed can further narrow the artery. In small (diameter < 1 mm) arteries in muscles another disease, *thrombo-angiitis obliterans*, may also occur. Both conditions place a restriction on flow *upstream* from the arterioles, so that in exercise the diseased vessel (rather than the arterioles) becomes the major site of resistance to flow, and no amount of arteriolar dilatation can cause the flow increase normally seen in exercise.

The consequences are that O_2 and foodstuff supplies are restricted and the muscles have to switch to glycolysis earlier. Weakness and pain set in after a few minutes of only moderate exercise, but as soon as the sufferer rests, the pain passes off again. A person with such disease will say that he can walk at his 'normal' pace for 400 or even only 200 metres and then cramp-like pain develops in his muscles. After resting, the pain goes and he can then set off again, only to be stopped after the same distance yet again. If he deliberately slows down his walking speed, he may manage to get much farther before the pain develops or even escape it altogether. The condition is called *intermittent claudication* because of the way the pain comes and goes.

It should be clear that treatments aimed at reducing arteriolar smooth muscle contraction (sympathectomy or α-receptor blockade) will have little effect on the condition, since the disease, and therefore the restriction of flow, is not in the arterioles but in the larger arteries. Diseased vessels have a great capacity for recanalisation, and insignificant collateral vessels that by-pass a damaged area can enlarge considerably and greatly improve blood flow. Both processes take time and may be encouraged by maintaining as high a blood flow as is possible through the affected region. It therefore seems that the best treatment is frequent exercise *to the limit of pain onset*, since this will raise the blood flow as far as the diseased vessel will allow.

EXTERNAL RESPIRATION AND EXERCISE

The changes in internal respiration must speedily be reflected in corresponding changes in external respiration. The increased O_2 use of the muscles, and their increased CO_2 production are accompanied by corresponding changes in gas exchange between blood and alveolar air, aided by rises in pulmonary blood flow and ventilation. As the whole of the right heart output passes to the lungs, the pulmonary blood flow inevitably increases as does the venous return and cardiac output. The actual gas exchange in the lungs has already been described in Chapter 18. It only remains here to describe the way in which pulmonary ventilation is adjusted to the changing gas exchange required in exercise.

Figure 23.7 portrays typical changes in ventilation that occur in periods of mild or severe exercise. In both forms there is an initial sharp increase in ventilation at the onset of exercise and an equally sharp fall at the end. This is followed by a gradual rise to a steady level of ventilation in mild exercise, the curve being remarkably similar to the curve for oxygen consumption at this exercise rate (Fig. 23.3). Observe, too, that the ventilation deficit is repaid after exercise exactly as is the oxygen deficit. In the case of severe exercise, although oxygen consumption (Fig. 23.4) reaches its peak value of 4 litres per minute by the end of the first $1\frac{1}{2}$ min, the ventilation goes on rising steadily throughout exercise. This further rise accompanies a progressive accumulation of lactic acid in the blood. It has already been noted that, until lactic acid production and the consequent release of H^+ provides an additional stimulus to pulmonary ventilation, this function is closely related to oxygen consumption and to oxidative CO_2 production. We will now discuss how this relationship might be determined.

The regulation of the changes in ventilation has constantly engaged the attention of research workers since the first decade of the 20th century, but the actual mechanism is still not certainly known. The chemoreceptors bathed in arterial blood, the carotid and aortic bodies, and the medullary CO_2 sensitive areas, were naturally thought to be concerned in the exercise response, but no sufficiently large changes have ever been detected in arterial blood O_2 or CO_2. Arterial blood CO_2 may actually fall in vigorous exercise, when lactic acid is being produced. The rise in acidity of blood that is thus caused may so stimulate ventilation that it actually results in a fall in the volume of oxygen taken up per unit volume of inspired air. Various other factors, such as limb movement, body temperature rise, adrenaline production and irradiation from other brain regions such as the alerting reaction centre, have all been investigated and may perhaps play a part in either the initial rapid rise in ventilation or the later gradual rise to the final level.

Recent work has turned attention back to CO_2 in the gradual change and final level phases of ventilation, for these have been found to follow, almost breath by breath, the changes in CO_2 production from the tissues and the mixed venous blood CO_2 content. A search for chemoreceptors on the course of the pulmonary artery has proved unavailing, but it is now suspected by some that the rapid changes in alveolar gas composition that must occur in exercise are reflected in the arterial blood levels of these cases, and that these rapid changes might stimulate the arterial chemoreceptors. There is also evidence that chemoreceptors, stimulated as the metabolic rate increases, may actually be present in skeletal muscle. In summary, just as with local blood flow changes, it is probably that there is no *one* stimulus to respiration, but that the response is the result of several changes, all of which occur progressively with rising exercise rates. All that can be said is that the total response in ventilation is nicely adjusted to the changing metabolic requirements of exercising muscle.

THE SUPPLY OF NUTRIENTS

MUSCLE GLYCOGEN

The muscle glycogen store is built up from the blood glucose, under the action of *insulin* (see Chapter 26), whenever glucose is being absorbed into the blood from the gut. Even after an overnight fast, some glucose is probably entering the muscle glycogen. In mild exercise, with the greatly raised blood flow and therefore raised glucose supply, the muscle uptake is increased, but usually not proportionally to the blood flow increase. As has been described already, most of the energy of exercise is not derived from glucose oxidation, and so most of the extra glucose entering muscle must be changed into glycogen.

The actual stimulus for mobilising the glycogen is probably a combination of local and systemic effects. Adrenaline, circulating in the blood in exercise, can initiate glycogen breakdown, as also can ADP, if this accumulates at all in the exercising muscle. However, as has already been emphasised, in mild and moderate exercise, this forms only a minor source of energy. It is only in vigorous exercise, with lactic acid formation, that muscle glycogen becomes an important direct energy source.

LIVER GLYCOGEN

This is undoubtedly used in prolonged moderate exercise, being converted into glucose which passes, via the blood, to the muscles where it is oxidised. A proportion of the energy used in continuous exercise must come from a carbohydrate source (glucose or glycogen), for pyruvic acid, essential for the mitochondrial oxidative system can only be obtained from carbohydrates. Without some carbohydrate breakdown, the products of partial breakdown of fatty acid chains accumulate in the body and can eventually interfere with physical and mental activity. These products are collectively known as *keto acids* and their accumulation as *ketosis*. Ketosis is a well-recognised feature of endurance-type of athletics. A mild ketosis is harmless, and always occurs when fatty acids are mobilised from the fat stores of the body, but severe ketosis, associated with an exhausted liver glycogen store and a fall in blood glucose, has been linked with physical and mental exhaustion or confusion, muscular weakness, even hypothermia and death in prolonged exercise in the cold.

ADIPOSE TISSUE

As already mentioned, the fatty acids metabolised by muscle are transported to it in the blood plasma, attached to plasma albumin. In adipose tissue, fats are present as neutral fat, a mixture of the glycerol esters of the fatty acids. The esters are hydrolysed into glycerol and free fatty acids. Glycerol is a water-soluble carbohydrate materal and can enter the glycolysis series of reactions or it can be oxidised in any tissues, being freely transported in the blood. Fatty acids are sparingly soluble in water and have to be attached to albumin, forming *lipoprotein* molecules before being taken to muscle and other tissues for oxidation.

The mechanisms for mobilising fat include circulating catecholamines and possibly direct sympathetic nerve stimulation. Secretion of growth hormone from the pituitary gland (Chapter 26) is also thought to play a part. Finally, as the amount of blood glucose available for oxidation falls, the liberation of fatty acid from adipose tissue rises.

When the fatty acids reach the muscle, the long chain molecules are successively broken into two carbon atom fragments and these fragments are either oxidised or combined in pairs to form the keto-acids. Oxidation of unit mass of fatty acid liberates about twice as much energy as does oxidation of the same mass of carbohydrate. Tissues can store fats nearly pure, but in all animal storage sites, glycogen is mixed with twice its weight of water. For these reasons one can readily see that fat is very much more suitable as a substrate for oxidation in exercising muscle than carbohydrate, especially when the exercise is being continued for a long time such as a day's walk in the hills or a marathon run.

DISPOSAL OF WASTE PRODUCTS

CARBON DIOXIDE

The CO_2 produced by metabolism, or by displacement from $NaHCO_3$ by lactic acid, is expelled from the body via the expired air, pulmonary ventilation being regulated in some way by the CO_2 produced to facilitate its elimination.

INORGANIC MATERIALS

In exercise, K^+ ions enter the tissue fluid spaces from the sarcoplasm. Some of this may be washed away in the venous blood.

The phosphate ions, released from CP and ATP as they provide energy, are present after exercise in the tissue fluid for the re-formation of CP when exercise ceases. Some of the phosphate enters the bloodstream and has to be retrieved by the muscle during the recovery from exercise.

HEAT

Heat production is an inevitable accompaniment of energy-transfer reactions. 50% of energy released in oxidation reactions appears as heat, and 50% is retained in ATP molecules. In muscular contraction, it is rare for more than 40% of the energy of ATP to be converted into external work (such as raising a weight against gravity). The rest emerges as heat. In this way, at least 80% of the energy input, in the form of oxidisable foodstuffs, appears as heat output. It is not surprising then, that we get hot when we exercise!

Heat is removed from the muscles by the blood, which distributes the heat throughout the body, warming it up. Receptors sensitive to a raised temperature are then stimulated and initiate the reactions of cutaneous blood-flow increase and of sweating, which serve to reduce body temperature (see Chapter 22).

EXERCISE FOR ALL

The general public, as well as the medical and allied professions, have become aware since the 1970s of the value of exercise and physical fitness. Two popular exercise crazes have been jogging and marathon running.

The only major objective benefit to be obtained from exercise is that lack of it is one of about six risk factors (p. 471) in the development of arterial disease, so that regular exercise should help to prevent this disease from developing. Otherwise, exercise keeps us fit so that we can take exercise. It is doubtful whether jogging exercise is enough to prevent arterial disease. In the person who starts too late, it may even precipitate trouble, either heart attacks or strokes. The heart rate in exercise is a useful indicator of its severity. Rates to be aimed at for people of different ages have now been worked out, as have rates that should not be exceeded.

Competitive marathon running is another matter. Running over a set course 26 miles in length imposes a variety of problems. Firstly, most marathon races are run on hard road surfaces and the repetitive impact of feet on these may lead to damage to articular cartilage surfaces. Special shoes with shock-absorptive soles must be worn to prevent this. Then, unless climatic conditions are optimal, the runner is likely to develop dehydration and salt-loss due to sweat secretion, part of the normal response to the extra heat production in the exercising muscles. Finally, the exercising muscles need a continuous supply of metabolic fuel. While the average human body's adipose tissue contains enough fuel for many marathon runs, and the endocrine mechanisms can mobilise this and the circulating blood can supply the fatty acids to the exercising muscles, these also require a constant small supply of carbohydrate if fat oxidation is to continue effectively. Their own glycogen store and that of the liver, are not sufficient for a marathon race, so the runner may develop both hypoglycaemia and ketosis. These together lead to fatigue and to mental confusion. During a marathon run, then, the runner, no matter how well trained, needs to replenish the water and salt loss in his sweat, and to take a sugar supplement during the race, if he is to avoid the consequences of dehydration and hypoglycaemia.

24 · The Digestive System

THE ANATOMY OF THE DIGESTIVE SYSTEM

The digestive system consists essentially of a continuous tube beginning at the mouth and ending at the anus, together with associated glands, some of which are very large. At its upper end, in fact, there are two cavities: the mouth, belonging to the digestive system, and the nasal cavity which is part of the respiratory system and has been described in Chapter 19. The two cavities are separated by the palate which itself is divided into an anterior bony or *hard palate* and a posterior mobile *soft palate*. The latter contains muscles which can elevate it, when the palate is elevated it is also tightened and its posterior surface comes into contact with the posterior wall of the pharynx. In this way the nasal cavity and mouth cavity are separated so that (a) during swallowing, food does not pass up into the nasal cavity and (b) so that during speech, pressure can be increased in the mouth and released suddenly by opening the lips, thus producing the explosive consonants such as 'p' and 'b'. Both these facilities are affected in paralysis of the palate or in the congenital defect of *cleft palate*. You can watch it at work by opening the mouth and saying 'Ah'. Both nasal and mouth cavities open posteriorly into the *pharynx*—into which you can see when the soft palate is elevated. Below the level of the inlet of the larynx, the digestive system consists of a single tube whose basic structure is similar throughout its length (Fig. 24.1). It consists of an outer *serous coat*, which is nothing more than the visceral layer of peritoneum (see below). Inside this is a *muscular layer* which mostly consists of an outer longitudinal and an inner circular coat of smooth muscle. Inside this again is a layer of loose connective tissue—the *submucosa*. Finally, lining the tube is the *mucous membrane* which is a layer of connective tissue covered by (mostly) columnar epithelium and containing glands which open into the lumen. The basic digestive tube structure is modified in different parts of the gut according to their function. There follows a systematic (but elementary) description of the various parts of the digestive system.

THE MOUTH

The mouth is lined by stratified squamous epithelium— a tough replaceable epithelium which is able to cope with the trauma inflicted by chewing hard toast and similar foods without being worn away. The epithelium is particularly thick and tough on the tongue but this has already been dealt with in Chapter 13 and will not be further discussed here. The teeth are 32 in number in the adult and consist of, in each jaw, 4 *incisors*, 2 *canines* (for biting), 4 premolars and 6 molars (for crushing and grinding). In the child, the milk teeth are 10

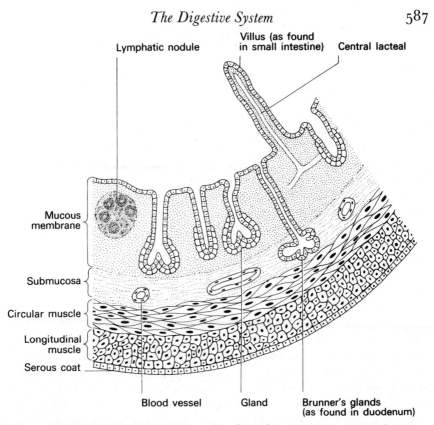

Fig. 24.1. Section through the intestine, to show its major components, some of which are only found in certain regions such as the duodenum.

in number in each jaw. The first milk tooth erupts at about 6 months of age and the first permanent tooth (the first molar) at 6 years.

The *pharynx* is also lined by stratified squamous epithelium and it leads down into the larynx (see Chapter 19) and the oesophagus.

THE OESOPHAGUS

This is the rather narrow tube which connects the pharynx to the stomach. It is about 25 cm long and is lined by stratified squamous epithelium. It has the usual muscle coats (although, curiously, the upper part has skeletal muscle in its wall). Food is passed down the oesophagus by muscle action (peristalsis) and not merely by gravity. It is possible to drink quite easily while upside down and, of course, giraffes always swallow water up-hill.

The oesophagus begins in the neck and then descends through the thorax, lying posteriorly near the vertebral column in the mediastinum. It then pierces the muscular part of the diaphragm (Fig. 19.3), and after a course of only $1-1\frac{1}{2}$ inches, enters the stomach at the cardiac orifice.

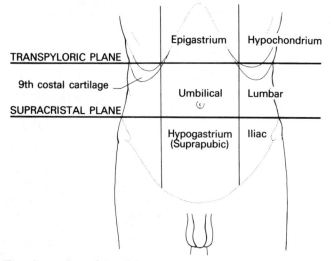

Fig. 24.2. The nine regions of the abdomen.

THE ABDOMINAL PLANES

Before describing the abdominal viscera, it will be helpful to mention here some of the surface features of the abdomen (Fig. 24.2). The anterior abdominal wall is bounded above by the costal margin and xiphisternum and below by the inguinal ligaments and the pubic symphysis. The *transpyloric plane* is a horizontal plane which transects the costal margin at the level of the tip of the ninth costal cartilage. Officially, this plane lies half-way between the suprasternal notch and the symphysis pubis, but much the easiest way to find it is to draw a line between the points where the lateral border of rectus abdominis crosses the costal margin on each side. Incidentally, on the right side, this point marks the fundus of the gall bladder. The pylorus (see below) lies at or a little below the level of the transpyloric plane; the plane itself lies at the level of the lower border of the first lumbar vertebra.

The *supracristal plane* lies at the level of the spine of the fourth lumbar vertebra and marks the level of the highest point on the iliac crests. The surface of the abdomen is divided into nine regions by these two horizontal planes and by two vertical lines running through the centres of the clavicles. The upper central region is the *epigastrium*, bounded by the two *hypochondria*. The middle central region is the *umbilical* region, flanked by the two *lumbar* regions, while below is the *hypogastrium* (or *suprapubic region*) with the iliac regions (or *iliac fossae*) on either side.

THE PERITONEAL CAVITY

Before describing the abdominal part of the digestive system, it will be necessary to mention the complex lining of the abdomen—the *peritoneum*. Peritoneum is a membrane consisting of a squamous epithelium placed on a layer of connective

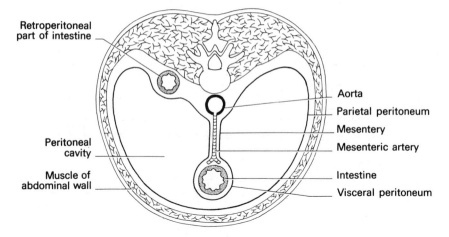

Retroperitoneal part of intestine

Peritoneal cavity

Muscle of abdominal wall

Aorta

Parietal peritoneum

Mesentery

Mesenteric artery

Intestine

Visceral peritoneum

Fig. 24.3. Diagram to illustrate the parietal and visceral layers of peritoneum and the formation of a mesentery.

tissue. It is therefore rather like the pleura (Chapter 19) and, like the pleura, it has parietal and visceral layers. The parietal layer lines the abdominal wall anteriorly and posteriorly, and also the pelvis and the under surface of the diaphragm. It is reflected in various places to form *mesenteries* (Fig. 24.3), but these are much more complicated than appear in the illustration and need not be described in detail. The greater part of the digestive tract has a mesentery but some parts, such as the duodenum, lie directly on the posterior abdominal wall and are covered on their anterior surface by parietal peritoneum. They are said to be *retroperitoneal*. The *visceral peritoneum* covers those parts of the gut which have a mesentery, and forms the serous coat. The parietal part of the peritoneum is part of the abdominal wall and is thus supplied by spinal nerves but the visceral layer is part of the gut wall and is supplied by the autonomic nervous system.

The main part of the peritoneal cavity is called the *greater sac*. It is the greater sac which is opened by the surgeon when he has made an incision through the abdominal wall. It appears to contain the intestines, stomach, liver, etc., but since these organs are themselves surrounded by visceral peritoneum they actually are outside the peritoneal cavity as a study of Figure 24.3 will show. An offshoot from the main peritoneal cavity is called the *lesser sac*. It lies behind the stomach, liver and transverse colon. It may best be imagined as a rubber hot water bottle, situated behind the stomach and with its orifice, pointing to the right, just above the beginning of the duodenum. The orifice represents the opening into the lesser sac from the greater sac and, apart from this opening, the lesser sac is a closed space. It functions as a sort of bursa so that the stomach can distend or contract freely by moving over the underlying structures. The lesser sac is important surgically because ulcers on the posterior wall of the stomach may rupture into it and because the pancreas lies in its posterior wall.

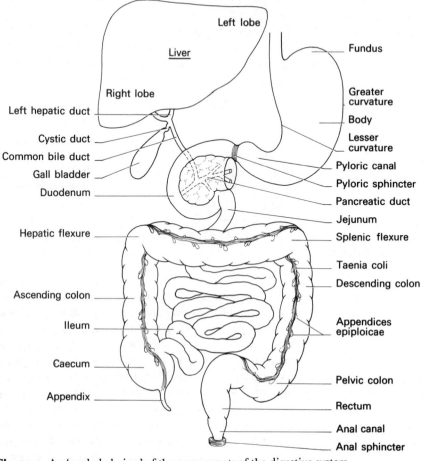

Fig. 24.4. An 'exploded view' of the components of the digestive system.

THE STOMACH

The cardiac orifice of the stomach lies about 1½ inches to the left of the midline behind the 7th costal cartilage. The other end of the stomach, the *pylorus*, is continuous with the duodenum. It lies on or a little below the transpyloric plane about 1 inch to the right of the midline.

The stomach itself is usually J-shaped, but its shape varies according to the body build. In short, stocky individuals it may lie almost transversely, but in tall, thin subjects the loop of the J may lie well below the iliac crests. The stomach has a *lesser* and a *greater* curvature (Fig. 24.4), a *fundus*, a *body* and a *pyloric canal*. To the greater curvature is attached a large 'apron' of peritoneum known as the *greater omentum*. This is laden with fat and is very vascular. It can help to localise infections within the abdomen by wrapping itself around the inflamed area and adhering to the surrounding structures so that the infection becomes sealed off from the rest of the peritoneal cavity. It has therefore been called the 'abdominal policeman'.

The muscular wall of the stomach has an additional layer of oblique muscle fibres between the longitudinal and circular muscle coats. The mucous membrane contains numerous glands which secrete mucus, and also hydrochloric acid and various digestive enzymes to be described later.

THE SMALL INTESTINE

The small intestine is about 20 feet in length (although during digestion it may be much shorter than this due to muscle contraction) and comprises the *duodenum*, the *jejunum* and the *ileum*. The pyloric end of the stomach opens into the beginning of the duodenum and the opening is surrounded by a thickening in the circular smooth muscle coat known as the *pyloric sphincter*. The duodenum is a C-shaped retroperitoneal tube, the concavity of the 'C' enclosing the head of the pancreas. The pancreatic duct and also the common bile duct, open into the vertical portion of the duodenum (see below). A little to the left of the midline, the duodenum passes over into the *jejunum* which comprises about two fifths of the length of this part of the small intestine, the remaining three-fifths being formed by the *ileum*. There is no sharp junction between them, merely a gradual change in the form of the intestine.

The small intestine, as a whole, has the important functions of digestion and absorption of foodstuffs. Its mucous membrane is lined by numerous glands which secrete mucus; the cells lining the intestine also produce digestive enzymes (see below). In order to help absorption, the internal surface area is increased in three ways. Firstly, the mucous membrane is thrown into a series of *circular folds* which run transversely around the lumen. These can be seen with the naked eye. Secondly, on a microscopic scale, the surface of the mucous membrane has a large number of villi. These are finger-like processes, covered by columnar epithelium and containing blood vessels and lymphatics (*lacteals*—see Chapter 20). These give the surface a velvety appearance. Finally on an ultramicroscopic scale, each epithelial cell has a *brush border* consisting of an enormous number of microvilli. These are only visible with an electron microscope. Another feature of the small intestine is the appearance of clusters of lymphatic tissue, similar in appearance to lymph nodes, in the mucous membrane.

All these features are present throughout the small intestine but there are quantitative differences. In the duodenum there are a large number of particularly long glands which penetrate deeply into the submucosa and are known as *Brunner's glands*. The circular folds of mucous membrane become less and less prominent along the small intestine until near the end of the ileum they are absent. The wall of the ileum therefore feels much thinner than the wall of the jejunum. The lymphatic nodules, however, increase in number and size; in the ileum they form areas an inch or more in diameter and are known as *Peyer's patches*.

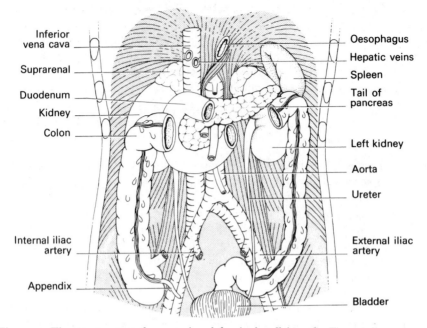

Inferior vena cava
Suprarenal
Duodenum
Kidney
Colon
Internal iliac artery
Appendix

Oesophagus
Hepatic veins
Spleen
Tail of pancreas
Left kidney
Aorta
Ureter
External iliac artery
Bladder

Fig. 24.5. The structures on the posterior abdominal wall (see also Fig. 21.1).

The small intestine has a mesentery along its whole length (except for the duodenum), through which numerous blood vessels and lymphatics travel to and from the intestine. Owing to its length, the intestine is thrown into a series of complex folds.

THE LARGE INTESTINE

In the right iliac region, the terminal ileum enters the *caecum* which in turn leads into the *colon*, the *rectum* and finally the *anal canal*.

The caecum is a blind-ended, thin-walled pouch, to which is attached the *appendix*. Both these structures are much more prominent in herbivorous animals but in man the appendix is a vestigial structure. The colon is subdivided into four parts. The *ascending colon* passes upwards on the posterior abdominal wall, from the right iliac region to the right upper abdomen where it becomes closely related to the liver. Here it changes direction suddenly at the *right colic* (or hepatic) *flexure* (Fig. 24.4) to become the *transverse colon*. This crosses the abdomen below the stomach and behind the greater omentum as far as the spleen which lies posteriorly in the left hypochondrium. Here it forms a *left colic* (or *splenic*) flexure and runs down the left side of the posterior abdominal wall as the *descending colon*. Finally, it forms a loop in the pelvis where it is known as the *pelvic colon*.

The transverse and pelvic colon each has a mesentery (*transverse mesocolon* and *pelvic mesocolon*) but the ascending and descending colon are retroperitoneal. The outer longitudinal muscle coat of the colon does not cover the whole circumference but takes the form of three longitudinal bands known as *taenia coli*. These have the effect of causing a puckering or sacculation of the colon which is particularly well seen on x-ray examination. The colon is also characterised on the outside by a number of large fatty tags called *appendices epiploicae*. The colon is thus easily distinguished from the small intestine. Internally, the mucous membrane does not have villi but it does contain numerous simple glands which secrete a good deal of mucus.

At the level of the third sacral vertebra, the pelvic colon gives way to the *rectum*, which, following the anteroposterior curve of the sacrum, terminates by turning abruptly downwards and backwards as the *anal canal*. The latter is surrounded by the *anal sphincter* which, like the sphincters of the urinary system, has components of both smooth and skeletal muscle.

THE LIVER

The liver is a large and important organ that lies in the upper right part of the abdominal cavity, mostly within the protection afforded by the rib cage. It is soft, and friable and extremely vascular. Its upper border can be marked out by a line through the xiphisternal joint passing slightly upwards and to the right to the 5th intercostal space near the nipple (in the male). On the left it extends to a point about 3 inches from the midline. The lower border follows the right costal margin but crosses the epigastrium from the tip of the 9th right costal cartilage to the 8th left costal cartilage. The anterior and upper surfaces of the liver are smooth and rounded but the lower border is sharp and can easily be felt when the liver is enlarged. On the inferior surface lies the 'gateway' to the liver—the *porta hepatis*— in which lie the *portal vein*, the *hepatic ducts* and the *hepatic artery*. Also attached to this surface is the *gall bladder*. The inferior vena cava is embedded in the back of the liver for a small part of its course.

The liver is divided primarily into right and left lobes, the former being by far the larger. The lobes are further subdivided but such details need not be described here except to say that the lobes of the liver, like the lobes of the lung, may be subdivided into segments, each of which is supplied by a branch of the portal vein and of the hepatic artery. Such segments are of importance to the surgeon.

The liver has a dual blood supply. The hepatic artery is derived from one of the major branches of the abdominal aorta and the portal vein (see Chapter 14), receives its blood from the intestine and the spleen. It is very much larger than the hepatic artery. Both vessels enter the porta hepatis, divide into right and left branches for the corresponding lobes, and then divide further into segmental branches. The large hepatic veins drain directly into the inferior vena cava.

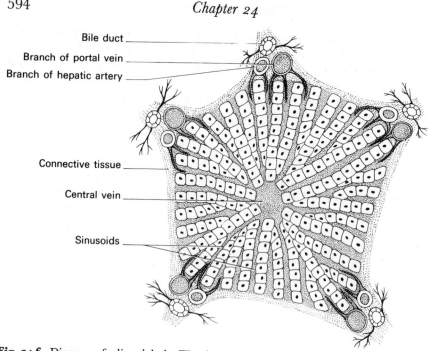

Bile duct

Branch of portal vein

Branch of hepatic artery

Connective tissue

Central vein

Sinusoids

Fig. 24.6. Diagram of a liver lobule. The sinusoids and the central vein are stippled.

The liver tissue itself is made up of a series of lobules (Fig. 24.6) composed of large liver cells arranged around a central vein. The columns of liver cells which form each lobule are separated by sinusoids through which blood can pass from the periphery of the lobule to the central vein. Between the lobules are patches of connective tissue containing branches of the hepatic artery and portal vein and also bile ducts.

The oxygenated blood from the hepatic artery and the venous blood from the portal vein mingle in the sinusoids and pass towards the central veins. You will remember that sinusoids are blood vessels which lack an endothelial lining so that the blood, which contains the products of digestion from the intestine, bathes the liver cells themselves so that the foodstuffs can be absorbed into the cells. From the central veins, the blood is collected into larger veins which finally empty via the hepatic veins into the inferior vena cava.

THE BILIARY APPARATUS

As you will see later in this chapter, one of the numerous functions of the liver cells is to secrete bile. The smallest biliary channels are the bile capillaries which lie between the liver cells themselves. These unite to form larger vessels which can be seen in the connective tissue between the lobules. These finally unite to form the right and left *hepatic ducts* which, at the porta hepatis, join to form the *common hepatic duct*. The *gall bladder* is a closed sac rather like a dilated test tube which is adherent to the under surface of the liver. Its closed end—the *fundus* —projects slightly beyond the lower border of the liver at the point where the lateral border of the right rectus muscle crosses the costal margin. Its other end tapers off into the *cystic duct*. This joins the common hepatic duct to form the *common bile duct* which passes behind the first part of the duodenum and enters the medial surface of the second (vertical) part in common with the pancreatic duct.

THE PANCREAS

The pancreas is about 6–8 inches long and lies on the posterior abdominal wall behind the stomach. Its head is enclosed within the C-shaped curve of the duodenum while its body and tail extend to the left as far as the hilum of the spleen (Fig. 24.5). The pancreatic duct runs along the whole length of the organ, collecting tributaries as it does so and finally joins the common bile duct and enters the duodenum.

The pancreas is both an endocrine and an exocrine gland. The endocrine portion is represented by the islets of Langerhans and will be described in Chapter 26. The exocrine part has a typical glandular structure with acini secreting into small ducts which, finally, join the main pancreatic duct.

THE DIGESTION AND ABSORPTION OF FOOD

INTRODUCTION

Most components of food cannot be absorbed into the body without alteration. Absorption is essentially the passage of single molecules of substances that are water-soluble through the cell membrane into the interior of the cells lining the intestine. Food comes in large solid masses which have to be broken up and then acted upon by the enzymes of the digestive juices which convert large, insoluble molecules into small soluble ones that can pass through cell membranes and cytoplasm.

THE MOUTH

The teeth, powered by the muscles of the jaw (supplied by Vth cranial nerve) break up the large food particles and, aided by the tongue, mix it with saliva into a pasty constituency. During this process the superficial facial muscles (supplied by the facial or VIIth cranial nerve) keep the mouth closed and the buccinator muscles (also supplied by the VIIth nerve) prevent food collecting in the cheek pouches. Saliva contains a starch splitting enzyme, which works best in an alkaline medium, so we find sodium bicarbonate also in saliva.

When these processes are complete, voluntary movements of the tongue project the mass (*bolus*) of chewed food into the oral part of the pharynx. The pharyngeal muscles (supplied mainly by the XIth cranial nerve) now take over, their contractions (and those of other muscles as described below) being reflexly stimulated by the bolus touching the pharyngeal mucous membrane. Although striated, we have no control over them—they are involuntary muscles. A contraction occurs immediately *above* the bolus, forcing it down the pharynx. As the bolus moves, the contracted region of the pharyngeal muscle moves with it, steadily pushing it down the pharynx, and then into and down the oesophagus. This progression of contraction along the tubular oesophagus, which is seen also in the stomach and small intestine, in the ureters between kidneys and bladder and in the uterine tube is called *peristalsis*. In the oeso-phagus it depends upon efferent nerve fibres in the Xth cranial nerve (vagus) and upon the local nerve network in the wall of the oesophagus. Sensory stimu-lation of the pharynx is needed to start both pharyngeal peristalsis and other muscle contractions essential in the initial part of swallowing. These latter are elevation of the soft palate to close off the nasal part of the pharynx and elevation of the larynx which, aided by the epiglottis and a closure of the vocal cords, prevents the bolus from entering the glottis and trachea. Application of a local anaesthetic agent to the pharyngeal mucous membrane abolishes the act of swallowing—thus showing its reflex nature. This mucous membrane is innervated by the IXth cranial nerve.

THE STOMACH

In man this acts both as a reservoir for ingested food and it starts the digestion of protein. It is while food is stored in the stomach before it becomes mixed with gastric secretion that the salivary starch-splitting enzyme can do most of its work. Beer drinkers will boast about how many pints their stomachs can ac-commodate, but the normal meal is probably around one litre in volume and a greater volume leads to some discomfort. The important secretion of the stomach, coming from the tubular glands of its body, contains a protein-splitting enzyme, *pepsin*, and hydrochloric acid. Pepsin splits some of the peptide links between adjacent amino acids of the protein chain molecules. It will only work well in an acid medium—hence the HCl secretion. Mucus is secreted from the necks of the tubular glands, from the epithelial surface of the whole of the stomach and from glands in the pyloric antrum.

The outlet of the stomach is guarded by a ring of smooth muscle, the *pyloric sphincter*. Only when peristaltic contraction waves raise the pressure in the antrum is the contraction of the pyloric sphincter overcome and a jet of acid *chyme* (mixed food and gastric secretion) from the stomach enters the duo-denum. Emptying begins soon after the food first enters the stomach, but may continue for many hours. Much fat in the chyme causes a slowing of the rate of gastric emptying.

REGULATION OF THE STOMACH

The stomach is supplied by the Xth cranial (vagus) nerves, except the pyloric sphincter which receives a motor supply from the sympathetic. The vagus nerves supply motor and secreto-motor fibres. Peristalsis is stimulated by the vagus nerves and the glands when neurally stimulated, produce a juice that is rich in pepsin, but of small volume. The vagus nerve fibres are reflexly stimulated by the sight, smell and taste of food, or even the thought of it. When food, particularly protein-containing food that has been partly digested by pepsin or altered by cooking, reaches the pyloric antrum, it causes the release there of a polypeptide molecule, *gastrin*, which enters the blood and is finally delivered by the blood back to the glands of the body of the stomach. Gastrin stimulates these glands to produce a large volume of HCl-containing secretion. Gastrin production depends also upon vagal nerve fibres, and one way of reducing gastric acid secretion in the treatment of peptic ulcers is by cutting the vagus nerves. This operation also reduces stomach movements and may thus cause further trouble.

Peristalsis and secretion by the stomach are inhibited when acid chyme, particularly when it also contains fat, enters the small intestine. A *gastric inhibiting polypeptide* (GIP) has been recently isolated, which is formed in the lining of the small intestine when sugar or fat is present. It enters the blood and is thus carried to the stomach, where it inhibits both motility and gland secretion.

THE PANCREAS AND THE LIVER (Digestive functions only)

Shortly after entering the small intestine the acid chyme is mixed with secretions from both these organs. The pancreas produces a secretion containing sodium bicarbonate, which neutralises the HCl from the stomach. This secretion also contains inactive precursors of the enzymes that split starch, fats and peptide bonds of proteins. The liver, via the bile, provides *bile salts*, which aid fat digestion by *emulsifying* (breaking up into very small particles) the fat droplets in the chyme. This enables the fat-splitting enzyme to work more effectively.

Although the pancreas is supplied with secreto-motor nerve fibres from the Xth cranial nerve, its secretion is also controlled by substances produced by the small intestinal wall when chyme is present. *Secretin* is one such material. It stimulates a copious flow of a watery bicarbonate solution. *Pancreozymin* produces a small volume of secretion that is very rich in the digestive enzymes. Pancreozymin has, now that it has been purified and chemically analysed, been found to be the same as a third material, *cholecystokinin*, which causes the gall bladder to contract and empty stored and concentrated bile into the intestine. Remember that the pancreas also has an endocrine function (Chapter 26). The other functions of the liver are described later in this chapter.

SECRETIONS AND ENZYMES OF THE SMALL INTESTINE

Although the simple tubular glands of the small intestine secrete 2 litres of fluid each day, contrary to accepted doctrine this secretion is now found to be virtually enzyme free. Most of the enzymes that were thought to be in this secretion, the *succus entericus*, are known to be contained in the *brush border* margins of the intestinal epithelial cells, and the last stages of the digestion processes actually occur there, as absorption begins. The one exception to the above statement about intestinal secretion of enzymes is *enterokinase*. This is produced in the duodenum and itself activates the protein splitting enzymes of the pancreatic juice. These must remain inactive when they are produced and until they actually enter the gut. Activation of the pancreatic digestive enzymes within the pancreas itself causes a major disaster known as *acute pancreatitis* which is frequently fatal.

Altogether, including saliva, gastric juice, bile, pancreatic juice and succus entericus, about 9 litres of fluid are poured into the digestive canal in each day of normal life. It is not surprising then that any obstruction or interference of absorption of this fluid causes disastrous results.

THE DIGESTIVE PROCESSES

Starch is a substance formed from molecules of glucose, a simple sugar or *monosaccharide*. Many hundreds or thousands of glucose molecules may be present in one starch molecule. The starch splitting enzyme breaks the links between adjacent glucose molecules, producing free glucose and some *maltose* which consists of two glucose molecules joined together (maltose is therefore called a *disaccharide*).

Fats, and vegetable oils, consist of molecules of *glycerol*, a simple sugar-like material, to each of which are attached three *fatty-acid* molecules. The digestion of fats consists of splitting off the fatty acids from the glycerol. This renders the molecules far more water-soluble and therefore readily absorbable by the intestinal epithelium.

Protein chain molecules are broken at their many peptide bond links into individual amino acids and some *dipeptides* (two amino acids joined by a peptide bond). Many protein digesting enzymes jointly work on a protein, some attacking each end and other at peptide bonds between particular types of amino acid along the chain.

Two disaccharides are commonly present in food. These are *sucrose* or common, cane, or beet sugar and *lactose*, the sugar of milk. These are not digested in the intestine but are absorbed, unchanged, into the brush borders of the intestinal epithelial cells.

ABSORPTION PROCESSES

In the brush borders, the three disaccharides, maltose, sucrose and lactose are split into monosaccharides. Maltose gives glucose alone, while sucrose is split into glucose and *fructose* and lactose into glucose and *galactose*. The fates of the three monosaccharices are described later. Special carrier systems exist in the epithelial cell membranes that enable them to absorb all the major products of digestion, and the salts added to the intestinal contents by the various secretions. Whenever any of these enter the cells the total concentration of all dissolved substances in the cells is raised, and that in the intestinal lumen lowered. This creates an osmotic pressure gradient, so water follows the absorbed materials. Most of the absorbed material enters the blood directly, but the fat, which is reconstituted within the intestinal cells, enters both the lymphatic vessels (lacteals) and the blood stream. Naturally, to perform the absorption of the products of digestion, and the reabsorption of the 9 litres of secreted water, a large surface area is needed, and constant mixing and moving of the intestinal contents. The villi and mucosal folds together give a surface area of about 150 sq metres. Alternate contraction and relaxation of the circular muscle of adjacent segments of the gut wall, called *segmentation* movements, continually mix the contents, and help to drive them along the intestine. Peristalsis does occur also, but a single peristaltic contraction wave only travels a short distance (10–20 cm) before dying out. Segmental movements are more rapid (12/min) near the duodenal end of the small intestine and peristaltic movements *always*, in the healthy gut, move away from the stomach and towards the large intestine. Individual villi also can wave about in the intestinal contents, independently of each other. This they do when absorbable substances are present.

THE LARGE INTESTINE

This part of the gut absorbs the last $\frac{1}{2}$–1 litres of water. It has little other function, except to act as a store for the unabsorbable material, until this can be disposed of at a convenient time and place. Despite much that has been said about malfunction of the large intestine, the breeding of noxious bacteria therein and absorption of 'toxins' made by these bacteria, these traditional bogies are mostly false. It is now believed that an over-lengthy stay of intestinal contents in the colon, with too great an absorption of water may require such powerful contractions of the large intestinal muscle to move the semi-solid contents that diverticula are formed from the lumen. These diverticula may then cause trouble if material becomes stuck in them and inflammation occurs, a condition called *diverticulitis*. This causative link is not proven, but it seems that diverticular disease can be arrested by a diet containing more of the non-digestible vegetable *fibre*. This adds to the bulk and water content of the material in the large intestine, which thereby remains fluid or soft and is easily propelled onwards and outwards.

METABOLISM OF FOODS

INTRODUCTION

Foods serve three basic functions. They provide a source of potential chemical energy, which can be liberated and utilised by living cells when the foodstuffs are oxidised. Secondly, they provide the materials needed for growth and maintenance of the body's structure, and finally there are both organic compounds, the *vitamins*, and inorganic elements needed for catalysing and regulating metabolic functions. All living cells, as already described in Chapter 2, must expend energy constantly in order to maintain their normal state, and the sum of these activities when the body is at complete rest is called the *basal metabolic rate* or BMR. This can be measured and, in the S.I. units of joules, is found to be about 8 million joules (8 MJ) per day. The figure is some 10% higher for men than for women and in both sexes declines as maturity advances. Muscular activity adds another 2–8 MJ per day onto the BMR. A vigorous 15-minute game of squash may add 0.6 MJ to the day's metabolic rate.

CARBOHYDRATE METABOLISM

Of the three monosaccharides absorbed from the gut, glucose is the most abundant and important. Not only is it the preferred source of energy for the nervous system, but it is also used in vigorous exercise by muscles, and can be used by all other tissues of the body. Indeed, all tissues have to use a small amount of glucose at all times, to keep the oxidation of fat proceeding satisfactorily. Much of the glucose of a meal will be stored in the liver as *glycogen*, a starch-like compound. Some enters skeletal muscle and is there also converted into glycogen. Liver glycogen can be returned to the blood as glucose, but muscle glycogen cannot. The subject of the regulation of blood glucose level and of glucose metabolism is dealt with more fully in Chapter 26.

Galactose is converted into glucose or glycogen in the liver and fructose can enter the same energy-yielding chain of reactions as does glucose, but principally in the liver and the gut.

Glucose (and fructose) metabolic breakdown proceeds in a series of about 15 separate steps, at first without the use of oxygen, and with the liberation of 10% of its total energy content. This part of the process is called *glycolysis* and its end product, if oxidation did not ensue, would be *lactic acid*. During glucose oxidation the remaining 90% of the energy is released, and equal amounts of oxygen and carbon dioxide are consumed and produced. Due to the stepwise degradation and oxidation of the glucose, about 50% of the released energy is trapped and stored in the high-energy compounds, adenosine triphosphate and creatine phosphate.

Galactose is present in one of the several materials that are found in the myelin sheaths of nerve fibres. To this extent only are the monosaccharides involved in growth of the body.

THE RESPIRATORY QUOTIENT

In the oxidation of carbohydrates, as oxygen is used, an equal volume of carbon dioxide is produced. We express this as a fraction or ratio, expressing it as CO_2 produced $\div O_2$ used. For carbohydrates the RQ is of course unity as the volumes are equal. Normally the RQ is determined from the volumes of O_2 removed from the inspired air and the CO_2 given off in the expired air, but there are occasions when this, the respiratory RQ or *respiratory exchange ratio*, as the experts call it, differs from the RQ of metabolising tissues. In vigorous excercise, (Chapter 23) when lactic acid from the muscles displaces CO_2 from sodium bicarbonate in the tissue fluids and blood, the respiratory exchange ratio may rise to 1.2 or higher, while the tissue RQ may still be about 0.8 or 0.9. Under steady state mild exercise or at rest, though, the respiratory RQ and the tissue RQ are the same.

FAT METABOLISM

Fats are the primary energy source for the human being, most tissues using them in preference to glucose when presented with both together. This is especially true of skeletal muscle, as is seen in the chapter on muscular exercise. Fatty acids are mostly chain molecules of carbon and hydrogen, containing 16, 18 or 20 carbon atoms. Oxidation starts by splitting off pairs of carbon atoms, and the fragments can enter then the same oxidation mechanism as glucose. In fat oxidation, about 7 molecules of CO_2 are produced for every 10 molecules of O_2 consumed (unlike the 1:1 ratio in carbohydrate oxidation) so the RQ fat oxidation is 0.7. If this system becomes overloaded the 2 carbon fragments combine in pairs, producing the keto-acids, *β-hydroxybutyric* and *aceto-acetic* acids. Accumulation of these ketone bodies and the related compound, *acetone*, produces the state of *ketosis*. While these compounds stimulate respiration because they are acidic, they depress other nervous system functions so a ketotic person is drowsy and confused or unconscious, but breathing deeply and rapidly even while at rest. Mild ketosis, short of CNS depression, is a regular accompaniment of prolonged exercise or starvation, in both of which fat is mobilised for oxidative metabolism from the fat stores. When drowsiness or confusion develops it indicates an urgent need for giving carbohydrate foods (and insulin if ketosis develops in a diabetic).

Fatty acids are also a component of all cell membranes, both intracellular and the bounding plasma membranes. They are mostly present in *phospholipids*, in which 2 fatty acids and 1 phosphoric acid molecule combine with glycerol. The fatty acids are of the type known as *polyunsaturated*. In these, 1, 2 or 3 pairs of adjacent carbon atoms are joined by double valency bonds, instead of the usual single bonds. The fats containing polyunsaturated fatty acids have recently come into prominence for they may be concerned in reducing the incidence of disease of coronary and other arteries.

Most of the body fats are present in the fat stores. These may contain enough fat to provide the energy needed to maintain a person for several weeks (we have enough glycogen to last one day). Excess food fat may enter the stores, and excess glucose certainly does, where it is converted into fat that contains fatty acids with no double bonds, i.e. *saturated* fatty acids (these *may increase* the liability to develop arterial disease).

AMINO ACID METABOLISM

The amino acids absorbed into the blood from the gut join the body's pool of these compounds, available for making new proteins and for other syntheses. Eight of the amino acids are called *essential*, for they cannot be formed in the human body, but otherwise there is a regular interchange of the NH_2 groups and other components of the amino acids between the different substances, so that the protein formation can always proceed with adequate supplies of all 20 of the amino acids. When these requirements have been met, the liver *de-aminates* excess amino acids, forming urea, a highly soluble, non-toxic waste product which is excreted from the body in the urine. The de-aminated residues of the amino acids are disposed of by the liver's oxidative mechanisms, some following the glucose and others the fatty-acid routes to complete oxidation. Since new protein manufacture and other synthetic processes needing amino acids continue at all times in all cells a constant supply of amino acids is essential, regardless of the food supply. The liver contains a modest store of usable protein, but this is far outweighed by the protein content of skeletal muscles. This will be drawn upon in periods of illness or of protein under-nutrition, with a consequent loss of muscle mass and power. During the recovery period from such events, the re-growth of the muscles will be enhanced by physical activity, using the muscles to their maximal ability.

VITAMINS

Most vitamins become built into the metabolic machinery as essential co-factors to the protein enzymes. It is frequently the case that the vitamin molecule itself reacts with and produces the change in the substance being handled by any particular enzyme. For example, Vitamin A is a component of *visual purple*, the substance that enables us to see in dim light (see below). There is some inevitable daily loss of all vitamins, though this may be only a small portion of the whole body's content and reserves of the substances. You should have sufficient ascorbic acid (vitamin C) in your body to last 3–6 months, unless some illness or accident, like a major limb bone fracture, makes an excessive demand on these stores.

INORGANIC MATERIALS

The metabolism of *iron* has already been mentioned with reference to haemoglobin, its most abundant form. Iron, like some of the vitamins, is also found as the co-factor in several enzymes, particularly in those that catalyse the union of hydrogen with oxygen in oxidative metabolism. Small losses of iron occur continually in urine and in sweat, and further losses in women in the menstrual flow or into a developing fetus in pregnancy. Stores of iron are present which can be drawn on to offset these losses, and iron absorption from the gut may be depressed when the stores are full and the very low normal concentration of iron in the blood rises towards its upper limit.

Calcium is required for making bone, and for normal blood clotting, muscle contraction, nerve impulse conduction and many other processes. As described in the chapter on endocrine glands (Chapter 26), the blood calcium level is maintained constant by alterations in absorption from the gut and absorption from or deposition in the bones. During growth, bone deposition of calcium is stimulated by other endocrine glands, and this may take precedence over other functions of calcium. Intestinal absorption of calcium will be stimulated during bone growth by 1,25 dihydroxycholecalciferol, provided enough of its precursor is present in the diet or is made in the skin by ultraviolet light.

Iodine metabolism is described in the section on the thyroid gland (Chapter 26). The metabolism of the many other inorganic materials found in the body has already been described in various places in this book, e.g. that of sodium and potassium in membrane functions and nerve impulse conduction.

THE FUNCTIONS OF THE LIVER

While the digestive and metabolic functions of the liver have been described piecemeal earlier in this chapter, some of its other functions have not been mentioned at all. For convenience, all the most important functions of the liver will be summarised in this section.

The chief digestive and absorptive role of the liver is in the production and secretion of the bile salts. These aid the digestion of fats and the absorption of fatty acids, retinol, cholecalciferol and vitamin K. The liver converts galactose and fructose to glucose, and stores excess carbohydrate as glycogen, which may be changed back to glucose and discharged into the blood at times of need. The liver is also the chief site of the de-amination of amino acids, converting the amine groups to the excretory waste product, urea.

Many compounds are inactivated or detoxified by the liver. Some are excreted in the bile, while in other cases, the inactive form passes via the blood to the kidneys for excretion. Detoxication involves either a chemical change in the substance, or its conjugation with other compounds such as sulphuric acid which renders it both inactive, stable and water-soluble. The liver is also involved in a wide range of synthetic activities, making compounds not present in the food, and also making many proteins. Conspicuous among these are the albumins of the blood plasma. These form about 70 % of normal blood protein, and are largely responsible for the colloid osmotic (oncotic) pressure of plasma. They are widely used as 'carrier' proteins, for many substances that are not readily dissolved in water (such as fat) are carried in blood attached to albumins. The liver is also the site of production of the proteins which cause blood to clot. For at least two of these, prothrombin being one, vitamin K is needed. Blood clotting may be impaired doubly by liver disease. Deficiency of bile salt production (or a bile-tract blockage) may prevent vitamin K absorption from the gut, and any disease of the liver may adversely affect prothrombin synthesis.

Finally, the liver acts as a store of many compounds of metabolic importance. Glycogen has already been mentioned. There is also a modest store of protein. Retinol, cholecalciferol, ascorbic acid, cobalamin and many other vitamins are stored in the liver, which also stores considerable amounts of iron.

The role of the von Kupffer cells in the liver, part of the reticulo-endothelial system, has already been described in Chapter 20.

NUTRITION

We must now consider some aspects of the food we eat, in particular, the dietary requirements for certain foodstuffs, without which we cannot maintain good health. It is conventional to consider this under three headings: *energy* requirements, *protein* required for growth and repair, and *vitamins* required to enable growth, repair and energy-producing processes to proceed in a normal manner.

ENERGY REQUIREMENTS

While the basal requirement for the average man at rest is 8 MJ/day, and for for the average woman 7.2 MJ/day, activity, fever or widespread neoplastic illness will raise these. A male office worker may expend 10.5 MJ and a house-wife 8.7. The average woman assistant in a department store 'on her feet all day' expends 9.4 MJ, though individuals may use as much as 12 MJ. The highest figures are achieved by coalminers and forestry workers. Both have an average expenditure of 15.4 MJ, with a range of 12–19 MJ/day.

As already seen, carbohydrates and fats in the food are the principle sources of metabolic energy. Each gramme of a sugar provides 16.6 kilojoules of energy, so something over 500 g (about 1¼ lb) of sugar would provide enough energy to keep a person going for a day. Each gramme of pure fat provides 37.5 kJ of energy. Therefore, about 250 g or ½ lb of fat would provide the daily energy needs. All diets must contain a sufficient weight of these substances to provide our energy needs, but for most people in Great Britain nowadays, the main problem is keeping the energy content of our food *down* to the level of daily needs. Excess of intake over need will cause an increase in fat stores of the body and thus of body weight. An increase in body weight leads directly to many other disturbances of our body function which reduce the quality and duration of life. The heart and circulation are constantly overloaded by the need to carry the extra weight. Respiratory excursions may be limited by the excessive fat stored within the abdomen. Weight-bearing joints, like the hips and knees, are exposed to excessive strain and may be damaged or develop osteoarthrosis. It is quite possible for dieticians to devise diets containing about half the normal daily energy need (4 MJ), but with all other known requirements fully met. On such a regime and with a moderate level of physical activity, the daily energy deficit might be as much as 8 MJ, which would have to be met by with-drawing 220 g or ½ lb of fat from the body fat stores. This would result in a weekly weight loss of 1½ kg or 3½ lb.

PROTEIN

Eight of the amino acids are essential and cannot be made in the human body, and so must be included in adequate amounts in the food. During growth another amino acid is also essential. One or two grammes of each essential amino acid are needed daily. Animal proteins contain a greater proportion of the essential amino acids than do vegetable proteins, most of which are entirely deficient in one or more of the essential amino acids. About 70 g of protein, containing at least 35 g of animal protein are usually regarded as the daily minimum, but judicious mixtures of vegetable proteins would enable both figures to be lowered considerably.

VITAMINS

Since there are about 15 different vitamins with different distributions among the potential foods available to us, it is very difficult to summarise statements of how much of which foods must be eaten to ensure an adequate intake of all vitamins. The picture is further complicated by the uncertainty that still exists concerning daily needs of so well-known a substance as ascorbic acid (vitamin C), or the extent to which ultra-violet irradiation of the skin may replace the dietary intake of cholecalciferol (vitamin D). A varied diet, containing wholemeal bread, cheese, fresh fruit (citrus and summer fruits, but not apples), lightly cooked or raw green vegetables, meat, liver and the fat fish, will provide an adequacy of all known vitamins and probably of all inorganic materials as well. The emphasis should be on the word *varied*, and this is as important as the earlier emphasis on the need to keep the energy content in balance with the energy need of the body, lest obesity overwhelm it.

CLASSIFICATION AND DESCRIPTION OF THE IMPORTANT VITAMINS

Some are described as *fat soluble* and others as *water soluble*. Among the latter, some are *heat stable* and so not affected by cooking, and others are described as *heat labile*, and may be partially or wholly destroyed by cooking. Before their chemical nature was identified they were given alphabetical letters, but modern practise is tending towards using the specific names, internationally accepted, for the different vitamins. In the following description, we will commence with fat-soluble substances, and then continue with the heat labile water-soluble vitamins and conclude with some of the more important heat-stable water-soluble vitamins.

Vitamin A, now known as *retinol*, was the first identified fat-soluble vitamin. Chemically it is closely related to the yellow plant pigments known as *carotenes* which occur in all green plant material and of course, in carrots. Carotenes are converted into retinol in the liver. Retinol is a component of the chemical *visual purple* (rhodopsin) essential for vision in dim light. It is also required for the normal maintenance of the delicate conjunctiva of the eye and of the respiratory epithelium. It is also required for normal bone development. Apart from its involvement in visual purple, we do not know how retinol acts.

Vitamin D, the other important fat-soluble vitamin, is now known as *cholecalciferol*, and is the precursor of 1.25 dihydroxycholecalciferol, a hormone secreted by the kidneys which regulates calcium metabolism. This stimulates calcium absorption from the gut and phosphate absorption from glomerular filtrate, thus raising the blood concentration of both and facilitating the calcification of bone. As has already been mentioned, cholecalciferol may be made by ultra-violet light acting on the precursor, 7 dehydrocholesterol, in the skin. It is present in small amounts in milk and milk products, but in significant amounts only in the livers of fish and the flesh of fat fish such as the herring, mackerel and Pacific salmon.

Vitamin C (ascorbic acid) was chemically identified in the 1930s, though its existence had been suspected for 200 years before that. It is water soluble, readily destroyed by heat, particularly when also exposed to the oxygen of room air. *Scurvy* results after several weeks or months of a diet free from ascorbic acid and the American expression 'Limey', for a Britisher, derives from the ration of lime juice which was added to the diet in the old sailing ships in order to prevent scurvy. It is widely distributed among fresh green vegetables, potatoes (but not true root vegetables) and fresh fruit. The human body normally contains several months' supply in its tissues, so that scurvy is rarely seen nowadays. The characteristic signs of the disease are an increased tendency for bleeding from minor injuries such as bruises or joint and ligament strains, and poor wound healing and bone formation. Ascorbic acid is essential for the metabolism of a component of collagen and without ascorbic acid, collagen production is defective. It is thought that this defect underlies the signs of scurvy.

Vitamin B1 or thiamine was the original heat-labile water-soluble vitamin. It forms part of the chemical structure of a key enzyme involved in glucose oxidation, and so is widely distributed in foods. It is found principally in grains and meats; pork and bacon contain significant amounts of thiamine. Thiamine lack affects the nervous system primarily (not surprising since the CNS is dependent upon carbohydrate oxidation for its energy), but the contractile power of the heart is also depressed in thiamine deficiency. *Beri-beri* is the name given to the disease, a neuropathy with or without oedema due to heart failure, that developed in the Far East when thiamine-containing outer layers of the rice grain were stripped off to provide the more desirable white rice with which we are so familiar. The same applies to wheat. White wheat flour and white bread contain much less thiamine than wholemeal flour and bread, but we eat sufficient meat to obtain our thiamine that way.

Nicotinic acid and riboflavine were identified in the 1930s as components of the heat-stable, water-soluble vitamin B_2. Both are found as components of enzymes concerned in hydrogen atom transfer pathways in metabolic oxidation processes. They are widely distributed therefore in foodstuffs, but, like thiamine, are concentrated in the grain seeds. Riboflavine deficiency produces no very striking disease state, but nicotinic acid deficiency causes *pellagra*, in which dermatitis, diarrhoea and dementia develop. It is now recognised that it is not so much dietary deficiency of nicotinic acid that causes pellagra, but its presence, in *maize* products in an indigestible form, if this food is a major constituent of the diet. Thiamine, nicotinic acid and riboflavine are all slowly lost from the body, the losses being proportional to the total energy consumption or carbohydrate oxidation. This has led to the policy, originally adopted in the 1940s by the United Kingdom government of adding the required amounts of these three vitamins to white flour, to ensure that the bread made from it has its energy content adequately 'covered' by these three vitamins. Removal of this governmental requirement would probably lead to re-emergence of vitamin deficiency states among the poorer and less aware sections of the population.

Folic acid, also called *pteroylglutamic acid*, and *cobalamin* or vitamin B_{12}, were identified in the late 1940s, though their existence had been suspected for 20–30 years before. Both are required for normal development of red blood cells and B_{12} also for normal function of the nervous system. Without any vitamin B_{12}, degeneration occurs in the long fibre tracts in *both* posterior and lateral columns of the spinal cord. While folic acid is present in moderate amounts in many vegetable and grain foods, and also in liver, vitamin B_{12} is *only found in animal foods*, especially in liver. Furthermore, the vitamin cannot be absorbed from the gut until after combination with a material, called *intrinsic factor*, secreted by the stomach (see Chapter 17).

Vitamin K is a fat-soluble material essential for forming some of the blood-clotting proteins. Although present in many foods in small amounts, the greater part of our requirement comes from the activity of intestinal tract bacteria. Its absorption depends upon bile salts. Vitamin K deficiency, then, may occur in infancy before the right bacteria have established themselves in the gut, if vigorous anti-bacterial therapy has sterilised the intestinal contents, or if liver disease or biliary tract obstruction cut off the supply of bile salts. Vitamin K deficiency may then lead to reduction in the clotting power of the blood.

25 · Regional Anatomy

There are two ways of studying anatomy and there are correspondingly two different types of anatomy textbooks. *Systematic anatomy* is the study of the body by systems. In systematic textbooks such as Gray's Anatomy, separate chapters are devoted to the cardiovascular system, the digestive system, the nervous system and so on. *Regional anatomy* is the study of the body by regions, so that in describing, for example, the gluteal region, the bones, muscles, blood vessels, nerves and lymphatics would all have to be discussed, as well as the relationships between them. Textbooks of surgical anatomy and operative surgery are always of this type. In the present volume, the anatomy sections are dealt with systematically since the physiotherapist is concerned very largely with the functional anatomy of the muscles and joints, while relatively little attention need be paid to the blood vessels, organs of digestion and other viscera. There are a few regions, however, where a little regional anatomy is helpful, because so many structures, important to the physiotherapist, are crowded together and are often palpable. This chapter will therefore deal with each of these regions as a whole, although all the structures that will be mentioned have already been described individually. The chapter may therefore be regarded as useful revision.

THE AXILLA

The *axilla*, or 'armpit', is important because it contains all the important nerves, blood vessels and lymphatics that pass to and from the upper limb, while its lymph nodes are also extremely important in the lymphatic drainage of the breast. It is roughly pyramidal in shape with the apex being at the triangular area bounded by the clavicle, the scapula and the outer border of the first rib (Fig. 9.19), while its concave base is formed by skin and fascia. The four sides of the pyramid are unequal, the lateral wall being very narrow.

THE BOUNDARIES OF THE AXILLA

The *anterior wall* of the axilla is formed by the pectoralis major and pectoralis minor muscles. It also includes the clavipectoral fascia which lies deep to the pectoralis major and runs between the lower border of the clavicle and the fascial floor of the axilla. It splits to enclose pectoralis minor (Fig. 25.1). Pectoralis major forms the thick lower border of the anterior wall (the *anterior fold*) which can be held between the fingers and thumb. The cephalic vein enters the axilla by passing between pectoralis major and deltoid and piercing the clavipectoral fascia before joining the axillary or subclavian veins.

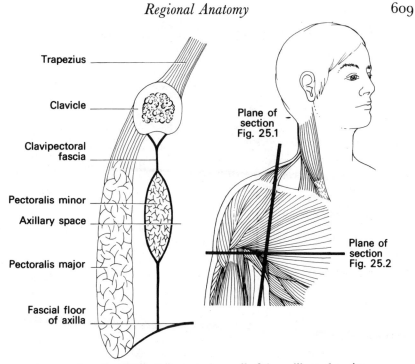

Fig. 25.1. A coronal section through the anterior wall of the axilla to show its components.

The *posterior wall* comprises the subscapularis backed up by the scapula itself, the teres major and the latissimus dorsi (Fig. 11.20). Of these, the latissimus dorsi has the lowermost origin and is the lowest muscle in the medial part of the axilla, but more laterally its tendon turns under teres major to be inserted higher up on the humerus, so that for the greater part of its extent the lower border of the posterior wall (the *posterior axillary fold*) of the axilla is formed by teres major. The axillary nerve leaves the axilla, along with a large artery (the posterior circumflex humeral) by passing backwards through the *quadrangular space*. The largest branch of the axillary artery (the subscapular artery) runs downwards and medially on the posterior wall, more or less following the lateral border of the scapula. A perpendicular line dropped halfway between the anterior and posterior axillary folds is called the *midaxillary line* and is a useful landmark, for example, in mapping out the lower borders of the lungs and pleura (Chapter 19).

The *medial wall* of the axilla consists of the upper ribs and intercostal muscles, covered by the serratus anterior. It can be palpated by running the hand up the side of the chest into the axilla.

The *lateral wall* is very narrow and is really nothing more than the floor of the intertubercular sulcus since the tendon of pectoralis major is inserted into the lateral lip of the sulcus and the teres major into the medial lip (Fig. 25.2). In the sulcus itself lies the tendon of the long head of biceps while in the angle

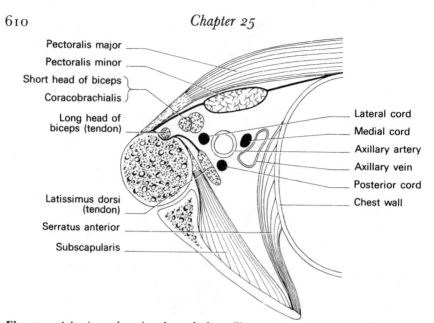

Pectoralis major
Pectoralis minor
Short head of biceps
Coracobrachialis
Long head of biceps (tendon)
Lateral cord
Medial cord
Axillary artery
Axillary vein
Posterior cord
Chest wall
Latissimus dorsi (tendon)
Serratus anterior
Subscapularis

Fig. 25.2. A horizontal section through the axilla along the line shown in Fig. 25.1.

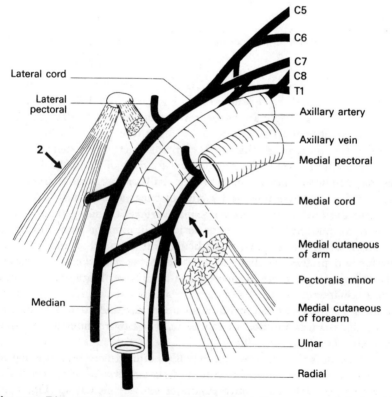

C5
C6
C7
C8
T1
Lateral cord
Lateral pectoral
2
Axillary artery
Axillary vein
Medial pectoral
Medial cord
1
Medial cutaneous of arm
Pectoralis minor
Median
Medial cutaneous of forearm
Ulnar
Radial

Fig. 25.3. Diagram to show the relationships of the axillary artery and the brachial plexus. The posterior cord is hidden behind the axillary artery. The arrows 1 and 2 refer to the position of the sections shown in Figure 25.4.

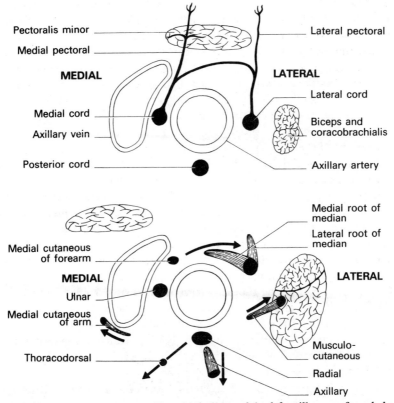

Fig. 25.4. The axillary artery and brachial plexus of the *left* axilla, seen from below. The upper figure shows the second part of the axillary artery and the lower shows the third part.

between the anterior and posterior walls are the short head of biceps and the coracobrachialis, travelling downwards from the coracoid process.

The major, but least important, content of the axilla is a large mass of loose fascia and fat. Embedded in this are the axillary lymph nodes, the axillary vessels and the cords and branches of the brachial plexus, the neurovascular bundle being enclosed in a tube-like sheath of fascia called the *axillary sheath*. Pectoralis minor is the key structure here. It is used to divide the axillary artery into three parts. Part one lies above pectoralis minor, part two behind it and part three below it. The artery begins at the outer border of the first rib and ends at the lower border of teres major. The three cords of the brachial plexus are related to the second part of the axillary artery in accordance with their names, i.e. the lateral cord is lateral, the medial cord is medial and the posterior cord is behind. Since the axillary vein lies medial to the axillary artery, the medial cord therefore lies between the two vessels (Figs. 25.3 and 25.4). The cords break up into branches which are arranged correspondingly around the third part of the axillary artery. Thus, the two major branches of the lateral cord—

the lateral root of the median and the musculocutaneous nerves—lie lateral to the third part of the artery. The ulnar and medial cutaneous nerve of the forearm and the medial root of the median nerve, all derived from the medial cord, lie between the artery and the vein but the medial root of the median then crosses the artery to join the lateral root so that the median nerve as a whole is lateral to the axillary and then the brachial artery in the upper half of the arm. The lateral and medial pectoral nerves from the lateral and medial cords respectively, travel forwards to supply the two muscles of the anterior wall. The remaining (very small) branch of the medial cord—the medial cutaneous nerve of the arm—passes medially behind the vein to supply a small area of skin on the medial side of the arm. The posterior cord lies directly on the posterior wall of the axilla so that it is easy for the axillary nerve to pass backwards through the quadrangular space. As soon as the radial nerve becomes clear of the axilla it inclines backwards, below teres major, to enter the radial groove. The two subscapular nerves are also on the posterior wall and supply two of its muscles (subscapularis and teres major). The remaining branch of the posterior cord—the thoracodorsal—descends near the edge of latissimus dorsi to supply the whole of this muscle. On the medial wall, two other nerves are seen. One of these is the long thoracic nerve which arises from the roots of the brachial plexus (C5, 6 and 7) and descends behind the axillary sheath and its contents, lying on the surface of serratus anterior, which it supplies. Emerging from the second intercostal space is the large lateral branch of the second intercostal nerve (the intercostobrachial nerve) which joins the medial cutaneous nerve of the arm and helps it to supply the skin.

The axillary lymph nodes are described in detail in Chapter 20. Briefly, the lateral group lie along the axillary vessels, the medial group in relation to pectoralis major and minor, the posterior group in relation the posterior wall of the axilla, the central group in the middle of the floor and the apical group around the axillary vessels at the apex of the axilla. The nodes receive lymphatic vessels not only from the upper limb but also from the chest wall, including the breast, and the back down as far as the iliac crest.

THE CUBITAL FOSSA

This is a triangular area in front of the elbow joint (Fig. 25.5) The base of the triangle is represented by a line joining the two epicondyles of the humerus. The lateral border is represented by the medial border of brachioradialis and the medial border by the lateral border of pronator teres so that the apex is where these two muscles meet. Its floor is formed by brachialis, which completely covers the front of the elbow joint. The depression between the muscles which forms the cubital fossa is roofed over by deep and superficial fascia, the latter containing the median antecubital vein which runs between the basilic and cephalic veins (Fig. 20.3). The medial and lateral cutaneous nerves of the forearm also lie in the superficial fascia on the medial and lateral sides respectively.

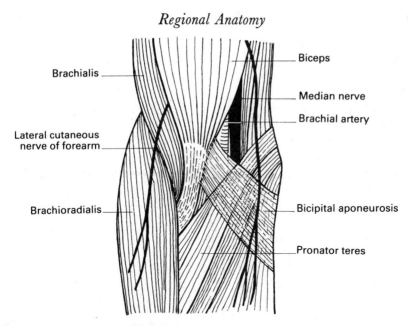

Brachialis

Biceps

Median nerve

Brachial artery

Lateral cutaneous
nerve of forearm

Brachioradialis

Bicipital aponeurosis

Pronator teres

Fig. 25.5. The right cubital fossa.

The principal structures within the fossa, all of which are palpable, are the tendon of biceps, the brachial artery and the median nerve. The biceps tendon is thick and tough and is most easily felt between finger and thumb. From it, the strong bicipital aponeurosis runs medially and its thin upper edge is often a prominent landmark. Just medial to this, the pulsations of the artery can be felt and the artery itself may be visible in a thin, elderly patient. The median nerve is medial to the artery, having crossed the brachial artery from the lateral to the medial side halfway down the arm.

Another large nerve is closely related to the cubital fossa, although rather deeply buried. This is the radial nerve, which lies at the bottom of the deep groove between the brachialis medially and the brachioradialis and extensor carpi radialis longus laterally. The ulnar nerve is, of course, behind the medial epicondyle and therefore does not enter the fossa.

THE FEMORAL TRIANGLE

The femoral triangle can be seen on the surface as a slight depression on the front and medial side of the thigh. Its boundaries are the inguinal ligament above, the sartorius laterally and the medial border of adductor longus medially. The gracilis muscle is closely related to the medial border (Fig. 25.6). The floor is formed by, from medial to lateral, the adductor longus, pectineus, psoas, and iliacus. The triangle is roofed over by the fascia lata (the deep fascia of the thigh) which is deficient below and lateral to the pubic tubercle, the gap forming the saphenous opening. Through this the great saphenous vein passes to

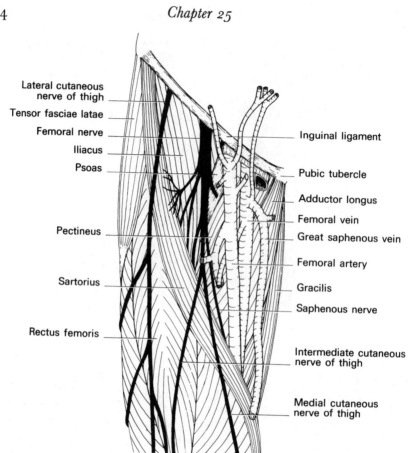

Lateral cutaneous
nerve of thigh

Tensor fasciae latae

Femoral nerve

Iliacus

Psoas

Pectineus

Sartorius

Rectus femoris

Inguinal ligament

Pubic tubercle

Adductor longus

Femoral vein

Great saphenous vein

Femoral artery

Gracilis

Saphenous nerve

Intermediate cutaneous
nerve of thigh

Medial cutaneous
nerve of thigh

Fig. 25.6. The femoral triangle.

join the femoral vein and in this region it is joined by various other superficial
tributaries from the lower abdomen, from the external genitalia and from the
lateral part of the thigh. The 'entrance' to the femoral triangle is the space
between the inguinal ligament and the hip bone (Fig. 25.7). Through this gap
pass the iliacus and psoas, on their way from their intra-abdominal origins and
pectineus, whose origin is from the pubis. The femoral artery and vein, as they
emerge from the abdomen (where they are called the external iliac artery and
vein) take with them some of the intra-abdominal fascia in the form of a fascial
sleeve called the *femoral sheath*. The vein is medial to the artery. Also within the
sheath, and lying medial to the vein, is the *femoral canal*. This is a space occupying the angle between the inguinal ligament and the pubis and bounded
medially by a curved band of fascia, called the *lacunar ligament* which fills in
the angle (Fig. 25.7). The femoral canal contains only some fat and loose fascia
and its function is to provide sufficient space to allow for enlargement of the
femoral vein when the venous return from the lower limb is increased during
exercise. It is important because it represents a weak spot in the abdominal

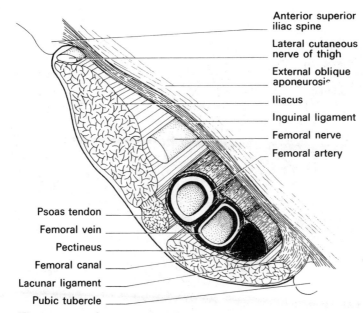

Anterior superior
iliac spine

Lateral cutaneous
nerve of thigh

External oblique
aponeurosis

Iliacus

Inguinal ligament

Femoral nerve

Femoral artery

Psoas tendon

Femoral vein

Pectineus

Femoral canal

Lacunar ligament

Pubic tubercle

Fig. 25.7. The structures that pass below the inguinal ligament.

wall through which a hernia may occur (*femoral hernia*). The femoral nerve lies lateral to the femoral artery and is outside the femoral sheath. Further laterally again, the lateral cutaneous nerve of the thigh emerges from the abdomen just medial to the anterior superior iliac spine. It lies on the iliacus.

The femoral artery and vein pass through the femoral triangle from base to apex, the vein gradually coming to lie behind rather than medial to the artery. The pulsations of the artery may be palpated just below the inguinal ligament, half-way between the anterior superior iliac spine and the midline. During its course the artery gives off a number of branches including the *profunda femoris*. This passes deep to adductor longus, along with its vein which lies in front of it. At the apex of the femoral triangle, therefore, the order of structures from before backwards is: femoral artery, femoral vein, adductor longus, profunda vein and profunda artery. A stab wound at this point will therefore injure the four largest vessels in the lower limb.

The femoral nerve only has a very short course in the femoral triangle. About one inch below the inguinal ligament, it breaks up into its numerous branches (see Chapter 12). Of these, the nerves to sartorius, rectus femoris, and vastus intermedius have only a short distance to travel. The nerve to vastus lateralis descends behind rectus and sartorius and then runs along the anterior border of its muscle before supplying it. The nerve to vastus medialis also descends the thigh, and enters the adductor (subsartorial) canal before supplying the muscle. It is closely accompanied by the saphenous nerve (one of the three cutaneous branches of the femoral nerve) which is about the same size but continues downwards to emerge from the canal near the knee. The remaining

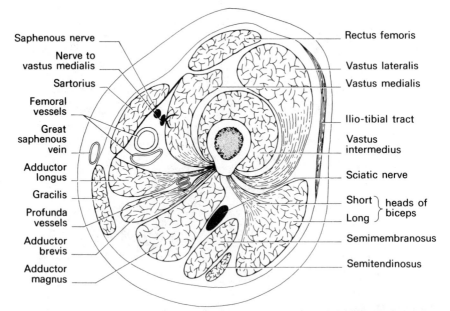

Fig. 25.8. A cross-section through the thigh to show the subsartorial (adductor) canal.

two cutaneous branches of the femoral nerve—the medial and intermediate cutaneous nerves of the thigh—pierce the deep fascia near the sartorius muscle in mid-thigh and in the lower third of the thigh respectively.

Several other cutaneous nerves supply the skin in the region of the femoral triangle. The lateral cutaneous nerve of the thigh emerges from the abdomen by passing under the inguinal ligament. It is derived from the lumbar plexus (L2 and 3) and supplies an area of skin down the lateral side of the thigh. A small area of skin over the saphenous opening is supplied by the femoral branch of the genitofemoral nerve and a small area of the medial side of the thigh by the ilio-inguinal. The obturator nerve gives an occasional cutaneous branch a little below the ilio-inguinal nerve.

The femoral triangle contains the important inguinal groups of lymph nodes. The *superficial inguinal nodes* lie along the upper part of the long saphenous vein and also along a line below and parallel to the inguinal ligament. They drain the superficial tissues of the front and medial sides of the lower limb and also the lower part of the abdominal wall, the external genitalia and the gluteal region.

THE ADDUCTOR (SUBSARTORIAL) CANAL

The reason that the femoral triangle is visible as a depression on the medial side of the front of the thigh is because the femur and the quadriceps muscle form a cylindrical unit which runs obliquely downwards while the adductor group

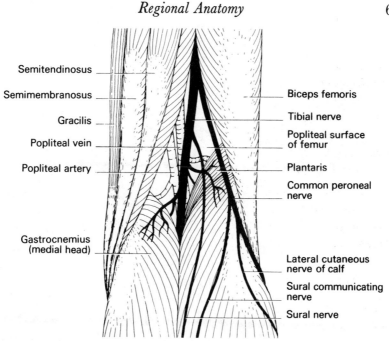

Semitendinosus

Semimembranosus

Gracilis

Popliteal vein

Popliteal artery

Gastrocnemius
(medial head)

Biceps femoris

Tibial nerve

Popliteal surface
of femur

Plantaris

Common peroneal
nerve

Lateral cutaneous
nerve of calf

Sural communicating
nerve

Sural nerve

Fig. 25.9. The right popliteal fossa.

of muscles lie on a more posterior plane since they are attached to the linea aspera (Fig. 25.8). This posterior position of the adductors relative to the quadriceps becomes even more evident lower down the thigh, where the angle between vastus medialis and the adductor longus is bridged over by a strong layer of fascia and by sartorius. The triangular space between these three muscle is the *adductor canal*, also known as the subsartorial, or Hunter's canal. The boundaries of the canal are thus the vastus medialis laterally, the sartorius anteromedially and the adductor longus behind. Below the lower border of adductor longus, the posterior wall is formed by adductor magnus since this is well below adductor brevis.

Entering the canal are the femoral artery and vein, the nerve to vastus medialis and the saphenous nerve (see above). The vein, which was lying posterior to the artery at the apex of the femoral triangle, now comes to lie lateral to it. The two vessels leave the canal by going through its posterior wall, i.e. they pass through the large opening in adductor magnus.

THE POPLITEAL FOSSA

The popliteal fossa is a diamond-shaped area behind the knee joint. It is only a fossa (i.e. a depression) when the knee is flexed since, on extension of the knee, the large amount of fat in the fossa, particularly in the female, produces a bulge rather than a fossa.

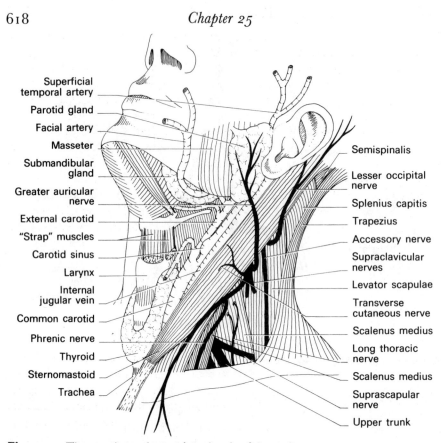

Superficial
temporal artery

Parotid gland

Facial artery

Masseter

Submandibular
gland

Greater auricular
nerve

External carotid

"Strap" muscles

Carotid sinus

Larynx

Internal
jugular vein

Common carotid

Phrenic nerve

Thyroid

Sternomastoid

Trachea

Semispinalis

Lesser occipital
nerve

Splenius capitis

Trapezius

Accessory nerve

Supraclavicular
nerves

Levator scapulae

Transverse
cutaneous nerve

Scalenus medius

Long thoracic
nerve

Scalenus medius

Suprascapular
nerve

Upper trunk

Fig. 25.10. The anterior and posterior triangle of the neck.

The fossa is bounded above and medially by the semitendinosus and semi-membranosus, the former being much thinner than and superficial to the latter (Fig. 25.9). Above and laterally is the tendon of biceps femoris on its way to the head of the fibula. All three of these structures are easily palpable when the knee is semiflexed. The two lower boundaries of the fossa are the medial and lateral heads of gastrocnemius respectively, the latter having plantaris as a close relation. The floor of the fossa is formed by the popliteal surface of the femur, the posterior part of the capsule of the knee joint and the popliteus muscle with its covering fascia. The fossa is roofed over by a layer of very tough deep fascia which is pierced by the short saphenous vein just before the latter joins the popliteal vein. The lower part of the posterior cutaneous nerve of the thigh lies in the superficial fascia.

As in the axilla, the most bulky but least important content of the fossa is fat. Running through the centre of the fossa from the upper to the lower angle are the popliteal vessels. The vein is, at first, lateral to the artery (see above) but in the fossa it crosses superficial to the artery to reach its medial side. The artery is thus the deepest structure in the fossa and is closely related to the femur. The artery gives off a number of genicular branches which take part in the peri-articular plexus around the knee joint.

Two important nerves pass through the fossa. These are the tibial and common peroneal nerves. The sciatic nerve usually divides into these two branches at the upper angle of the fossa but occasionally the division occurs higher up (see Chapter 12) so that the nerves enter the fossa independently. The tibial nerve runs down the midline, being superficial to both artery and vein (Fig. 25.9). In the fossa it gives genicular branches to the knee joint, muscular branches to gastrocnemius, soleus, plantaris and popliteus and the sural nerve which is cutaneous. The common peroneal follows the upper lateral boundary of the fossa and can usually be felt just alongside the tendon of biceps. It follows biceps to the head of the fibula and then winds round the neck before dividing into superficial and deep peroneal nerves. Whilst in the fossa, it gives genicular branches to the knee joint and two cutaneous branches—the sural communicating and the lateral cutaneous nerve of the calf.

THE TRIANGLES OF THE NECK

There are two principal anatomical triangles in the neck, separated by sternomastoid. The *posterior triangle* is bounded by the posterior border of sternomastoid in front, the anterior border of trapezius behind and the middle third of the clavicle below (Fig. 25.10). The apex of the triangle is where sternomastoid and trapezius meet at the superior nuchal line of the skull. The floor of the triangle is formed by the scalenus anterior, scalenus medius and posterior, levator scapulae, splenius capitis and a small part of semispinalis capitis near the apex. The triangle is roofed over by the deep fascia of the neck. In spite of its name, the posterior triangle is *not* on the back of the neck. In fact its lower part is anterior, since it is just above the clavicle, and the triangle, seen in three dimensions, then spirals round the neck so that its apex is near the back of the skull. In thin persons the lower part of the triangle can be seen on the surface as a fairly deep depression—the supraclavicular fossa.

A number of important nerves pass through the triangle. There are also a number of small arteries and veins but these are not important. The spinal part of the accessory nerve passes from a point half-way down the posterior border of sternomastoid to a point on the anterior border of trapezius two fingers' breadth above the clavicle. It supplies both these muscles. The trunks of the brachial plexus emerge from between scalenus anterior and scalenus medius. The upper, and perhaps the middle, trunks are easily seen in a dissection, and may be palpable, but the lower trunk is down on the first rib and is rather inaccessible. The suprascapular nerve comes from the upper trunk and crosses the triangle towards the suprascapular notch. The long thoracic nerve emerges on the surface of scalenus medius before descending on that muscle to reach serratus anterior (see above). The remaining supraclavicular branches of the brachial plexus are small.

The subclavian artery is closely related to the posterior triangle since it, too, emerges from between scalenus anterior and scalenus medius but since it is normally behind the clavicle it is, strictly speaking, outside the triangle.

The lower deep cervical group of lymph nodes lies in the lower anterior part of the posterior triangle, i.e. in the angle between sternomastoid and the clavicle. They drain a number of important structures (Chapter 20).

The anterior triangle of the neck is bounded by the anterior border of sternomastoid, the lower border of the mandible and the midline (Fig. 25.10). It contains the larynx and upper part of the trachea and the pharynx and the upper part of the oesophagus. Further up is the submandibular region, including the submandibular salivary gland and a number of groups of lymph nodes, including the upper deep cervical group (Chapter 20).

The most important structures to traverse the anterior triangle are the common and internal carotid arteries, the internal jugular vein and the vagus nerve. All three structures are enclosed in a tube of deep fascia called the *carotid sheath*. The common carotid is medial to the internal jugular vein and it runs from behind the sternoclavicular joint (where it is under cover of sterno-mastoid) to the level of the upper border of the thyroid cartilage of the larynx. At this point, it divides into internal and external carotid arteries, the former being dilated at its origin to form the carotid sinus. The external carotid gives a series of branches to supply structures in the neck and face.

Closely related to the lower part of the larynx and the trachea is the thyroid gland. Its isthmus lies over the 2nd, 3rd and 4th rings of the trachea and its lateral lobes extend up alongside the larynx (Chapter 26). It is partly covered by some thin, flat muscles (the 'strap' muscles) but is easily visible when enlarged, and sometimes when normal. Like the larynx, it moves upwards on swallowing.

26 · The Endocrine Glands

The complex multicellular organism, with its different organs and systems each in its own peri-cellular environment and each with its own specific cell types and functions, requires a communication system enabling it to respond to, or to produce, changes in function if it is to survive. The nervous system is one such communicating system, employing a network of specific conductors and exchanges for the routing of impulses from receptor to effector organs within the body. In this respect the nervous system is analagous to a country's telephone system, albeit with but one vast 'exchange', the central nervous system.

The endocrine glands and their secretions, *hormones*, form a second communicating system. *The secretions of these glands pass directly into the blood and are carried therein to all parts of the body.* (The endocrine glands are therefore also known as the *ductless glands*.) The effects of the hormones may be widespread or only on a few *target cells* or tissues. This system is perhaps analogous to the older postal system, in which a common carrier (mail coaches and railways) is employed to convey discrete packages or circulars from a single distribution point (the endocrine gland), via the blood to single or multiple receivers, the target cells. If you remember how many circulars, received through the letter post, are simply dumped without evoking any response, you will realise that this is an inefficient system of communication. Perhaps you might be led to think the same of the endocrine glands. Be that as it may, we inherited the system from our biological forebears and we must now study how it is used in the human body.

We must consider three main areas, firstly, how the hormones act, secondly, how their own secretion is controlled and finally, the functions they regulate. In practice it is difficult wholly to separate the second and third aspects of this study and after a brief introduction to the principles of control of endocrine glands, both aspects will be considered together, gland by gland.

THE CHEMISTRY OF HORMONES

Before progressing to these three areas of study, something must be said about the chemical nature of hormones. They fall into two groups, one of which comprises the *steroid* hormones. These have a chemical nucleus made of three 6-carbon atom rings and one 5-carbon atom ring, known to the chemists as the *cyclopento-phenanthrene ring* (Fig. 26.1). Addition to this nucleus of different side groups, or opening of one of the rings, produces an array of different compounds with different actions, some of them hormones, others playing different roles in the living body. Cholesterol (Fig. 26.1b) is readily synthesised in the body and from it are derived the bile acids and the steroid hormones. Irradiation of the skin with ultra-violet light produces vitamin D from the very similar compound 7-dehydrocholesterol. The steroid hormones are produced by the adrenal cortex, the gonads and the kidneys (Figs. 26.1d–f).

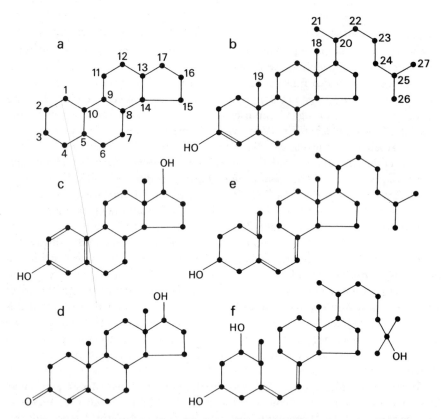

Fig. 26.1. Some steroid molecules: (a) the nucleus with carbon atoms numbered; (b) cholesterol; (c) estradiol; (d) testosterone; (e) cholecalciferol (vitamin D); (f) the active form of vitamin D which potentiates calcium absorption in the gut.

The other main group is formed by the polypeptides or amino acid derivatives. A single amino acid, modified chemically to produce adrenaline is the simplest of these, while insulin, formed of a chain of 21 amino acids joined by two cross-links to another chain of 30 amino acids, represents a more complex polypeptide material.

The steroid molecules, being small and lipid-soluble, can readily pass through cell membranes. The polypeptide molecules, because of their larger size and lipid-insolubility, do not penetrate cells but act on their surface membranes as will now be described.

THE MODE OF ACTION OF HORMONES

ON CELL MEMBRANES

Some hormones, usually the larger polypeptide materials (such as insulin), as has just been mentioned, cannot penetrate cell membranes and have to work wholly from outside living cells. Their action is typically on cell membrane carrier systems. They thus enhance or reduce the rate at which materials are transported through cell membranes.

ON CELL CYTOPLASM—ENZYME ACTIVATION

Some hormones, again while themselves remaining outside the cell, can activate enzymes already present within the cell's cytoplasm. They do this by a chain of reactions discovered in the years before 1970. The hormone itself activates an enzyme present in the target cell's membrane, called *adenyl cyclase*. This enzyme converts ATP into *cyclic adenosine monophosphate*, cyclic AMP for short. The cyclic AMP then diffuses through the cytoplasm and can activate, directly or indirectly, enzymes in the cell and so the chemical reactions performed by the activated enzymes are themselves started. At least 12 hormones use cyclic AMP as their *second messenger* in this way. What we do not yet know is how the cyclic AMP produced by one hormone produces its specific effect, uncomplicated by the different effects produced by this compound in response to stimulation of the cell by other hormones.

ON CELL NUCLEI—ENZYME INDUCTION

Steroid hormones can penetrate cell membranes and enter the cytoplasm. There they combine with specific receptor proteins. The hormone–protein complex then moves into the cell's nucleus and there combines with specific DNA molecules or genes. Transcription of the DNA to messenger RNA and the subsequent formation of new protein enzymes within the cell's cytoplasm then follow. To illustrate some of the puzzles that remain in this story, we would here interpose some of the known facts concerning hair growth.

Presumably, all cells in the body have inherited the genes responsible for producing the enzymes that are needed to make the keratin molecules of hair. Shortly after birth, only those cells that are in the skin of the scalp and eyebrows commence spontaneously to produce coarse hair (as opposed to the fine downy hair that is found over most of the skin). The genes have become activated in these cells only. At puberty, due to the production of gonadal steroid hormones, hair growth starts in other regions of the body. But note how this is restricted. Axillary hair growth commences in both sexes. In girls also, a triangular area of hair appears in the pubic region. In boys the pubic hair growth spreads up the abdominal skin surface and further areas of hair growth appear on the trunk, limbs and on the face. We do not know what factors permit the activation

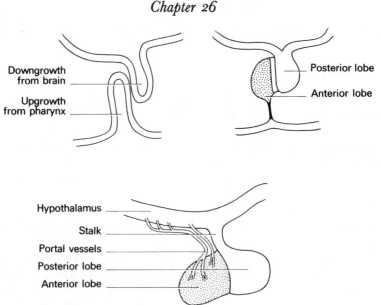

Fig. 26.2. The development of the pituitary gland. The anterior lobe is connected to the brain by blood vessels and the posterior lobe by nervous tissue.

of the hair growth genes only in axillary and pubic skin in girls when their gonads become active, why the hormones of the boy's testes should have a more widespread effect, nor why in both sexes the gene activation is confined so rigidly to certain epidermal cells. In all other cells of the body, repression of hair growth genes is complete and absolute, yet they are bathed in just as much steroid as are cells of the skin. Moreover, in some cells the gonadal steroids produce other, equally specific, responses when they appear at puberty.

CONTROL OF HORMONES' SECRETION

In the many cases in which hormones are playing a part in maintaining the constancy of some bodily function, or preserving some aspect of the internal environment, the control mechanism is some form of negative feedback loop control. Thus, if a hormone's action is required to raise the concentration of substance X to its optimal value, then any fall in X will stimulate further production of the hormone and an excess of X will inhibit the hormone's production. Sometimes we find two hormones at work, one raising X and inhibited by its rise, and the other lowering X and inhibited by a fall in X, but stimulated when X rises. X may be the concentration of a specific component of extracellular fluids, the volume of these fluids, the rate of cellular metabolism in general or the metabolism of a specific substance. In some cases, X itself is a hormone, for in man we find that some hormones control the production of other hormones as we shall see when we consider the pituitary gland.

THE INDIVIDUAL ENDOCRINE GLANDS

THE PITUITARY GLAND

This is an extremely important gland although it is only about the size of a large pea. It is in a highly inaccessible part of the body, being situated in a cavity—the *pituitary fossa*—in the middle of the base of the skull (Fig. 9.6). It is attached to the base of the brain below the hypothalamus by a thin stalk which contains large bundles of nerve fibres. Once again, the development of this gland helps to explain its structure. It develops from two rudiments: an upgrowth from the roof of the embryonic pharynx and a downgrowth from the base of the fore-brain (Fig. 26.2). The upgrowth separates itself from the pharynx but the downgrowth remains attached to the brain. The pituitary thus consists of two principal lobes. The anterior lobe has a typical endocrine structure, being composed of columns of cells of three principal types, separated by thin-walled blood vessels (sinusoids). The posterior lobe, however, being a derivative of the nervous system, consists of bundles of nerve fibres and supporting cells, although here again, thin-walled sinusoids are present. As can be seen in Fig. 26.2, the posterior lobe is connected to the hypothalamus by the nerve fibres of the stalk but there is no direct connection between the anterior lobe and the hypothalamus. There is, however, a vascular connection since capillaries in the lower part of the hypothalamus coalesce to form larger vessels which run down the stalk and break up into sinusoids again in the anterior lobe. You will recognise this as a *portal system* (see Chapter 14).

The pituitary gland has been termed the Master or Leader of the Endocrine Orchestra by those who enjoy seeking out analogies. *Leader* is the most appropriate term, for the site of the *conductor* lies elsewhere. The pituitary gland produces at least eight known hormones. Of these, four control other endocrine glands,

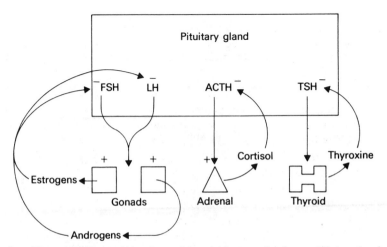

Fig. 26.3. The controlling mechanisms of the pituitary and other endocrine glands.

in each case the pituitary hormone stimulating the production of hormone by the target gland. Thus, production by ovaries and testes of oestrogens, progesterone and androgens (see Chapter 27) is stimulated by follicle stimulating (FSH) and luteinising (LH) hormones of the pituitary. Thyroxine production from the thyroid gland and cortisol production from the adrenal cortex are similarly stimulated by thyroid stimulating hormone (TSH) and adrenocorticotrophic hormone (ACTH). All four of these hormones come from the anterior lobe of the pituitary gland. In all four cases the production of the pituitary hormone is inhibited by a rise in blood levels of the hormones of the target glands and stimulated by a fall in these levels. In this way, a constant production of the hormones of the target glands would be maintained (Fig. 26.3)

But this is not necessarily desirable, for changing bodily states may require changes in the activity of the target glands. We need a conductor of the orchestra and I wonder if it is not too fanciful to suggest that, as in the 18th century the musical orchestra became too large and complex for adequate control by the first violinist, so in biological evolution, the endocrine gland control became too complex to be handled by the pituitary and so another boss had to be found. In the body this is the *hypothalamus*, the portion of the brain that is, in fact, nearest to the pituitary gland. During the years since 1950, we have learned how the hypothalamus controls the anterior part of the pituitary gland. The axons of some hypothalamic neurones terminate on the capillaries in the lowest part of the hypothalamus. When impulses pass down these neurones, chemical transmitter-like substances are released from the terminals. These substances enter the local blood vessels, are carried down the portal vessels into the anterior lobe of the pituitary gland and finally delivered to the secretion-forming cells of the gland. For each of the six known hormones produced by the anterior pituitary, there is evidence now of either a hypothalamic *releasing factor* or a *release inhibiting factor*, liberated into the blood vessels of the pituitary stalk from the terminals of axons of hypothalamic neurones when these are activated (Fig. 26.4).

Table 26.1 the hypothalamic regulation of the anterior pituitary via the median eminence and the pituitary stalk. Gaps in the table indicate lack of evidence at present for the existence of any substance.

	Release promotion	Release inhibition
Growth hormone	Not identified	Somatostatin (14-peptide)
Prolactin		Dopamine
Gonadotrophins	10-peptide	
Adrenocortico-trophic hormone	41-peptide	
Thyrotrophic hormone	3-peptide	

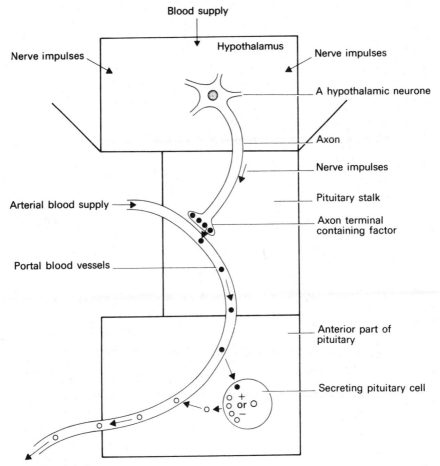

Blood supply

Hypothalamus

Nerve impulses

Nerve impulses

A hypothalamic neurone

Axon

Nerve impulses

Pituitary stalk

Arterial blood supply

Axon terminal
containing factor

Portal blood vessels

Anterior part of
pituitary

Secreting pituitary cell

+
or
−

Venous blood drainage

Fig. 26.4. Shows how the hypothalamus controls the anterior lobe of the pituitary gland.

A final point that must be made, now that we know of this link between hypothalamus and pituitary, is that the negative feedback element in the control of the gland may be applied, not only to the cells of the gland itself but also to the hypothalamic neurones where it would be one of many competing stimulatory and inhibitory influences that may, at any time, be acting on these nerve cells. Can we stretch our musical analogy just a little further; the conductor is not only aware of the markings of the composer's score, but also of the volume of sound produced by the players. His control must then be sensitive to both the composer's intentions (themselves influenced by many cultural factors) and to the responding efforts of the players, synthesising all into a harmonious whole. In the same way, perhaps, the hypothalamus has come to controlling the output of the endocrine orchestra, integrating information from many aspects of bodily function, one of which is undoubtedly the products of the endocrine target glands themselves. Table 26.1 indicates our knowledge in 1984 concerning the existence of the controls exerted by the CNS on the anterior pituitary gland through the mechanisms depicted in Fig. 26.4.

Apart from the four trophic hormones already described, the anterior lobe of the pituitary gland secretes two further hormones. All six are polypeptide in nature and although only two types of secreting cell, *eosinophil* (ie.. stainable with acid dyes such as eosin) and *basophil* (i.e. stainable with basic dyes such as haematoxylin), can be identified in the gland there is evidence that each of the six hormones is produced by its own line of cells. A third cell type, called *chromophobes* because their cytoplasm is hardly stained at all. The two hormones whose actions remain to be described are called *prolactin* and *growth hormone*.

Prolactin is secreted by eosinophil cells. It acts on breast tissue, causing it to secrete milk, when growth has already been stimulated in pregnancy by the combined action of oestrogen and progesterone. It is a large polypeptide hormone. It also opposes the actions of the gonadotrophins on the ovary and may completely block ovulation for as long as breast-feeding continues. Its production is stimulated by tactile contact at the nipple.

Growth hormone is also a polypeptide hormone. Its sequence of amino acids is close to that of prolactin, and it is not surprising therefore that it is difficult to prepare growth hormone with no prolactin-like activity. How this large protein molecule exerts its actions is not fully understood, though there is evidence that it stimulates the formation in the liver of smaller peptides, called *somatomedins*. Whether directly or not, bone growth is promoted by growth hormone, and an excess of this hormone will cause *gigantism* in the young before epiphyses have fused and *acromegaly* (overgrowth of bone in hands, feet and jaw) in adult life. It affects protein synthesis, and stimulates glucose output from the liver. It also stimulates fatty acid output from the adipose tissues. These last two actions probably aid in the provision of extra metabolic fuel for muscles during prolonged periods of exercise, for exercise is a potent stimulus to growth hormone production. Its greatest effect on growth is exerted between ages 2–9 years, after which it has progressively less effect. Growth is also affected by other hormones, especially those of thyroid gland, pancreas and gonads (see below). It is difficult to ascribe specific functions to the various types of cell found in the anterior lobe, but broadly speaking, the eosinophilic cells produce growth hormone and prolactin while the basophils produce the remaining hormones.

In man, two materials are liberated from the posterior pituitary gland. One of these allows water reabsorption from the collecting ducts of the kidneys, and in higher doses causes arteriolar constriction. It is properly known as *antidiuretic hormone* (ADH), but is commonly misnamed *vasopressin*. The other is called *oxytocin*. It causes uterine muscle contraction, also contraction of the myo-epithelial cells of the ducts in mammary glands. These hormones are released and enter the sinusoids in the posterior lobe but this is not a gland in the true sense of the word. The hormones are, in fact, produced in the nerve cells of the hypothalamus (a process known as *neurosecretion*). They then pass along the axons of the pituitary stalk and are released at the nerve endings in the posterior lobe.

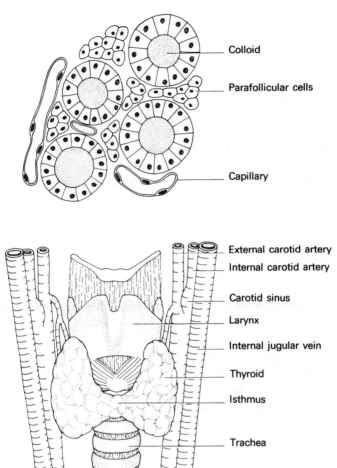

Colloid

Parafollicular cells

Capillary

External carotid artery

Internal carotid artery

Carotid sinus

Larynx

Internal jugular vein

Thyroid

Isthmus

Trachea

Fig. 26.5. The upper diagram shows a histological section of the thyroid gland while the lower diagram shows the whole gland *in situ*.

Production of both hormones occurs naturally as a reflex response. A rise in osmotic pressure of blood plasma stimulates hypothalamic *osmoreceptors*. Impulses from these stimulate cells whose axons run to the posterior pituitary where the terminals liberate ADH. This causes water reabsorption in the kidneys and a consequent fall in solute concentration and osmotic pressure of plasma. Conversely, a rise in central venous blood volume or in arterial blood pressure inhibit ADH production. Oxytocin is reflexly released when mechanoreceptors in the uterine cervix or in the nipple are stimulated. The uterus, in particular, needs to be sensitised to oxytocin, which is most effective *after* labour has already started. In lactation, oxytocin is the cause of the *let-down* of milk, known both to nursing mothers and those familiar with the hand milking of cattle and goats, etc. This reflex may be evoked or inhibited by other sensory stimuli.

THE THYROID GLAND

This gland is situated in the midline at the root of the neck. It is rather H-shaped (Fig. 26.5) with the two *lateral lobes* connected across the trachea by the *isthmus*. It is folded around the trachea and between the two common carotid arteries. Like the larynx, it moves upwards on swallowing and may sometimes be recognised on inspection of the neck in females when they swallow, even when it is normal. It is an extremely vascular organ. Embedded in the back of each of the lateral lobes of the thyroid are two parathyroid glands (see later).

Histologically, the thyroid consists mainly of spherical vesicles lined by cubical or low columnar cells, the lumen of the vesicles being filled with *colloid*. In between the vesicles are small clumps of clear cells, the *parafollicular cells* which produce a hormone known as *calcitonin*. This will be described later. The cells that line the vesicles manufacture tri-iodothyronine (T_3) and thyroxine (T_4) from iodine and the amino acid tyrosine. The active hormones are either released or stored in the colloid. In childhood they are responsible for normal growth and development, at all ages they are concerned both with overall levels of metabolism and with mental function. If iodine is lacking in the food, then less thyroid hormone can be made, but colloid accumulates and the thyroid enlarges to produce one form of *goitre*. In certain parts of Great Britain, iodine is lacking in the water and some table salt is iodinated to provide the vital element. In Canada all table salt is iodinated by law, because much natural water in the Great Lakes region is iodine deficient. Any deficiency in circulating hormone results in a raised production by the anterior pituitary gland of TSH, due to the negative feedback control loop already described. TSH causes, amongst other things, an enlargement of the thyroid gland, so that it can better extract iodine from the blood and make more of its hormones. Thyroid deficiency (whether due to iodine lack, or some other cause) results in a low metabolic rate, poor growth in early childhood and mental deficiency or dementia. The deficiency states are known as *cretinism* in childhood, *myxoedema* in adults. An over-active thyroid causes a high basal metabolic rate, excitability, heart failure and a deposition of fat behind the eye and in the external ocular muscles, and a contraction of the levator muscle of the upper eyelid so that the eyes appear enlarged and protruding. This condition is called *thyrotoxicosis* and the ocular condition is known as *exophthalmos*. Two main types exist. In one, one or more nodules of tissue are present in the gland that produce hormone at a high rate, even without any stimulus from TSH. Since the thyroid hormone levels in the blood become raised, TSH production from the pituitary ceases and the rest of the gland around the secreting nodule(s) become atrophied. In the other type of thyrotoxicosis, it is excessive production of TSH from the pituitary, or the presence of a 'long acting thyroid stimulant' (LATS), that causes the *whole gland* to become enlarged and to secrete excessive hormone. It is curious to note that LATS is actually an anti-thyroid antibody, mistakenly produced by the body's immune system. In these two types of hyperthyroidism, TSH production is stopped by the excessive thyroid hormone production, so surgical removal of most of the over-active gland may be followed by a hypo-thyroid hormone production crisis. In the third type, excess TSH production itself causes the over-secretion of thyroid gland hormones. One can thus see how the various thyroid diseases illustrate the way in which anterior pituitary and thyroid glands control each other—a veritable model of the negative feedback control principle.

Concerning the physiological function of the thyroid hormones, we really know very little. It's all very well to say they regulate the metabolic activity of all cells, but we don't know why this regulator exists or what changes in body function may produce changes in the regulator. There is now a little evidence that it has something to do with regulating body temperature by varying the rate at which the body produces heat, but the evidence is by no means sufficient to enable us to say that this is the primary function of the thyroid gland.

We will meet this gland again when we consider the hormonal regulation of calcium metabolism.

THE SUPRARENAL GLANDS

The suprarenal glands (also known as the adrenal glands) are each of them really two glands (in some animals they are indeed separate), the cortex being a true gland, surrounding the medulla which is really a modified sympathetic ganglion. The glands lie at the upper poles of the kidneys, in close relation to the diaphragm and to various upper abdominal viscera (Fig. 24.5). They weigh about 5 g each and are therefore quite small. The cortex consists of columns and bands of cells arranged in three principal layers and separated by sinusoids. The medulla again contains sinusoids but the cells arranged around them are derived embryologically from the nervous system and are analogous to sympathetic ganglion cells—in fact the medulla may be regarded as an enormous sympathetic nerve ending in which the transmitter substance, adrenaline, is produced in such large quantities that it enters the circulation. The causes of release of adrenaline and its actions were described in Chapter 6. The suprarenal cortex produces two main steroid hormones, *aldosterone* and *cortisol*. Aldosterone enhances sodium reabsorption by the renal tubules and thus affects, somewhat indirectly, the volumes of extracellular fluids, including blood volume. It is produced by the outermost layer of the adrenal cortex, the *zona glomerulosa*, and its production is regulated, indirectly by a hormone produced in the kidneys when a lowered blood volume (or other change) alters renal function. This hormone is called *renin*. It is a protein-splitting enzyme and produces, from a plasma globulin, an 8 amino-acid polypeptide, *angiotensin*, which itself stimulates aldosterone production and is the most powerful known constrictor of blood vessels.

Cortisol's actions are still something of a mystery. It clearly stimulates protein breakdown in tissues, and it is likely that the rise in blood glucose produced by cortisol is the result of protein breakdown and the consequent elevation in plasma amino acid content, for the liver can convert excess amino acids into glucose. But, quite distinct from these actions, cortisol exerts a profound effect upon various aspects of inflammatory and immune reactions. It reduces the inflammatory response to bacterial infections or injury and reduces the production or activity of antibodies. It also stabilises the membranes of lysosomes, the intracellular bodies concerned in the cell's defence mechanisms. It can thus be used to 'damp down' unwanted or excessive reactions to tissue injury, and states of heightened immune reaction. One of the present authors knows, from personal experience of having his *tennis elbow* most effectively treated, how cortisol reduces the chronic inflammation that may surround a small injury to the tissues around a joint, and we see many people in whom abnormally active immune reactions cause diseases like *rheumatoid arthritis* which are treated with derivatives of cortisol.

Cortisol's production is controlled by ACTH, which causes an increase in size and output of the adrenal cortices. ACTH is itself regulated by the cortisol levels in the blood. Extra ACTH and cortisol are produced in times of psychological or physiological stress, such as examinations or acute infections like a boil or acute appendicitis.

THE OVARIES AND THE TESTES

Details of the hormones secreted by these glands (*oestrogens* and *progesterone* from the ovaries, *testosterone* from the testes), their control and effects will be described in the next chapter. Since these hormones are all steroids (as are aldosterone and cortisol) (Fig. 26.1), the metabolic paths of their formation from cholesterol share many steps. It is not surprising, therefore, that the adrenal glands in both sexes produce oestrogen and testosterone in small amounts. Disturbance, due to an absent enzyme, in one pathway may lead to abnormal amounts of inappropriate sex hormones being produced and a person may acquire physical characteristics of the wrong sex. This almost never alters the person's mental sense of identity of his or her 'proper' role in society. Such changes in personal identity, when they occur, are not caused or accompanied by changes in the person's oestrogen or testosterone production.

So far in our description of individual endocrine glands, we have been concerned with organs directly or indirectly under the control of the nervous system. In the remainder, there is either no nervous control or this is of relatively minor importance. (Aldosterone production from the adrenal cortex is one case wherein the nervous system has no part to play.)

THE ISLETS OF LANGERHANS

As was mentioned in Chapter 24, the pancreas is both an endocrine and an exocrine gland, the former component being represented by the islets of Langerhans. These are small clumps of cells embedded in the main mass of the pancreas. Their relation to diabetes mellitus was shown in 1889 after the first successful operation for removal of the pancreas. In man there are 1–2 million islets, varying from 20 to 300 μm in diameter. They contain two main types of granular cells, a and β. The a cells produce *glucagon* when the blood glucose level falls. Glucagon raises blood sugar by increasing the breakdown of liver glycogen to give glucose. The β cells produce *insulin* when the blood glucose level is raised, and insulin causes an increased uptake of glucose from the blood by most metabolising tissues of the body and by adipose tissue, wherein the glucose is converted to fat. Thus we have, with respect to blood glucose concentration, two control systems, each of them operating in a negative feedback manner and both serving to maintain a constant blood glucose concentration. The importance of this concentration lies probably in the nearly complete dependence of the CNS upon glucose as a source of metabolic energy. Without adequate glucose the brain will not work properly and the fall in blood glucose concentration, due to exhausted liver glycogen stores, may contribute to mental confusion and collapse after long-continued vigorous exercise without adequate food intake. Insulin release in response to raised blood glucose is enhanced by Gastric Inhibitory Polypeptide (GIP), which is normally secreted as food leaves the stomach and is being absorbed from the small intestine. Insulin release is inhibited by *somatostatin*, which is present in the islets' delta cells, as well as in the median eminence (where it controls growth hormone production).

Other hormones play a part in blood glucose maintenance. Adrenaline enhances glycogen breakdown on the liver. The action of cortisol in raising blood glucose has already been mentioned. Because the fatty acid released by growth hormone provides an alternative source of energy, this hormone indirectly depresses glucose utilisation by bodily tissues and thus its accumulation in the blood.

In diabetes mellitus an imbalance develops between these actions and that of insulin, or the tissues become less responsive to the presence of insulin, or again, the islet tissue's production of insulin becomes defective. We find an elevation of blood glucose concentration, and a shift away from glucose to fat in metabolism. Excess glucose is lost in the urine, taking with it water and salt. Fat metabolism may become impaired and *ketosis* develop. The combination of water and salt loss and of ketosis cause the development of diabetic *coma* and will cause the patient's death unless insulin is given to restore a normal glucose metabolism and blood concentration. It must be repeated, though, that diabetes is not solely caused by insulin lack, but by any one of several factors that results in an elevation of blood glucose concentration. Perhaps the most important of these is eating too much, especially the rapidly absorbed disaccharide, sucrose. Equally it should be realised that diabetes is not treated by just giving insulin, but by seeking to restore the normal balance between glucose entry into the blood from intestine and liver and its removal by metabolising tissues and into glycogen stores.

Diabetes produces *complications*. Among the more important of these are impairment of blood supply to tissues, particularly to skin and muscles in the legs, damage to sensory nerve fibres, again most prominent in the legs, and an increased liability to local infections. Trivial damage to the feet or toes goes unnoticed because of the sensory loss. It heals slowly, because the blood supply is poor, and it readily becomes infected with pus-forming bacteria, or the more dangerous gas-gangrene organisms. Second only to cigarette smoking, these complications of diabetes are a major cause of loss of the lower limb(s) by amputation.

THE PARATHYROID GLANDS AND THE HORMONAL REGULATION OF
CALCIUM METABOLISM

Embedded in the deep surface of the thyroid gland are two pairs of *parathyroid* glands. These secrete the hormone *parathormone* which has two chief actions. One is to potentiate the removal of calcium from bone and the other to raise the rate of phosphate excretion in the urine. Both actions cause the blood calcium concentration to rise, and the principal cause of secretory activity in the parathyroids is a reduction in blood calcium concentration.

Two other endocrine glands are involved in calcium metabolism. The parafollicular cells of the thyroid gland produce *calcitonin* which has the opposite action to parathormone on bone. It stimulates the laying down of more calcium salt in bone, causing the blood calcium level to fall. It is produced when the blood calcium level is raised above normal. Parathormone and calcitonin, then, operate in a manner on calcium metabolism in a manner similar to that of insulin and glucagon on glucose metabolism.

The third hormone is known as *1,25-dihydroxy-cholecalciferol* and the cells which produce it have yet to be discovered. The hormone is produced in the kidneys from 25-hydroxy cholecalciferol, when blood calcium concentration falls. Cholecalciferol is better known as *vitamin D₃*. This is formed when ultra-violet light falls upon 7-dehydrocholesterol which is present, along with cholesterol, in many body tissues (see Chapter 22). The action of the hormone is to raise the absorption of calcium from the intestinal contents, if sufficient absorbable calcium is available. And here is another twist to the calcium story: cereals contain calcium, but they also contain, in the outer part, a compound which prevents absorption of this calcium, and an enthusiast for whole oatmeal or whole-wheat flour may render his dietary calcium unabsorbable, even though he has plenty of vitamin D and be producing 1,25-dihydroxy-cholecalciferol in the amounts required.

Shortage of calcium is particularly important when bones are growing and the disease of *rickets* is the result of calcium or vitamin D shortages. Bones become soft and bend easily. They become swollen at the sites of growth and where bone formation should be progressing actively, especially at the epiphyseal plates of cartilage in long bones. *Osteomalacia* is a very similar disease sometimes seen in childbearing women who are either calcium- or vitamin D-deficient.

Decalcification of bones, *osteoporosis*, is commonly seen in the elderly. The cause of this is not clearly defined, but may be due to imbalance between parathormone and calcitonin or to a reduced absorption of Ca^{2+} from the food despite adequate 1,25-hydroxy vitamin D_3. Its importance lies in the ease with which these bones break and become deformed and may press on nerves causing pain in the elderly.

OTHER HORMONAL FUNCTIONS OF THE KIDNEYS

Both renin and erythrogenin have already been described, both are hormones in the sense that they are formed in the kidneys, released into the blood stream and produce their effects on other organs. Both, however, are proteolytic enzymes. Renin causes the production of angiotensin from a plasma protein, and the angiotensin acts on the adrenal cortex stimulating it to produce aldosterone, and is itself an extremely powerful vasoconstrictor and consequently may raise the arterial blood pressure. Erythrogenin also acts on a plasma protein, producing *erythropoietin*, which itself acts on undifferentiated bone-marrow cells, causing them to develop into red blood cells. Renin is produced, as already described, when sodium contents of the body are low, and erythrogenin when the oxygen content of blood entering the kidneys is low.

HORMONES OF THE GUT

These have been described in the chapter on the digestive system, so nothing more need be said of them here.

27 · Reproduction, Growth and Development

Reproduction of new members of the human species begins in the act of fertilisation, the union of a spermatozoon from the male of the species with an ovum from the female. In common with other mammalian species the fertilised ovum is retained within the female body for the initial stages of development.

Before we study the growth and development of the fertilised ovum, we should first see the origin of the male and female members of the species, and the way their bodies differ in order to fulfill their different requirements in the reproductive process.

SEX DETERMINATION

The sex of a member of the species is determined at the time of fertilisation, by the chromosomes found in the nucleus of spermatozoon and ovum. In Chapter 2 we described the chromosomes of human cells; how there are 23 pairs, that in one pair there is the sex-determining difference, all females having two 'X' chromosomes and all males having one 'X' and one 'Y' chromosome, and how in the formation of ova and spermatozoa the total chromosome numbers were halved. All ova contain 23, one of which is an X chromosome. All spermatozoa also contain 23 chromosomes but half of their number include one X and the other half one Y chromosome. In fertilisation when the full number of chromosomes is restored, there are thus equal chances of an embryo having two X chromosomes and becoming female or having an X and a Y chromosome and thus becoming a genetic male. It is the presence of the Y chromosome in the male and the two X chromosomes in the female that later in development lead to the different development of the male and female bodies. This will be described later. Here we will briefly describe the structures and their functions as found in adult humans.

ANATOMY OF HUMAN REPRODUCTIVE SYSTEMS

THE MALE REPRODUCTIVE SYSTEM

The male sex gland or *gonad* is the *testis* and in the adult male the pair of testes occupy the *scrotum* which keeps them at a temperature below that of the body (see below). The testis itself consists of a number of coiled *seminiferous* tubules in between which lie *interstitial cells;* the latter secrete important hormones. The tubules are lined by the cells which, by their division, produce the male sex cells, the *spermatozoa.* From the testis, the spermatozoa, which are still not fully mature, pass through a series of tubules into the beginning of the *epididymis*, which itself consists of a tightly coiled fine tubule about 20 feet long (the same as the length of the small intestine). From here, they pass into the *ductus* (or *vas*) deferens, a thick-walled muscular tube which leads out of the scrotum, through the inguinal canal and into the pelvis (Fig. 27.1). It runs to the back of the

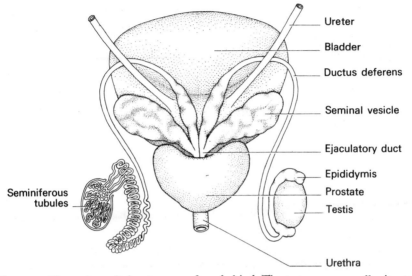

Fig. 27.1. The male genital system seen from behind. The testes are normally situated in the scrotum.

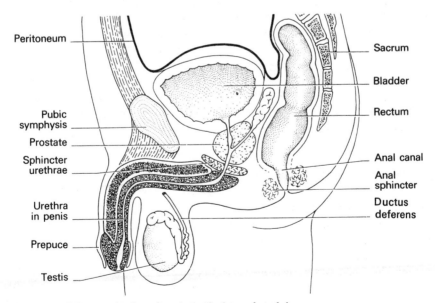

Fig. 27.2. Midline sagittal section through the male pelvis.

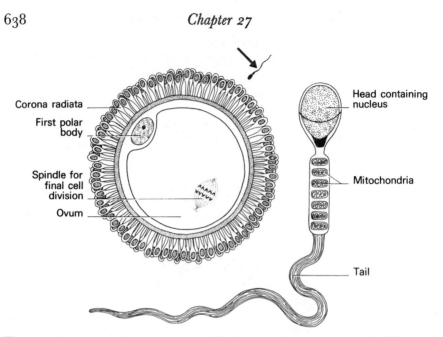

Fig. 27.3. An ovum and a spermatozoon. Highly magnified, but not equally. The arrow indicates the size of a spermatozoon in relation to the ovum.

bladder neck where it joins the duct of the *seminal vesicle* to form the *ejaculatory duct*.

The contents of the male pelvis are shown in Figure 27.2. Immediately below the neck of the bladder is the *prostate gland*. This is an exocrine gland that produces prostatic fluid, the main vehicle for carrying the spermatozoa to the outside world. The prostatic ducts, 15–20 in number, open into the prostatic urethra. Behind the prostate and bladder are the two *seminal vesicles* that produce a secretion which also forms part of the *semen*. From each seminal vesicle a duct runs downwards for a short distance before being joined by the corresponding ductus deferens to form the ejaculatory duct which, in turn, opens on each side into the prostatic urethra. The *penis* consists mainly of erectile tissue. This is a very vascular tissue composed mainly of large venous channels into which arteries open directly, the blood flow being controlled by smooth muscle sphincters. The end of the penis—the *glans*—is covered by the foreskin, or *prepuce*.

The structure of a spermatozoon is shown in Fig. 27.3. The head contains the nuclear material. Behind this is the source of energy, in the form of a coil of mitochondria, and behind this again is the long mobile tail.

THE FEMALE GENITAL SYSTEM

The female gonad is the *ovary*. The ovaries lie against the side wall of the pelvis, and are each about the size of a date. Each ovary contains a very large number of *primordial ova* which are awaiting full development and each ovum is sur-

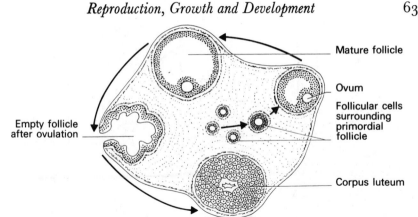

Fig. 27.4. Diagram to show the maturation of an ovarian follicle and its corpus luteum.

rounded by layers of *follicular cells*. There are also present a number of *immature follicles* and perhaps a *corpus luteum* which will be described later (Fig. 27.4).

In the midline of the pelvis, lying behind the bladder and in front of the rectum, is the uterus. This has a neck, or *cervix* (part of which projects into the *vagina*), a *body*, a *fundus* and a pair of *uterine* (or *Fallopian*) *tubes* (Fig. 27.5). The uterus is normally anteverted, i.e. pointing forwards with its long axis more or less at right angles to the vagina. The cervix and body have extremely thick (over 1 cm) walls of smooth muscle and are lined by the *endometrium*. The latter contains numerous glands but it is not possible to give a straightforward description of its structure here since it undergoes important changes during the menstrual cycle. These changes will therefore be discussed later in this chapter.

Each uterine tube passes laterally to reach the ovary, and each ends by opening directly into the peritoneal cavity in the vicinity of the ovary. The opening is surrounded by finger-like processes or *fimbria*. It is important to understand that when an ovum is shed from the ovary, it is shed into the peritoneal cavity and is then picked up by the fimbriated end of the uterine tube. The tubes are lined by a ciliated columnar epithelium and they have walls of smooth muscle.

THE FEMALE BREAST

Since this is an organ that undergoes marked changes at puberty and during pregnancy it is included in this Chapter. The actual milk-producing organ is known as the *mammary gland*, the breast being a term that includes this and the large quality of fat and fibrous tissue that makes up the characteristic shape. The gland consists of 15–20 lobes, subdivided into lobules and embedded in dense connective tissue. From each lobe a *lactiferous duct* passes towards the nipple on the summit of which it opens. The breast lies on pectoralis major but overflows onto serratus anterior and the external oblique aponeurosis.

The lymph drainage of the breast is mainly laterally and upwards into the lymph nodes in the axilla. It also drains through the thoracic wall to lymph nodes inside the thorax.

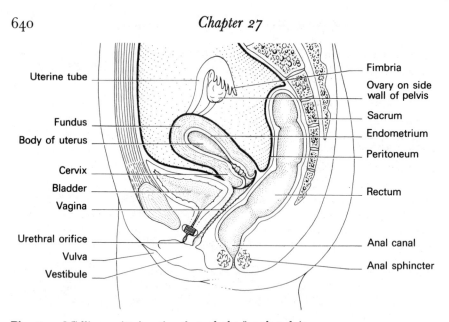

Fig. 27.5. Midline sagittal section through the female pelvis.

FUNCTIONS OF THE HUMAN REPRODUCTIVE SYSTEMS

TESTES

These serve the dual roles of secreting the hormones that cause, in other organs, the development needed for the male functions, and that of producing the spermatozoa. *Androgenic hormones* are steroid compounds which, like most other biologically active steroids, are produced from the parent compound cholesterol. A series of steps, each mediated by an enzyme, causes the production in the testes of *testosterone*. These processes are carried out in the interstitial cells of the testes, which secrete their product directly into the blood. The production is regulated by a hormone from the anterior pituitary, called (because of its action in the female ovary) *luteinising hormone*, or LH for short. LH stimulates testosterone formation, but its production by the pituitary gland is reduced by a rising blood level of testosterone—yet another example of negative feedback control.

Testosterone production begins early in embryonic life, and is responsible for the differentiation of the external genitalia into the male form and the descent of the testes into the scrotum. This low level production continues through childhood and has been thought to cause the behaviour characteristic of little boys. It is more likely however, that such behaviour is learned from other people than determined by hormones. At puberty its production increases and the *secondary sexual* characteristics of the adult male emerge. These include the development of facial hair and (in some) the later recession of scalp hair, enlargement of the larynx with deepening of the voice, a marked increase in muscle mass, unrelated to any increase in physical exercise or 'training' but undoubtedly enhancing muscular strength and thus athletic performance, and, after an initial increase in the rate of growth of bones, a fusion of epiphyses and so a cessation of bone growth. Enlargement of the testes and accessory glands occurs. Spermatozoa formation and glandular secretions are initiated and the penis grows to adult size.

Spermatogenesis is the other main function of the testes. This occurs in the seminiferous tubules. The cells lining these tubules undergo two cell divisions, during which the chromosome number is halved (meiosis). The resulting cells lose much of their cytoplasm, organise the remainder into a tail and are now mature spermatozoa. The whole process takes about 6 weeks and is dependent upon the testis being at a lower temperature than the normal body temperature of 37°C. For this reason, in most mammals, the testes are found outside the body in the scrotum. A testis that fails to leave the high temperature of the abdomen and descend into the scrotum does not form normal spermatozoa.

Further maturation occurs during a number of days spent in the epididymis and the spermatozoa become mobile. They enter and are driven along the ductus deferens and mixed with prostatic and seminal vesicle secretions by contractions of its wall during coitus. The added secretions provide nutrients and factors that increase the mobility of spermatozoa and the final semen has, at ejaculation, a volume of 3–5 ml containing 50–100 million spermatozoa per ml.

OVARIES

Again, these serve the dual roles of hormone and ova production or *ovulation*, but the two processes are more closely linked than in the testes. In the complete absence of ovulation, the ovaries maintain, together with the adrenal cortices, a continuous basal level of secretion of *oestrogens*. These hormones, of which the most important is *oestradiol*, cause the characteristic deposition of fat in the subcutaneous tissues of the body, the development of breasts and nipples, the *growth spurt*, usually earlier in girls than in boys (p. 651) but ceasing when on the face, trunk and limbs. It is surprising that one hormone can 'turn on' the genetic mechanism for hair growth on the scalp, and suppress the same mechanism on the face. This basal secretion begins at the onset of puberty, when it initiates a growth spurt, usually earlier in girls than in boys (p. 000) but ceasing when epiphyses fuse, which also happens earlier in the female. Bone growth in face, hands and feet does not occur, and oestrogens do not stimulate any increase in muscle growth.

Ovulation is, in contrast to spermatogenesis, a very economical process. About 400,000 primordial ova are present in the two ovaries at birth, of which only about 400 ever reach maturity. This should be compared with the 300 million sperm produced at each act of ejaculation of semen. One mature ovum is shed from one or other of the two ovaries at monthly intervals and since this process is linked with changes in secretion of hormones by the ovary and further changes in both the pituitary gland and the uterus, the whole will be described together, under the term *Menstrual cycle* (Fig. 27.6).

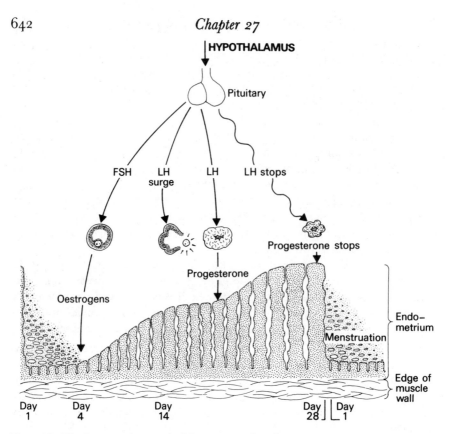

Fig. 27.6. The hormonal control of the menstrual cycle.

THE MENSTRUAL CYCLE

The endometrium that lines the uterus undergoes periodic changes so that it is necessary to describe its structure at different stages of the cycle. The cycle normally lasts about 28 days but wide variation can occur. It is customary to use the first day of menstrual bleeding as a baseline from which to time events in the cycle since this is an easily recognised and accurately timed occurrence. The bleeding is caused by the almost complete breakdown of the endometrium which, in the previous stage of the cycle, was thick and highly vascular. After ovulation, a number of primordial ova begin to develop but usually only one completes the course. The developing ovum becomes surrounded by several layers of *follicular cells*. These produce *oestrogens* (particularly *oestradiol*) and the blood oestrogen level rises to a peak a few days before ovulation. The oestrogens act on the remnants of the endometrium and stimulate it to regenerate and to grow in thickness—the *proliferative phase* (Fig. 27.6). The endometrium at the end of this phase is 2–3 mm thick and consists of tubular glands embedded in a loose connective tissue stroma. The glands are slightly tortuous, lined by columnar epithelium which is partly ciliated, and which secrete a little serous fluid. The arteries of the endometrium increase in length so that they become spiral in form.

When the successful follicle is fully mature, the nucleus of the contained ovum undergoes the first of its two meiotic divisions in which its chromosome number is halved (see Chapter 2). The two cells thus produced are unequal in size, most of the cytoplasm remaining in the ovum while the other cell forms the small *first polar body*. At this time the ovum is 100–150 μm in diameter, just visible to the naked eye. The ovum is contained in and attached to the inner wall of the mature follicle which is now filled with fluid and about 1 cm in diameter (Fig. 27.4). It protrudes from the surface of the ovary so that when the exposed wall of the follicle breaks down and the ovum detaches itself, it passes straight into the fimbriated opening of the uterine tube. The cells of the follicle now proliferate to form a large yellowish mass of cells called the *corpus luteum*. These cells continue to produce some oestrogens but their most important function is to produce *progesterone*. This hormone prepares the endometrium for implantation of the ovum. Under its influence the endometrium enters the *secretory phase*. The thickness increases to 5–6 mm, mainly due to increasing oedema of the stroma, the glands become more tortuous and dilated, and their secretions pour into the lumen. The spiral arteries become even longer and the vascularity increases.

If the ovum is not fertilized, all this preparation becomes useless, the blood levels of oestrogens and progesterone fall, the spiral arteries contract and the endometrium becomes ischaemic. As a result, the endometrium, except for its basal layer, is shed and when the arteries open up again the remains of the endometrium, along with a good deal of blood, is discharged as the menstrual flow. This continues for about 4 days whereupon the endometrium begins a new proliferative phase under the influence of a steadily rising oestrogen production from the ovary where a new follicle is ripening.

It now remains to describe the mechanisms whereby these processes are controlled. The production of oestrogens and progesterone from the follicular cells and corpus luteum are under the control of hormones from the pituitary which are appropriately named *follicle stimulating hormone* (FSH) and *luteinizing hormone* (LH). Since they act on the gonad (the ovary) they are known collectively as *gonadotrophins*. The pituitary gland is not the primary source of the rhythm, however, since it itself, is controlled by a *gonadotrophic releasing hormone* (GnRH) produced by the hypothalamus, the part of the brain that lies above the pituitary. Even the hypothalamus may be affected by higher authority, since impulses from the cerebral cortex or alterations in the sleep-awake rhythm may affect its function so that the menstrual cycle may be affected by anxiety or fear of pregnancy, while night nurses may also find a change in their cycle.

It is now possible to explain (in a rather simplified way) the control of the menstrual cycle. The 'timekeeper' is in the hypothalamus which sends 'pulses' of GnRH to the pituitary via their connecting blood vessels. The pituitary responds by producing FSH and LH, which, in turn stimulate the ovary to produce oestrogens and progesterone. The steadily rising blood level of oestrogens acts on the endometrium to produce its proliferative phase until, shortly before ovulation, there is a sudden rise in LH (*the LH surge*) to very high levels which in some way triggers off ovulation. The level of oestrogens is prevented from rising too rapidly by a negative feedback effect on the pituitary but as the level gets higher and higher, this becomes converted to a positive feedback and the output of both FSH and LH rise, the latter becoming the LH surge.

The LH surge reduces oestrogen production by the ovary but stimulates the formation of the corpus luteum and hence of progesterone which, in turn produces the secretory phase in the corpus luteum. Should the ovum be fertilized, the corpus luteum and the production of progesterone continue so that the embryo can become embedded in the succulent endometrium but if fertilization does not occur, a negative feedback mechanism operated by progesterone leads to a rapid fall in FSH and LH levels and menstruation begins.

Work on the hormonal control of the menstrual cycle is continuing but has already produced useful results. Gonadotrophins may be used to control the cycle if necessary and, in particular, a form of LH can induce ovulation in infertile women; the main difficulty is that this 'fertility drug' may be too successful and produce multiple births.

It must be remembered that similar hormones are found in the male, but in the absence of a high level of oestrogens to act on the pituitary there is no LH surge. FSH helps to stimulate the production of spermatozoa while LH stimulates the interstitial cells of the testis to produce testosterone (p. 640).

TIMING OF OVULATION

One systemic effect of progesterone is that of causing a rise in body temperature of about 0.5°C. This happens 1–2 days after ovulation, when the corpus luteum is most active, and it persists for 12–13 days, until the failure of progesterone production from the corpus luteum initiates the menstrual flow.

In the whole menstrual cycle, while the luteal phase is usually constant for the individual person, the follicular phase may vary in length from cycle to cycle. This makes it impossible accurately to time ovulation from the preceding menstrual period. It can only be determined from the *subsequent* rise in body temperature (1–2 days later) or the onset of menstruation (14 days later). A more accurate timing of ovulation may be obtained from the LH surge. This can be determined by analysis of the urine, and ovulation occurs 20–24 hours later. This is the method used for extracorporeal fertilization.

THE MENOPAUSE

Diminution of reproductive capacity in the male is a gradual process and no definite age can be given at which it ceases entirely. In the female however, it is marked by the *menopause*. The menopause, when menstruation becomes irregular, usually occurs between 45 and 55 years; the average being age 48. Oestrogen production falls, with consequent loss of the LH surge, ovulation ceases and finally the cycles stop. The endometrium and the vaginal mucosa become atrophied and various symptoms such as hot flushes, insomnia and depression may occur. Some of these symptoms may be due to side-effects of the continued production of gonadotrophin, released from the negative feedback control of oestrogen.

The reduction in oestrogen production is associated with various physiological changes. Osteoporosis develops after the menopause, but also changes in bone growth so that the face and hands becomes more like those of a male. Hair growth may begin in the male distribution and scalp hair growth is reduced. The larynx may enlarge and the voice deepen. Atheroma may begin to develop in arterial walls if sufficient of the other risk factors are present.

FERTILISATION AND INTRA-UTERINE DEVELOPMENT

Having entered the fimbriated end of the uterine tube, the ovum is carried along it by a combination of muscular and ciliary action. Fertilisation normally occurs in this situation, the spermatozoa having travelled from the vagina mainly by their own mobility. It is probable that the spermatozoa have to accomplish fertilisation within 24–48 hours as they cannot live after this time. As soon as one spermatozoon has penetrated the ovum, the latter becomes impenetrable to others, otherwise the fertilised ovum would have too many chromosomes.

The result of fertilisation is the restoration of the full number of chromosomes (as a result of the union between the male and female germ cells, each of which carries only 23 chromosomes). The fertilised ovum divides repeatedly to produce a tiny globule of cells—the *blastocyst*—and this finally reaches the uterus and becomes embedded in the endometrium about a week after fertilisation. In some women, the uterine tubes will not allow the passage of ova, though in all other respects ovarian and uterine function are normal. It is now possible to retrieve the ova as they are shed from the Graafian follicles, to fertilise them *in vitro* and then, after a period of growth, to return them to the uterine cavity, ensuring that the endometrium has been primed by ovarian hormones and is ready to receive the developing embryo. This enables an otherwise sterile woman to carry her own baby to term. Fertilized ova can be preserved by deep-freezing: by appropriate hormonal treatment a woman's ovaries can be encouraged to produce several ova at one time; 'surrogate' mothers' uteri can have embryos implanted in them. Thus, one does not know the lengths to which man might go in these directions, were it not for moral proscriptions.

The whole period of pregnancy lasts for 40 weeks from the date of the first day of the last menstrual period, and is subdivided into two periods. For the first 8 weeks, a process of *differentiation* occurs. This is the *embryonic period* when the cells of the blastocyst, which are all very similar, divide rapidly over and over again to produce not only a bigger embryo but one in which all the multiplicity of cell types which have been described in Chapter 3 are present. The first separation of cell types is into those which are going to form the supporting tissues such as the placenta and fetal membranes and those which will form the fetus itself. When all these cells and tissues have become differentiated and the embryo is recognisable as a tiny human being, the second or *fetal period* begins which is mainly one of *growth*. Growth, of course, continues until adulthood but at 40 weeks a number of changes take place in the body at birth. The development of the embryo and fetus will now be considered in a little more detail.

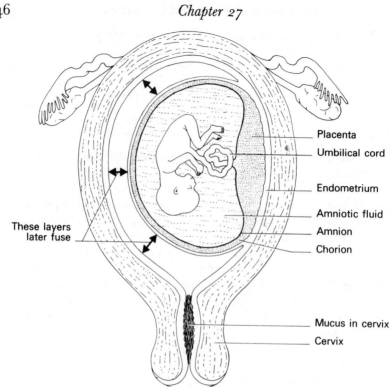

Placenta

Umbilical cord

Endometrium

Amniotic fluid

Amnion

Chorion

Mucus in cervix

Cervix

These layers later fuse

Fig. 27.7. The fetus in utero during early pregnancy.

The blastocyst, having implanted itself deep within the endometrium, proceeds to grow rapidly while its cells become differentiated into *embryonic* and *supporting cells*. The embryonic cells become further differentiated to form the rudiments of the various systems of the body. The nervous system develops very early and the brain and spinal cord can soon be recognised. As the embryo grows, it needs a vascular system to carry oxygen and nutrients to all parts of the body so that the heart starts to beat surprisingly early in development and it propels plasma and, a little later, blood cells through a primitive vascular system. The lungs, of course, are useless in the uterus as is the digestive system so that their development is delayed. Even so, by the end of 8 weeks of development, all the organs can be recognised quite easily and the embryo, from now on to be called the *fetus* is about 3 cm long. At 4 months the hair develops and at 5 months the mother can feel fetal movements (*'quickening'*). At about 7 months the fetus begins to develop subcutaneous fat and, if born prematurely, it can survive. With modern intensive care methods, even younger fetuses may be viable. Finally at 40 weeks, the baby is ready to be born.

Meanwhile, the supporting cells of the blastocyst have formed the *fetal membranes* (Fig. 27.7). The fetus is contained within a thin-walled sac filled with fluid. The sac wall consists of two layers—an inner *amnion* and an outer *chorion*. The sac contains about 700 ml of *amniotic fluid* at full term; this is a valuable

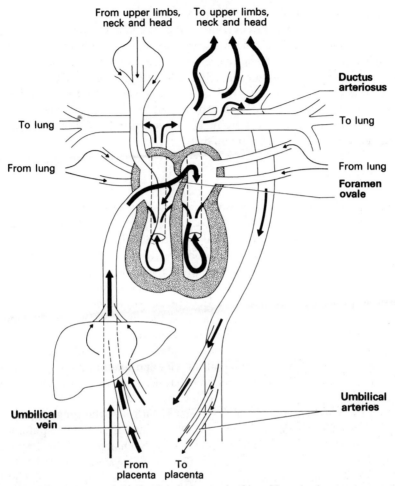

Fig. 27.8. Diagram to explain the circulation in the fetus. Note the four segments of the circulation which shut down at birth: the umbilical arteries and vein, the foramen ovale and the ductus arteriosus. (See also Fig. 14.1.)

protection for the fetus both during its development and during the early stages of labour. The chorion is directly in contact with the endometrium and a part of it grows enormously to produce the *placenta*. This is about 15 cm in diameter, 4–5 cm thick and weighs about 500 grammes. The placenta is essentially a vascular organ. It is connected to the fetus by the *umbilical cord* which transmits two umbilical arteries from fetus to placenta and one umbilical vein from placenta to fetus. These vessels are connected to a capillary plexus which occupies the interior of a huge number of *chorionic villi* within the placenta. Like the villi in the intestine (Chapter 24), they increase the internal surface area of the placenta which is important because the villi occupy a large cavity through which circulates the mother's blood. The maternal arteries actually open into this cavity and the veins drain it, so that the placenta is an exception to the general statement that blood is enclosed within the endothelial lining of blood vessels (Chapter 14). Since the coverings of the chorionic villi are extremely thin, the fetal and maternal blood come into very close contact without actually mixing (although a small number of fetal red cells may pass into the maternal circulation (p. 485), but the contact is close enough to allow the passage of oxygen and food substances to the fetus and of carbon dioxide and waste materials from fetus to mother. The placenta may thus be regarded as the 'lungs' and 'digestive system' of the fetus as well as being an endocrine gland which produces certain hormones essential for the maintenance of pregnancy.

BIRTH

The smooth muscle of the uterine wall begins to contract at the onset of labour and the thick wall of the cervix is taken into the body of the uterus so that a wide birth canal is opened up. As cervical dilatation begins and especially after the fetal head enters the birth canal, receptors are stimulated which reflexly cause secretion of oxytocin from the posterior pituitary gland. This hormone further stimulates uterine muscle to contract, which causes even more stimulation of the cervical receptors, a positive feedback state being established. Uterine muscle is also powerfully stimulated by prostaglandins, though how their production in labour is altered is not known. The fetal membranes (amnion and chorion) rupture and the amniotic fluid escapes. The uterine muscle, aided by voluntary contractions of the mother's abdominal muscles and diaphragm, now expel the baby, usually head first. This can sometimes happen quite quickly but is usually a prolonged process, particularly with a first baby.

When the baby emerges it takes its first breath and the lungs expand, at least partially. The alveoli open up, and at the same time the full pulmonary circulation begins so that the baby's colour rapidly changes to a healthy pink. At birth, important changes also occur in the vascular system (Fig. 27.8). In the fetus, oxygenated blood from the placenta, via the umbilical vein, enters the inferior vena cava after traversing the liver and thence reaches the right atrium. Only a little of this blood enters the right ventricle from the atrium and of this, nearly all never reaches the lungs because it is shunted from the pulmonary artery straight into the aorta through a channel called the *ductus arteriosus* (it would not be helpful for oxygenated blood to circulate through the collapsed and airless lung of the fetus). Most of the oxygenated blood entering the right atrium passes through an opening in the interatrial septum (the *foramen ovale*) to the left atrium and thence to the left ventricle. This oxygenated blood is thus ready to be pumped out into the systemic vessels. At birth, the smooth muscle in the walls of the umbilical arteries and vein contracts strongly so that the lumens of these vessels, and that of the channel through the liver, become closed off. The foramen ovale closes by a flutter-valve action and the ductus arteriosus also closes down by a contraction of its smooth muscle. The blood from the right atrium now all passes into the right ventricle and thence via the pulmonary arteries to the lungs. Since the lungs now contain air, the blood is oxygenated and returned to the left atrium via the pulmonary veins. In fact, the normal adult circulation is usually established within seconds or minutes of birth. As might be expected, this complicated mechanism does not always function properly, two of the most common developmental errors being failure of the foramen ovale to close (*atrial septal defect*) and failure of the ductus arteriosus to close (*patent ductus arteriosus*).

As soon as the blood flow through the umbilical vessels has ceased—and the midwife can tell this by noting the cessation of pulsation in the umbilical arteries—the umbilical cord is tied and cut and the baby separated from the mother. It is a mistake to tie and cut the umbilical cord too soon, since about $\frac{1}{3}$ of the fetal blood may then be trapped in the placenta and not available to the new-born infant which has, with the expansion of its lungs, an increased circulatory capacity. After a few more uterine contractions, the membranes and the placenta are expelled from the uterus, which contracts down tightly so as to prevent bleeding from the severed ends of the arteries in the endometrium.

POST-NATAL GROWTH AND DEVELOPMENT

Most of the basic life-supporting systems in the human body, once they have adjusted to independent existence after birth, are able to function just as in the adult. Thus the cardiovascular and respiratory systems can, almost immediately, ensure an adequate blood supply to all tissues and the adequate oxygenation of this blood, though, owing to the horizontal direction of the ribs, respiration is almost entirely diaphragmatic. The kidneys excrete urea, excess salt and water, though they do not concentrate urine as well as in the adult. Since the pelvic bones are very small, the bladder is an abdominal organ and less well protected than in the adult. The digestive system is at once capable of digesting most foodstuffs, though it is simplest (and probably best) not to tax it with adult-type foods for a few months, but to rely on that which evolution has provided! In the central nervous system the great growth spurt that began before birth continues after birth with much axonal growth, synapse formation and myelin sheath development. These processes probably continue for 2–5 years. They are somewhat dependent upon adequate nutrition and there is evidence that deficient neuronal growth and myelination in the first 2 years can never be fully made good in later life. Since the pyramidal tract, for example, is not fully myelinated until the infant is a year old, the plantar response during this time is extensor, instead of flexor as in the healthy adult (Chapter 5). The brain during this period increases in mass from 300 to 900 g, and reaches the adult weight of 1,200 g at 6 years of age. It is widely conjectured that the development and learning of motor skills accompanies the formation of rich synaptic connection between neurones, some people even asserting that unless the skill is attempted, the synapses will not form. It is difficult to see which comes first in this egg and chicken-type argument! During the first year of post-natal life, the kidneys lose their lobulation and the large fetal adrenal cortex shrinks away.

Birth weight may be tripled or quadrupled by the end of one year. During this time the maxilla and mandible grow even faster than does the cranium, and the milk teeth begin to appear, all 20 of them erupting by 2 years.

GROWTH IN GENERAL

Growth during childhood and adolescence is most easily studied from measurements of the weight and height at different ages. More sensitive estimations are given by measuring the *rate (velocity) of growth*. Such indicators show four well-defined periods of growth:

1　From birth to 2 years there is a marked slowing down of growth rate.
2　For the next 10 years, growth rate diminishes rather more slowly.
3　At adolescence, there is a sudden and very marked growth spurt.
4　The adolescent growth spurt is followed by a gradual deceleration of growth rate until maturity.

Growth, however, is not uniform throughout all the body tissues. While the internal organs such as the liver, spleen and intestines follow a similar growth curve to that of weight and height, the nervous system, as has been seen, undergoes most of its growth in the later part of fetal life and in the first 6 years of childhood. The lymphatic system grows rapidly during childhood and reaches the peak of its development at adolescence, after which it slowly declines. The reproductive system, of course, does not reach its full development until puberty.

The maturity of the skeletal system is usually estimated by taking an X-ray of the wrist and hand and estimating a *bone age* from the state of ossification of the bones. This does not necessarily correspond to the chronological age and can be used in the diagnosis of certain growth disturbances. The final mature height of a child can be estimated fairly accurately from the bone age, the present height, and the height of the parents—this estimation is often called for by ballet schools.

GROWTH IN INFANCY

During the first two years of life the infant is weaned from breast or bottle-feeding and develops some immunological protection. It learns to stand and walk and begins to talk and develop some aspects of social behaviour. The most important factor in retarding infant growth is malnutrition.

During this period, the normal adult curves in the spine are developed. In the new born baby the whole spine has an anterior concavity and, if unsupported, the head will fall forward onto the chest. At about 3 months of age, the infant begins to raise its head and look about and the cervical forward convexity develops (Fig. 10.9). Round about the age of one year the infant begins to stand up and then to walk, and during this period the lumbar forward convexity develops. This causes the forward rotation of the pelvic girdle so that the pubic symphysis drops relative to the sacrum. The hip and knee joints become extended and the feet everted so that the sole comes into contact with the ground.

In the toddler, the centre of gravity is high, so that the stance has a wide base and falls are frequent. By the age of 5 the pelvis has become deeper and the bladder and uterus descend to become pelvic rather than abdominal organs. At this age, too, ribs descend and respiration becomes thoracic as well as diaphragmatic.

At birth, red marrow is found in all bones but during childhood this becomes reduced and in the adult it is only found in the vertebrae, limb girdles, skull, ribs and sternum and in the ends of some of the long bones.

GROWTH IN CHILDHOOD

The fontanelles (p. 170) have become fully ossified at 2 years and the growth of the brain, and therefore the skull, slows down. The mandible and maxilla become deeper to accommodate the permanent dentition (from 6 years onwards) as well as the maxillary sinus. The size of the face therefore increases relative to that of the cranium.

GROWTH IN ADOLESCENCE

The *adolescent growth spurt* begins at about 10 years in girls and 12 years in boys and it lasts about 3 years in both sexes. Boys, however, gain more in height than girls. Growth in height ceases when most of the epiphyses fuse, usually at about the age of 15 in boys and 16 in girls.

It is during the adolescent growth spurt that the reproductive organs and secondary sexual characteristics develop with the onset of puberty. In boys, beginning about $11\frac{1}{2}$ years, the penis, scrotum and other parts of the genital system enlarge, and pubic and axillary hair develop. The larynx enlarges, the voice 'breaks', and as the genital system develops, semen is produced. In girls, from 11 years or so, the breasts begin to form, the female distribution of fat can be recognised and the uterus and vagina enlarge. The onset of menstruation (*menarche*) occurs at about 13 years although the first periods may be irregular and unaccompanied by ovulation. They may sometimes be profuse and can be the cause of an iron deficiency anaemia in young women.

Sexual maturation, as measured by the occurrence of ovulation and spermatogenesis, occurs at about the same age in both sexes, i.e. between 14–15 years.

Growth disturbances during childhood and adolescence may be the result of under-nutrition, severe illness, genetic factors including such conditions as Down's syndrome, endocrine deficiencies, congenital heart conditions causing lack of oxygen, and psycho-social disturbances. In many of these conditions, however, the stunting of growth is not permanent because of the phenomenon of *catch-up growth*. If, for example a child is severely under-nourished for a time, so that his growth curve falls below the normal level, when he returns to a normal diet, his rate of growth will increase until he returns to the normal height and weight for his age or may even overshoot. Unfortunately such catch-up does not occur in the central nervous system if, through infantile under-nutrition, it has not grown normally.

Index

Abdominal planes 588
Abdominal wall muscles 290–6
Abductor digiti minimi
  foot 357
  hand 281, 326
Abductor hallucis 357
Abductor pollicis brevis 324
Abductor pollicis longus 320, 321
Absorption of food 595, 598–9
Accessory ligaments 222
Accessory nerve 371, 619
Acetabulum 199, 202, 203, 204, 205, 249
Aceto-acetic acid 601
Acetone 601
Acetylcholine (A.Ch.) 52, 64–7, 68, 114, 130,
  141, 466
Acetylcholinesterase 65, 68
Acetyl Co-A 569–70
Achilles tendon 215, 349–50
Acini, of glands 36
Acromegaly 628
Acromioclavicular joint 186, 187, 237–8
Acromion process 186, 187, 190
Actin 135, 138, 141–4, 148, 569
Action potential 58
  cardiac muscle 439–44
    time course 442–3
  skeletal muscle 141
  see also Nerve impulses
Actomyosin 142
Adaptation in receptor function 60, 99
Adductor brevis 357
Adductor canal 615, 616–17
Adductor hallucis 359
Adductor longus 204, 337, 613, 615, 617
Adductor magnus 207, 208, 337–8, 344, 617
Adductor pollicis 324–6
Adenosine diphosphate 21, 25
  in muscle contraction 142, 148–9, 569–71
Adenosine triphosphate 21, 22, 23, 25, 600
  in muscle contraction 142–3, 148–51, 568–71
Adenyl cyclase 623
Adipose tissue 38, 39, 583–4
Adolescent growth spurt 651
Adrenal glands see Suprarenal glands
Adrenaline 26–7, 64, 133, 582–3, 622, 631
  effect on cardiac muscle 451
Adrenocorticotrophic hormone (ACTH) 626
Afterdischarge, in reflex activity 73
Agglutination of red blood cells 485
Agranulocytes 479, 480

Air, composition of 524–6
  alveolar 525–6
  atmospheric 524–5
  expired 526
Alarm reaction 108, 131–2, 133, 578
Albumins 481–2
Aldosterone 475, 631, 632, 635
Alerting reaction centres 108, 131–2, 133, 578
Alpha receptors 461, 464, 575
Alveoli (acini) of glands 36
Alveoli of lungs 514–15
Amino acids 22, 598
  deamination 602, 603
  essential 602, 605
  genetic regulation of sequence in proteins
    30–1
  metabolism 602
Amnion 646, 648
Amniotic fluid 646–7, 648
Amputation neuroma 160
Anaemia 496
  aplastic 478
  sickle cell 497
Anal canal 592, 593
Anastomoses 148, 419, 422, 423–4
  arteriovenous 424, 559
Anatomical dead space volume (lungs) 521
Anatomical position 4
Anatomical snuffbox 194, 198, 321, 322
Anconeus 317–18
Androgenic hormones 640
Aneurysm 429
Angina pectoris 276
Angiotensin 463–4, 631, 635
Angle of Louis 185
Anisotropic bands 136
Ankle jerk 117, 124
Ankle joint 263–5
Annular ligament, elbow joint 242
Annulospiral nerve endings 100, 102, 115–16
Annulus fibrosus 224–5
Anorectal ring 297
Anterior fontanelle 170
Anterior horn cells 89, 91, 140–1
Anterior nerve roots 90–1
Antibodies 537, 539, 544
Antidiuretic hormone (ADH) 475–6, 552, 553,
  554, 628–9
Antigens 537, 539, 544
Antihaemophiliac globulin (AHG) 483–4
Aorta 8, 422, 426

Aortic bodies 531
Aortic valve 418
Apex beat 420
Aphasia 80
Apneustic centre 529
Apocrine glands 559
Aponeurosis 274
   bicipital 309, 613
   palmar 322–4
   plantar 356, 364
Appendices epiploicae 593
Appendix 592
Aqueduct 76, 96
Aqueous humour 403, 475, 476, 490
Arachnoid mater 76
Arachnoid villi 97, 488
Arches of foot 363–5
Archicerebellum 88, 113, 118
Arcuate artery, of foot 435
Areolar connective tissue 37–9
Arterial diseases 601
   effect on blood flow 471, 580
   exercise 580–1, 585
   investigation of arteries 471–2
Arteries 8, 420
   elastic 420–1
   muscular 421
   see also names of individual arteries
Arterioles 8, 9, 411, 421, 424
Articular facets of vertebrae 175, 178
Ascorbic acid (vitamin C) 478, 602, 604, 606
Association areas 80
Association fibres 76
Atheroma 471, 580, 645
Athetosis 113, 114, 122
Atlanto-occipital joint 227–8
Atlas 177
Atria 416–17
Atrial septal defect 416, 649
Atrio-ventricular (a-v) node 418, 442
Atropine, effect on nervous system 66
Auditory nerve *see* Vestibulocochlear nerve
Auricle (of heart) 415, 416
Autonomic nervous system 53–4, 75, 125–6
   anatomy 126–9
   functions 129–33
   see also Parasympathetic nervous system;
      Sympathetic nervous system
Autoregulation of blood flow 465, 470
Axilla 608–12
Axillary artery 430, 432, 609, 611
Axillary nerve 382, 609, 612
Axillary sheath 611, 612
Axillary vein 433, 608
Axis 177, 178
Axolemma 47, 52
Axon 44, 47, 75
   connector ('preganglionic') 125–9
   effector ('postganglionic') 125–9
   hillock of 62

Axon reflex 560
Axonotmesis 153, 155, 160
Axoplasm 47

Babinski response 121, 123–4
Bainbridge reflex 454
Balance, sense of 406–8
Baroreceptor reflex 8, 452–4, 461–2, 463
Basal metabolic rate (BMR) 600
Basal nuclei of cerebral hemispheres 83, 113–14
Basement membrane 34, 35
Basilar artery 429
Basilic vein 432, 433, 612
Basis pedunculi 84
Basophil leucocytes 479, 480
Bell's palsy 284, 369
Beri-beri 606
Beta receptors 461, 464, 576
Betz cells 110
Bicarbonate ions 498, 499
Biceps brachii 189, 194, 309, 609, 610, 613
Biceps femoris 344, 618
Bicipital aponeurosis 309, 613
Bifurcate ligament 267
Bile salts 597, 603
Biliary apparatus 594–5
Bilirubin 496
Biliverdin 496
Biological variations 6–7
Birth 648–9
Bladder 546–9
Blastocyst 645, 646
Bleeding time 484
Blood 37, 473, 476
   cells 476–81
   circulation of *see* Circulation of blood
   coagulation (clotting) 481, 482–5
   groups 485–6
   plasma 481–2
   platelets 481
Blood pressure, arterial 7–8, 455–6
   diastolic 455–6
   and exercise 579
   factors affecting 459–60
   hypertension 463–4
   regulation of 26, 460–4
   systolic 455, 456
Blood vessels 420–5
   abdomen 426–7
   head and neck 427–9
   lower limb 433
   sympathetic innervation 126, 460–1, 575–6
   thorax 426–7
   upper limb 430–3
   see also Arteries; Veins; *names of individual*
      vessels
Body fluids
   composition 474–5
   volumes, measurement of 473–4

volumes, regulation of 475–6
see also specifically named body fluids
Body temperature control 559, 562–6, 584
Bohr shift 494, 496
Bolus, of food 595–6
Bone, 37, 162
    cancellous 162, 168
    compact 162, 168
    development 165–7
    structure 162–4
    see also Bones *and names of individual bones*
Bone age 650
Bones
    individual, study of 170
    long, structure of 168–70
    shape of 167–8
    see also names of individual bones
Boutons terminaux (synaptic bags) 52, 60–1
Bowman's capsule 550
Brachial artery 430–1, 432, 612, 613
Brachial plexus 92, 371, 373–4, 398, 611, 619
    lateral cord branches 374–9, 611–12
    medial cord branches 379–81, 611–12
    posterior cord branches 381–6, 611–12
Brachialis 195, 309–10, 612
Brachiocephalic artery 426
Brachiocephalic vein 427, 433
Brachioradialis 317, 612
Bradykinin 465
Brain 76–8
    brain-stem 77, 78, 82, 88
    cerebellum 76–7, 88, 113, 118
    cerebral hemispheres 76, 77–8, 78–83
    development 76–7
    medulla oblongata 76, 84–8
    midbrain 76, 82, 83–4
    pons 76, 84–8
Brain-stem 77, 78, 82, 88
Breast, female 639
Bronchi 513–14, 515–18, 519
Bronchiectasis 516
Bronchioles 514
Bronchopulmonary segments 515–18
Brown-Séquard's disease 120
Brunner's glands 591
Brush border 591, 597–8
Buccinator muscle 595
Buffering reactions 498–500
Bundle of His (atrioventricular bundle) 481,
    442–3
Bursae 223, 272, 274, 339
    iliopsoas muscle 332
    knee joint 259, 260, 335
    subacromial 239–40, 307
    subscapular 239
    tendo calcaneus 350

Caecum 592
Calcaneo-cuboid joint 266–8

Calcaneofibular ligament 264
Calcaneus 213, 214–15
Calcitonin 164, 165, 630, 634
Calcium
    in bone 164–5
    metabolism 164–5, 603
        hormonal regulation 634
    in muscle contraction 130, 141–4
    in nerve impulse transmission 65
    role of 164–5
Calmodulin 165
Capillaries 421, 424–5
Capitate bone 196
Capitulum 192
Capsular ligaments 220, 222
    ankle 263
    elbow 242
    hip 249
    knee 255
    shoulder 238, 239, 240
    vertebrae 227
    wrist 245
Carbamino compounds 497, 498
Carbohydrate metabolism 600, 604
Carbon dioxide
    cerebral circulation 466–7
    disposal 584
    measurement of production 449
    transport 491–2, 497–500
Carbon dioxide narcosis 533
Carbonic anhydrase 476–7, 498, 499
Cardiac cycle 438, 445–9
Cardiac impulses 420
Cardiac muscle 134, 438
    contraction 433–4
    electrical activity 439–45
    histology 438–9
Cardiac output 448, 455, 459
    measurement of 448–9
    venous return 458–9
Cardiac plexus 419–20
Carotenes 605
Carotid arteries
    common 426, 427, 620
    external 427, 620
    internal 427, 428, 620
Carotid bodies 532
Carotid sheath 620
Carotid sinus 620
Carpal bones 196–8
Carpal tunnel 197, 311–12
Carpal tunnel syndrome 378
Carpo-metacarpal joints 246–7
    first 247–8
Carriers in haemophilia 484
Carrying angle 192, 243
Cartilage 37, 42–3
    articular 42, 168, 220–2
    elastic 42, 43
    fibrocartilage 42

hyaline 42–3, 168, 219, 220–2
Catalysts 22
Catch-up growth 651
Cauda equina 89, 182
Caudate nucleus 83, 113, 114
Cause-effect processes 8–9
Cells 2
  differentiation 10, 31–2
  division 10, 12, 15–19, 643
  electron microscopy 11
  energy production 19–22
  function 19–23
  light microscopy 10–11
  membrane 12, 22, 23–4
    nuclear 11
  nucleus 11–12, 15–19
    nuclear membrane 11
  regulation of activity 24–5
    chemical messengers (hormones) as factors
      in 26–7
    development and growth regulation 31–2
    genetic mechanism 27–32
    nervous system as factor in 26
  structure 10–19
Central canal of spinal cord 76
Central nervous system (CNS) 75–6
  effects of diseases of 119–24
Cephalic vein 432–3, 608, 612
Cerebellar peduncles 77, 84
Cerebellum 76–7, 86, 88, 113, 118
Cerebral arteries
  anterior 428
  middle 428
  posterior 429
Cerebral circulation, regulation of 466–7
Cerebral cortex 75, 78–82
  motor 80, 109–10
  representation of body on 80, 104, 109
  sensory 80, 104–5
Cerebral hemispheres 76, 77–8, 78–83
  dominant hemisphere 80
Cerebral peduncle 83, 84
Cerebral vascular accident 82 *see also* Stroke
Cerebrospinal fluid (CSF) 76, 96–7, 402, 488–9
  composition 475
  control of volume 476
  functions 489–90
  raised pressure of 489
Cervical plexus 371
Cervical rib 175–7
Cervix, of uterus 639
Charcot's joint 121
Chemoreceptors 97
  aortic and carotid bodies 531–2
  brain-stem 532
  peripheral 533–4
Chest wall 503–5
Chloride ions, and membrane potential
  nerve cells 54–5, 61, 63, 74
  skeletal muscle fibres 140

Chloride shift 499–500
Cholecalciferol 561–2, 603, 604, 605, 606, 634
  see also Vitamins D; D₃
Cholecystokinin 597
Cholesterol 561, 621, 632
Cholinergic autonomic vasodilatation 466
Chondrocytes 42
Chordae tendineae 417, 445
Chorea 113, 114, 122
  Huntingdon's 114
Chorion 646, 647, 648
Chorionic villi 647, 648
Choroid, of eye 401
Choroid plexuses 96, 488
Chromatids 16, 17
Chromatin 12, 15
Chromatolysis 46, 154–5
Chromosomes 15–19, 29, 636
Chyme 596, 597
Cilia 34–5
Ciliary body 402–3
Circle of Willis 429
Circulation of blood 7, 411, 455
  arterial diseases affecting 471, 580–1
  blood flow, local regulation 465–6
  blood pressure, arterial 455–6
    factors affecting 459–60
    regulation of 460–4
  blood volume control 464
  cerebral, regulation of 466–7
  collateral 422, 423
  coronary, regulation of 467–8
  cutaneous, regulation of 469–70, 559–60
  kidneys 427, 470, 546, 551
  peripheral resistance 456–7
  portal 412–13, 470–1, 625
  pulmonary 411, 472
  skeletal muscle 276, 468–9, 575–9
  systemic 411
  venous return 425, 457–8, 577–8
    and cardiac output 458–9
Cisterna chyli 487
Claudication 276
  intermittent 581
Clavicle 186–7, 236
  movements 238
Clavipectoral fascia 608
Claw hand 280, 381
Clonus 118
Close-packed position of joints 222, 265, 335
Clotting of blood 481, 482–5
Clotting time 484–5
Cobalamin 478, 604, 607
Coccyx 173, 179
Cochlea 405, 406
Colic 106–7
Collagen fibres 38, 39, 42–3
Collecting duct 551
Colliculi of midbrain 83
Colon 592–3

Commissural fibres 76, 80
Common bile duct 595
Common hepatic duct 594, 595
Conchae, of nose 408, 512, 518
Conjunctiva 401
Connective tissue 37–9
Connective tissues 33, 37–43
  see also specified types
Conoid ligament 237–8
Contraction of skeletal muscle 140–4, 279–83
  energy sources 148–51
  heat production and energy efficiency 151
  isometric 146, 282
  small muscle masses 579
  isotonic 146, 282
  mechanical effects of 145–8
  pre-loaded and after-loaded 146–7
  refractory period 144
  summation in 144, 147–8
  twitch contraction 142–4
  see also Exercise; Muscle, skeletal, action
Coraco-acromial arch 239, 240
Coraco-acromial ligament 239
Coracobrachialis 309, 611
Coracoclavicular ligament 186, 187, 190, 237–8
Coracohumeral ligament 239
Coracoid process 187, 189–90
Core body temperature 564, 565–6
Cornea 401
Corona radiata 82
Coronary arteries 418–19, 426
Coronary circulation, regulation of 467–8
Coronary sinus 416, 419
Coronary thrombosis 419
Coronoid process 192, 195
Corpora quadrigemina (colliculi) 83
Corpus callosum 80
Corpus luteum 639, 643, 644
Corpus spongiosum 549
Corpus striatum 83
Corticopontine fibres 86
Corticospinal tract (pyramidal tract) 76, 83, 84, 86, 87, 110
Cortisol 631–2
Costal cartilages 182, 503–4
Costal margin 185
Costoclavicular ligament 186, 237, 238
Costodiaphragmatic recess 509
Costovertebral joints 231–2
Co-transmitters 67
Cough reflex 513
Coughing 513, 516, 519, 520
Countercurrent multiplier system 553–4
Cranial fossae 173
Cranial nerves 75, 88, 91, 368–71
Creatine phosphate 21–2, 600
  in muscle contraction 148–9, 569, 574
Cretinism 630
Crossed extensor reflex 71, 73, 74, 123
Cruciate ligaments 256–7

Crus, of diaphragm 505
Crutch palsy 385
Cubital fossa 312, 431, 612–13
Cuboid bone 214, 216
Cuneiform bones 216
Cutaneous nerve of arm
  lateral, lower 384
  lateral, upper 382
  medial 379, 612
  posterior 384
Cutaneous nerve of calf, lateral 394, 619
Cutaneous nerve of forearm
  lateral 374–5, 612
  medial 379–80, 612
  posterior 384
Cutaneous nerve of thigh
  intermediate 616
  lateral 387, 615, 616
  medial 616
  posterior 390, 618
Cyclic adenosine monophosphate 623
Cystic duct 595
Cytoplasm 11

Deamination of amino acids 602, 603
Decussation of lemnisci 87
Decussation of pyramids 87
7-Dehydrocholesterol 164, 561, 606, 622, 634
Deltoid ligament 263
Deltoid muscle 192, 304–6, 608
Dendrites 47
Dens of axis 177
Dense connective tissue 39
Deoxyribonucleic acid (DNA) 11, 12, 29
Depolarisation of cell membranes 54, 56, 438
Dermal papillae 558
Dermatomes 92, 93–5, 121, 396–8
Dermis 556, 558
Detrusor muscle 555
Deuterium in body water measurement 474
Diabetes mellitus 632, 633
Diaphragm 505–7
  nerve supply 372
Diastole 420, 438
Dicrotic pressure wave in cardiac cycle 448
Digestion of food 595–9
Digestive system
  anatomy 586 see also named parts
  nerve supply 132–3
Digital arteries 432
Digital nerves
  foot 393
  hand 378
1, 25 Dihydroxycholecalciferol 603, 606, 634
Disaccharides 598
Distal convoluted tubule 551
Diverticula 599
Diverticular disease 599
Diverticulitis 599

Dopamine, as neurotransmitter 67, 114
Dorsal artery of foot (dorsalis pedis) 435, 436
Dorsal scapular nerve 374
Dorsiflexion of foot 345
Duchenne's muscular dystrophy 123
Ductus arteriosus 648, 649
Ductus (vas) deferens 637–8, 641
Duodenum 591
Dupuytren's contracture 324
Dura mater 76
Dural venous sinuses 429
Dust cells 511, 543

Ear 404–8
  inner 405–8
  middle 404–5
  outer 404
Ejaculatory duct 638
Ejection phase, cardiac cycle 447
Elastic fibres 38
Elbow joint 242–3
Electrocardiography 444–5
Electromyography 159
Embryo 646
End arteries 423
  functional 419, 423
End-diastolic volume of heart 448, 451, 455
End-expiratory position (chest) 520
End-expiratory volume (lungs) 521
End-plate potential 141
End-systolic volume of heart 448, 451, 455
Endocardium 415
Endocrine glands 36, 621–4
  see also names of individual endocrine glands
Endolymph 406–7
Endometrium 639, 642–4
Endomysium 100, 270
Endoneurium 49, 152–7, 160
Endoplasmic reticulum 13–15
Endothelium 34, 415
Energy production 19/22
  and exercise 149–50, 569–70, 571
  food as energy source 599–601
  muscle contraction 148–51
Energy requirements 604–5
Enkephalins 67, 88, 107, 108
Enterokinase 598
Enzymes 20, 22–3
  lysosomal 15
  production, by cells 15
Eosinophil leucocytes 479, 480
Epicardium 413
Epicondyles
  femur 207, 208
  humerus 192, 193
Epidermis 556
Epididymis 637, 641
Epigastrium 588
Epiglottis 513
Epimysium 270

Epineurium 49
Epiphyseal plate 166, 167
Epiphyseal scar 167, 168
Epiphyses 166, 167
  fusion of 167
Epithelium 33–4, 36
  columnar 34, 35
  cubical 34, 35
  keratinised 35–6
  olfactory 409
  pseudostratified 35
  simple 34–5
squamous 34
  stratified 35–6
  stratified squamous 35
  transitional 36
Equilibrium potential 55–6, 140
Erb's paralysis 398
Erector spinae 288–90, 294
Erythrocytes 476–9
Erythrogenin 478, 634–5
Erythropoietin 478, 635
Eustachian tube 404–5
Evans' blue, in plasma volume measurement 474
Excitatory post-synaptic potential (EPSP) 61–2, 63, 71–2, 74
Exercise 567
  and arterial disease 580–1, 585
  blood pressure 579
  body temperature control 566
  circulatory changes 575–81
  disposal of waste products 584–5
  energy sources 149–50
  external respiration 581–2
  heart in 578–9
  heat production and removal 584
  internal respiration 572–4
  jogging 585
  marathon running 584, 585
  metabolism in skeletal muscle 568–75
  motor units in 567–8
  muscle blood flow 575–7
  nutrients, supply of 582–4
  peripheral resistance in blood circulation 579
  venous return 577–8
Exocrine glands 36, 37
  see also names of individual exocrine glands
Exophthalmos 630
Expiratory centre 529, 530
Expiratory reserve volume (lungs) 522–3
Extensor carpi radialis brevis 318–19
Extensor carpi radialis longus 318–19
Extensor carpi ulnaris 320
Extensor digiti minimi 320
Extensor digitorum 319–20
Extensor digitorum brevis 347–8
Extensor digitorum longus 347
Extensor hallucis longus 347
Extensor indicis 322

Extensor pollicis brevis 320, 321
Extensor pollicis longus 321
Extensor retinaculum 312, 322
External auditory meatus 171
External oblique 290–1, 294
External occipital protuberance 171, 173
Exteroceptors 131
Extrafusal fibres 100
Extrapyramidal system 83, 111
    lesions 121–2
Eye 399–404

Facial artery 427–8
Facial nerve 369–70
Facilitation, in reflex activity 73
Factor VIII 483–4
Fallopian tubes 639
Fascia
    deep 39, 41, 556, 612
    superficial 39–40, 556, 612
Fascia lata 41, 613
Fasciculation 122, 157, 158
Fasciculi
    muscle 270
    nerve 49
Fat cells 38
Fat metabolism 601, 604
Fatigue 567–8, 572
Fatty acids 149, 569–70, 583, 598, 601
    polyunsaturated 601
    saturated 601
Feedback 8
    negative 8, 25–6, 452
        glucose concentration 633
        hormones, action of 624
        menstrual cycle 644
        pituitary gland 627
        testosterone production 640
        thyroid gland 631
Femoral artery 427, 433, 436, 614, 615, 617
Femoral canal 614–15
Femoral hernia 615
Femoral nerve 93, 389, 615–16
Femoral sheath 614
Femoral triangle 332, 337, 613–16
Femoral vein 436, 437, 614, 615, 617
Femur 205–8
Fertilization 17–18, 636, 645
Fetus 646–8
Fibre in food 599
Fibrillation of muscle 157–8
Fibrin 482
Fibrinogen 482
Fibroblasts 38, 39
Fibrocartilage 42
Fibrous flexor sheath 198–9, 276, 314, 354
Fibula 209, 212–13
Fibular collateral ligament 256
Fick's method 448–9

Fight or flight response 131–2
    see also Alarm reaction
Fimbria, of uterine tubes 639
Final common pathway 141
First polar body 643
Fixators 281
Flaccid paralysis 122, 157
Flare, in skin (in triple response) 106, 108,
    469–70, 560
Flexor carpi radialis 312–13
Flexor carpi ulnaris 281, 313–14
Flexor digiti minimi brevis
    foot 359
    hand 326
Flexor digitorum accessorius 358
Flexor digitorum brevis 357
Flexor digitorum longus 349, 353–4, 357, 358
Flexor digitorum profundus 315–16
Flexor digitorum superficialis 314–5
Flexor hallucis brevis 359
Flexor hallucis longus 214, 215, 349, 353, 357
Flexor pollicis brevis 324
Flexor pollicis longus 315
Flexor reflex 70, 74, 123
Flexor retinaculum 197, 245, 311–12
Flower spray nerve endings 100
Folic acid (pteroylglutamic acid) 477, 607
Follicle stimulating hormone (FSH) 626, 643,
    644
Follicular cells of ovary 639, 642
Fontanelle 170
Foods, metabolism of *see* Metabolism of foods
Foot drop 345, 396
Foramen magnum 171
Foramen ovale 648, 649
Foramen transversarium 175, 177, 178
Forced vital capacity (lungs) 524
Forebrain 76, 82
Forefoot, joints of 268–9
Fossa ovalis 416
Frontal bone 78, 170
Frontal lobe 78, 80, 170
Frontal sinus 408
Frontopontine fibres 84, 86
Fructose 598, 600
Functional residual capacity (lungs) 523

Galactose 598, 600
Gall bladder 593, 594–5
Gametes *see* Ova; Spermatozoa
Gamma-amino butyric acid (GABA), as
    neurotransmitter 67
Ganglia 75–6
    autonomic nervous system 125, 126–7
    posterior root 90
Gastric inhibiting polypeptide (GIP) 597, 633
Gastrin 27, 597
Gastrocnemius 349–51, 618
Gemelli muscles 343, 344

Genes 27, 29
Genetic code 12, 30
Genetic regulation of cell activity 27–32
Genitofemoral nerve 387, 616
Gigantism 628
Glands 36–7
  apocrine 559
  endocrine (ductless) 36, 621–4
    see also names of individual endocrine glands
  exocrine 36, 37
    see also names of individual exocrine glands
Glans 638
Glaucoma 403
Glenohumeral ligaments 239
Glenoid fossa 187, 189, 190, 238, 240
Glenoidal labrum 238
Globulins 481, 482
Globus pallidus 114
Glomerular filtrate 552
Glomerulus 550
Glossopharyngeal nerve 370
Glucagon 632
Glucose 571, 573, 598
  CNS dependence on 633
  diabetes mellitus 633
  hormone control 632–3
  liver glycogen 583
  metabolism 600
  muscle glycogen 25–6, 582
Glutamic acid, as neurotransmitter 67
Gluteal nerve
  inferior 390–1
  superior 390
Gluteus maximus 41, 204, 205, 273, 339–40
Gluteus medius 207, 340–2
Gluteus minimus 207, 340–2
Glycerol 598
Glycine, as neurotransmitter 67
Glycogen 25–6
  in liver 583, 600
  in muscle 150, 570–1, 582–3, 600
Glycolysis 570–1, 573, 574, 580, 600
Goblet cells 35
Goitre 630
Golgi apparatus 15
Golgi bottle neurones 116
Golgi tendon organ 74, 99, 101, 116
Gonad 636, 638 *see also* Ovaries; Testes
Gonadotrophic releasing hormone (GnRH) 643
Gonadotrophins 643, 644
Gracilis 338–9, 613
Granulocytes 479, 480
Greater omentum 590–1
Greater sac 589
Grey matter 75
Growth 649–51
Growth hormone 628
Gustatory cells 410
Gyri, cerebral 78
  post-central 80, 104–5

precentral 80

Haemocytoblasts 477, 481
Haemoglobin 476, 477, 478, 492, 498
  abnormal forms 496–7
  formation and fate 496
  oxygenation of 493, 498–9
  structure 493–4
Haemolysis 485, 486
Haemophilia 483–4, 490
Haemostasis 484
Hair cells 406–7
Hairs 558–9
  growth of hair 623–4
Hamate bone 196, 197, 198
Hamstrings 203, 220, 344–5
  actions and functions 345
Hand movements 328–30
Haversian system 162–3
Heart 413–20
  apex 416, 420
  atria 416–17
  block 451
  blood supply 418–19
  conducting system 418, 439–44
  examination of 449, 454
  and exercise 578–9
  nerve supply 419–20
  rate 448, 455
  regulation of 449–54
  sounds 445
  surface markings 420
  valves 417, 418, 445
  ventricles 417–18
  see also Cardiac . . .
Heat loss 562–3, 584
Heat production 151, 584
Hemiplegia 118, 122
Heparin 483
Hepatic artery 593, 594
Hepatic ducts 593, 594
Hepatic veins 593, 594
Hering-Breuer reflex 529, 530
Hernia 294–5, 615
Herpes zoster (shingles) 120, 121
High energy phosphate molecules 21–2
Hilton's Law 224, 251, 277, 389
Hindbrain 76, 84
Hip bone (os innominatum) 199–205
Hip joint 249–54
Histamine 560
Homans' sign 351
Homeostatic control (homeostasis) 8–9
Hook grip 329
Hormones 621, 640–1
  chemistry 621–2
  mode of action 623–4
  regulation of cellular activity 26–7
  secretion, control of 624

see also names of individual hormones
Humerus 190–3
Hunter's canal (adductor canal) 615, 616–17
Huntingdon's chorea 114
Hydrocephalus 489
Hydrochloric acid 596, 597
β-Hydroxybutyric acid 601
5-Hydroxytryptamine (5-HT), as
        neurotransmitter 67
Hyperalgia 108
Hypertension 463–4
Hyperthyroidism 630
Hypochondrium 588
Hypogastrium 588
Hypoglossal nerve 371
Hypothalamus 82–3, 625, 626, 643
Hypothenar eminence 322

Ileum 591
Iliac arteries
  common 427
    external 427, 433
    internal 427, 433
Iliac crest 203, 204
Iliac fossa 203, 588
Iliacus 332, 613, 614
Iliofemoral ligament 207, 251
Iliohypogastric nerve 387
Ilio-inguinal nerve 387, 616
Iliolumbar ligament 233
Iliolumbar nerve 373
Iliopsoas 207, 332
Iliotibial tract 332
Ilium 199, 202–3
Immune reaction 537, 544
Immunity
  cell-mediated 544–5
  humoral 544–5
Immunoglobulins 544
Immunology 543–5
Incus 405
Infarct 423
Inflammatory reaction 106, 108, 469–70, 560
Infraspinatus 190, 192, 240, 307
Inguinal hernia 295
Inguinal ligament (Poupart's ligament) 204,
        290–1, 613
Inhibitory post-synaptic potential (IPSP) 63, 74
Inspiratory capacity (lungs) 521–2, 523
Inspiratory centre 529, 530
Inspiratory reserve volume (lungs) 521–2
Insula 78
Insulin 26, 582, 601, 622, 632–3
Intention tremor 113
Intercalated discs 438–9
Interclavicular ligament 237
Intercostal membranes 504
Intercostal muscles 504–5
Intercostal nerves 372–3, 612

Intercostal spaces 504
Intercostobrachial nerve 373, 612
Intermittent claudication 581
Internal arcuate fibres 87
Internal capsule 82, 83, 119
Internal elastic lamina, of arteries 421
Internal oblique 291, 294
Interossei
  foot 359–61
  hand 327–8, 329–30
Interosseous membrane, radio-ulnar 245
Interosseous nerve, posterior 383, 384–5
Interphangeal joints
  fingers 249
  toes 269
Interspinous ligament of vertebrae 227
Intertransverse ligaments of vertebrae 227
Intervertebral discs 175, 178, 224–6
  herniation 225–6
Intervertebral foramina 174, 179
Intervertebral joints 224–31
Intervertebral ligaments 226–7
Intestine
  large 592–3, 599
  small 591–2
    secretions and enzymes 597–8
Intracellular water 475, 487–8
Intrafusal fibres 100, 115, 117
Intra-pleural pressure 519–20
Intra-pulmonary pressure 520
Intrinsic factor 478, 607
Iodine 630
Iris 401
Iron 478, 496, 602
Irradiation, in reflex activity 74
Ischiofemoral ligament 251
Ischium 199, 202, 203
Islets of Langerhans 595, 632–3
Isometric contraction
  cardiac muscle 447
  skeletal muscle 146, 282
Isometric relaxation phase, in cardiac cycle 448
Isotonic contraction
  cardiac muscle 447
  skeletal muscle 146, 282
Isotropic bands

Jejunum 591
Jogging 585
Joint mice 222
Joints 218
  ball and socket 220
  cartilaginous 219
  close-packed position of 222, 265, 335
  ellipsoid (condyloid) 220
  fibrous 218
  hinge 220
  pain from injuries 102, 224
  pivot 220

plane (gliding) 220
sellar 220
spine 224–9
    movements of spine 229–31
synovial 219–24
    blood supply and nerve supply 224
    see also names of individual joints
Jugular vein
    external 429
    internal 427, 429, 620

Karyotype 17
Keratin 36, 556, 557
Keto acids 583, 601
Ketosis 583, 601, 633
Kidneys 546
    blood circulation 427, 470, 546, 551
    functions 549, 552–4, 634–5
    structure 549–51
Klumpke's paralysis 398
Knee jerk 68, 73, 116–17, 124
Knee joint 254–62, 335–6
    locking 261, 335, 352
    movements 260–2
von Kupffer cells 543, 604

L-dopa 114
Labrum acetabulare 249
Labyrinth, of ear
    membranous 406
    osseous 405–6
Lacrimal gland 400–1
Lacteals 487, 537, 591, 599
Lactic acid 150, 571, 572–4, 600
Lactiferous ducts 639
Lactose 598
Lacunar ligament 614
Lamina of vertebrae 174
Laminectomy 174
Larynx 513
Latent heat of vaporisation 563
Latissimus dorsi 192, 299–300, 609, 612
Lemniscus, medial 84, 86–7, 104
Lens, of eye 402–3
Lenticulostriate arteries 82
Lentiform nucleus 83, 114
Lesser sac 589
Leucocytes 476, 479–81
Leucocytosis 480
Levator anguli oris 283
Levator ani 296
Levator scapulae 302, 619
Lifting heavy weights 225
Ligaments 39
    see also specified ligaments
Ligamentum flavum 39, 227
Ligamentum nuchae 227
Ligamentum patellae 208, 210, 255, 260, 335

Ligamentum teres 205, 251, 252
Line of gravity 6, 254, 365–7
Linea alba 290
Linea aspera 207
Linea semilunaris 293
Lipoprotein 482, 583
Liver 593–4
    functions 597, 603–4
Long acting thyroid stimulant (LATS) 630
Long thoracic nerve 374, 612, 619
Longitudinal ligaments
    anterior 227
    posterior 227
Loop of Henlé 550–1, 553
Loose connective tissue 37–9
Lordosis 253
Lower motor neurone 110
    diseases of 122–3
    lesions 122
Lumbar plexus 92, 386–9
Lumbar puncture 89, 178
Lumbocostal arches 505
Lumbo-sacral trunk 389
Lumbricals
    foot 357–8, 360, 362
    hand 326–7
Lunate bone 194, 196
Lungs 510–11, 517–18
    diffusion capacity of 528
    volumes and capacities 521–4
        measurement of 524
Luteinising hormone (LH) 626, 640, 643
    LH surge 643, 644
Lymph 473, 487, 536, 538
Lymph nodes 536, 537–42
    cervical 542, 620
    lower limb 542, 616
    upper limb 540–1, 611, 612
Lymphangitis 537
Lymphatic duct, right 539, 540, 542
Lymphatic follicles 538
Lymphatic system
    anatomy 537–42
        spleen 542–3
        thymus 542–3
    functions 536–7, 543–5
Lymphatics 537–40
    lower limb 542
    upper limb 540–1, 612
Lymphoblasts 481
Lymphocytes 479, 480, 481, 536, 538
    bursa equivalent (B) 544–5
    thymic processed (T) 544–5
Lysosomes 15

Macrocytes 478
Macro motor units 158 161
Macrophages 38 543
Maculae 407–8

Main en griffe (claw hand) 280, 381
Malleolus
  lateral 211, 212, 213
  medial 211, 212, 213
  third 265
Malleus 405
Malpighian body 550
Maltose 598
Mammary glands 639
Mandible 171, 173
  angle of 171, 173
Mandibular nerve 369
Manubriosternal joint 219, 232
Manubrium sterni 184, 185
Marathon running 584, 585
Marrow 168
Masseter 236, 283, 369
Mast cells 38
Mastoid antrum 405
Mastoid process 171
Maxilla 171
Maxillary nerve 369
Maxillary sinus (antrum) 408
Maximal voluntary ventilation test 524
McArdle's disease 572, 576
Median antecubital vein 612
Median nerve 197, 362, 613
  lateral root 376–9, 612
  medial root 376–9, 379, 612
Mediastinum 413
Medulla oblongata 76, 84–8
Medullary cavity, of bone 166, 168
Megakaryocytes 481
Megaloblasts 478
Meiosis 17–19, 643
Meissner's corpuscles 98
Melanin 558
Melanoblasts 558
Membrane potential 54–6, 140, 438
Menarche 651
Meninges 76, 97
Menisci (knee joint) 257–9
Menopause 644–5
Menstrual cycle 641, 642–4
Mesaxon 47
Mesenteries 589, 593
Metabolism of foods 569–74, 599–607
  amino acids 602
  carbohydrates 600, 604
  fat 601
  vitamins 605–7
Metacarpal arteries, palmar 432
Metacarpals 198
Metacarpophalangeal joints 248–9
Metatarsal arteries
  dorsal 435
  plantar 436
Metatarsals 213, 216–17
Metatarsophalangeal joints 269
Micturition 549, 554–5

Midaxillary line 609
Midbrain 76, 82, 83–4
Mid-carpal joint 245–6
Milieu interne 33, 549
Mitochondria 12–13, 25
Mitosis 15–17
Mitral valve (bicuspid valve) 417
Molar solution 474
Monosaccharides 598
Monocytes 479, 480, 481
Motor area of cerebral cortex 80, 109–10
Motor end plates 53, 68
Motor pathways 110–11
  lesions 121–4
Motor point 138, 158 159
Motor units 138, 140–1, 277
  in exercise 567–8
  see also Macro motor units
Mouth 586–7, 595–6
Movements
  terminology 4–6
  voluntary, control of 112–14
Muscle, cardiac see Cardiac muscle
Muscle, skeletal 134,135, 270–2
  action 279–83
    antagonists 280
    concentric 281–2
    eccentric 281–2
    fixators 281
    insufficiency of muscle 282–3
    prime movers 280
    synergists 281
    see also Contraction of skeletal muscle
  architecture 277–9
  attachments 272–3
  blood circulation regulation 276, 468–9,
      575–9
  chemical nature 136–9
  contraction see Contraction of skeletal muscle
  metabolism, in exercise 568–75
  microscopic appearance 136–9
  nerve supply 276–7
  neurovascular hilus 276
  red fibres 136–7, 567, 571
  tendons and tendon sheaths 274–6
  white fibres 136–7, 567, 571
  see also Muscles
Muscle, smooth 130, 134–5, 461
Muscle pump 425 458 578
Muscle spindles (neuromuscular spindles)
      99–100, 115–16
Muscle tone 97, 113
  control of 115–18
  and stretch reflex 118
  abdominal wall 290–6
  eye 399–400
  facial expression 283–4, 369
  head and neck 283–7
  lower limb 330
  adductor group 336–9

calf muscles 349–55
  extensor group 345–8
  foot, sole of 355–62
  gluteal region 339–44
  hamstrings 344–5
  lateral muscles 348–9
  thigh, front of 331–6
mastication 283, 369
pelvis 296–7
trunk 287–90
  abdominal wall 290–6
  respiration 502–8
  spine 288–90
upper limb 297
  arm 309–11
  forearm, extensors 317–22
  forearm, flexors 311–17
  hand 322–30, 362
  shoulder and scapular regions 192, 240, 297–308
see also Muscle, skeletal; *and names of individual muscles*
Musculocutaneous nerve 374–5, 612
Myasthenia gravis 67, 68
Myelin sheath 47–9
Myeloblasts 481
Myelocytes 481
Myocardium 413–15
Myofibrils 137–8
Myoglobin 497, 574–5
Myosin 135, 138, 141–4, 148, 569
Myotomes 396–8
Myxoedema 630

Naso-lacrimal duct 400–1
Navicular 213, 216
Neck, triangles of 619–20
Nélaton's line 208
Neocerebellum 86, 88, 113, 118
Neostigmine 67
Nephron 550
Nerve, peripheral *see* Peripheral nerves
Nerve cell *see* Neurone
Nerve-effector organ junctions 53–4, 67–8, 140
Nerve fibres 44, 47–9
  classification 50
  size of 50
Nerve impulses 44–5, 54–8
  initiation 58–60
  motor nerve fibres 140
  refractory period 58
  saltatory conduction 48, 58
  see also Neurotransmitters; Reflex(es); Synapses
Neural arch, of vertebra 173, 175
Neurapraxia 152–3
Neurocentral joints (joints of Luschka) 227
Neurofilaments 46
Neuroglia 75

Neuromodulators 67
Neuromuscular junctions 53–4, 67–8, 140
Neuromuscular spindles (muscle spindles) 99–100, 115–16
Neurone 44
  structure 46–54
neurosecretion 628
Neurotmesis 153, 155–6, 157, 160
Neurotransmitters 52, 60, 63, 64–8, 114, 130
Neurovascular bundles 41, 272, 276
Neurovascular hilus 276
Neutrophil leucocytes 479, 480
Nexuses, in smooth muscle 461
Nicotine, effects on nervous system 66
Nicotinic acid 606–7
Nissl substance (Nissl granules) 46, 154
Nodes of Ranvier 48, 58
Noradrenaline 52, 64–5, 66–7, 130
Noradrenergic terminals 66–7
Normoblasts 477, 478
Nose 409–10, 512, 518–19
Nuclear bag fibres 100
Nuclear chain fibres 100
Nucleic acids 11, 12, 29, 30–1
Nucleotides 29–30
Nucleus, of cell 11–12, 15–19
  nuclear membrane 11
Nucleus cuneatus 87, 104
Nucleus gracilis 87, 104
Nucleus pulposus 225
Nutrient foramen in long bone 168
Nutrition 604–5
Nystagmus 408

Oblique popliteal ligament 256
Obstructive lung disease 519
Obturator externus 343, 344
Obturator foramen 204
Obturator internus 203, 296, 343, 344
  nerve to 390
Obturator nerve 388
Occipital bone 78, 170
Occipital lobe 78, 80
Occlusion phenomenon, in reflex activity 72
Oedema, lymphatic 536
Oesophagus 587, 596
Oestradiol 32, 641, 642
Oestrogens 32, 632, 641–4
Ohm's Law 457
Olecranon process 192, 193, 195
Olfactory epithelium 409
Olfactory nerve 368, 409
Oligodendrocytes 48
Olivary nucleus 87
Omentum, greater 590–1
Ophthalmic artery 428
Ophthalmic nerve 369
Opponens digiti minimi 326
Opponens pollicis 324

Optic chiasma 403–4
Optic disc 402, 489
Optic foramen 400
Optic nerve 368, 400, 402, 403
Orbicularis oculi 284, 401
Orbicularis oris 284
Orbits 173
Organ of Corti 406
Organelles 11–15
Osmoreceptors 554, 629
Osmotic pressure 554, 599, 629
Ossicles, of ear 405
Ossification
  in cartilage 165–7
  in membrane 165, 166
Osteoblasts 165–8
Osteoclasts 167
Osteocytes 163, 167
Osteomalacia 634
Osteoporosis 165, 634, 645
Otoliths 407
Ova 17, 18, 636, 638–9, 641–4
Oval window 405, 406
Ovaries 632, 638–9, 641
Ovulation 566, 641, 642
  timing of 644
Oxygen debt 150, 573–4
Oxygen haemoglobin dissociation curve 493–4
Oxygen transport 491–7
  delivery to tissues 495–6
  uptake in lung, factors affecting 494–5
Oxytocin 628–9, 648

Pacemaker, cardiac 440–1
Pacinian corpuscles 51, 60, 98, 99, 102
Pain 105–9
  hyperalgia 108
  joint injuries 102, 224
  muscles, joints and ligaments 102
  receptors and perception 105–6
  referred 106, 107, 109
  variations in sensation 108
Palaeocerebellum 88, 113
Palate 586
Palmar aponeurosis 322–4
Palmar arch
  deep 432
  superficial 431, 432
Palmar spaces 324
Palmaris longus 313
Pancreas 37, 595, 597, 632–3
Pancreatitis, acute 598
Pancreozymin 597
Papillae, of tongue 410
Papillary muscles 417, 445
Papilloedema 489
Paraesthesia 49
Parafollicular cells of thyroid 630, 634
Paranasal sinuses 408, 409, 512

Paraplegia 122
Parasympathetic nervous system 126–9, 131–3
Parathormone 164–5, 634
Parathyroid glands 630, 634
Parietal bone 78, 170
Parietal lobe 78, 170
Parkinson's disease 113–14, 117
Parotid gland 369
Partial pressure of respiratory gases 524–5, 527
Patella 208–9, 255, 259, 260, 335
Patent ductus arteriosus 649
Pectineus 204, 337, 613
Pectoral nerve
  lateral 375, 612
  medial 379, 612
Pectoralis major 192, 302–3, 608, 609
Pectoralis minor 303–4, 608, 611
Pedicle of vertebrae 174
Pellagra 606–7
Pelvis 199
  muscles 296–7
Penis 549, 638
Pennate type muscles 277–9
Pepsin 596, 597
Peri-articular plexus, of arteries 422
Pericardium 413
Perichondrium 42
Pericytes 421
Perilymph 406
Perimysium 270
Perineal body 296–7
Perineurium 49
Periosteum 163, 166, 168
Peripheral nerves 49, 75, 368
  classification of fibres 50
  damage 152
    anatomical effect 152–5
    physiological effects 157–61
    regeneration 155–7, 160–1
  dermatomes 92, 93–5, 121, 396–8
  electric stimulation 158–9
  myotomes 396–8
  see also Cranial nerves; Spinal nerves; *and
    names of individual nerves*
Peristalsis 134, 587, 596, 599
Peritoneal cavity 588–9
Peritoneum 588–9
Peroneal nerve
  common 391–2, 394–6, 619
  deep 394, 395, 619
  superficial 394–5, 619
Peroneus brevis 213, 215, 217, 349
Peroneus longus 213, 215, 216, 349, 359
Peroneus tertius 347
Peyer's patches 536, 591
pH of blood 482, 498–500, 549, 552–3
Phagocytosis 15, 154, 480
  reticulo-endothelial system 536, 538, 539, 543
Phalanges

fingers 198–9
toes 213, 217
Phantom limb sensation 160
Pharyngotympanic tube 404–5
Pharynx 512, 586, 587
Phospholipid 482, 601
Phrenic nerve 372, 506
Physostigmine 67, 68
Pia mater 76
Piriformis 343, 344
    nerve to 390
Pisiform bone 196, 197, 198
Piso-triquetral joint 246
Pituitary fossa 173, 625
Pituitary gland 625–9
Placenta 647–8
Plantar aponeurosis 356, 364
Plantar arch, deep 435, 436
Plantar arteries
    lateral 436
    medial 436
Plantar calcaneo-navicular ligament 214, 264,
    266
Plantar flexion, of foot 345
Plantar nerves
    lateral 362, 392, 394
    medial 362, 392, 393–4
Plantaris 349, 351, 618
Plasma 481
    proteins 481–2
Plasma cells 38, 545
Platelets (thrombocytes) 481
Pleura 508–10
Plexus
    cardiac 419–20
    peri-articular (of arteries) 422
Plexuses of spinal nerves 92–3
    brachial 92, 371, 373–4, 398, 611, 619
        lateral cord branches 374–9, 611–12
        medial cord branches 379–81, 611–12
        posterior cord branches 381–6, 611–12
    cervical 371
    lumbar 92, 386–9
    sacral 92, 389–96
Pneumotaxic centre 529, 530
Polarisation of cell membranes 54
Poliomyelitis 122–3, 160
Polycythaemia 477, 478
Polymorphonuclear leucocytes 479, 480
Pons 76, 84–8
Pontine nuclei 84, 86
Popliteal artery 433–4, 436, 618–19
Popliteal fossa 207, 617–19
Popliteal vein 433, 436, 437, 618
Popliteus 349, 351–3, 618
Porphyrin 493
Porta hepatis 593
'Porta pedis' 393
Portal system 412–13, 470–1
    hepatic 412, 470

hypophyseal 413, 470–1, 625
Portal vein 427, 593, 594
Posterior column pathway 104
Posterior nerve roots 90–1
Posterior root ganglion 90
Post-herpetic neuralgia 120, 121
Postural drainage 516–17
Posture 115, 118–19, 365–7
Potassium ions, and electrical potential
    cardiac muscle cells 439, 441–2
    nerve cells 54–7, 61, 63, 74
    skeletal muscle fibres 140, 141
Poupart's ligament 204, 290–1, 613
Power grip 329
Precision grip 320, 328, 329–30
Pregnancy 645–8
Prepuce 638
Presynaptic inhibition 63, 74
Pro-erythroblasts 477
Profunda femoris artery 433, 615
Profunda femoris vein 437, 615
Progesterone 32, 632, 643, 644
Projection fibres 76
Prolactin 628
Pronator quadratus 316
Pronator teres 194, 312, 612
Propagated action potential *see* Nerve impulses
Proprioception 97, 102
Proprioceptive neuromuscular facilitation
    (PNF) 119
Prostacyclin 482
Prostaglandins 482, 554, 648
Prostate gland 549, 638
Proteins 22–3, 605
    functions 23
    genetic control of production 29–30
    metabolism 598, 602
    in muscle 150
    production, by cells 13–15
Prothrombin 482
Proximal convoluted tubule 550
Psoas 332, 613, 614
Ptosis 401
Pubic symphysis 199, 204, 219, 234
Pubis 199, 202, 204–5
Pubofemoral ligament 251, 252
Puborectalis 297
Pudendal nerve 390
Pulmonary arteries 411, 426, 515
Pulmonary circulation 411, 472
Pulmonary trunk 411, 422, 426, 515
Pulmonary valve 418
Pulmonary veins 411, 515
Pulse pressure 459
Pupil 401
Purkinje cells 418, 443
Pyloric sphincter 591, 596
Pylorus 590
Pyramid 87, 110
Pyramidal tract 76, 83, 84, 86, 87, 110

Pyruvic acid 571, 583

Quadrangular space 311, 609
Quadratus femoris 207, 343, 344
Quadratus lumborum 290
Quadriceps femoris 208, 209, 333–6
Quadriceps lag 335
Quadriplegia 122
Quickening 646

Radial artery 431, 432
Radial collateral ligament 242, 245
Radial nerve 93, 382–6, 612, 613
Radiocarpal joint 245
    movements 245–6
Radio-ulnar joint
    inferior 244–5
    superior 220, 242–3
Radius 192, 194
Rami communicantes 92, 126
Reabsorption, in kidneys 552
Reaction of degeneration (motor nerves) 158–9
Receptors 44, 50–1, 97–102
    alpha 461, 464, 575
    baroreceptors 8, 452–4, 461–2, 463
    beta 461, 464, 576
    body temperature control 563–4, 584
    chemoreceptors 97, 531–2, 533–4
    exteroceptors 131
    function 58–60
    neuromuscular spindles (muscle spindles)
        99–100, 115–16
    osmoreceptors 554, 629
    pain, from stimulation of 105–6
    proprioception 102
    skin 98–9
    tendon and joint 101–2
    volume (cardiovascular system) 454
Reciprocal innervation (reciprocal inhibition)
    74
Recruitment, in nerve stimulation 73
Rectum 592, 593
Rectus abdominis 290, 293, 294, 296
Rectus femoris 334–6
Red blood cells 476–9
Red nucleus 84
Red reaction, in skin 560
    see also Triple response
Referred pain 106, 107, 109
Reflex(es) 9, 44, 45, 97
    arc 45, 68–9
        monosynaptic 69
        polysynaptic 69
    autonomic nervous system 125–6, 129–30,
        131–3
    axon 560
    baroreceptor 8, 452–4, 461–2, 463
    blood pressure control 461–2

body temperature control 563–5
cardiac
    baroreceptor 8, 452–4, 461–2, 463
    others 454
circulatory 461–3
cough 513
crossed extensor 71, 73, 74, 123
flexor 70, 74, 123
Hering-Breuer 529, 530
response 44, 45, 68–74
stretch 68, 69, 74, 115–17, 123
    brain and 117–18
    gamma (γ)-efferent control of 115, 117
    muscle tone 118
    posture 118–19
    see also Tendon jerks
Refractory period
    in muscle contraction 144
    in nerve impulse conduction 58
Regeneration in nerve damage 155–7, 160–1
Reinforcement in reflex activity 73, 117
Release inhibiting factor, hypothalamic 626
Releasing factor, hypothalamic 626
Renal arteries 427, 546, 551
Renin 463, 464, 631, 634–5
Reproductive systems
    anatomy
        female 638–9
        male 636–8
    functions 636, 640–9
Residual volume (lungs) 522–3
Respiration 502
    control of 528–34
        chemical 530–2
        intrinsic brain-stem mechanism 529–30
        neural control of brain-stem mechanism
            530
        responses to blood gas composition changes
            532–3
    and exercise 572–4, 581–2
    external 502, 534–5, 581–2
    internal 502, 534–5, 572–4
        study of, from external respiration
            measurements 534–5
    movements of 507–8
    see also Respiratory system
Respiratory exchange ratio 573, 574, 601
Respiratory minute volume 526
Resiratory quotient 572–3, 574, 600–1
Respiratory system 132, 502
    air, composition of 524–6
    anatomy 502
        air passages 512–15, 518–19
        bronchopulmonary segments 515–18
        chest wall 503–5
        diaphragm 505–7
        lungs 510–11
        pleura 508–10
        pulmonary artery 515
    exchange of gases between alveolar air and

blood 527–8
pressure and volume changes 519–24
upper respiratory tract functions 518–19
see also Respiration
Resting membrane potential 55, 140, 438–42
Reticular fibres 38
Reticular formation 88, 104, 108
Reticulocyte 477
Reticulo-endothelial system 536, 538, 539, 543
Reticulo-spinal tracts 88, 408
Reticulum of lymph nodes 538–9, 543
Retina 401–2, 428
Retinacula 41, 274–5
    extensor (leg) 345
    femoral 249, 251
    fingers 314–15
    patellar 255–6, 335
    peroneal 349
    wrist 311–12
Retinol (vitamin A) 602, 603, 604, 605
Retrograde degeneration in nerve damage 154, 155
Rhesus factor (Rh) 485–6
Rheumatoid arthritis 632
Rhomboids 302
Riboflavine 606, 607
Ribonucleic acids (RNA) 11, 29
    messenger RNA (mRNA) 30–1
    transfer RNA (tRNA) 30
Ribosomes 13, 30
Ribs 182–4
    action in respiration 503–5
    rib cage 185
Rickets 634
Rods and cones 401–2
Root values, of nerves 93, 94
Rotator cuff 192, 240, 308, 344
Rubro-spinal tract 408
Running
    action of muscles in 367
    marathon 584, 585

Saccule 407
Sacral plexus 92, 389–96
Sacro-iliac joint 179, 199, 203, 204, 232–4
Sacro-iliac ligaments 232, 234
Sacrospinous ligament 232–3, 234
Sacrotuberous ligament 232–3, 234
Sacrum 173, 178–9
Sagittal sinus, superior 97, 429
Saliva 595
Saltatory conduction of nerve impulse 48, 58
Saphenous nerve 389, 615, 617
Saphenous veins
    great (long) 436, 613–14, 616
    small (short) 436, 618
Sarcolemma 136
Sarcomere 137
Sarcoplasm 136–7

Sarcoplasmic reticulum 136
Sartorius 332–3, 613, 617
Saturated water-vapour pressure (S.V.P.) 525
Saturday night paralysis 153, 385
Scalene muscles 286, 619
Scaphoid 194, 196, 197, 198
Scapula 187–90, 236
    movements 238, 301–2
    winged 302, 374
Schwann cells 47–9
    in nerve damage 154–7
Sciatic nerve 391–2, 619
Sclera 401
Scrotum 636, 641
Scurvy 484, 606
Sebaceous glands 558
Sebum 558
Secondary sexual characteristics 640–1, 651
Secretin 597
Segmentation movements of gut 599
Semen 638
Semicircular canals 405, 406
Semicircular ducts 406
Semimembranosus 211, 344, 618
Seminal vesicle 638
Seminiferous tubules 636
Semispinalis capitis 288, 619
Semitendinosus 345, 618
Sensation 97–102
    see also Pain; Receptors; Sensory pathways
Sensory association area of cerebral cortex 104–5
Sensory neurone disease 120–1
Sensory pathways 102–5
Sensory system lesions 120
Septum
    interatrial 416, 648, 649
    interventricular 416–17
Serratus anterior 189, 301–2, 609, 612
Serum 482–3
Servo mechanism 25, 26
Sesamoid bones 167–8
    foot 216, 269, 353, 359
    patella 208, 335
    thumb 324, 326
Sex chromosomes 17, 18, 19, 636
Sex determination 636
Shingles 120, 121
Shoulder joint 238–42
    movements 240–2
Sickle cell anaemia 497
Sinu-atrial (s-a) node 418, 441–2
Sinusoids 412, 470, 594, 625
Skin 556
    blood supply 469–70, 559–60
    functions 561–6
    receptors 98–9
    structure 556–9
    triple response 106, 108, 469–70, 560
Skull 170–3

'Slipped disc' 226
Smell, sense of 408–9
Sodium bicarbonate 595, 597, 601
Sodium ions, and electrical potential
  cardiac muscle cells 439, 441–2
  nerve cells 24, 54–7, 61, 74
  skeletal muscle fibres 140, 141
Sole plate 53
Soleus 209, 349, 351
Somatomedins 628
Somatostatin 633
Spasmodic torticollis 286
Spastic paralysis 121
Spatial summation 62, 71, 72
Speech centre 80, 110
Spermatogenesis 641
Spermatozoa 17, 18, 636–7, 638, 644, 645
Sphincter 273
  anal 270, 593
  pre-capillary 425
  pyloric 591, 596
  urethrae 270, 549, 555
  vesicae 549, 555
Sphygmomanometer 456
Spinal cord 88–9
  section of 123–4
Spinal nerves 75, 89–93, 371–3
  see also Plexuses of spinal nerves
Spinal shock 123
Spine of scapula 187, 190
Spine (spinous process) of vertebra 174, 175, 178
Spine (vertebral column) 88, 173, 179–82
  movements 182, 229–30
Spinocerebellar tract 113
Spinothalamic tracts 76, 103, 107–8
Spirometer 524, 534
Splanchnic nerves 126
Spleen 542–3
Splenius capitis 619
Spring ligament (plantar calcaneo-navicular ligament) 214, 264, 266
Standing, action of muscles in 365–7
Stapes 405
Starch 598
Starling's law of the heart 443, 450–2, 454, 579
Stato-acoustic nerve *see* Vestibulocochlear nerve
Sternal angle (angle of Louis) 185
Sternebrae 184
Sternoclavicular joint 185, 186, 187, 236–7, 238
Sternoclavicular ligaments 237
Sternocleidomastoid (sternomastoid) 285–6, 619, 620
Sternocostal joints 232
Sternum 184–5
Steroid hormones 621–2, 623
Stomach 590–1, 596–7
Stratum basale 556a–7, 561
Stratum corneum 557, 559, 561
Stratum granulosum 557

Stratum lucidum 557, 559
Stratum spinosum 557
Strength-duration (SD) curve, in muscle contraction 159
Stretch reflex 68, 69, 74, 115–17
  and brain 117–18
  gamma (γ)-efferent control of 117
  and muscle tone 118
  and posture 118–19
  see also Tendon jerks
Striate arteries 429
Stroke 82, 429, 471
Stroke volume, cardiac 448, 451, 455
Styloid process
  radius 194
  ulna 195
Subarachnoid space 76, 97
Subclavian arteries 426, 430, 620
Subclavian vein 427, 429, 433, 608
Subcostal angle 185
Subliminal fringe activity, in reflex activity 72
Subsartorial canal (adductor canal) 615, 616–17
Subscapular artery 609
Subscapular nerves, upper and lower 382, 612
Subscapularis 189, 191, 240, 306–7, 609, 612
Substance P 67, 107, 560
Substantia nigra 83, 113–14
Subsynaptic region 52
Subtalar joint 266
Succus entericus 597
Sucrose 598
  in body fluid volume measurement 473–4
Sulci, cerebral 78
  central sulcus 80
Superior nuchal line 171
Supinator 195, 320–1
Supraclavicular fossa 619
Supracristal plane 588
Suprarenal glands 631–2
  medulla 133, 631
Suprascapular nerve 374, 619
Supraspinous ligament, of vertebrae 227
Supraspinatus 190, 192, 240, 307
Suprasternal notch 185
Sural communicating nerve 392, 394, 619
Sural nerve 392
Surfactant, secretion in air passages 514
Sustentaculum tali 213, 214
Sutures 170, 218
Swallowing 595–6
Sweat, in body temperature control 563, 564–5
Sweat glands 559, 563
Sympathetic nervous system 126, 128, 131–3
  blood flow in skeletal muscle 575–6
  innervation of blood vessels 126, 460–1, 575–6
  stimulation of heart 442, 451–2
Symphysis pubis 199, 204, 219, 234
Synapses 44, 51–3, 60–3

Synaptic bags 52, 60–1
Synaptic gap (cleft) 51–2
Synaptic inhibition 63
Synaptic transmission 60–3
Synaptic vesicles 52
  dense-cored 52
Synchondrosis 219
Syndesmosis 218, 262
Synostosis 219
Synovial fluid 222, 223, 475, 490
Synovial membrane 220, 222, 223, 224
Synovial sheaths 223, 316–17, 322, 354–5
Systole 420, 438

T-tube system in muscle fibre 136
Taenia coli 593
Talco-calcaneal ligaments 266
Talo-calcaneo-navicular joint 266
Talofibular ligaments 264
Talus 213
Tarsus 213
Taste, sense of 409–10
Taste buds 410
Teeth 586–7, 595
Temporal artery, superficial 428
Temporal bone 78–80, 170–1
Temporal lobe 78–80
  functions associated with 80
Temporal summation 62, 71
Temporalis 236, 283, 369
Temperature control *see* Body temperature
  control
Temporomandibular joint 171, 235–6
Temporopontine fibres 84, 86
Tendo calcaneus (Achilles tendon) 215, 349–50
Tendon jerks 68, 73, 116–17, 124
Tendon and joint receptors 101–2
Tendon sheaths 275–6
Tendons 274–6
Tensor fasciae latae 343
Teres major 192, 299, 306, 609, 612
Teres minor 192, 240, 307–8
Terminology, anatomical 4–6
Testes 632, 636, 640–1
Testosterone 632, 640, 644
Tetanus 144
  complete 144
  partial 144, 148
Tetany 144
Thalamus 82, 104
Thenar eminence 322
Thiamine (vitamin B₁) 606, 607
Thoracic duct 539, 540, 542
Thoracodorsal nerve 382, 612
Thoraco-lumbar fascia 288
Thrombo-angiitis obliterans 580
Thrombocytes 481
Thrombosis 471
Thromboxane 482

Thrombus 482
Thumb movements 247, 329–30
Thymus 542–3
Thyroid gland 620, 630–1, 634
Thyroid stimulating hormone (TSH) 626, 630
Thyrotoxicosis 630
Thyroxine (T₄) 478, 630
Tibia 209–12
Tibial arteries
  anterior 434–5, 436
  posterior 434, 435–6
Tibial collateral ligament 256
Tibial nerve 391–4, 619
Tibial tuberosity 210, 212
Tibialis anterior 346–7
Tibialis posterior 211, 216, 349, 353, 359
Tibiofibular joints 218, 220, 262–3
Tidal volume (lungs) 521
Tinel's sign 161
Tissue fluids 473, 486–7
Tissues 33
  see also specified types
Tongue 409–10
Total lung capacity 521, 523
Trabeculae, in bone 162
Trabeculae carneae 417
Trachea 513, 518, 519
Transcellular fluids 473, 475, 476, 488–90
Transneuronal (transynaptic) degeneration 155
Transpyloric plane 588
Transversalis fascia 293
Transverse processes, of vertebrae 174, 175,
    177, 178
Transverse tarsal joint 268
Transversus abdominis 293, 294, 504
Transversus thoracis 373, 504
Trapezium bone 196, 197, 198
Trapezius muscle 297–9, 619
Trapezoid bone 196, 197
Trapezoid ligament 237–8
Tremor
  intention 113
  in Parkinsonism 122
  in Parkinson's disease 113
Trendelenberg's sign 342
Triceps 189, 310–11
Tricuspid valve 417
Trigeminal nerve 104, 369
Trigone, of bladder 546
Tri-iodothyronine (T₃) 630
Triple response 106, 108, 469–70, 560
Triquetral bone 196
Tritium in body water measurement 474
Trochlea 192
Trophic ulcers 121
Tropomyosin 138, 141, 142
Troponin 138, 141
Tunica adventitia 421
Tunica intima 421
Tunica media 421

Two-point discrimination 98
Tympanic membrane 404, 405

Ulna 192, 194–5
Ulnar artery 431, 432
Ulnar collateral ligament 242, 245
Ulnar nerve 380–1, 612
Ulnar paradox 381
Umbilical cord 647, 649
Umbilical hernia 295
Undifferentiated cells in areolar connective
    tissue 38
Upper motor neurone 110
    lesions 121
Ureters 546
Urethra 546, 549
Urine 549
Uterine (Fallopian) tubes 639
Uterus 639
Utricle 407

Vagina 639
Vagus nerve 126, 127–9, 132, 371, 620
    cardiac action of 442, 451, 452–4
    oesophageal peristalsis 596
    stomach innervation 596–7
Varicose veins 437
Vasomotor centre 461, 462–3
Vasomotor nerve supply of blood vessels 126,
    460–1, 575–6
Vasopressin *see* Antidiuretic hormone (ADH)
Vasovagal attack 454, 467
Vastus intermedius 334–6
Vastus lateralis 334–6
Vastus medialis 334–6, 617
Veins 411, 421–2, 425
    communicating 425
    head and neck 429
    lower limb 436–7
    thorax and abdomen 427
    upper limb 432–3
    valves 458
    see also names of individual veins
Vena cava
    inferior 411, 416, 425, 427
    superior 411, 416, 425, 427
Venae comitantes 425, 433, 437
Venous admixture 494
Venous return 425, 457–9, 577–8
Venous tone 457
Venous valves 458
Ventricles of brain
    fourth 77, 84–6, 96
    lateral 76, 95–6
    third 82, 96
Ventricles of heart 417–18
Ventricular septal defect 416–17
Venules 411, 421
Vertebrae 173

cervical 173, 175–8
coccygeal 173, 179
lumbar 173, 178
sacral 173, 178–9
thoracic 173–5
Vertebral artery 175, 177, 428–9
Vertebral canal 173, 175
Vertebral column 88, 173, 179–82
    movements 182, 229–30
Vestibular nuclei 86, 407
Vestibule, of ear 405, 406
Vestibulocochlear nerve 86, 370, 406–8
Vestibulospinal tract 407, 408
Vibration sense 102
Villi
    arachnoid 97, 488
    chorionic 647, 648
    intestinal 591, 599
Vis a fronte 458
Vis a tergo 457
Visual pathway 403
Visual purple (rhodopsin) 602, 605
Vital capacity (lungs) 524
Vital centres 78
Vitamins 600, 602, 605–7
    A (retinol) 602, 603, 604, 605
    B₁ (thiamine) 606, 607
    B₁₂ 478, 604, 607
    C (ascorbic acid) 478, 602, 604, 606
    D 164, 605, 606, 622
        see also Cholecalciferol
    D₃ 634
        see also Cholecalciferol
    K 482, 603, 607
        see also Folic acid; Nicotinic acid; Riboflavine
Vitreous humour 403

Walking, actions of muscles in 367
Wallerian degeneration 153–5
Water, of the body 473
Wheal 106, 470, 560
White blood cells 476, 479–81
White fibrous tissue 39
White matter 75
Winging of scapula 302, 374
Wrist drop 386
Wrist joint (radiocarpal joint) 245
    movements 245–6

X chromosomes 17, 18, 19, 636
Xiphoid process (xiphisternum) 184

Y chromosomes 17, 18, 636

Zygomatic arch 171, 173
Zygomatic bone 171
Zymogen granules 15